T0140603

Many-body Approaches at Different Scales

G. G. N. Angilella · C. Amovilli
Editors

Many-body Approaches at Different Scales

A Tribute to Norman H. March on the Occasion of his 90th Birthday

Editors
G. G. N. Angilella
Dipartimento di Fisica e Astronomia
Università di Catania
Catania
Italy

and

IMM-CNR, UdR Catania
Catania
Italy

and

INFN, Sez. Catania
Catania
Italy

and

Scuola Superiore di Catania
Università di Catania
Catania
Italy

C. Amovilli
Dipartimento di Chimica e Chimica Industriale
Università di Pisa
Pisa
Italy

ISBN 978-3-030-10191-6 ISBN 978-3-319-72374-7 (eBook)
https://doi.org/10.1007/978-3-319-72374-7

Softcover re-print of the Hardcover 1st edition 2018

Printed on acid-free paper

This Springer imprint is published by the registered company Springer International Publishing AG part of Springer Nature
The registered company address is: Gewerbestrasse 11, 6330 Cham, Switzerland

Preface

This volume is dedicated to the 90th birthday of Prof. Norman H. March. Already in the title, *Many-body Approaches at Different Scales*, the book endeavours to epitomize Prof. March's very broad range of interests, which spans both chemistry and physics, with an emphasis on condensed matter and many-body theory. The majority of the contributions to this book come from Norman's closest collaborators and friends around the world.

Norman was born in 1927. He obtained the B.Sc. in Physics (1st Class) and the Ph.D. from London University, UK (MA Oxford). He conducted research for the Ph.D. degree directly under the supervision of Prof. Charles A. Coulson. He became Professor of Physics at the University of Sheffield in 1961 and Professor of Theoretical Solid-State Physics at Imperial College, London, in 1972. After the untimely death of Prof. Coulson, in 1976 the Coulson Chair of Theoretical Chemistry was established at the University of Oxford. Norman March was the first holder of that chair, to which he was appointed in 1976 and in which he remained until retirement in 1994. Since 1994, Norman March is Emeritus Professor of the University of Oxford. He received the Honorary Doctorate in 1974 from Chalmers University (Gothenburg, Sweden) and the Laurea Honoris Causa in Physics in 2003 from the University of Catania (Italy) (see Fig. 1). In 2000–2001, he enjoyed an inter-university foreign Franqui Chair between the Free University of Brussels (VUB) and Antwerp University (UA), Belgium. He also served as Chair of the Solid-State Advisory Committee at the International Centre for Theoretical Physics (ICTP) in Trieste, Italy, from 1980 to 1991.

Norman has continued to be scientifically active after his formal retirement, maintaining affiliations in Belgium (U. Antwerp) and in Italy (Abdus Salam International Center, Trieste). A true globetrotter (frequently using an InterRail ticket across Europe, to the astonishment of his much younger fellow travellers), his 'fixed points' included Cardiff (UK), Catania (Italy), Debrecen (Hungary), Donostia/San Sebastian (the Basque Country, Spain), Galveston (Texas), Heidelberg (Germany), Pisa (Italy), Szeged (Hungary) and Valladolid (Spain), in addition to Trieste and Antwerp (which he uses to dub as his 'hub'). For decades, he has been a regular attendant at the Sanibel Symposium on Quantum Chemistry, Dynamics, Condensed Matter and Materials Physics and Biological Applications at St. Simon's Island (Florida, USA).

His activity covers more than six decades. Since the 1950s, Prof. Norman March has been a leading scientist in atomic, molecular and condensed state physics. He has written more than 1150 papers and has written or edited more than 30 monographs, alone or in collaboration with more than 300 coauthors, including a long list of young people he mentored during his academic career. Since decades, he has been Editor of 'Physics and Chemistry of Liquids', a truly international journal covering most aspects of the chemistry and physics of liquids.

The early years of Norman's career were focused on Thomas–Fermi theory [1–3]. The Thomas–Fermi statistical method laid the foundations of the theory of the density functional that only with the Hohenberg–Kohn theorems in 1964 was put on a firm theoretical footing [4] (see Ref. [5] for a review). Norman has made a pioneering contribution in the 1950s and early 1960s in this context. Two papers on variational methods involving one particle density and density matrix for particles moving in a linear harmonic oscillator are worthy of mentioning [6–8]. The idea of taking into account first the mathematically treatable models to understand the essential aspects of the physics has been one of the main features in the scientific approach of Norman March. The Thomas–Fermi model was, in that period, an innovative method in tackling many-body problems in quantum mechanics not only for atoms but also for molecules and solids. In order to emphasize the tremendous impact that March's early papers on Thomas–Fermi theory [9, 10] had among his contemporaries (see Ref. [11] for a review), let us mention the fact that the original paper was translated into the Chinese language by a colleague, and circulated as a manuscript.

Norman computed in the 1950s the momentum distribution and the electron density in atoms and simple organic molecules. He introduced a model, again based on a Thomas–Fermi (TF) approach, to determine the bond length in tetrahedral and octahedral molecules [12]. The so-called March model has been applied more recently in boron and carbon cages including fullerene [13, 14]. Norman applied for the first time the TF model to a variety of systems in going from atoms to solid metals.

The condensed state is another field of major interest for Norman March. Starting from the statistical theory of the electron gas and reaching to metals and semiconductors including the presence of impurities, he studied a wide class of phenomena by interpreting experimental results in terms of concise and elegant theories. After the discovery of high-T_c superconductivity, Prof. March has been attracted by the challenging exotic properties of these materials, including their unconventional pairing mechanism. In particular, he and his collaborators gave contributions to describe their phase diagram in terms of universal properties, such as the superconducting coherence length and effective mass, which enabled to embrace other superconducting material classes, including the heavy-Fermion materials and the ruthenates. Professor March's insight and earlier results within linear response theory on the behaviour of an electron liquid around isolated impurities in solids [15] stimulated more recent work of younger scientists in similar problems, but in unconventional superconductors and in other low-dimensional materials, such as graphene. Professor March also gave contributions to the study of the properties of matter under extreme conditions, one example being molecular iodine under high pressure [16]. His efforts in this direction then kindled more research towards compressed hydrogen and the alkalis.

Another critical phenomenon that has been studied in many works by Norman is Wigner crystallization. Wigner predicted in 1934 the crystallization of a uniform electron gas when the density is lower than a critical value [17]. Ceperley and Alder found this critical density by means of Quantum Monte Carlo simulations in 1980 [18]. Norman March made his first contribution in this area in 1958 through the virial theorem by partitioning the total correlation energy per particle into kinetic and potential energy contributions [19]. Electron crystallization of jellium, the uniform electron gas, is then treated in a paper with Care in 1975 [20, 21]. In such a low-density regime, electron correlation takes a dominant role even for a finite number of electrons (see Ref. [22] for a review). In recent years, Norman considered also, more in general, an assembly of electrons in a quantum dot or confined by an extremely weak external potential where Wigner molecules can be formed [23].

The present volume endeavours to lay a track through the diverse interests covered by Prof. March's long career in science, by presenting both novel results and overall afterthoughts. As suggested by the title, one *fil rouge* through the milestones set by Norman in his studies is provided by many-body correlations, which determine the often unconventional properties of large assemblies of particles, as their number or scale increases. This goes through electrons in atoms (Chaps. 16–18, 20), atoms in molecules and molecules in clusters (Chap. 19), condensed matter in both the liquid (Chaps. 12, 13) and the solid states (Chap. 8), with specific reference to the role of impurities and disorder (Chaps. 6, 9, 10, 24), superconductivity (Chaps. 1, 5), nuclear and subnuclear matter (Chap. 23), and even the whole Universe and the fabric of spacetime (Chap. 29).

While we are aware that we have possibly omitted several topics to which Prof. March has given contributions, we hope this volume will stand as a testimony and serve as a grateful and affectionate tribute to Norman's 'quiet' way of doing science.

Catania, Italy
Pisa, Italy
July 2017

G. G. N. Angilella
C. Amovilli

References

1. L.H. Thomas, Math. Proc. Cambridge Phil. Soc. **23**, 542 (1926). https://doi.org/10.1017/S0305004100011683
2. E. Fermi, Rendiconti dell'Accademia Nazionale dei Lincei **6**, 602 (1927)
3. E. Fermi, Z. Physik **48**, 73 (1928). https://doi.org/10.1007/BF01351576
4. P.C. Hohenberg, W. Kohn, Phys. Rev. **136**, B864 (1964). https://doi.org/10.1103/PhysRev.136.B864
5. R.G. Parr, W. Yang, *Density Functional Theory of Atoms and Molecules* (Oxford University Press, Oxford, 1989). ISBN 9780195092769
6. N.H. March, W.H. Young, Proc. Phys. Soc. **72**(2), 182 (1958). https://doi.org/10.1088/0370-1328/72/2/302
7. N.H. March, W.H. Young, Phil. Mag. A **4**(39), 384 (1959), Reprinted in Ref. [24]. https://doi.org/10.1080/14786435908233351
8. W.H. Young, N.H. March, Proc. R. Soc. A **256**(1284), 62 (1960), Reprinted in Ref. [24]. https://doi.org/10.1098/rspa.1960.0093
9. N.H. March, Acta Cryst. **5**(2), 187 (1952), Reprinted in Ref. [24]. https://doi.org/10.1107/S0365110X52000551
10. N.H. March, Adv. Phys. **6**(21), 1 (1957). https://doi.org/10.1080/00018735700101156
11. N.H. March, in *Theory of the Inhomogeneous Electron Gas*, ed. by S. Lundqvist, N.H. March (Springer, New York, 1983), Chap. 1, p. 1. https://doi.org/10.1007/978-1-4899-0415-7_1
12. N.H. March, Proc. Cambridge Philos. Soc. **48**, 665 (1952), Reprinted in Ref. [24]. https://doi.org/10.1017/S0305004100076441
13. F. Siringo, G. Piccitto, R. Pucci, Phys. Rev. A **46**, 4048 (1992). https://doi.org/10.1103/PhysRevA.46.4048
14. F.E. Leys, C. Amovilli, N.H. March, J. Chem. Inf. Comput. Sci. **44**(1), 122 (2004). https://doi.org/10.1021/ci0200624
15. J.C. Stoddart, N.H. March, M.J. Stott, Phys. Rev. **186**, 683 (1969). https://doi.org/10.1103/PhysRev.186.683
16. F. Siringo, R. Pucci, N.H. March, Phys. Rev. B **37**, 2491 (1988). https://doi.org/10.1103/PhysRevB.37.2491
17. E. Wigner, Phys. Rev. **46**, 1002 (1934). https://doi.org/10.1103/PhysRev.46.1002
18. D.M. Ceperley, B.J. Alder, Phys. Rev. Lett. **45**, 566 (1980). https://doi.org/10.1103/PhysRevLett.45.566
19. N.H. March, Phys. Rev. **110**, 604 (1958), Reprinted in Ref. [24]. https://doi.org/10.1103/PhysRev.110.604
20. C.M. Care, N.H. March, J. Phys. C: Solid State Phys. **4**(18), L372 (1971), Reprinted in Ref. [24]
21. C.M. Care, N.H. March, Adv. Phys. **24**(1), 101 (1975), Reprinted in Ref. [24]. https://doi.org/10.1080/00018737500101381
22. G. Senatore, N.H. March, Rev. Mod. Phys. **66**, 445 (1994), Reprinted in Ref. [24]. https://doi.org/10.1103/RevModPhys.66.445
23. C. Amovilli, N.H. March, Phys. Rev. A **83**, 044502 (2011). https://doi.org/10.1103/PhysRevA.83.044502
24. N.H. March, G.G.N. Angilella (eds.), *Many-body theory of molecules, clusters, and condensed phases* (World Scientific, Singapore, 2009). ISBN 9789814271776

Fig. 1 Professor Norman H. March (right) with Professor Renato Pucci (left), in Catania, Italy, 2007

Contents

Contributors

F. Aliotta CNR-IPCF, Messina, Italy

Julio A. Alonso Departamento de Física Teórica, Atómica y Óptica, Universidad de Valladolid, Valladolid, Spain

Lydia Alonso Departamento de Física Teórica, Atómica y Óptica, Universidad de Valladolid, Valladolid, Spain

C. Amovilli Dipartimento di Chimica e Chimica Industriale, University of Pisa, Pisa, Italy

G. G. N. Angilella Dipartimento di Fisica e Astronomia, Università di Catania, Catania, Italy; Scuola Superiore di Catania, Università di Catania, Catania, Italy; IMM-CNR, UdR Catania, Catania, Italy; INFN, Sez. Catania, Catania, Italy

P. W. Ayers Department of Chemistry and Chemical Biology, McMaster University, Hamilton, ON, Canada

Robert Balawender Institute of Physical Chemistry of the Polish Academy of Sciences, Warsaw, Poland

M. Baldo Istituto Nazionale di Fisica Nucleare, Sezione di Catania, Catania, Italy

G. Baskaran The Institute of Mathematical Sciences, Chennai, India; Perimeter Institute for Theoretical Physics, Waterloo, ON, Canada

M. Bercx EMAT, Department of Physics, University of Antwerp, Antwerpen, Belgium

A. Cabo Montes de Oca Departamento de Física Teórica, Instituto de Cibernética, Matemática y Física, Havana, Cuba

D. Chakraborty Department of Chemistry and Chemical Biology, McMaster University, Hamilton, ON, Canada

M. Chan Department of Chemistry and Chemical Biology, McMaster University, Hamilton, ON, Canada

M. E. Charro Faculty of Education, University of Valladolid, Valladolid, Spain

F. Claro Pontificia Universidad Católica de Chile, Instituto de Física, Santiago, Chile

D. Corradini American Physical Society, Ridge, NY, USA

R. Cuevas-Saavedra Department of Chemistry and Chemical Biology, McMaster University, Hamilton, ON, Canada

Frank De Proft General Chemistry (ALGC), Vrije Universiteit Brussel (Free University of Brussels, VUB), Brussels, Belgium

F. Flores Departamento de Física Teórica de la Materia Condensada and IFIMAC, Universidad Autónoma de Madrid, Madrid, Spain

F. M. Floris Dipartimento di Chimica e Chimica Industriale, University of Pisa, Pisa, Italy

G. Forte Dipartimento di Scienze del Farmaco, Università di Catania, Catania, Italy

P. Fulde Max-Planck-Institut für Physik Komplexer Systeme, Dresden, Germany

P. Gallo Dipartimento di Matematica e Fisica, Università "Roma Tre", Roma, Italy

Paul Geerlings General Chemistry (ALGC), Vrije Universiteit Brussel (Free University of Brussels, VUB), Brussels, Belgium

P. V. Giaquinta Dipartimento di Scienze Matematiche e Informatiche, Scienze Fisiche e Scienze della Terra, Università degli Studi di Messina, Messina, Italy

M. L. Glasser Donostia International Physics Center, San Sebastián, Spain; Department of Physics, Clarkson University, Potsdam, NY, USA; Department of Theoretical Physics, University of Valladolid, Valladolid, Spain

E. C. Goldberg Instituto de Física del Litoral (CONICET-UNL), Santa Fe, Argentina; Departamento Ingeniería de Materiales, Facultad de Ingeniería Química, Universidad Nacional del Litoral, Santa Fe, Argentina

A. Grassi Dipartimento di Scienze del Farmaco, Università di Catania, Catania, Italy

Andrzej Holas Institute of Physical Chemistry of the Polish Academy of Sciences, Warsaw, Poland

A. Kamenev W. I. Fine Theoretical Physics Institute and School of Physics and Astronomy, University of Minnesota, Minneapolis, MN, USA

D. Lamoen EMAT, Department of Physics, University of Antwerp, Antwerpen, Belgium

G. M. Lombardo Dipartimento di Scienze del Farmaco, Università di Catania, Catania, Italy

María J. López Departamento de Física Teórica, Atómica y Óptica, Universidad de Valladolid, Valladolid, Spain

M. Martin Conde Dipartimento di Matematica e Fisica, Università “Roma Tre”, Roma, Italy

C. C. Matthai Department of Physics and Astronomy, Cardiff University, Cardiff, UK

Ch. B. Mendl Stanford Institute for Materials and Energy Sciences, SLAC National Accelerator Laboratory and Stanford University, Menlo Park, CA, USA

K. Morawetz Münster University of Applied Sciences, Steinfurt, Germany; International Institute of Physics (IIP), Campus Universitário Lagoa Nova, Natal, Brazil; Max-Planck-Institute for the Physics of Complex Systems, Dresden, Germany

Á. Nagy Department of Theoretical Physics, University of Debrecen, Debrecen, Hungary

I. Nagy Department of Theoretical Physics, Institute of Physics, Budapest University of Technology and Economics, Budapest, Hungary; Donostia International Physics Center, San Sebastián, Spain

L. M. Nieto Departamento de Física Teórica, Atómica y Óptica, Universidad de Valladolid, Valladolid, Spain; Instituto de Investigación en Matemáticas (IMUVA), Facultad de Ciencias, Universidad de Valladolid, Valladolid, Spain

D. Oriti Max Planck Institute for Gravitational Physics (Albert Einstein Institute), Potsdam-Golm, EU, Germany

S. Palpacelli Hyperlean S.r.l, Ancona, Italy

B. Partoens CMT group, Department of Physics, University of Antwerp, Antwerp, Belgium

F. M. D. Pellegrino NEST, Istituto Nanoscienze-CNR, Modena, Italy; Scuola Normale Superiore, Pisa, Italy

M. Piris Donostia International Physics Center (DIPC), Donostia, Euskadi, Spain; Euskal Herriko Unibertsitatea (UPV/EHU), Donostia, Euskadi, Spain; Basque Foundation for Science (IKERBASQUE), Bilbao, Euskadi, Spain

M. Polini Graphene Labs, IIT, Istituto Italiano di Tecnologia, Genova, Italy

R. C. Ponterio CNR-IPCF, Messina, Italy

R. Pucci Dipartimento di Fisica e Astronomia, Università di Catania, Catania, Italy; IMM-CNR, UdR Catania, Catania, Italy

P. Pugliese Dipartimento di Matematica e Fisica, Università “Roma Tre”, Roma, Italy

P. Robles Escuela de Ingeniería Eléctrica, Pontificia Universidad Católica de Valparaíso, Valparaíso, Chile

M. Rovere Dipartimento di Matematica e Fisica, Università "Roma Tre", Roma, Italy

F. Saija CNR-IPCF, Messina, Italy

R. Saniz CMT group, Department of Physics, University of Antwerp, Antwerp, Belgium

F. Siringo Dipartimento di Fisica e Astronomia, Università di Catania, and INFN Sezione di Catania, Catania, Italy

S. Succi Istituto Applicazioni Calcolo, CNR, Roma, Italy; Institute for Applied Computational Science, John Paulson school of Engineering and Applied Sciences, Harvard University, Cambridge, MA, USA

R. H. Squire Department of Natural Sciences, West Virginia University, Institute of Technology, Beckley, WV, USA

I. Torre NEST, Istituto Nanoscienze-CNR, Modena, Italy; Scuola Normale Superiore, Pisa, Italy; Graphene Labs, IIT, Istituto Italiano di Tecnologia, Genova, Italy

Christian Van Alsenoy Department of Chemistry, University of Antwerp, Antwerp, Belgium

Zhidong Zhang Shenyang National Laboratory for Materials Science, Institute of Metal Research, Chinese Academy of Sciences, Shenyang, People's Republic of China

Part I
Condensed Matter Theory

Chapter 1
Correlations in the Superconducting Properties of Several Material Classes

G. G. N. Angilella and R. Pucci

Abstract Complex phenomena, as those involving many particles, may still exhibit simple patterns, usually expressed by simple relations among relevant quantities. We briefly recount how Professor Norman March led our way through superconducting experiments for several material classes, including the high-T_c cuprates and the heavy-Fermion materials, in the unending quest for simple paths among intricated data.

1.1 Introduction

Looking for correlations among different physical variables is certainly a powerful, yet quite difficult and sometimes even deceiving, way to assess a theory, or even a physical law, concerning a given physical system. This has to do with the difficulty of distilling those (hopefully) few physical variables which almost certainly relate to the effect under consideration, and which may belong to quite different scales. In the case of many-body systems, one intrinsic source of difficulty in looking for correlation among physical quantities comes however from the inherent complexity of the system: the correlation one is after might not be a linear one, as is e.g. the case for strain and applied stress in Hooke's law of elasticity, but an intrinsically nonlinear (and often complicated) correlation. This is for instance the case of the correlation between superconducting temperature T_c, electron-phonon coupling

G. G. N. Angilella (✉) · R. Pucci
Dipartimento di Fisica e Astronomia, Università di Catania, Catania, Italy
e-mail: giuseppe.angilella@ct.infn.it

R. Pucci · G. G. N. Angilella
IMM-CNR, UdR Catania, Via S. Sofia, 64, 95123 Catania, Italy
e-mail: renato.pucci@ct.infn.it

G. G. N. Angilella
INFN, Sez. Catania, Via S. Sofia, 64, 95123 Catania, Italy

G. G. N. Angilella
Scuola Superiore di Catania, Università di Catania, Via Valdisavoia, 9, 95123 Catania, Italy

G. G. N. Angilella and C. Amovilli (eds.), *Many-body Approaches at Different Scales*, https://doi.org/10.1007/978-3-319-72374-7_1

constant λ, and Debye frequency ω_D in conventional superconductors. Even at the mean-field level, where many-body correlations are treated at the lowest (albeit self-consistent) level, BCS theory shows that the three quantities above are correlated via $k_B T_c \propto \hbar\omega_D e^{-1/\lambda N(0)}$, where $N(0)$ is the single-particle density of states (DOS) at the Fermi level. Although the superconducting energy scale $k_B T_c$ is linearly correlated with the phonon energy scale $\hbar\omega_D$, these two energies can differ by orders of magnitude in several superconducting materials. Moreover, the functional correlation between $k_B T_c$ and the dimensionless coupling strength $\lambda N(0)$ is nonlinear, and even non-analytic. It has been emphasized that this precludes the application of perturbation theory to the problem of superconductivity, despite pairing instability in the presence of a filled Fermi sea was predicted by Cooper to occur for any weak coupling [1]. In fact, the best approximation available to date for a theory of superconductivity, viz. BCS theory, firmly rests on a variational ansatz [2, 3].

1.2 Preformed Pairs Scenario in High-T_c Superconductors

Looking for (and often finding) correlations among relevant physical quantities in many-body phenomena is an art in which Professor Norman March has been a rather skilful artist. One example concerning superconductivity in the high-T_c (HTSC) cuprates is the search for a crossover between the normal and superconducting state. Using two-dimensional Fermi liquid theory, as discussed by Kohno and Yamada [4], March, Pucci and Egorov [5] (see also [6–8]) were able to relate the product of electrical resistivity R and nuclear spin-lattice relaxation time T_1 to temperature T via the remarkably simple correlation law

$$RT_1 \propto T. \tag{1.1}$$

This linear correlation is indeed obeyed by $YBa_2Cu_4O_8$ for temperatures $T \gtrsim T_m$, where $T_m \approx 150\,K$ is a crossover temperature below which such correlation is lost [5, 9] (Fig. 1.1). March, Pucci and Egorov [5] then conclude that $E_b \sim k_B T_m$ sets the energy scale at which the system departs from a Fermi liquid theory of monomers, and rather enters a phase characterized by charged dimers ($2e$ bosons). At that time, it was tempting to identify these dimers either with Cooper pairs, or with localized, strongly-bound composite bosons. A theoretical description of the crossover between BCS pairing and Bose-Einstein condensation, with the coupling strength and density as parameters driving such crossover, was emerging, following the seminal work of Nozières and Schmitt–Rink [10] (see also Refs. [11, 12]). The physical nature of such bosons, and therefore the microscopic mechanism at the basis of their binding, was at that time (and presently still is!) the subject of an open debate, one likely possibility being that of bisolitons [13] or bipolarons [14, 15], or of resonating valence bonds [16–19] (see also Baskaran [20], in this volume), or tunneling pairs between adjacent CuO_2 layers [21–26]. Compelling experimental evidence, first from calorimetric measurements [27], then from angle-resolved photoemission spectroscopy (ARPES)

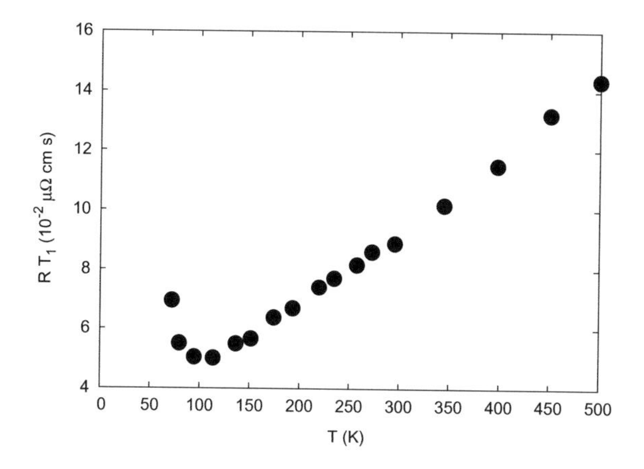

Fig. 1.1 Product of electrical resistivity R times spin-lattice relaxation time T_1 versus temperature T for $YBa_2Cu_4O_8$. Data from Bucher et al. [9]. Redrawn after Ref. [5]

[28], and from scanning tunneling microscopy [29], eventually pointed towards the existence of a 'pseudogap' phase below a characteristic temperature $T^* \gg T_c$ in the underdoped cuprates. Such a pseudogap phase is characterized by preformed pairs indeed, which do not however yet participate in superconductivity, as they lack phase coherence [30].

1.3 Towards a Fermionic Mechanism for Exotic Superconductivity: The Case of the Heavy-Fermion Materials

One source of evidence that HTSC may originate from a non-phononic mechanism was provided by Uemura et al. [31, 32], who recognized that a universal correlation (again) between T_c and the ratio of superconducting density n_s to fermionic effective mass m^* exists for different superconducting material classes, thereby supporting the idea of a fermionic nature of the superconducting mechanism. This idea was generalized by Angilella, March and Pucci [33] to embrace the so-called heavy-Fermion materials, a class of exotic superconductors characterized by magnetic correlations and relatively large value of the effective to bare electronic mass ratio, m^*/m_e, although their relatively low superconducting temperature T_c does not qualify them for 'high-T_c' superconductors. In Ref. [33] (see also Refs. [34, 35]), then, Angilella, March and Pucci indeed reported an empiric correlation between the superconducting condensation energy $\sim k_B T_c$ and a fermionic 'characteristic energy' $\varepsilon_c = \hbar^2/m^*\xi^2$, where ξ is the superconducting coherence length. Such a correlation, however, proved to be nonlinear, with $k_B T_c$ saturating at large values of the characteristic energy ε_c. Angilella et al. [36] then developed a model, based on the Bethe–Goldstone equation for the Cooper problem in non-s-wave superconductors, explicitly providing an expression for the universal correlation

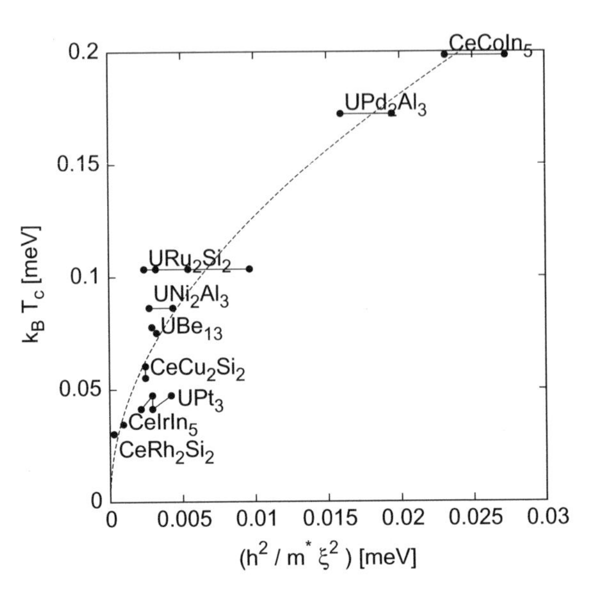

Fig. 1.2 Experimental values of the superconducting transition temperatures T_c for several heavy-Fermion materials plotted against the characteristic energy $\varepsilon_c = \hbar^2/m^*\xi^2$. Data for T_c, m^*, ξ have been taken from Table 1 in Ref. [34]. The dashed curve is a phenomenological fit based on Eq. (1.2), as discussed in Ref. [36] (see also Ref. [37])

$$k_B T_c = f\left(\frac{\hbar^2}{m^*\xi^2}\right) \tag{1.2}$$

already empirically demonstrated to hold for the heavy-Fermion materials (Fig. 1.2). Such expression (see also Ref. [37] for a recent review) parametrically depends on ℓ, with $\ell = 0, 1, 2$ for an s-, p- or d-wave symmetry order parameter, which therefore enables one to embrace conventional, heavy-Fermion, and cuprate superconductors, respectively.

One open question remains of course the physical nature of the pairing interaction in the HTSC. Several experimental findings seem to disfavour a purely phononic mechanism of superconductivity in the cuprates, although the lattice and its deformation is still expected to play an important role (see e.g. Ref. [38], where the correlation between T_c and elastic constants in various superconducting elements is investigated). Applied strain, either external of 'chemical', may indeed affect the kinetic sector of the many-body Hamiltonian, through a modification of the band parameters and/or doping, rather than the coupling sector thereof. This novel kind of correlation was indeed recognized by Pavarini et al. [39], who established a non-monotonic dependence of the optimal T_c on the next-nearest to nearest hopping parameter ratio, $r = t'/t$, across several cuprate compounds. A model for this dependence was then provided by Angilella et al. [40, 41], in terms of the proximity to a strain-induced electronic topological transition (ETT) [42, 43].

In the cuprates, characterized by a relatively short superconducting coherence length ξ, various kinds of short-range spectroscopies, including optical and neutron ones, pointed strikingly towards the existence of antiferromagnetic fluctuations, which of course originate in the antiferromagnetic texture of their parent (i.e.

undoped) compounds [44]. Such fluctuations were therefore claimed to be responsible of pairing in the superconducting phase: since the pairing interaction which eventually enable electrons to get bound in Cooper pairs is not of phononic origin, but rather has an electronic one, it has been said that 'electrons pair themselves' [45]. Evidence of antiferromagnetic fluctuations has also been gathered for several heavy-Fermion materials [46, 47], where also ferromagnetism can coexist with spin-triplet superconductivity [48–50], and for Sr_2RuO_4 [51, 52].

Angilella, March and Pucci [53] then exploited such evidence for superconductivity mediated by antiferromagnetic fluctuations to infer a correlation between the spin-fluctuation energy scale $k_B T_{sf}$ and the pairing characteristic energy $\varepsilon_c = \hbar^2/m^*\xi^2$ defined above, and entering Eq. (1.2) for the universal law of superconduting correlation.

1.4 Summary

We have briefly reviewed some research activity spanning several decades of close acquaintance and collaboration with Professor Norman March. The results presented here are related to superconductivity in several material classes, including the cuprates and the heavy-Fermion materials. The guiding light in Professor March's quest has always been the search for simple correlation among 'factual data': these always come before any model or theory, but have certainly fostered new and stimulating views at how relevant physical quantities actually correlated in complex phenomena involving many particles.

Acknowledgements It is a pleasure and a honour to acknowledge scientific guidance, fruitful collaboration, and intimate friendship with Professor Norman March during the last several decades. RP wishes to thank the Department of Theoretical Chemistry, Oxford University, for much hospitality during his yearly visits for over a decade in the 1990s. Similarly, GGNA wishes to thank the Department of Physics, University of Antwerp, for providing a fruitful and stimulating environment where it was possible to meet and interact with Professor March for almost a decade in the 2000s. In particular, thanks are due to Professors I. Howard, D. Lamoen and K. Van Alsenoy, as well as to Dr F. E. Leys.

References

1. L.N. Cooper, Phys. Rev. **104**, 1189 (1956). https://doi.org/10.1103/PhysRev.104.1189
2. J. Bardeen, L.N. Cooper, J.R. Schrieffer, Phys. Rev. **106**, 162 (1957). https://doi.org/10.1103/PhysRev.106.162
3. J. Bardeen, L.N. Cooper, J.R. Schrieffer, Phys. Rev. **108**(5), 1175 (1957). https://doi.org/10.1103/PhysRev.108.1175
4. H. Kohno, K. Yamada, Prog. Theor. Phys. **85**(1), 13 (1991). https://doi.org/10.1143/ptp/85.1.13
5. N.H. March, R. Pucci, S.A. Egorov, Phys. Chem. Liq. **28**(2), 141 (1994). https://doi.org/10.1080/00319109408029549

6. S.A. Egorov, N.H. March, Phys. Chem. Liq. **27**(3), 195 (1994). https://doi.org/10.1080/00319109408029525
7. S.A. Egorov, N.H. March, Phys. Chem. Liq. **30**(1), 59 (1995). https://doi.org/10.1080/00319109508028433
8. G.G.N. Angilella, N.H. March, R. Pucci, Phys. Chem. Liq. **38**, 615 (2000). https://doi.org/10.1080/00319100008030308
9. B. Bucher, P. Steiner, J. Karpinski, E. Kaldis, P. Wachter, Phys. Rev. Lett. **70**, 2012 (1993). https://doi.org/10.1103/PhysRevLett.70.2012
10. P. Nozières, S. Schmitt-Rink, J. Low Temp. Phys. **59**(3), 195 (1985). https://doi.org/10.1007/BF00683774
11. F. Pistolesi, G.C. Strinati, Phys. Rev. B **49**, 6356 (1994). https://doi.org/10.1103/PhysRevB.49.6356
12. M. Randeria, in *Bose Einstein Condensation*, ed. by A. Griffin, D. Snoke, S. Stringari (Cambridge University Press, Cambridge, 1995), p. 355
13. A.S. Davydov, Phys. Rep. **190**(4), 191 (1990). https://doi.org/10.1016/0370-1573(90)90061-6
14. A.S. Alexandrov, *Theory of Superconductivity: From Weak to Strong Coupling* (IOP Publishing, Bristol, 2003)
15. A. La Magna, R. Pucci, Phys. Rev. B **55**, 14886 (1997). https://doi.org/10.1103/PhysRevB.55.14886
16. P.W. Anderson, Science **235**, 1196 (1987). https://doi.org/10.1126/science.235.4793.1196
17. P.W. Anderson, G. Baskaran, Z. Zou, T. Hsu, Phys. Rev. Lett. **58**, 2790 (1987). https://doi.org/10.1103/PhysRevLett.58.2790
18. G. Baskaran, Z. Zou, P.W. Anderson, Solid State Commun. **63**(11), 973 (1987). https://doi.org/10.1016/0038-1098(87)90642-9
19. G. Baskaran, P.W. Anderson, Phys. Rev. B **37**, 580 (1988). https://doi.org/10.1103/PhysRevB.37.580
20. G. Baskaran, in *Many-Body Approaches at Different Scales: A Tribute to Norman H. March on the Occasion of his 90th Birthday, chap. 5*, ed. by G.G.N. Angilella, C. Amovilli (Springer, New York, 2018), p. 43. (This volume). https://doi.org/10.1007/978-3-319-72374-7_5
21. P.W. Anderson, Phys. Rev. B **42**, 2624 (1990). https://doi.org/10.1103/PhysRevB.42.2624
22. P.W. Anderson, Phys. Rev. Lett. **64**, 1839 (1990). https://doi.org/10.1103/PhysRevLett.64.1839. Reprinted in [24]
23. S. Chakravarty, A. Sudbø, P.W. Anderson, S. Strong, Science **261**(5119), 337 (1993). https://doi.org/10.1126/science.261.5119.337
24. P.W. Anderson, *The Theory of Superconductivity in the High-T_c Cuprates* (Princeton University Press, Princeton NJ, 1997). ISBN 9780691043654
25. G.G.N. Angilella, R. Pucci, F. Siringo, A. Sudbø, Phys. Rev. B **59**, 1339 (1999). https://doi.org/10.1103/PhysRevB.59.1339
26. G.G.N. Angilella, A. Sudbø, R. Pucci, Eur. Phys. J. B **15**(2), 269 (2000). https://doi.org/10.1007/s100510051125
27. J.W. Loram, K.A. Mirza, J.R. Cooper, W.Y. Liang, Phys. Rev. Lett. **71**, 1740 (1993). https://doi.org/10.1103/PhysRevLett.71.1740
28. M. Randeria, J. Campuzano, in *Models and Phenomenology for Conventional and High-Temperature Superconductivity*, ed. by G. Iadonisi, J.R. Schrieffer, M.L. Chiofalo (IOS, Amsterdam, 1999), Proceedings of the CXXXVI International School of Physics E. Fermi (Varenna, Italy, 1997), pp. 115–139. ISBN 9781614992219
29. M. Kugler, Ø. Fischer, C. Renner, S. Ono, Y. Ando, Phys. Rev. Lett. **86**, 4911 (2001). https://doi.org/10.1103/PhysRevLett.86.4911
30. V.J. Emery, S.A. Kivelson, Nature **374**(6521), 434 (1995). https://doi.org/10.1038/374434a0
31. Y.J. Uemura, G.M. Luke, B.J. Sternlieb, J.H. Brewer, J.F. Carolan, W.N. Hardy, R. Kadono, J.R. Kempton, R.F. Kiefl, S.R. Kreitzman, P. Mulhern, T.M. Riseman, D.L. Williams, B.X. Yang, S. Uchida, H. Takagi, J. Gopalakrishnan, A.W. Sleight, M.A. Subramanian, C.L. Chien, M.Z. Cieplak, G. Xiao, V.Y. Lee, B.W. Statt, C.E. Stronach, W.J. Kossler, X.H. Yu, Phys. Rev. Lett. **62**, 2317 (1989). https://doi.org/10.1103/PhysRevLett.62.2317

32. Y.J. Uemura, L.P. Le, G.M. Luke, B.J. Sternlieb, W.D. Wu, J.H. Brewer, T.M. Riseman, C.L. Seaman, M.B. Maple, M. Ishikawa, D.G. Hinks, J.D. Jorgensen, G. Saito, H. Yamochi, Phys. Rev. Lett. **66**, 2665 (1991). https://doi.org/10.1103/PhysRevLett.66.2665
33. G.G.N. Angilella, N.H. March, R. Pucci, Phys. Rev. B **62**, 13919 (2000). https://doi.org/10.1103/PhysRevB.62.13919
34. G.G.N. Angilella, N.H. March, R. Pucci, Supercond. Sci. Technol. **18**, 557 (2005). https://doi.org/10.1088/0953-2048/18/4/028
35. G.G.N. Angilella, N.H. March, R. Pucci, Phys. Chem. Liq. **39**, 405 (2001). https://doi.org/10.1080/00319100108031673
36. G.G.N. Angilella, F.E. Leys, N.H. March, R. Pucci, Phys. Lett. A **322**(5–6), 375 (2004). https://doi.org/10.1016/j.physleta.2003.12.066
37. G.G.N. Angilella, in *Correlations in Condensed Matter Under Extreme Conditions: A Tribute to Renato Pucci on the Occasion of his 70th Birthday, chap. 3*, ed. by G.G.N. Angilella, A. La Magna (Springer, Berlin, 2017), pp. 31–46. https://doi.org/10.1007/978-3-319-53664-4_3. ISBN 9783319536637
38. G.G.N. Angilella, N.H. March, R. Pucci, Eur. Phys. J. B **39**(4), 427 (2004). https://doi.org/10.1140/epjb/e2004-00213-y
39. E. Pavarini, I. Dasgupta, T. Saha-Dasgupta, O. Jepsen, O.K. Andersen, Phys. Rev. Lett. **87**(4), 047003 (2001). https://doi.org/10.1103/PhysRevLett.87.047003
40. G.G.N. Angilella, E. Piegari, A.A. Varlamov, Phys. Rev. B **66**(1), 014501 (2002). https://doi.org/10.1103/PhysRevB.66.014501
41. G.G.N. Angilella, G. Balestrino, P. Cermelli, P. Podio-Guidugli, A.A. Varlamov, Eur. Phys. J. B **26**(1), 67 (2002). https://doi.org/10.1140/epjb/e20020067
42. I.M. Lifshitz, Sov. Phys. JETP **11**, 1130 (1960); [Zh. Eksp. Teor. Fiz. **38**, 1569 (1960)]
43. Ya.M. Blanter, M.I. Kaganov, A.V. Pantsulaya, A.A. Varlamov, Phys. Rep. **245**(4), 159 (1994). https://doi.org/10.1016/0370-1573(94)90103-1
44. J.P. Carbotte, E. Schachinger, D.N. Basov, Nature **401**(6751), 354 (1999). https://doi.org/10.1038/43843
45. J. Orenstein, Nature **401**(6751), 333 (1999). https://doi.org/10.1038/43801
46. G.R. Stewart, Z. Fisk, J.O. Willis, J.L. Smith, Phys. Rev. Lett. **52**, 679 (1984). https://doi.org/10.1103/PhysRevLett.52.679
47. G.R. Stewart, Rev. Mod. Phys. **56**(4), 755 (1984). https://doi.org/10.1103/RevModPhys.56.755
48. S.S. Saxena, P. Agarwal, K. Ahilan, F.M. Grosche, R.K.W. Haselwimmer, M.J. Steiner, E. Pugh, I.R. Walker, S.R. Julian, P. Monthoux, G.G. Lonzarich, A. Huxley, I. Sheikin, D. Braithwaite, J. Flouquet, Nature **406**(6796), 587 (2000). https://doi.org/10.1038/35020500
49. D. Aoki, A. Huxley, E. Ressouche, D. Braithwaite, J. Flouquet, J. Brison, E. Lhotel, C. Paulsen, Nature (London) **413**(6856), 613 (2001). https://doi.org/10.1038/35098048
50. T. Hattori, Y. Ihara, Y. Nakai, K. Ishida, Y. Tada, S. Fujimoto, N. Kawakami, E. Osaki, K. Deguchi, N.K. Sato, I. Satoh, Phys. Rev. Lett. **108**, 066403 (2012). https://doi.org/10.1103/PhysRevLett.108.066403
51. P. Monthoux, G.G. Lonzarich, Phys. Rev. B **59**, 14598 (1999). https://doi.org/10.1103/PhysRevB.59.14598
52. Ar Abanov, A.V. Chubukov, A.M. Finkel'stein, Europhys. Lett. **54**(4), 488 (2001). https://doi.org/10.1209/epl/i2001-00266-0
53. G.G.N. Angilella, N.H. March, R. Pucci, Phys. Rev. B **65**(9), 092509 (2002). https://doi.org/10.1103/PhysRevB.65.092509

Chapter 2
All-Electrical Scheme for Hall Viscosity Measurement

F. M. D. Pellegrino, I. Torre and M. Polini

Abstract In highly viscous electron systems such as, for example, high quality graphene above liquid nitrogen temperature, a linear response to applied electric current becomes essentially nonlocal, which can give rise to a number of new and counterintuitive phenomena including negative nonlocal resistance and current whirlpools [1]. Moreover, in a fluid subject to a magnetic field the viscous stress tensor has a dissipationless antisymmetric component controlled by the so-called Hall viscosity. We propose an all-electrical scheme that allows a determination of the Hall viscosity of a two-dimensional electron liquid in a solid-state device.

2.1 Introduction

Hydrodynamics [2, 3] is a powerful non-perturbative theory forgraphene the description of transport in materials where the mean free path ℓ_{ee} for electron-electron (e-e) collisions happens to be much smaller than the sample size W and the mean free path ℓ for momentum non-conserving collisions, i.e. $\ell_{ee} \ll \ell, W$. Despite the abundance of theoretical works [4–24], clear-cut experimental evidence of hydrodynamic transport in the solid state has been lacking until recently, with the exception of early longitudinal transport experiments in electrostatically defined wires in the two-dimensional (2D) e lectron gas in (Al,Ga)As heterostructures [25, 26]. The latter work reported the observation of negative differential resistance, which was interpreted as the Gurzhi effect [12] arising due to an increase in electron temperature due to current heating.

F. M. D. Pellegrino (✉) · I. Torre
NEST, Istituto Nanoscienze-CNR, Modena, Italy
e-mail: francesco.pellegrino@sns.it

F. M. D. Pellegrino · I. Torre
Scuola Normale Superiore, Piazza dei Cavalieri, 7, Pisa, Italy

I. Torre · M. Polini
Graphene Labs, IIT, Istituto Italiano di Tecnologia, Via Morego 30, 16163 Genova, Italy
e-mail: marco.polini@iit.it

G. G. N. Angilella and C. Amovilli (eds.), *Many-body Approaches at Different Scales*, https://doi.org/10.1007/978-3-319-72374-7_2

In graphene [27], hydrodynamic flow was originally predicted [10, 17, 18] to occur at the charge neutrality point (CNP), where thermally-excited electrons and holes undergo frequent collisions due to poorly-screened Coulomb interactions [28]. In this regime, the authors of Ref. [29] have recently reported experimental evidence of the violation of the Wiedemann–Franz law, which is consistent with the occurrence of highly-frictional electron-hole flow.

In the future, the strongly-interacting 2D electron-hole liquid in undoped graphene may enable investigations of solid-state nearly-perfect fluids [18], i.e. fluids with very low values of the shear viscosity (in unit of the entropy density) and therefore minimal dissipation [30]. At the CNP, however, carrier density inhomogeneities due to long-range disorder are unavoidable [31] and should be taken into account for a reliable description of the physics [14].

Microscopic calculations [32–34] suggest that also doped graphene sheets can display hydrodynamic behavior above liquid-nitrogen temperatures and for typical carrier concentrations. The reason is easy to understand. In the conventional Fermi-liquid regime, i.e. for $T \ll T_{\rm F} \equiv E_{\rm F}/\hbar$, where $E_{\rm F}$ is the Fermi energy, Pauli blocking is responsible for a very small rate of quasiparticle collisions and very long *e-e* mean free paths. In doped graphene [32–34], $\ell_{ee} \propto -[T^2 \ln(T)]^{-1}$ for $T \ll T_{\rm F}$. As temperature increases, however, the Fermi surface 'softens', Pauli blocking is not as effective, and ℓ_{ee} quickly decays, reaching a sub-micron size with an approximate power law $\ell_{ee} \propto T^{-2}$. Furthermore, in 2D crystals where momentum-non-conserving collisions are dominated by acoustic phonon scattering, ℓ decays like T^{-1}, thereby guaranteeing the existence of a temperature window where the hydrodynamic inequalities $\ell_{ee} \ll \ell, W$ can be satisfied, where W represents the typical size of the sample.

Doped graphene systems display very weak inhomogeneities due to the screening exerted on the long-range scattering sources by the electron liquid itself. Moreover, doped systems are characterized by large viscosities [34, 35] and values of ℓ_{ee} that can be comparable to ℓ, thereby offering an ideal platform to access a hydrodynamic regime in which quantum corrections to the Navier–Stokes equation are necessary, e.g. in finite magnetic fields.

A recent experimental study [35] of ultra-clean single- and bilayer graphene encapsulated between boron nitride crystals has indeed demonstrated that the 2D electron system in doped graphene displays hydrodynamic flow. For completeness, let us also mention recent reports on hydrodynamic transport in narrow quasi-2D channels of palladium cobaltate [36].

The setup that will be analyzed in this work is sketched in Fig. 2.1. It is a half-plane geometry with a single current injector, the simplest setup one can possibly imagine for the identification of viscosity-related features in nonlocal transport. The half-plane setup is conceptually very instructive.

The present chapter is based on Ref. [37] and is organized as follows. In Sect. 2.2 we review the theory of magneto-hydrodynamic transport in viscous 2D electron systems. In Sect. 2.3 we present the solution of the magneto-hydrodynamic equations in the case of a half-plane setup. Finally, in Sect. 2.4 we summarize our principal findings and draw our main conclusions.

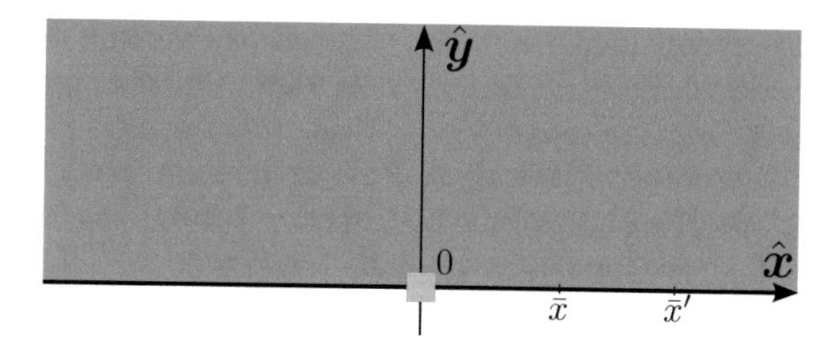

Fig. 2.1 A sketch of the nonlocal transport setup analyzed in this work. In this geometry, current is injected into a single electrode at the origin

2.2 Hydrodynamic Theory

In the presence of time-reversal symmetry breaking (e.g. due to an external magnetic field), a dissipationless term, controlled by the so-called Hall viscosity [38–49] $\eta_{\rm H}$, appears in the viscous stress tensor [3] σ'_{ij}. In two spatial dimensions one has

$$\sigma'_{ij} = \sum_{k,\ell} \eta_{ij,k\ell} v_{k\ell}, \tag{2.1}$$

where i, j, k and ℓ denote Cartesian indices, $v_{k\ell} \equiv (\partial_k v_\ell + \partial_\ell v_k)/2$, and $\eta_{ij,k\ell}$ is a rank-4 tensor, usually called the 'viscosity' tensor [40],

$$\eta_{ij,k\ell} \equiv \zeta \delta_{ij}\delta_{k\ell} + \eta(\delta_{ik}\delta_{j\ell} + \delta_{i\ell}\delta_{jk} - \delta_{ij}\delta_{k\ell}) + \eta_{\rm H}(\delta_{jk}\epsilon_{i\ell} - \delta_{i\ell}\epsilon_{kj}). \tag{2.2}$$

In Eq. (2.2), $\eta_{\rm H}$ parametrizes the portion of $\eta_{ij,k\ell}$ which is antisymmetric with respect to the exchange $ij \leftrightarrow k\ell$ and is non-zero only when time-reversal symmetry is broken.

In the linear-response and steady-state regimes, electron transport in the hydrodynamic regime in the presence of a static magnetic field $\mathbf{B} = B\hat{\mathbf{z}}$ is described by the continuity equation

$$\nabla \cdot \mathbf{J}(\mathbf{r}) = 0, \tag{2.3}$$

and the Navier–Stokes equation

$$-\nabla P(\mathbf{r}) + \nabla \cdot \boldsymbol{\sigma}'(\mathbf{r}) + e\bar{n}\nabla\varphi(\mathbf{r}) - \frac{e}{c}\mathbf{J}(\mathbf{r}) \times \mathbf{B} = \frac{m}{\tau}\mathbf{J}(\mathbf{r}). \tag{2.4}$$

Here, $\mathbf{J}(\mathbf{r}) = \bar{n}\mathbf{v}(\mathbf{r})$ is the particle current density, $\mathbf{v}(\mathbf{r})$ is the fluid element velocity, $\bar{n}$ is the ground-state uniform density, $P(\mathbf{r})$ is the pressure, $\boldsymbol{\sigma}'(\mathbf{r})$ is the viscous stress tensor whose Cartesian components have been explicitly reported in Eqs. (2.1)–(2.2), $\varphi(\mathbf{r})$ is the 2D electrostatic potential in the plane where electrons move, $-e$ is the electron charge, m is the electron effective mass, and τ is a phenomenological transport time describing momentum-non-conserving collisions [23] (e.g. scattering of electrons against acoustic phonons). The gradient of the pressure is proportional to the gradient of the density via $\nabla P(\mathbf{r}) = (\mathscr{B}/\bar{n})\nabla n(\mathbf{r})$, where $\mathscr{B} = \bar{n}^2/\mathscr{N}_0$

is the bulk modulus of the homogeneous electron liquid, $\mathcal{N}_0$ being the density of states at the Fermi energy. It is useful to define the electrochemical potential as $\phi(\mathbf{r}) = \varphi(\mathbf{r}) + \delta\mu(\mathbf{r})/(-e)$, where $\delta\mu(\mathbf{r}) = [n(\mathbf{r}) - \bar{n}]/\mathcal{N}_0$ is the chemical potential measured with respect to the equilibrium value, e.g. $\bar{\mu} = \hbar v_{\rm F}\sqrt{\pi\bar{n}}$ for the case of single-layer graphene [28] and $\bar{\mu} = \hbar^2\pi\bar{n}/(2m)$ for bilayer graphene [28]. Since experimental probes are usually sensitive to $\phi(\mathbf{r})$, from now on we will focus our attention on the electrochemical potential rather than on $\varphi(\mathbf{r})$.

We now note that the viscous stress tensor in Eqs. (2.1)–(2.2) can be written in the following compact form

$$\boldsymbol{\sigma}' = (\eta + i\eta_{\rm H}\tau_y)[(\partial_x v_x - \partial_y v_y)\tau_z + (\partial_x v_y + \partial_y v_x)\tau_x] + \zeta\nabla\cdot\mathbf{v}, \tag{2.5}$$

where τ_i with $i = x, y, z$ are standard 2×2 Pauli matrices acting on Cartesian indices. As in Eq. (2.4) above, in the linear-response and steady-state regimes we can write $\mathbf{v}(\mathbf{r}) = \mathbf{J}(\mathbf{r})/\bar{n}$. We then note that the bulk viscosity ζ couples to $\nabla\cdot\mathbf{J}$, which vanishes because of the continuity equation (2.3). The bulk viscosity term in the viscous stress tensor therefore drops out of the problem at hand. In summary, Eq. (2.5) simplifies to:

$$\boldsymbol{\sigma}' = m(\nu + i\nu_{\rm H}\tau_y)[(\partial_x J_x - \partial_y J_y)\tau_z + (\partial_x J_y + \partial_y J_x)\tau_x], \tag{2.6}$$

where $\nu \equiv \eta/(m\bar{n})$ is the kinetic shear viscosity and $\nu_{\rm H} \equiv \eta_{\rm H}/(m\bar{n})$ is the kinetic Hall viscosity. Making use of Eq. (2.6) in Eq. (2.4) and introducing the electrochemical potential $\phi(\mathbf{r})$, we can write the Navier–Stokes equation (2.4), as

$$\frac{\sigma_0}{e}\nabla\phi(\mathbf{r}) = (1 - D_\nu^2\nabla^2)\mathbf{J}(\mathbf{r}) + \omega_c\tau\left(1 + D_{\rm H}^2\nabla^2\right)\mathbf{J}(\mathbf{r}) \times \hat{\mathbf{z}}, \tag{2.7}$$

where $\sigma_0 = ne^2\tau/m$, $D_\nu \equiv \sqrt{\nu\tau}$ has been introduced in Refs. [1, 23, 35], $D_{\rm H} \equiv \sqrt{-\nu_{\rm H}/\omega_c}$, and $\omega_c \equiv eB/(mc)$ is the usual cyclotron frequency. As we will see below, $\nu_{\rm H}$ and ω_c have opposite signs so that $D_{\rm H}$ is a well defined length scale. Notice that the Hall viscosity parametrizes a correction to the ordinary Lorentz force due to the spatial dependence of the velocity $\mathbf{v}(\mathbf{r})$.

Since the setup in Fig. 2.1 is translationally-invariant in the $\hat{\mathbf{x}}$ direction, it is useful to introduce the following Fourier Transforms [1, 23] (FTs) with respect to the spatial coordinate x: $\tilde{\phi}(k, y) = \int_{-\infty}^{+\infty} dx\ e^{-ikx}\phi(\mathbf{r})$ and $\tilde{\mathbf{J}}(k, y) = \int_{-\infty}^{+\infty} dx\ e^{-ikx}\mathbf{J}(\mathbf{r})$. The three coupled partial-differential equations (2.3)–(2.4) can be combined into a 4×4 system of first-order ordinary differential equations:

$$\partial_y\mathbf{w}(k, y) = \mathcal{M}(k)\mathbf{w}(k, y), \tag{2.8}$$

where $\mathbf{w}(k, y)$ is a four-component vector, i.e.

$$\mathbf{w}(k, y) = [k\tilde{J}_x(k, y), k\tilde{J}_y(k, y), \partial_y\tilde{J}_x(k, y), e\bar{n}\tilde{\phi}(k, y)/(m\nu)]^\top, \tag{2.9}$$

and

$$\mathcal{M}(k) = k \begin{pmatrix} 0 & 0 & 1 & 0 \\ -i & 0 & 0 & 0 \\ 1 + 1/(kD_\nu)^2 & \nu_r + \omega_c\tau/(kD_\nu)^2 & i\nu_r & -i \\ (\nu_r - \omega_c\tau)/(kD_\nu)^2 & 1 + \nu_r^2 + (1 + \nu_r\omega_c\tau)/(kD_\nu)^2 & i(1 + \nu_r^2) & -i\nu_r \end{pmatrix}, \tag{2.10}$$

where $\nu_r \equiv \nu_{\rm H}/\nu$. The matrix $\mathcal{M}(k)$ has four eigenvalues: $\lambda_{1/2}(k) = \pm|k|$ and $\lambda_{3/4}(k) = \pm q$, where we have introduced the shorthand notation

$$q \equiv \sqrt{k^2 + 1/D_\nu^2}. \tag{2.11}$$

The corresponding eigenvectors are:

$$\mathbf{w}_{1/2}(k) = \begin{pmatrix} i \\ \pm\text{sgn}(k) \\ \pm i\,\text{sgn}(k) \\ \dfrac{1 \mp i\,\text{sgn}(k)\omega_c\tau}{D_\nu^2 k^2} \end{pmatrix}, \quad \mathbf{w}_{3/4}(k) = \begin{pmatrix} \pm\dfrac{k}{q} \\ -i\dfrac{k^2}{q^2} \\ 1 \\ \dfrac{(\nu_{\mathbf{r}} - \omega_c\tau)}{D_\nu^2 q^2} \end{pmatrix}. \tag{2.12}$$

Note that the eigenvalues are independent of the cyclotron frequency and Hall viscosity, while the eigenvectors explicitly depend on them. The general solution of equation (2.8) can be therefore written as a linear combination of exponentials of the form $\sum_{j=1}^{4} a_j(k)\mathbf{w}_j(k)\exp(\lambda_j y)$. The four coefficients $a_j(k)$ can be determined from the enforcement of suitable boundary conditions (BCs).

2.3 Half-Plane Setup

We consider a single current injector in a half-plane setup with infinite length in the $\hat{\mathbf{x}}$ direction. A current injector is mathematically described by the usual point-like BC for the component of the current density perpendicular to the $y = 0$ edge:

$$J_y(x, 0) = -I\delta(x)/e, \tag{2.13}$$

where I in the dc drive current [50]. The solution of the viscous problem requires an additional BC on the tangential component of the velocity at the $y = 0$ edge. We work with no-slip boundary conditions

$$J_x(x, 0) = 0. \tag{2.14}$$

Finally, we also impose the following BCs at $y = +\infty$: $J_x(x, y \to +\infty) = 0$ and $J_y(x, y \to +\infty) = 0$.

In FT with respect to x, the BCs become

$$\tilde{J}_y(k, 0) = -I/e, \tag{2.15a}$$

$$\tilde{J}_x(k, 0) = 0, \tag{2.15b}$$

$$\tilde{J}_y(k, +\infty) = 0, \tag{2.15c}$$

$$\tilde{J}_x(k, +\infty) = 0. \tag{2.15d}$$

The FT of the electrochemical potential along the edges reads as follows:

$$\tilde{\phi}_+(k) = I[\tilde{r}_+(k) - i\rho_{\rm H}/k + 2i\rho_{\nu_{\rm H}} k D_\nu^2 + \tilde{r}_{+A}(k)], \tag{2.16}$$

where $\bar{k} = kD_\nu$, $\bar{q} = qD_\nu$, $q = \sqrt{k^2 + 1/D_\nu^2}$, $\rho_0 = m/(\bar{n}e^2\tau)$, $\rho_{\rm H} = -m\omega_c/(\bar{n}e^2)$, and $\rho_{\nu_{\rm H}} = m\nu_{\rm H}/(\bar{n}e^2 D_\nu^2)$. In particular, $\rho_{\rm H} = B/(-e\bar{n}c)$ is the usual Hall resistivity. The terms $\tilde{r}_+(k)$ and $\tilde{r}_{+A}(k)$ are expressed as

$$\tilde{r}_+(k) = \rho_0 \left[\frac{1}{|k|} + D_\nu^2 (q + |k|)\right], \tag{2.17a}$$

$$\tilde{r}_{+A}(k) = i\rho_0 \frac{\nu_{\rm H}}{\nu} D_\nu^2 {\rm sgn}(k) (q - |k|). \tag{2.17b}$$

Equation (2.17) can be transformed back in real space analytically as

$$r_+(x) = -\rho_0 \left[\frac{1}{\pi} \ln\left(\frac{|x|}{D_\nu}\right) + \frac{D_\nu^2}{\pi x^2} + \frac{D_\nu}{\pi |x|} K_1\left(\frac{|x|}{D_\nu}\right)\right], \tag{2.18a}$$

$$r_{+A}(x) = \rho_0 \frac{\nu_{\rm H}}{\nu} \frac{D_\nu}{2x} \left[-I_1\left(\frac{|x|}{D_\nu}\right) + \mathbf{L}_1\left(\frac{|x|}{D_\nu}\right)\right], \tag{2.18b}$$

where $I_1(x)$ ($K_1(x)$) is the modified Bessel function of first (second) kind and order one, and $\mathbf{L}_1(x)$ is the modified Struve function of order one.

We now introduce the 'transverse' resistance, which is measured in the setup sketched in Fig. 2.1 as

$$R_T(x) \equiv \frac{\phi(x, 0) - \phi(-x, 0)}{I} = \rho_{\rm H}{\rm sgn}(x) + 4\rho_{\nu_{\rm H}}\delta'\left(\frac{x}{D_\nu}\right) + 2r_{+A}(x). \tag{2.19}$$

We note that $R_T(x) \to \rho_{\rm H}{\rm sgn}(x)$ for $|x| \gg D_\nu$, since, in the same limit, $r_{+A}(x) \to 0$. In order to have a clear signature of the Hall viscosity it is therefore convenient to perform two measurements of the transverse resistance R_T, i.e. one at position $0 < \bar{x} \lesssim D_\nu$ and a second one at position $\bar{x}' \gg D_\nu$. The difference

$$\Delta R_T(\bar{x}) \equiv R_T(\bar{x}) - \lim_{\bar{x}' \to \infty} R_T(\bar{x}') = 2r_{+A}(\bar{x}), \tag{2.20}$$

is independent of $\rho_{\rm H}$ and non-zero only in the presence of a finite Hall viscosity. In particular, for $\bar{x}$ very close to the injector one finds

$$\Delta R_T(\bar{x} \to 0^+) \to -\frac{\rho_0}{2\nu}\nu_{\rm H}, \tag{2.21}$$

which makes it clear that a measurement of ΔR_T yields immediately the value of the Hall viscosity.

2.4 Conclusions

In summary, we have proposed an all-electrical scheme that allows a determination of the Hall viscosity $\nu_{\rm H}$ of a two-dimensional electron liquid in a solid-state device.

We have demonstrated that the transverse geometry in Fig. 2.1 is particularly suitable for extracting $\nu_{\rm H}$ from experimental data. Indeed, we have shown that a measurement of $\Delta R_T(x)$, as given by Eq. (2.20), yields immediately the value of the Hall viscosity, provided τ is measured from the ordinary longitudinal resistance [35] ρ_{xx} and $\nu(B = 0)$ from one of the protocols discussed in Refs. [35, 51].

References

1. P.J.W. Moll, P. Kushwaha, N. Nandi, B. Schmidt, A.P. Mackenzie, Science **351**(6277), 1061 (2016). https://doi.org/10.1126/science.aac8385
2. M. Mendoza, H.J. Herrmann, S. Succi, Phys. Rev. Lett. **106**, 156601 (2011). https://doi.org/10.1103/PhysRevLett.106.156601
3. M.J.M. de Jong, L.W. Molenkamp, Phys. Rev. B **51**, 13389 (1995). https://doi.org/10.1103/PhysRevB.51.13389
4. L.D. Landau, E.M. Lifshitz, *Fluid mechanics*, *Course of theoretical physics*, vol. 6 (Pergamon, New York, 1987). ISBN 9780080339337
5. U. Briskot, M. Schütt, I.V. Gornyi, M. Titov, B.N. Narozhny, A.D. Mirlin, Phys. Rev. B **92**, 115426 (2015). https://doi.org/10.1103/PhysRevB.92.115426
6. M.I. Dyakonov, M.S. Shur, Phys. Rev. B **51**, 14341 (1995). https://doi.org/10.1103/PhysRevB.51.14341
7. L. Levitov, G. Falkovich, Nat. Phys. **12**(7), 672 (2016). https://doi.org/10.1038/nphys3667
8. A. Lucas, J. Crossno, K.C. Fong, P. Kim, S. Sachdev, Phys. Rev. B **93**, 075426 (2016). https://doi.org/10.1103/PhysRevB.93.075426
9. M. Mendoza, H.J. Herrmann, S. Succi, **3**, 1052 EP (2013). https://doi.org/10.1038/srep01052
10. M. Müller, S. Sachdev, Phys. Rev. B **78**, 115419 (2008). https://doi.org/10.1103/PhysRevB.78.115419
11. B.N. Narozhny, I.V. Gornyi, M. Titov, M. Schütt, A.D. Mirlin, Phys. Rev. B **91**, 035414 (2015). https://doi.org/10.1103/PhysRevB.91.035414
12. D. Svintsov, V. Vyurkov, S. Yurchenko, T. Otsuji, V. Ryzhii, J. Appl. Phys. **111**(8), 083715 (2012). https://doi.org/10.1063/1.4705382

13. L.W. Molenkamp, M.J.M. de Jong, Solid-State Electron. **37**(4), 551 (1994). https://doi.org/10.1016/0038-1101(94)90244-5
14. V.N. Kotov, B. Uchoa, V.M. Pereira, F. Guinea, A.H. Castro Neto, Rev. Mod. Phys. **84**, 1067 (2012). https://doi.org/10.1103/RevModPhys.84.1067
15. J. Crossno, J.K. Shi, K. Wang, X. Liu, A. Harzheim, A. Lucas, S. Sachdev, P. Kim, T. Taniguchi, K. Watanabe, T.A. Ohki, K.C. Fong, Science **351**(6277), 1058 (2016). https://doi.org/10.1126/science.aad0343
16. T. Schäfer, D. Teaney, Rep. Prog. Phys. **72**(12), 126001 (2009). https://doi.org/10.1088/0034-4885/72/12/126001
17. M. Polini, G. Vignale, in *No-Nonsense Physicist: An Overview of Gabriele Giuliani's Work and Life*, ed. by M. Polini, G. Vignale, V. Pellegrini, J.K. Jain (Scuola Normale Superiore, Pisa, 2016), pp. 107–124, arXiv:1404.5728 [cond-mat.mes-hall], https://doi.org/10.1007/978-88-7642-536-3_9
18. A. Principi, G. Vignale, M. Carrega, M. Polini, Phys. Rev. B **93**, 125410 (2016). https://doi.org/10.1103/PhysRevB.93.125410
19. D.A. Bandurin, I. Torre, R.K. Kumar, M. Ben Shalom, A. Tomadin, A. Principi, G.H. Auton, E. Khestanova, K.S. Novoselov, I.V. Grigorieva, L.A. Ponomarenko, A.K. Geim, M. Polini, Science **351**(6277), 1055 (2016). https://doi.org/10.1126/science.aad0201
20. N. Read, E.H. Rezayi, Phys. Rev. B **84**, 085316 (2011). https://doi.org/10.1103/PhysRevB.84.085316
21. I.V. Tokatly, G. Vignale, Phys. Rev. B **76**, 161305 (2007). https://doi.org/10.1103/PhysRevB.76.161305
22. I.V. Tokatly, G. Vignale, J. Phys. Condens. Matter **21**(27), 275603 (2009). https://doi.org/10.1088/0953-8984/21/27/275603
23. D.A. Abanin, S.V. Morozov, L.A. Ponomarenko, R.V. Gorbachev, A.S. Mayorov, M.I. Katsnelson, K. Watanabe, T. Taniguchi, K.S. Novoselov, L.S. Levitov, A.K. Geim, Science **332**(6027), 328 (2011). https://doi.org/10.1126/science.1199595
24. R. Krishna Kumar, D.A. Bandurin, F.M.D. Pellegrino, Y. Cao, A. Principi, H. Guo, G.H. Auton, M. Ben Shalom, L.A. Ponomarenko, G. Falkovich, K. Watanabe, T. Taniguchi, I.V. Grigorieva, L.S. Levitov, M. Polini, A.K. Geim, Nat. Phys. **advance online publication** (2017). https://doi.org/10.1038/nphys4240
25. R.N. Gurzhi, Sov. Phys. Uspekhi **11**(2), 255 (1968). https://doi.org/10.1070/PU1968v011n02ABEH003815
26. S. Das Sarma, S. Adam, E.H. Hwang, E. Rossi, Rev. Mod. Phys. **83**(2), 407 (2011). https://doi.org/10.1103/RevModPhys.83.407
27. M. Müller, J. Schmalian, L. Fritz, Phys. Rev. Lett. **103**, 025301 (2009). https://doi.org/10.1103/PhysRevLett.103.025301
28. I. Torre, A. Tomadin, A.K. Geim, M. Polini, Phys. Rev. B **92**, 165433 (2015). https://doi.org/10.1103/PhysRevB.92.165433
29. L. Fritz, J. Schmalian, M. Müller, S. Sachdev, Phys. Rev. B **78**, 085416 (2008). https://doi.org/10.1103/PhysRevB.78.085416
30. C. Hoyos, D.T. Son, Phys. Rev. Lett. **108**, 066805 (2012). https://doi.org/10.1103/PhysRevLett.108.066805
31. A.O. Govorov, J.J. Heremans, Phys. Rev. Lett. **92**, 026803 (2004). https://doi.org/10.1103/PhysRevLett.92.026803
32. A.K. Geim, K.S. Novoselov, Nat. Mater. **6**(3), 183 (2007). https://doi.org/10.1038/nmat1849
33. P.S. Alekseev, Phys. Rev. Lett. **117**, 166601 (2016). https://doi.org/10.1103/PhysRevLett.117.166601
34. J.E. Avron, R. Seiler, P.G. Zograf, Phys. Rev. Lett. **75**, 697 (1995). https://doi.org/10.1103/PhysRevLett.75.697
35. R. Bistritzer, A.H. MacDonald, Phys. Rev. B **80**, 085109 (2009). https://doi.org/10.1103/PhysRevB.80.085109
36. Q. Li, S. Das Sarma, Phys. Rev. B **87**, 085406 (2013). https://doi.org/10.1103/PhysRevB.87.085406

37. F.M.D. Pellegrino, I. Torre, M. Polini, (2017), to appear, arXiv:1706.08363 [cond-mat.mes-hall]
38. G. Falkovich, *Fluid Mechanics: A Short Course for Physicists* (Cambridge University Press, Cambridge, 2011)
39. A.V. Andreev, S.A. Kivelson, B. Spivak, Phys. Rev. Lett. **106**, 256804 (2011). https://doi.org/10.1103/PhysRevLett.106.256804
40. M.I. Dyakonov, M.S. Shur, IEEE Trans. Electron Devices **43**, 380 (1996). https://doi.org/10.1109/16.485650
41. M. Dyakonov, M. Shur, Phys. Rev. Lett. **71**, 2465 (1993). https://doi.org/10.1103/PhysRevLett.71.2465
42. A. Tomadin, M. Polini, Phys. Rev. B **88**, 205426 (2013). https://doi.org/10.1103/PhysRevB.88.205426
43. A. Tomadin, G. Vignale, M. Polini, Phys. Rev. Lett. **113**, 235901 (2014). https://doi.org/10.1103/PhysRevLett.113.235901
44. B. Bradlyn, M. Goldstein, N. Read, Phys. Rev. B **86**, 245309 (2012). https://doi.org/10.1103/PhysRevB.86.245309
45. A. Cortijo, Y. Ferreirós, K. Landsteiner, M.A.H. Vozmediano, 2D Mater. **3**(1), 011002 (2016). https://doi.org/10.1088/2053-1583/3/1/011002
46. F.D.M. Haldane, Phys. Rev. Lett. **107**, 116801 (2011). https://doi.org/10.1103/PhysRevLett.107.116801
47. N. Read, Phys. Rev. B **79**, 045308 (2009). https://doi.org/10.1103/PhysRevB.79.045308
48. T. Scaffidi, N. Nandi, B. Schmidt, A.P. Mackenzie, J.E. Moore, Phys. Rev. Lett. **118**, 226601 (2017). https://doi.org/10.1103/PhysRevLett.118.226601
49. M. Sherafati, A. Principi, G. Vignale, Phys. Rev. B **94**, 125427 (2016). https://doi.org/10.1103/PhysRevB.94.125427
50. F.M.D. Pellegrino, I. Torre, A.K. Geim, M. Polini, Phys. Rev. B **94**, 155414 (2016). https://doi.org/10.1103/PhysRevB.94.155414
51. I. Torre, A. Tomadin, R. Krahne, V. Pellegrini, M. Polini, Phys. Rev. B **91**, 081402 (2015). https://doi.org/10.1103/PhysRevB.91.081402

Chapter 3
Computer Simulations of the Structure of Nanoporous Carbons and Higher Density Phases of Carbon

Lydia Alonso, Julio A. Alonso and María J. López

Abstract The most stable form of solid carbon is graphite, a stacking of graphene layers in which the carbon atoms show sp^2 hybridization which leads to strong intra-layer bonding. Diamond is a denser phase, obtained at high pressure. In diamond the carbon atoms show sp^3 hybridization. Metastable solid carbon phases can be prepared also with lower density than graphite (in fact, densities lower than water); for instance the carbide-derived carbons. These are porous materials with a quite disordered structure. Atomistic computer simulations of carbide-derived carbons indicate that the pore walls can be viewed as curved and planar nanographene ribbons with numerous defects and open edges. Consequently, the hybridization of the carbon atoms in the porous carbons is sp^2. Because of the high porosity and large specific surface area, nanoporous carbons find applications in gas adsorption, batteries and nanocatalysis, among others. We have performed computer simulations, employing large simulation cells and long simulation times, to reveal the details of the structure of the nanoporous carbons. In the dynamical simulations the interactions between the atoms are represented by empirical many-body potentials. We have also investigated the effect of the density on the structure of the disordered carbons and on the hybridization of the carbon atoms. At low densities, typical of the porous carbide-derived carbons formed experimentally, the hybridization is sp^2. On the other hand, as the density of the disordered material increases, a growing fraction of atoms with sp^3 hybridization appears.

Work dedicated to Professor N. H. March.

L. Alonso · J. A. Alonso (✉) · M. J. López
Departamento de Física Teórica, Atómica y Óptica, Universidad de Valladolid, 47011, Valladolid, Spain
e-mail: jaalonso@fta.uva.es

M. J. López
e-mail: maria.lopez@fta.uva.es

G. G. N. Angilella and C. Amovilli (eds.), *Many-body Approaches at Different Scales*, https://doi.org/10.1007/978-3-319-72374-7_3

3.1 Introduction

Nanoporous carbons exhibit a quite disordered structure with internal pores of nanometric size. These materials have low densities and high specific surface areas, properties which make them attractive in applications to gas adsorption and hydrogen storage, among others. The carbide-derived carbons (CDC) form an interesting family of nanoporous carbons. CDCs can be easily produced from metal carbides [1, 2], ZrC for instance, by selectively extracting the metal atoms through a chemical chlorination process performed at temperatures between 600 and 1200 °C. The structural characteristics and the properties of these materials can be tuned by selecting the production conditions, in particular the reaction temperature, and the appropriate post-treatments of the samples. The microstructure of the CDCs has been investigated by Raman spectroscopy, X-ray diffraction and high-resolution transmission electron microscopy [3]. The elucidation of the structure of nanoporous carbons is greatly aided by computer simulations. Atomistic computer simulations of carbide-derived carbons reveal a disordered structure of connected pores of different size and shape [4, 5]. The pore walls can be viewed as curved and planar graphene fragments with numerous defects and some open edges. Consequently, the hybridization of the carbon atoms in this family of porous materials is that typical of graphene, that is, sp^2. The increase of the experimental reaction temperature, and the increase of the temperature and annealing time in the computer simulations, both induce a progressive repair of the long range disorder and a substantial increase of the degree of graphitization in the system [6]. A main characteristic of the porous carbons is their small densities (number of atoms per unit volume), and this is the reason for the easy formation of a network of interconnected planar and curved two dimensional graphene-like layers. In the other extreme, diamond is a solid phase of carbon denser than graphite, produced under extreme high pressure and temperature conditions in the earth mantle or by synthetic methods in the laboratory [7–9]. The atomic coordination around each carbon atom in diamond is four, and the electronic hybridization is sp^3. Although diamond is metastable (the most stable solid phase of carbon is graphite), the rate of conversion of diamond to graphite is negligible at ambient conditions.

Early in 1977, Norman March and coworkers pioneered the investigation of the structure of amorphous carbon [10, 11]. Their conclusion was that the X-ray and electron diffraction intensities cannot be explained by simply considering a model structure and the appropriate atomic scattering factor, and that an explicit modelling of the chemical (covalent) bonds, is required. Amorphous carbon obtained by different experimental methods shows values of the interatomic distances and coordination numbers in between those for graphite and diamond [12, 13]. An early proposal for the structure of amorphous carbons was a disordered mixture of small two-dimensional graphitic fragments (in which the carbon atoms have sp^2 hybridization) linked by tetrahedrally bonded (sp^3) atoms [14]. However, neutron diffraction revealed little tetrahedral bonding [12]. This is easy to understand, because the densities are not in the high density regime favorable to tetrahedral bonding. In fact, our simulations of porous carbons for samples with more than sixty thousand atoms [4] revealed disordered structures formed by planar and curved graphitic fragments connected in a way leaving abundant empty spaces (pores), and the analysis of the structures indicated a very small amount of tetrahedrally coordinated carbon, certainly below one per cent. The resulting material can be viewed as low density (0.77 g/cm^{-3}) disordered carbon. Despite the highly disordered structure this type of carbons are not clasified as amorphous carbons. According to the recommendations of IUPAC [15], the term amorphous carbon is restricted to carbon materials having short-range order only (no long-range crystalline order) with deviations of the interatomic distances and/or interbonding angles with respect to the graphite lattice as well as to the diamond lattice. So, the term amorphous carbon is not applicable to materials with two-dimensional structural elements such as polyaromatic layers with a nearly ideal interatomic distance and an extension greater than one nanometer. Using density-functional molecular dynamics and simulation cells containing 216 atoms, Deringer and Csanyi [16] have found that the structure of amorphous carbon is a mixture of threefold (sp^2) and fourfold (sp^3) bonded atoms, the proportion of these depending sensitively on the density. The densities investigated by these authors are between 1.5 and 3.5 g/cm^{-3} (notice that the densities of graphite and diamond are 2.1–2.2 g/cm^{-3} and 3.5 g/cm^{-3}, respectively). Motivated by all the above works we present here the results of an investigation of the structure of non crystalline carbons (including amorphous carbon) over a wide range of densities, between 0.77 and 2.87 g/cm^{-3}. One of the main purposes is to investigate the transition between disordered porous carbons and amorphous carbons, and to detect the critical density for which the amount of atoms with tetrahedral (sp^3) coordination begins to be noticeable. Because the proportion of atoms with tetrahedral coordination is expected to be small, except at high densities, it is necessary to work with simulation cells having a large enough number of atoms.

3.2 Theoretical Models and Dynamical Simulations

Molecular dynamics simulations are performed to investigate the structural features of a variety of disordered carbon materials spanning a broad range of densities (0.7–2.8 g/cm^3). CDCs are porous carbon materials placed on the low density corner of pure carbon materials with densities smaller than that of water. Experimentally, these materials are produced from precursor metal carbides or silicon carbide, by selective chemical extraction (through chlorination) of the metal or silicon atoms. Upon removal of the metal or silicon, the carbon sublattice becomes very unstable and collapses internally. The transformation experienced by the material is a conformal transformation, that is, a transformation without change of the macroscopic piece shape or size, giving rise to the porous carbon structure. To mimic the formation proccess of the CDCs, we start the simulations after the removal of the metal atoms or silicon (the chlorination procedure is out of the scope of the present study) with the C atoms in the same structure as in the corresponding carbides, and perform molecular dynamics simulations at constant number of particles N, constant volume V, and constant temperature T, at several temperatures. Thus we simulate the structural transformation of the carbon skeleton of the carbide keeping constant the density of the carbon material along the simulation, similarly to the experimental conformal transformation. Two carbides have been considered: ZrC, which has a structure of two fcc interpenetrated lattices, and α-SiC, which has a 6H-hexagonal structure. The densities of the C atoms in the carbides are 0.77 and 0.96 g/cm^3, respectively. Large simulation cells are used within the periodic boundary conditions scheme to represent appropriately the disordered CDC porous structures. A $12 \times 12 \times 12$ cubic cell of 56.376 Å of side, containing 6912 C atoms is used to simulate the CDCs derived from ZrC, and a $10 \times 6 \times 18$ tetrahedral cell of dimensions $53.36 \times 55.458 \times 45.3$ Å, containing 6480 C atoms is used for the CDCs derived from SiC. Amorphous carbons, on the other hand, are disordered high density carbons, with densities between the density of graphite 2.1–2.2 g/cm^3 and the density of diamond, 3.5 g/cm^3. As the density of carbon increases from the densities typical of CDCs to the densities typical of amorphous carbons, the structure of the material is expected to experience a profound transformation. To assess the dependence of the final carbon structures on the density and on the initial configuration of the carbon atoms we have considered two types of materials, M1 and M2, having the initial structure of the carbon squeleton in ZrC and SiC, respectively, but with scaled C–C distances to produce the desired densities in the range of 0.77–2.87 g/cm^3. Notice that the CDC derived from ZrC is the M1 material with density 0.77 g/cm^3 and the CDC derived from SiC is the M2 material with density 0.96 g/cm^3.

The selected temperatures for the simulations,[1] $T = 350$, 2100 and 3010 K, are higher than the temperatures of the chlorination process of about 600–1500 K. The reason being that the former determine the speed of the structural changes in the material whereas the latter correspond to the chemical procedure for extraction of the metal or silicon [4]. Moreover the simulation time, of the order of a few hundred picoseconds, is orders of magnitude smaller than the experimental times of several minutes or hours. Thus, the higher temperatures of the simulations spead up the structural changes and therefore compensate for the shorter simulation times. An empirical correlation can be established between the temperatures of the simulations and the experimental ones [4, 17]. To start the simulations, the initial velocities of the carbon atoms are set to a Maxwellian distribution corresponding to the selected simulation temperature. The Nosé-Hoover thermostat is used to keep constant the temperature along the simulations. The equations of motion are integrated with a time step of 0.5 fs. Simulation runs of 300 ps lead to well converged final structures, as it has been checked by performing simulation runs of 600 ps for some selected cases. The final structures obtained in the simulations are quenched down to 0 K in order to perform the structural analysis. The structures are then characterized on the basis of (i) the local ordering as measured by the coordination of the C atoms, which

[1]Notice that the Tersoff potential gives a melting temperature for carbon of about 6000 K whereas the experimental value is about 4300 K. Therefore the temperatures given in this paper have been scaled by a 0.7 factor.

reflect the sp^2 or sp^3 hybridization of C, and the ring pattern of the C network, and (ii) global quantities such as the specific surface area of the pore walls, the porosity (ratio of empty to total volume) and the size distribution of pores. In the simulations, the interatomic C-C interactions are mimicked through the Tersoff potential [18–20], that appropriately represents the covalent bonds between carbon atoms in graphite and diamond, plus a weak interaction potential [21] which is added to represent the interaction between layers in graphite.

3.3 Results

We start investigating the formation of CDCs derived from ZrC and SiC through NVT molecular dynamics simulations. The chemical (chlorination) procedure used to extract the metal from the carbides is not simulated here. The simulation starts just after the removal of the metal atoms from the carbide and before the carbon atoms had time to move. Thus, the initial structure for the simulations consists in a carbon skeleton in which the carbon atoms are placed at the same positions that they occupied in the corresponding carbide lattice. The NVT molecular dynamics is aimed to simulate the subsequent conformal trasformation experienced by the carbon network. Thus, the density of the carbon material remains fixed in the simulations and is equal to the density of the carbon atoms in the original carbide.

Clearly, after the removal of the metal atoms from the carbide the structure of the carbon atoms is highly unstable, and the internal collapse of the structure occurs very fast. C–C bonds begin to form in the material, and after 0.5 ps most atoms are linked together forming a highly disordered carbon network. Figure 3.1 shows several snapshots at different times of the simulation of the formation of a CDC derived from SiC at a simulation temperature of 3010 K. After about 5 ps, the structure evolves forming an incipient porous structure which fully develops at longer times. After 300 ps, the final porous structure is almost reached, the pore walls are graphitic layers with defects, interconnected with each other. It is clearly apparent that the carbon atoms are not uniformly distributed in the material. Longer simulation times of the order of 600 ps only lead to minor modifications of the nanoporous structure of the CDC. This can be explained by the high thermal stability of the graphitic nanostrips [22]. Quite similar formation pathways and final structures have been obtained for CDCs derived from the other precursor carbide, ZrC, at the same ($T = 3010$ K) simulation temperature [4].

However, different structures are produced for carbon materials with higher densities, of the order or higher than the density of graphite. As an example, Fig. 3.2 shows snapshots along the dynamics, of the formation of the M1 material (initial fcc configuration of the C atoms) with a density of 2.2 g/cm^3. The internal collapse of the carbon structure also occurs very fast, in about 0.5 ps, as in the previous case. However, at this relatively high density, the subsequent evolution of the carbon network is quite different. In contrast with the CDC's, the high density M1 material does not develop a porous structure and it is not possible to identify in the structure

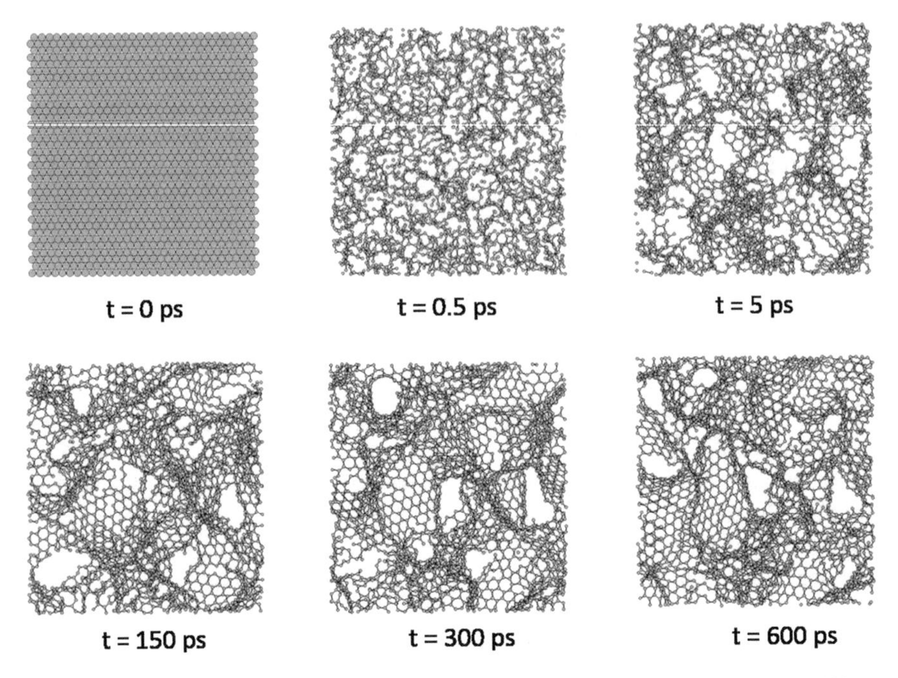

Fig. 3.1 Snapshots, for several times, of the formation of the CDC structure derived from SiC at a simulation temperature of 3010 K. A slab of 25 Å of depth is shown for clarity

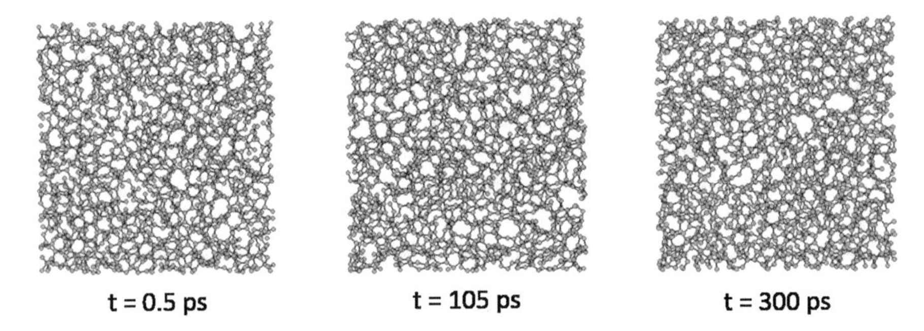

Fig. 3.2 Snapshots, for several times, of the formation of the M1 material with a density of 2.2 g/cm^3, at a simulation temperature of 3010 K. A slab of 10 Å of depth is shown for clarity

flakes of graphitic layers. The C atoms form, instead, a disordered three-dimensional C network.

The local structure of the carbon network and the level of graphitization of the pore walls can be analized by investigating the coordination of the carbon atoms. The coordination number is defined as the number of nearest neighbors of an atom. Three-fold coordinated carbon atoms with sp^2 hybridization are characteristic of two-dimensional graphitic layers. The edges of finite size graphitic flakes give rise

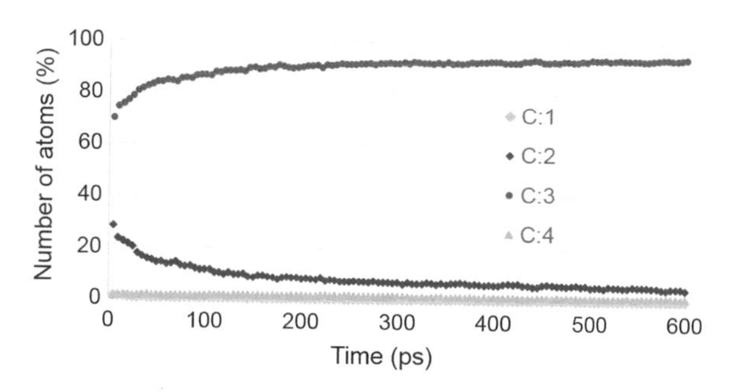

Fig. 3.3 Ratio of carbon atoms with coordinations one, two, three and four as a function of the simulation time, for the CDC structure derived from SiC at a simulation temperature of 3010 K

to the appearance of two-fold coordinated atoms. On the other hand, coordination four (sp^3 hybridization) leads to three dimensional structures and is characteristic of diamond. Figure 3.3 shows the time evolution along the simulation of the coordinations of the C atoms in the CDC structure obtained from the SiC precursor at a temperature of 3010 K. Coordinations two an three rise up within a few picoseconds. In about 5 ps, 70% of the C atoms become three-fold coordinated and about 30% become two fold coordinated. Coordinations one and four, on the other hand, are only marginaly present in the structure. As the time proceeds, the number of three-fold coordinated atoms increases slowly at the expense of the two-fold coordinated atoms. This indicates an increase of the size of the graphitic flakes accompanied by the corresponding reduction of the edges. After about 600 ps the structure does not change any further and the different coordinations become almost flat as a function of time. The ratio of the number of C atoms with different coordinations is well converged within 1%. A slightly relaxed convergence criterion of 2–3% is fulfilled with simulation times of about 300 ps (see Fig. 3.3). A similar convergence with time has been found for the simulated M1 and M2 structures generated at different temperatures and for different densities of carbon. Therefore, we will limit the simulations to 300 ps in order to reduce the computational cost of this study. The final structures obtained in the simulations are then quenched down 0 K in order to perform the structural analysis of the materials at equilibrium conditions.

For the low density materials (M1 and M2 with densities below 1.7 g/cm^3), the number of three-fold coordinated atoms increases with the simulation temperature indicating that the level of graphitization of the structure improves with temperature. However, even at the highest simulation temperature (T = 3010 K), the two-fold coordinated atoms do not disappear completely; this shows the imposibility of removing all the edges of the graphitic flakes and therefore the impossibility of fully graphitize the sample. As a representative example, the left panel of Fig. 3.4 shows the coordinations as a function of temperature of the CDCs produced using SiC as the precursor.

Figure 3.4 also shows, in the right panel, the coordinations of the C atoms in the M1 and M2 materials as a function of the C density. The number of three-fold coordinated atoms increases and the number of two-fold coordinated atoms decreases with increasing density, up to approximately the density of graphite (2.2 g/cm^3). This

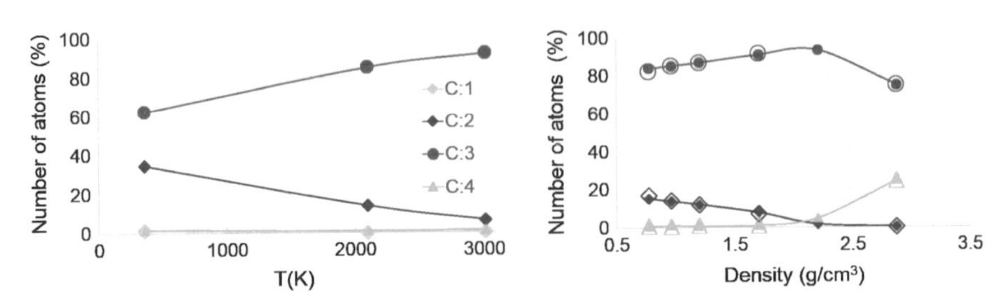

Fig. 3.4 Number (in percent) of carbon atoms with coordinations one, two, three and four. Left panel gives the coordinations as a function of the simulation temperature for the CDCs derived from SiC. Right panel gives the coordinations as a function of the density for the M1 (open symbols) and M2 (solid symbols) materials produced at $T = 2100$ K

increase of 3-fold coordinated atoms, however, is not indicative of an improvement in the level of graphitization of the material, as can be clearly seen from the snapshots of Fig. 3.2. The graphitic layers become smaller and more interconnected among them giving rise to a highly disordered three-dimensional network of sp^2 C atoms. For larger densities, above the density of graphite, the trend changes and the number of three-fold coordinated atoms decreases in favor four-fold coordinated atoms, with sp^3 hybridization, which are characteristic of diamond-like structures. Thus, the materials with a density of 2.7 g/cm^3 are formed by a mixture of three- and four-fold coordinated C atoms with a ratio of about three (75%) to one (25%), respectively. This mixture of coordinations is also found in amorphous carbon materials [16] although one should notice that a broad range of densities and ratio of coordinations is included under the generic name of amorphous carbons. The structures obtained for a given density and at a given simulation temperature for the two types of materials investigated here, M1 and M2, are very similar, what leads us to conclude that the initial configuration of the C atoms for the simulations (cubic or hexagonal) has little effect in the final structure of the simulated materials. The more relevant parameter being the density of carbon atoms.

The ring pattern of the carbon network is also used to characterize the structure of the simulated materials. In a perfect graphitic layer the carbon atoms are three-fold coordinated and form a honeycumb network of hexagonal rings. The presence of pentagonal rings in graphitic layers gives rise to closed carbon structures as in the fullerenes, whereas heptagonal and octagonal rings lead to open structures as in schwarzites. On the other hand, non planar hexagonal rings are also found in diamond like structures but, in this case, the C atoms form a non-layered three-dimensional network of four-fold coordinated atoms. At the low carbon densities of the CDCs derived from ZrC and from SiC, the walls of the porous structure of these materials are mainly formed by hexagonal rings, defining graphitic layers. Pentagonal and smaller rings are completely absent and, therefore, closed pores can not be formed. However, there is a non negligible number of heptagons and octagons in the layers which lead to the interconnection of the pore walls and to the open character of the pores in these materials. The number of hexagonal rings increases with increasing simulation

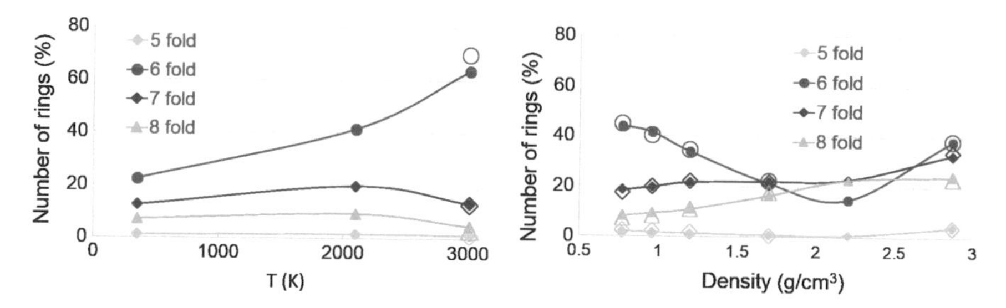

Fig. 3.5 Number (in percent with respect to the ideal number of hexagonal rings in a graphitic layer) of five-, six-, seven-, and eight-fold rings. Left panel: number of rings as a function of the simulation temperature for the CDCs derived from SiC. The solid and empty symbols correspond to simulation times of 300 and 600 ps, respectively. Right panel: number of rings as a function of the density for the M1 (open symbols) and M2 (solid symbols) materials produced at $T = 2100$ K

temperature, as it is shown in Fig. 3.5 (left panel) for the CDCs derived from SiC (a similar behaviour is observed for the CDCs derived from ZrC). This confirms the observed improvement with temperature of the level of graphitization of the pore walls. However, the number of heptagons and octagons, although smaller than the number of hexagons, remains almost constant with temperature; these rings do not disappear even at the highest simulation temperature, what prevents the formation of large graphitic layers.

The right panel of Fig. 3.5 reveals that with increasing carbon density, up to approximately the density of graphite, the number of hexagons decreases whereas the number of heptagons and octagons increases. As a consequence, the graphitic structure and the local order of the structure disappear and, as it is observed in Fig. 3.2, a three-dimensional disordered network of sp^2 C atoms develops. For densities above the density of graphite, the number of hexagonal rings increases again and the number of heptagons also increases. The hexagonal rings, however, now contain four-fold coordinated carbon atoms and are the signature of the development of diamond like structures. Thus, at high densities (below the density of diamond, 3.53 g/cm^3) the materials become amorphous with highly disordered structures and formed by a mixture of three- and four-fold coordinated C atoms, as discussed above.

A relevant quantity to characterize nanoporous materials is the specific surface area (SSA), which it is correlated, in general, with the capacity of the material to adsorb gases. The geometrical evaluation of this quantity from the simulated structures is quite difficult, because the walls of the pores are formed by small, non planar layers interconnected among them. Thus, it has been proposed [4] to approximate the SSA by the sum of the areas of the rings each considered as a perfect planar ring. Although the pore walls are one layer thick, the two sides of the walls are not fully available for gas adsorption, mainly due to the interconnection between layers. From the simulated structures one can estimate that, in average, only about one side and half of the other side of the layers are exposed for gas adsorption. Thus to calculate the SSA, we multiply the surface of one side of the layers by the empirical factor of 1.5. Figure 3.6 shows the SSA for the CDCs derived from ZrC and SiC as

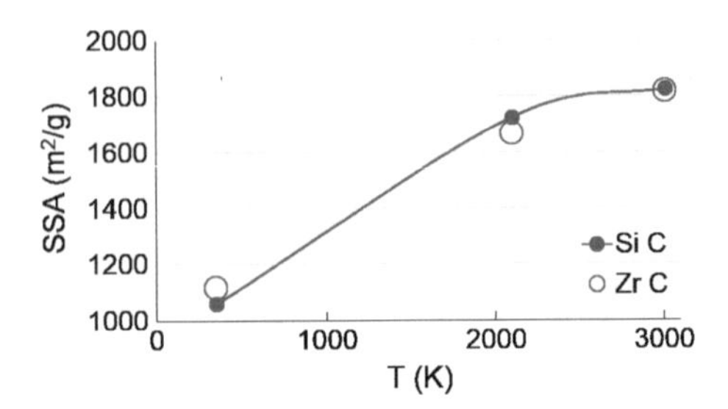

Fig. 3.6 Specific surface area (SSA) as a function of the simulation temperature for the CDCs derived from ZrC (empty symbols), and from SiC (solid symbols)

a function of temperature. The SSA increases moderately with the temperature of the simulation following a similar trend as the experimentally determined SSA [23] for various CDCs as a function of chlorination temperature. The simulated values match rather well the experimental values. Moreover we found no much difference between the two low density CDC's investigated, derived from ZrC and from SiC, respectively. For the high density simulated materials, the geometrical determination of the SSA is not physically meaningful since the number of rings does not correlate with the surface available for gas adsorption, as it has been seen above.

Of great interest is the porosity of the materials, that is, the empty volume that can be used to adsorb molecules or for catalysis. We have calculated the porosity from the simulated geometrical structures, considering an effective exclusion volume (volume that can not be used to adsorbe gases) of each C atom given by a sphere of radius equal to the Van der Waals radius of carbon (1.7 Å). The porosity increases weakly with the simulation temperature, as it is shown in Fig. 3.7. The porosity of the CDCs derived from ZrC is larger than that determined for SiC, due to the density difference between the two materials (the CDCs derived from ZrC have a lower density). However, calculated values of the porosity for the CDCs derived from ZrC are slightly smaller than the ones obtained in previous simulations [4]. This is due to the smaller simulation cells used here. Since the porosity is a global quantity of the material, large cells are required for an accurate determination of the porosity although the trends with temperature and density can be already revealed with smaller simulation

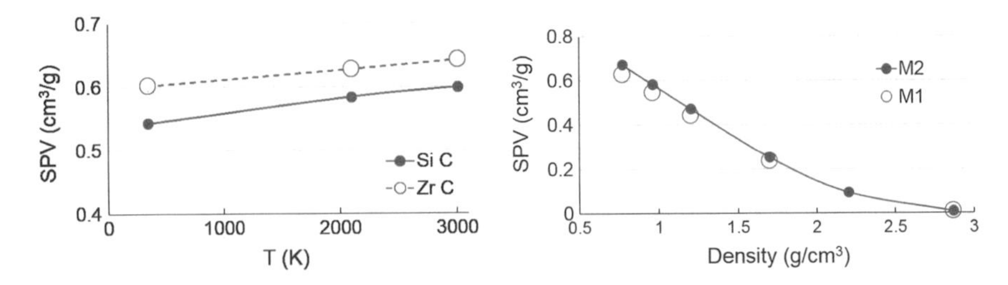

Fig. 3.7 Specific pore volume, SPV. Left panel: SPV as a function of the simulation temperature for the CDCs derived from ZrC (empty symbols), and from SiC (solid symbols). Right panel: SPV as a function of the density for the M1 (open symbols) and M2 (solid symbols) materials produced at $T = 2100$ K

cells as the ones used here. Moreover, the experimentally determined values exhibit a broader range of variation of the porosity with the chorination temperature, most probably due to a better level of annealing of the structures. The porosity, however, depends sensitively on the density of the material (see Fig. 3.7). With the geometrical definition given above, the porosity of graphite results in a negligible value, as is confirmed from adsorption experiments. The simulations lead to small values ot the porosity for both M1 and M2 materials for densities of the order of the density of graphite, and zero porosities for larger densities. The simulated M1 and M2 materials exhibit the same porosities for the same densities, confirming once more the weak dependence on the initial configuration of the C atoms and the more relevant dependence on the density of the material.

The gas adsorption capacity of a material is related with its specific surface area and its porosity. However, it has been shown both, from the experimental [24, 25] and theoretical [26] sides, that the adsorption capacity does not perfectly correlate with these two quantities but it depends on the size of the pores. For instance, it has been determined that the optimum pore size for hydrogen adsorption at room temperature is in the range of 7–11 Å [27]. Thus, to fully characterize the porous structure of a material and its adsorption characteristics it is convenient to determine the pore size distribution function (PSD), that measures the total volume contained in pores of a given size. From the simulation side, PSD is a very demanding quantity to compute since large simulation cells are required to obtain meaningful distributions. The simulation cells considered in this work (of about 50 Å wide) are sufficiently large to reproduce the distribution functions for pores smaller than about 20 Å. Larger pores are escarcely produced in these cells and, therefore, the distribution functions are not accurate for pores larger than about 20 Å. On the other hand, since the pores have no regular shapes and are interconnected among them, one has to introduce a model of pores to determine their size distribution. Based on our simulated structures we found appropriate to use a geometrical model of nonoverlapping spheres [4] to represent the pores. Figure 3.8 shows the pore size distribution function, PSD, of the CDCs obtained using SiC as the precursor, for several simulation temperatures. The CDCs produced at low temperatures ($T = 350$ K) exhibit a relatively narrow distribution of pore sizes around an average value of about 8 Å. With increasing simulation temperature, the distribution becomes wider and the average pore size increases up to a value of about 11 Å for $T = 3010$ K. Although the maximum size of the pores (about 20 Å) is limited by the size of the simulation cell, this does not affect the general trends of the PSDs below that size. A similar behaviour has been obtained for the CDCs derived from ZrC as a precursor. The trends in the simulated PSD functions are in good agreement with the PSDs determined from experimental adsorption isotherms for CDCs produced from several precursor carbides [3].

On the other hand, similarly to the porosity, the PSD depends strongly on the density of the material. The PSD becomes narrower with increasing density and the size of the pores decreases. At approximately the density of graphite, the network of C atoms does not leave empty space and therefore the material does not contain pores any more (see Fig. 3.8 for M2 materials; a similar behaviour has been found for M1 materials).

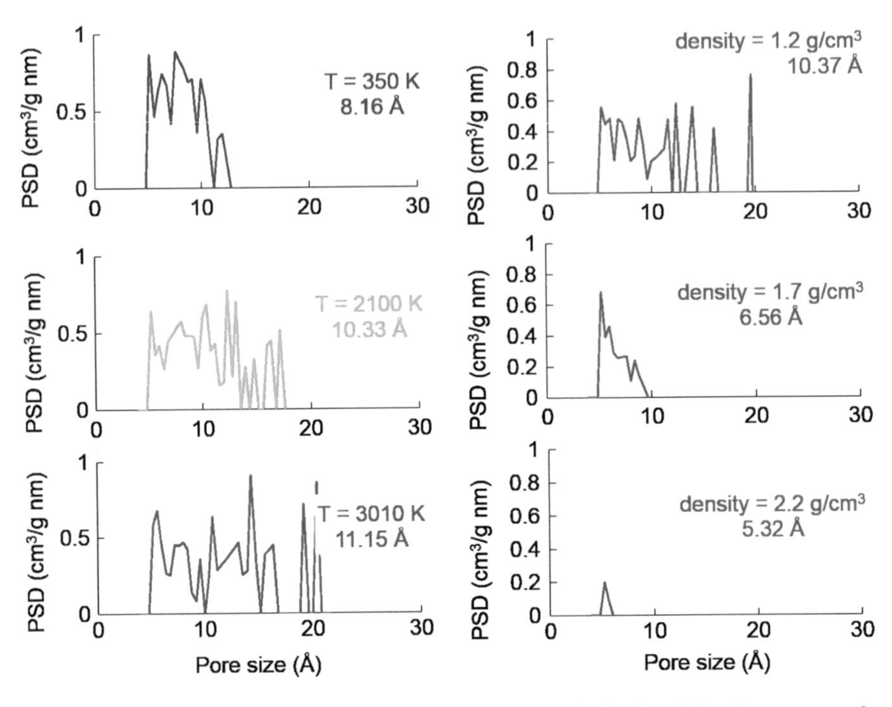

Fig. 3.8 Pore size distribution function, PSD of the CDCs obtained using SiC as the precursor, for several simulation temperatures (left panels). PSD of M2 materials of various densities produced at a temperature of $T = 3010$ K (right panels). Average pore size is indicated in each graph

3.4 Conclusions

We have performed atomistic molecular dynamics simulations to investigate the formation and structural characteristics of disordered carbon materials as a function of the C density, the initial carbon structure and the simulation temperature. Carbon materials ranging from CDCs porous carbons with densities of 0.77–0.96 g/cm^3 to amorphous carbons with densities of 2.8 g/cm^3 have been considered. The simulations have been designed to closely mimic the experimental process of formation of CDCs from metal carbides once the extraction of the metal has taken place, and generalized for materials with different densities. Thus, the initial configurations for the NVT dynamical simulations were the structures of the C atoms in two carbides, ZrC an SiC. Moreover, materials with different densities are generated using those two initial configurations but scaling the C–C distances to produce the desired densities. Upon removal of the metal from the carbide, the collapse of the structure takes place very fast, in a few ps, giving rise to a C network of sp^2 C atoms. A porous structure of open interconnected pores appears; the pore walls are one atom thick graphitic-like layers interconnected among them. The level of graphitization of the pore walls is assessed in terms of the coordinations of the C atoms and the ring structure of the C network. For the CDCs and the low density C materials with densities below 1.7 g/cm^3, there is a prevalence of three-fold coordinated C atoms and hexagonal rings, confirming the graphitic-like structure of the pore walls. However, there is a non negligible number of two-fold coordinated C atoms, corresponding to the edges of the graphitic flakes, and heptagonal and octagonal rings that lead to open pores. With increasing temperature, the level of graphitizacion improves, although the edges and the heptagonal/octogonal defects do not disappear completely. At higher densities, of the order or higher than the density of graphite, there is still a prevalence of the coordination three, but four-fold coordinated C atoms appear and its percentage increases with increasing density. The carbon network loses progressively its graphitic character with the appearance of diamond-like tetrahedral structures that mark the transition towards amorphous carbons. Thus, the high density materials are formed by a combination of sp^2 and sp^3 C atoms and the three dimensional C network does not enclose pores.

The specific surface area (SSA), the porosity, and the pore size distribution function (PSD) characterize the adsorption properties of a material. In agreement with experimental determinations, we found that the SSA of the simulated CDCs increases with increasing temperature, reaching values close to 2000 m^2/g. The porosity of the CDCs and the average size of the pores also increase with increasing temperature and the PSD becomes wider in close analogy with the experimental trends. However, with increasing carbon densities, the porosity drops dramatically and, as expected, for densities of the order and higher than the density of graphite the porosity vanishes.

In summary, we have found that the structure of the simulated carbon materials is not sensitive to the initial configuration of the C atoms. However, the structure is strongly dependent on the density, observing a transition from porous to amorphous structures at approximately the density of graphite. High temperatures favor higher level of graphitization in lower density, CDCs carbons.

Acknowledgements This work was supported by MINECO of Spain (Grant MAT2014-54378-R) and Junta de Castilla y León (Grant VA050U14). The authors thankfully acknowledge the facilities provided by Centro de Proceso de Datos - Parque Científico of the University of Valladolid.

References

1. W.A. Mohun, in *Proceedings of the 4th Biennial Conference on Carbon*, ed. by Y. Zhou (Pergamon Press, Oxford, 1960), pp. 443–453
2. Y. Gogotsi, A. Nikitin, H. Ye, W. Zhou, J.E. Fischer, B. Yi, H.C. Foley, M.W. Barsoum, Nat. Mater. **2**(9), 591 (2003). https://doi.org/10.1038/nmat957
3. G. Yushin, R. Dash, J. Jagiello, J. Fischer, Y. Gogotsi, Adv. Funct. Mater. **16**(17), 2288 (2006). https://doi.org/10.1002/adfm.200500830
4. M.J. López, I. Cabria, J.A. Alonso, J. Chem. Phys. **135**(10), 104706 (2011). https://doi.org/10.1063/1.3633690
5. J.A. Alonso, I. Cabria, M.J. López, J. Mater. Res. **28**(4), 589 (2013). https://doi.org/10.1557/jmr.2012.370

6. C. de Tomás, I. Suarez-Martinez, F. Vallejos-Burgos, M.J. López, K. Kaneko, N.A. Marks, Carbon **119**, 1 (2017). https://doi.org/10.1016/j.carbon.2017.04.004
7. J.C. Angus, C.C. Hayman, Science **241**(4868), 913 (1988). https://doi.org/10.1126/science.241.4868.913
8. F.P. Bundy, W.A. Bassett, M.S. Weathers, R.J. Hemley, H.U. Mao, A.F. Goncharov, Carbon **34**(2), 141 (1996). https://doi.org/10.1016/0008-6223(96)00170-4
9. J. Narayan, A. Bhaumik, J. Appl. Phys. **118**(21), 215303 (2015). https://doi.org/10.1063/1.4936595
10. B. Stenhouse, P.J. Grout, N.H. March, J. Wenzel, Philos. Mag. **36**(1), 129 (1977), Reprinted in Ref. [28]. https://doi.org/10.1080/00318087708244453
11. B.J. Stenhouse, P.J. Grout, J. Non-Cryst. Solids **27**(2), 247 (1978). https://doi.org/10.1016/0022-3093(78)90127-8
12. D.F.R. Mildner, J. Carpenter, J. Non-Cryst. Solids **47**(3), 391 (1982). https://doi.org/10.1016/0022-3093(82)90215-0
13. C.A. Majid, J. Non-Cryst. Solids **57**(1), 137 (1983). https://doi.org/10.1016/0022-3093(83)90416-7
14. T. Noda, M. Inagaki, Bull. Chem. Soc. Jpn. **37**(10), 1534 (1964). https://doi.org/10.1246/bcsj.37.1534
15. E. Fitzer, K. Kochling, H.P. Boehm, H. Marsh, Pure Appl. Chem. **67**, 473 (1995), (IUPAC Recommendations 1995). https://doi.org/10.1351/pac199567030473
16. V.L. Deringer, G. Csányi, Phys. Rev. B **95**, 094203 (2017). https://doi.org/10.1103/PhysRevB.95.094203
17. C. de Tomas, I. Suarez-Martinez, N.A. Marks, Carbon **109**, 681 (2016). https://doi.org/10.1016/j.carbon.2016.08.024
18. J. Tersoff, Phys. Rev. B **37**, 6991 (1988). https://doi.org/10.1103/PhysRevB.37.6991
19. J. Tersoff, Phys. Rev. Lett. **61**, 2879 (1988). https://doi.org/10.1103/PhysRevLett.61.2879
20. P.A. Marcos, J.A. Alonso, A. Rubio, M.J. López, Eur. Phys. J. D **6**(2), 221 (1999). https://doi.org/10.1007/s100530050304
21. K. Nordlund, J. Keinonen, T. Mattila, Phys. Rev. Lett. **77**, 699 (1996). https://doi.org/10.1103/PhysRevLett.77.699
22. M.J. López, I. Cabria, N.H. March, J.A. Alonso, Carbon **43**(7), 1371 (2005). https://doi.org/10.1016/j.carbon.2005.01.006
23. R.K. Dash, G. Yushin, Y. Gogotsi, Micropor. Mesopor. Mat. **86**(1), 50 (2005). https://doi.org/10.1016/j.micromeso.2005.05.047
24. A. Linares-Solano, M. Jordá-Beneyto, D.L.C.M. Kunowsky, F. Suárez-García, D. Cazorla-Amorós, in *Carbon Materials: Theory and Practice*, ed. by A. Terzyk, P. Gauden, P. Kowalczyk (Research Signpost, Kerala, India, 2008), pp. 245–281
25. Y. Gogotsi, R.K. Dash, G. Yushin, T. Yildirim, G. Laudisio, J.E. Fischer, J. Am. Chem. Soc. **127**(46), 16006 (2005). https://doi.org/10.1021/ja0550529
26. I. Cabria, M.J. López, J.A. Alonso, Carbon **45**(13), 2649 (2007). https://doi.org/10.1016/j.carbon.2007.08.003
27. I. Cabria, M.J. López, J.A. Alonso, Int. J. Hydrogen Energy **36**(17), 10748 (2011), *International Conference on Hydrogen Production (ICH2P)-2010*. https://doi.org/10.1016/j.ijhydene.2011.05.125
28. N.H. March, G.G.N. Angilella (eds.), *Many-body Theory of Molecules, Clusters, and Condensed Phases* (World Scientific, Singapore, 2009)

Chapter 4
Graphene-Like Massless Dirac Fermions in Harper Systems

F. Claro and P. Robles

Abstract It is shown that systems described by Harper's equation exhibit a Dirac point at the center of the spectrum whenever the field parameter is a fraction of even denominator. The Dirac point is formed by the touching of two subbands at a single point in momentum space, and the physics around such point is characterized by the relative field only, as if the effective field were null at the reference value. Such behavior is consistent with the nesting property conjectured by Hofstadter, and its experimental verification would give support to such hypothesis.

The relative simplicity with which graphene—carbon single layer sheets—may be made and handled in the laboratory has drawn much attention to the physics of massless Dirac particles [1]. In this material, electrons moving in two dimensions (2D) near the Fermi level are subject to an effective energy dispersion law proportional to momentum rather than the usual momentum squared. The dynamics is similar to that of photons and phonons, except that in graphene the particles are charged fermions. They interact among themselves through the Coulomb force and with external electric and magnetic fields, which makes them amenable to a varied palette of experimental manipulation and possible applications.

In graphene, carbon atoms are arranged in a two dimensional hexagonal lattice, obtained as the overlap of two identical triangular lattices displaced one respect to the other. As a result of the lattice symmetry, the valence and conduction bands have two inequivalent degenerate points about which the dispersion is linear, the so called Dirac points [1–3]. No such points are found in the square lattice arrangement. As we discuss below, however, for certain values of a perpendicular magnetic field the square lattice also supports a single Dirac point where two subbands meet. Such special values are defined by a magnetic flux through the unit cell that is a rational fraction of

F. Claro (✉)
Pontificia Universidad Católica de Chile, Instituto de Física,
Avda. Vicuña Mackenna 4860, Santiago, Chile
e-mail: fclaro@uc.cl

P. Robles
Escuela de Ingeniería Eléctrica, Pontificia Universidad Católica de Valparaíso,
Avenida Brasil 2147, Valparaíso, Chile

G. G. N. Angilella and C. Amovilli (eds.), *Many-body Approaches at Different Scales*, https://doi.org/10.1007/978-3-319-72374-7_4

even denominator, in units of flux quanta. Furthermore, in the neighborhood of every one of these values Landau levels emerge from the Dirac point as they do from zero magnetic field in the hexagonal graphene lattice. At the reference field, the massless Dirac particles behave as if there was no magnetic field and in its neighborhood, they appear to respond only to the difference field much as composite fermions do at major Landau level filling fractions of even denominator [4]. These results are a property of Harper's model and therefore generic to all systems governed by such relation [5]. Besides electrons in a square lattice and a perpendicular magnetic field, the equation has appeared in several contexts, including the quantum Hall effect [6, 7], superconducting networks [8], nonperiodic solids [9], electrons in superlattices [10], and ultracold atoms in optical lattices [11–13].

The dynamics of Bloch electrons in a square lattice of lattice constant a and a perpendicular magnetic field B may be described by Harper's equation [5],

$$f_{n+1} + f_{n-1} + 2\cos(2\pi\phi n + \nu) f_n = \varepsilon f_n, \tag{4.1}$$

where f_n is the amplitude of a Wannier state localized at site na along the x-axis, n any integer, ϕ is the magnetic flux traversing a plaquette in units of the flux quantum hc/e, $\nu = k_y a$ is the dimensionless wave number along the y-axis, and ε is the energy in units of the hopping integral t. The usual Landau gauge $\mathbf{A} = B(0, x, 0)$ has been used. When the flux parameter is a rational $\phi = p/q$, p and q integers prime to each other, then the diagonal term in Eq. (4.1) has period q and the set may be closed by selecting solutions with the property $f_{n+q} = \exp(iq\mu) f_n$, with μ a real number.

The spectrum is known to have $\varepsilon = 0$ as solution at all field values [14]. When $\phi = p/q$, q odd, this root is at the center of a subband whereas if q is even it corresponds to a subband edge. We are interested in this latter case, which we assume in what follows. The condition for the existence of solutions of the resulting q equations for the amplitudes $f_1, f_2, \ldots f_q$ is then that the determinant of the coefficients,

$$D(\varepsilon, \mu, \nu) = P_q(\varepsilon) - 2(\cos q\nu + \cos q\mu) + 4(-1)^{\frac{q}{2}}, \tag{4.2}$$

vanishes [15]. Here, $P_q(\varepsilon)$ is a polynomial of degree q in ε, of even parity and having the coefficient of highest power ε^q equal 1, with all other coefficients dependent on p and q but not on μ and ν. It has the additional property that $P_q(0) = 0$. For each value of μ and ν Eq. (4.2) has q roots which, as the phases cover their range, span the q subbands present in the spectrum at that value of the field. The solution $\varepsilon = 0$ is a band edge as can be easily verified from Eq. (4.2), where the sum of cosines acquires an extremum value. In fact, if $q = 4s$, s an integer, $\varepsilon = 0$, $\nu = \mu = 0$ solves Eq. (4.2). Likewise, $\varepsilon = 0$, $\nu = \pm\pi/q$, $\mu = \pm\pi/q$ are solutions when $q = 2s$, s odd. These solutions correspond to the edges of two separate subbands that meet at a single critical point in the Brillouin zone, its center if $q/2$ is even and the four equivalent corners if odd. The bands may only touch, never overlap [16].

That the dispersion near $\varepsilon = 0$ is linear in the phases μ, ν follows from the property $P_q(-\varepsilon) = P_q(\varepsilon)$ for all q even. Near zero energy, one has $P_q(\varepsilon) \approx$

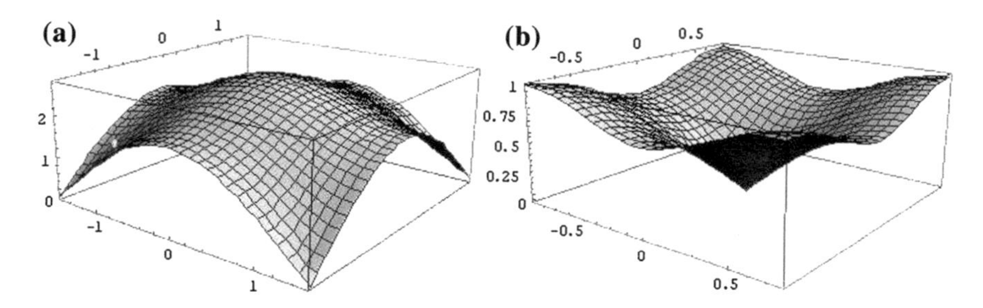

Fig. 4.1 Energy dispersion for (**a**) $q = 2$, and (**b**) $q = 4$. The energy (vertical axis) is in units of the band parameter t, while the perpendicular plane represents the first Brillouin zone

$-(-1)^{\frac{q}{2}} A(p,q)\varepsilon^2$, where $A(1,2) = 1$, $A(1,4) = 8$, $A(1,6) = 24$, $A(1,8) = 96 - 32\sqrt{2}$, $A(3,8) = 96 + 32\sqrt{2}$ and so on, are all positive constants [17]. To order ε^2, the condition on the determinant, Eq. (4.2), becomes

$$A(p,q)\varepsilon^2 + 2(-1)^{\frac{q}{2}}(\cos q\nu + \cos q\mu) - 4 = 0. \tag{4.3}$$

Near the Brillouin zone center this gives for $q/2$ even

$$\varepsilon(\mu, \nu, \phi) = \pm C(p,q)\sqrt{\nu^2 + \mu^2}, \tag{4.4}$$

where $C(p,q) = qA(p,q)^{-1/2}$ is a field dependent velocity in units $ta/\hbar$. A similar relation is obtained near each Brillouin zone corner for $q/2$ odd, with the phases μ, ν measured with respect to the appropriate zone corner.

Figure 4.1a shows a Dirac point placed at the zone corners corresponding to $p = 1$, $q = 2$, in which case the form Eq. (4.3) is exact. This case has been discussed previously [18]. Figure 4.1b is for $q = 4$ and exemplifies a Dirac point placed at the center of the zone. Specular reflection with respect to the $\varepsilon = 0$ plane gives the dispersion for negative energy, corresponding to holes. We note that in the case of a 2D triangular lattice a single Dirac point occurs at the special field values $\phi = n \pm 1/6$, n an integer [19].

As shown by Hofstadter, the spectrum of Harper's equation has a recursive subband structure arranged in such a way that, displayed over the fundamental cell $0 \leq \phi \leq 1$, resembles a butterfly with open wings [14]. He conjectured a nesting property that makes each subband in the spectrum a replica at some recursive level of the field-free band, about which a geometrically distorted version of the whole graph develops. One consequence is that, for instance, at $\phi' = \phi + \delta\phi$ a cyclotron frequency $\omega = q\delta\phi/\hbar g(\phi, \varepsilon)$ may be defined in the neighborhood of any subband pertaining to the spectrum at flux ϕ, where $g(\phi, \varepsilon)$ is the density of states at energy ε. Note that this expression scales as the departure from the field at the subband under consideration, just as the low field semiclassical cyclotron frequency departs from zero magnetic field [20].

To study the special situation when two subbands touch at energy $\varepsilon = 0$ we take advantage of the effective Hamiltonian formalism [21]. In this theory, the original problem of an electron moving in the presence of a square lattice potential and an external magnetic field B' in the neighborhood of a given subband belonging to the spectrum at field B, is replaced by the Hamiltonian problem defined by

$$H_r = \varepsilon_r\left(\frac{1}{\hbar}[\mathbf{p} + \frac{e}{c}\Delta\mathbf{A}(\mathbf{r})], \phi\right), \tag{4.5}$$

where $\varepsilon_r(\mathbf{k}, \phi)$ is the dispersion law in subband r, and $\Delta\mathbf{A}(\mathbf{r}) = \delta B(0, x, 0)$ is a vector potential for the departure magnetic field $\delta B = B' - B$. If the integer $r = 1, 2, \ldots q$ labels the subbands in order of increasing energy, then $\varepsilon_{q/2}$ and $\varepsilon_{q/2+1}$ touch at zero energy, near which the dispersion obeys Eq. (4.3). Recalling that $\nu = k_y a$ is the dimensionless crystal momentum in the y-direction and making the similar association of μ with the crystal momentum along the x-axis $k_x a$ [22], one can use Eqs. (4.4) and (4.5) to solve for the spectrum near zero energy using standard methods of quantum mechanics [23, 24]. One then obtains the sequence of eigenenergies

$$E_n = \mathrm{sgn}(n)2qt\sqrt{\frac{\pi}{A(p,q)}|n\delta\phi|}, \tag{4.6}$$

where $\delta\phi = \phi' - \phi = e\delta Ba^2/hc$ is the flux traversing a unit cell measured with respect to the reference value $\phi = p/q$, and $n = 0, \pm 1, \pm 2, \ldots$ is a Landau level index for electrons and holes. Figure 4.2 shows this expression evaluated in the neighborhood of flux 1/2 up to $n = 4$ (solid lines), together with the associated spectrum given by roots of Eq. (4.2) at a few rational values of the flux in that neighborhood (dots). In the latter, the Landau levels have a width and possibly internal structure, only that so narrow that it is not resolved in the scale of the figure. The agreement is excellent at low Landau levels, though it deteriorates slowly as the Landau index increases and the relative flux grows. The figure shows the positive quadrant only and it repeats for negative flux and negative energy, specularly reflected over the proper axes. It is worth noting that Eq. (4.6) leads to Landau energies with the same functional form as those obtained for Dirac fermions in graphite [3].

The number of states in each Landau level may be obtained from the gap labeling theorem, according to which the statistical weight—number of states per unit cell of the crystal—below any gap in the spectrum is given by a continuous function of the field, of the form $W = M\phi + N$, with M, N integers [16, 25]. The value $\varepsilon = 0$ divides the spectrum in two specularly symmetric halves, so at that energy $W = 1/2$ a condition that requires $M = mq/2$, $N = (1 - mp)/2$, m an odd integer. Using these facts one readily finds that the number of states of any level in the Landau fan emerging from the Dirac point at flux $\phi = p/q$, q even, is simply $D = q|\delta\phi|$. For instance, for $\phi = 1/2$, the number of states per cell below a gap reaching the apex at $\varepsilon = 0$ has the form $W_m = m\phi + (1 - m)/2$. The number of states between two neighboring gaps is then $|W_{m+2} - W_m| = 2\phi - 1 = 2|\delta\phi|$, in accordance with the general result just described. The number of states grows linearly with the relative flux

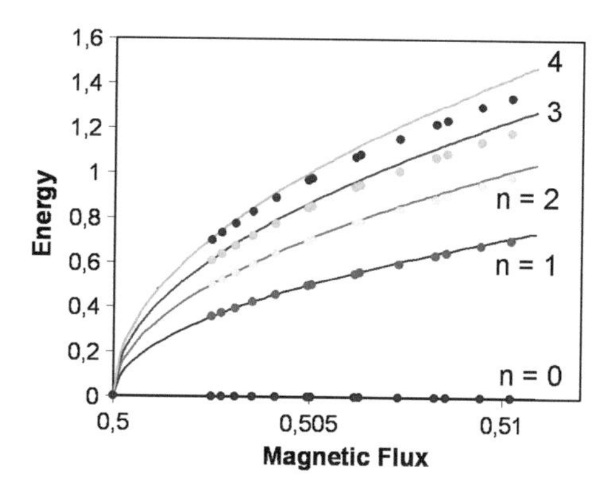

Fig. 4.2 Landau levels emerging from the Dirac point at 0.5 flux quanta per unit cell (solid lines), and solutions to Harper's equation in the same neighborhood (dots)

as it does for free 2D electrons in a magnetic field, yet the total number is q times larger. Since the index M may be identified with the dimensionless conductance [6, 26], in the neighborhood of a Dirac point at flux p/q one expects the Hall conductance to be quantized according to

$$\sigma_{\mathrm{H}} = qm\frac{e^2}{2h}. \tag{4.7}$$

Here, e is the electron charge and h Planck's constant. Because q is even, this quantity will always involve integer multiples of the quantum of conductance e^2/h even when the spin degeneracy is fully resolved. In the simple case $q = 2$ the sequence of multiples for electrons in the latter case is of the form $2n + 1$, n an integer, but if the Zeeman energy is small the sequence is the same as that observed in graphene, *e.g.* of the form $2(2n + 1)$ [1, 27].

Harper's model arises in the motion of 2D crystalline electrons in the presence of a perpendicular magnetic field if couplings up to nearest neighbors in reciprocal space are kept only. The spectrum it gives rise to, the so called Hofstadter butterfly, has been gradually confirmed experimentally [28–31]. As more details of the spectrum are unveiled, the presence of Dirac points may be tested. It is unlikely that they be observed in this physical system, however, since in more realistic models the degeneracy of neighboring bands at the center of the spectrum may be lifted [32]. Recent experiments with cold atoms and a fictitious magnetic field, described also by Harper's equation, may be more suitable to probe the presence of Dirac points in the spectrum [11–13].

Acknowledgements When one of us (FC) completed his PhD under Professor Gregory Wannier at the University of Oregon in 1972 and was planning to return to his home country, Chile, a well known physicist—actually, a Nobel Prize winner—remarked to him that if he wanted to build an academic career it was not sensible to try it in a country with no research tradition, nor established groups with appropriate funding and support as was the case of Chile then. Yet, at about the same time, a nuclear physicist of Hungarian origin, Michael Moravcsik, told him about the International Centre of Theoretical Physics (ICTP) in Trieste, Italy, precisely devoted to help young phycisists in developing countries building their scientific careers within their native community. Strongly commited to go back, he then returned to Chile, got in touch with the ICTP and travelled over the years to Italy several times to join the prestigious local Summer Workshop in Condensed Matter Physics. A most influential leader in this event was Professor Norman March, a well known scientist from England who spent every summer at the ICTP supporting the workshop organization and supervising research by attendants from all over the world, with exemplary devotion and commitement. He was widely respected by the community, and his constant availability, valuable counseling and advice helped making those events a remarkable opportunity for meeting first rate scientists, learning and doing quality research, to all attendants. It is then time to say, thank you Professor March.

References

1. K.S. Novoselov, A.K. Geim, S.V. Morozov, D. Jiang, M.I. Katsnelson, I.V. Grigorieva, S.V. Dubonos, A.A. Firsov, Nature **438**, 197 (2005). https://doi.org/10.1038/nature04233
2. J.W. McClure, Phys. Rev. **104**(3), 666 (1956). https://doi.org/10.1103/PhysRev.104.666
3. G. Li, E.Y. Andrei, Nat. Phys. **3**(9), 623 (2007). https://doi.org/10.1038/nphys653
4. J. Jain, *Composite Fermions* (Cambridge University Press, Cambridge, 2007)
5. P.G. Harper, Proc. Phys. Soc. A **68**(10), 874 (1955). https://doi.org/10.1088/0370-1298/68/10/304
6. D.J. Thouless, M. Kohmoto, M.P. Nightingale, M. den Nijs, Phys. Rev. Lett. **49**, 405 (1982). https://doi.org/10.1103/PhysRevLett.49.405
7. F. Claro, G. Huber, Phys. Today **57**(3), 17 (2004). https://doi.org/10.1063/1.1712487
8. B. Pannetier, J. Chaussy, R. Rammal, J.C. Villegier, Phys. Rev. Lett. **53**, 1845 (1984). https://doi.org/10.1103/PhysRevLett.53.1845
9. J.B. Sokoloff, Phys. Rep. **126**(4), 189 (1985). https://doi.org/10.1016/0370-1573(85)90088-2
10. R.R. Gerhardts, D. Weiss, U. Wulf, Phys. Rev. B **43**, 5192 (1991). https://doi.org/10.1103/PhysRevB.43.5192
11. M. Aidelsburger, M. Atala, M. Lohse, J.T. Barreiro, B. Paredes, I. Bloch, Phys. Rev. Lett. **111**, 185301 (2013). https://doi.org/10.1103/PhysRevLett.111.185301
12. P. Roushan, Bull. Am. Phys. Soc. **62**(4), Y23.00001 (2017), (APS March Meeting 2017)
13. M. Aidelsburger, M. Lohse, C. Schweizer, M. Atala, J.T. Barreiro, S. Nascimbene, N.R. Cooper, I. Bloch, N. Goldman, Nat. Phys. **11**(2), 162 (2015), Letter. https://doi.org/10.1038/nphys3171
14. D.R. Hofstadter, Phys. Rev. B **14**, 2239 (1976). https://doi.org/10.1103/PhysRevB.14.2239
15. D. Langbein, Phys. Rev. **180**, 633 (1969). https://doi.org/10.1103/PhysRev.180.633
16. F. Claro, G.H. Wannier, physica status solidi (b) **88**(2), K147 (1978). https://doi.org/10.1002/pssb.2220880262
17. Y. Hasegawa, P. Lederer, T.M. Rice, P.B. Wiegmann, Phys. Rev. Lett. **63**, 907 (1989). https://doi.org/10.1103/PhysRevLett.63.907
18. P. Delplace, G. Montambaux, Phys. Rev. B **82**, 035438 (2010). https://doi.org/10.1103/PhysRevB.82.035438
19. F.H. Claro, G.H. Wannier, Phys. Rev. B **19**, 6068 (1979). https://doi.org/10.1103/PhysRevB.19.6068
20. F. Claro, Physica Status Solidi (b) **97**(1), 217 (1980). https://doi.org/10.1002/pssb.2220970124
21. M. Taut, H. Eschrig, M. Richter, Phys. Rev. B **72**, 165304 (2005). https://doi.org/10.1103/PhysRevB.72.165304
22. G.H. Wannier, Physica Status Solidi (b) **100**(1), 163 (1980). https://doi.org/10.1002/pssb.2221000116
23. G.W. Semenoff, Phys. Rev. Lett. **53**(26), 2449 (1984). https://doi.org/10.1103/PhysRevLett.53.2449
24. I. Affleck, J.B. Marston, Phys. Rev. B **37**(7), 3774(R) (1988). https://doi.org/10.1103/PhysRevB.37.3774

25. G.H. Wannier, Physica Status Solidi (b) **88**(2), 757 (1978). https://doi.org/10.1002/pssb.2220880243
26. P. Streda, J. Phys. C **15**(36), L1299 (1982). https://doi.org/10.1088/0022-3719/15/36/006
27. Y. Zhang, Y. Tan, H.L. Stormer, P. Kim, Nature **438**(7065), 201 (2005). https://doi.org/10.1038/nature04235
28. T. Schlösser, K. Ensslin, J.P. Kotthaus, M. Holland, Europhys. Lett. **33**(9), 683 (1996). https://doi.org/10.1209/epl/i1996-00399-6
29. C. Albrecht, J.H. Smet, K. von Klitzing, D. Weiss, V. Umansky, H. Schweizer, Phys. Rev. Lett. **86**, 147 (2001). https://doi.org/10.1103/PhysRevLett.86.147
30. C.R. Dean, L. Wang, P. Maher, C. Forsythe, F. Ghahari, Y. Gao, J. Katoch, M. Ishigami, P. Moon, M. Koshino, T. Taniguchi, K. Watanabe, K.L. Shepard, J. Hone, P. Kim, Nature **497**(7451), 598 (2013), Letter. https://doi.org/10.1038/nature12186
31. B. Hunt, J.D. Sanchez-Yamagishi, A.F. Young, M. Yankowitz, B.J. LeRoy, K. Watanabe, T. Taniguchi, P. Moon, M. Koshino, P. Jarillo-Herrero, R.C. Ashoori, Science **340**(6139), 1427 (2013). https://doi.org/10.1126/science.1237240
32. F. Claro, Physica Status Solidi (b) **104**(1), K31 (1981). https://doi.org/10.1002/pssb.2221040151

Chapter 5
Silicene and Germanene as Prospective Playgrounds for Room Temperature Superconductivity

G. Baskaran

Abstract Combining theory and certain striking phenomenology, we suggest that silicene and germanene are *elemental Mott insulators* and abode of doping induced high-T_c superconductivity. In our theory, a three-fold reduction in π-π^* bandwidth in silicene, in comparison to graphene, and short range Coulomb interactions enable Mott localization. Recent experimental results are invoked to provide support for our Mott insulator model: (i) a significant π-band narrowing, in silicene on ZrB_2 seen in ARPES, (ii) a superconducting gap appearing below 35 K with a large $2\Delta / k_B T_c \sim 20$ in silicene on Ag, (iii) emergence of electron like pockets at M points, on electron doping by Na adsorbent, (iv) certain coherent quantum oscillation like features exhibited by silicene transistor at room temperatures, and (v) absence of Landau level splitting up to 7 T, and (vi) superstructures, not common in graphene, but ubiquitous in silicene. A synthesis of the above results using theory of Mott insulator, with and without doping, is attempted. We surmise that if competing orders are taken care of and optimal doping achieved, superconductivity in silicene and germanene could reach room temperature scales; our estimates of model parameters, t and $J \sim 1$ eV, are encouragingly high, compared to cuprates.

5.1 Introduction

Nearly 12 years ago, Professor Norman H. March and collaborators pointed out [1, 2] (see also [3]) the possibility of Mott localization of electrons in hexasilabenzene (Si_6H_6, a Si analogue of benzene), resulting from an expanded equilibrium Si–Si bond length. This article (see also [4]) presents some profound consequences of such localization in silicene (Si analogue of graphene) and germanene. It gives me great pleasure to contribute this article honouring Professor Norman March, on the occasion of his 90th birthday. My first meeting with Professor March was in early

G. Baskaran (✉)
The Institute of Mathematical Sciences, Chennai 600113, India
e-mail: baskaran@imsc.res.in

G. Baskaran
Perimeter Institute for Theoretical Physics, Waterloo, ON N2L2Y5, Canada

G. G. N. Angilella and C. Amovilli (eds.), *Many-body Approaches at Different Scales*, https://doi.org/10.1007/978-3-319-72374-7_5

1976 at the International Center for Theoretical Physics (ICTP), Trieste, Italy. He helped Professor Abdus Salam, in developing condensed matter theory activities at ICTP, for more than two decades, in the formative years. One of the focus of Professor March at ICTP was to nurture and mentor young minds from third world countries; I was one of the beneficiaries.

A wealth of activity in the field of graphene, following the seminal work of Novosolev and Geim [5–7], has paved the way for silicene [8–11], a silicon analogue of graphene. This new entrant might have a potential to begin another fertile direction in condensed matter science and technology. Replicating a rich graphene physics has been a part of recent efforts. Stable silicene layer has been created on a few metallic substrates, Ag, ZrB_2 and Ir. However, synthesis of free standing silicene remains a challenge. Interesting angular resolved photoemission spectroscopy (ARPES) and scanning tunneling microscopy (STM) results are available [12–19]. Silicene based field effect transistor has been also fabricated [20, 21]. There have been successful attempts to synthesize germanene and related systems [22–27] on certain metallic substrates.

Aim of the present article is to provide a low energy model, a rather unexpected one, for electrical and magnetic properties of silicene. Finding a suitable low energy model for strongly interacting quantum matter continues to be a challenge. The very experimental results we wish to understand guide us to correct theoretical modeling. Theory in turn guides experiments. The synergy continues. This is true from Standard Model building in elementary particle physics to Standard Model building for cuprate. Silicene is no exception.

The currently prevalent view is that neutral silicene is a Dirac Metal, a semimetal qualitatively similar to graphene [11]. Purpose of the present paper is to offer a different view point, that *silicene is not a carbon copy of graphene—it is an elemental Mott insulator*. If proved correct, our provocative proposal will make silicene different from semi metallic graphene in a fundamental fashion and open new avenues for physics and technology, arising from strong electron correlation effects.

It is well appreciated now, thanks to the path breaking discovery of high-T_c superconducting cuprates by Bednorz and Müller [28] and subsequent resonating valence bond theory by Anderson and collaborators [29–33], that Mott insulators and strong electron correlations are seats of a variety of rich physics and phenomena. In addition to superconductivity it includes, quantum spin liquids, emergent fermions with Fermi surfaces, gauge fields, quantum order, topological order and so on. Further, inspired by certain recent theoretical development [34] there is a debate and search for quantum spin liquids in honeycomb lattice Hubbard model [35–37]. We believe that silicene and germanene will fit well into the discussion as real candidate materials, albeit with added novel features.

Several ab initio calculations [8–10], many-body theory [38], quantum chemical calculations and insights [1, 2, 39–42] are available for silicene. Interestingly there is a differing view, which has not been well appreciated. Existing solid state many-body calculations predict stable free standing semimetallic silicene that is a similar to graphene. However, quantum chemical methods and insights doubt existence of a stable free standing silicene [39–41] because of radicalization/reactivity and reduced aromaticity, arising from a weakened p-π bond.

It is clear that theory of silicene is challenging and less understood compared to graphene, because of a growing importance of electron electron interaction and a soft c-axis displacement (puckering) degree of freedom, arising from an easy sp^3 mixing. Our model and theory is aimed to initiate new discussion, theoretical and experimental studies.

The present article is organized as follows. We interpret certain existing theoretical results, Coulomb screening argument for Mott transition and quantum chemical insights as providing support for our proposal of a narrow gap Mott insulating state for neutral silicene. We also identify and discuss a set of about 6 anomalous experimental results that point to a Mott insulating state.

A Heisenberg model, containing additional multispin interactions, is introduced to describe spin dynamics in a small gap Mott insulator, and a t-J model for the spin-charge dynamics of doped Mott insulator. Then we discuss the aforementioned anomalous experimental results in the light of our model.

Prospects for high-T_c superconductivity, within our model, is discussed next. In view of larger and more favourable t-J parameters, in comparison to layered cuprates, there is a distinct possibility of a T_c *approaching the room temperature scale,* provided competing interactions are taken care of. Using existing theoretical works we come to the conclusion that superconductivity is likely to be a chiral spin singlet $d + id$ superconductivity.

Superstructures, not common in graphene but ubiquitous in silicene grown on substrates, are interpreted as arising from a strong response of c-axis deformable Mott localized p-π electrons, to substrate perturbations, through site dependent sp^3 mixing. In a recent ab initio calculation with Vidya [43] we have found indirect evidence for Mott localization, through presence of sizable Kekulé (valence bond) order and a weak antiferromagnetism. Further we have interpreted experimentally seen reconstructions as Kekulé or Valence Bond order.

The present paper assumes that t-J model in 2D describes Kosterlitz–Thouless superconductivity for a range of doping. While there is no rigorous proof for this, it is well certified by a body of analytical, numerical and experimental efforts available for the square lattice cuprates, ever since Bednorz and Müller's discovery and the beginning of resonating valence bonds (RVB) theory. Existing studies on honeycomb lattice t-J model indicate that spin singlet superconductivity continues to be a dominating phase over a range of doping, but with $d + id$ order parameter symmetry.

Our present work is a natural extension of our earlier work on graphene [44–46]. It has been our view that graphite and graphene should show interesting electron correlation effects, even though electron-electron interaction strengths are moderate compared to the band width. According to us, *reduced 2-dimensionality of graphene causes an amplification of electron correlation effects*. Our earlier prediction of spin-1 collective mode in neutral graphene [45] and very high-T_c superconductivity in doped graphene [44, 46] are based on use of a moderate electron repulsion strength. Our main message in the present article is that 2D silicene and germanene, having a third of graphene band width, but two thirds of on site coulomb interaction strength of graphene should exhibit more pronounced electron correlation effects.

In a recent article we have suggested a *five-fold way* to new high-T_c superconductors [47] one of the ways is the *graphene route*. Silicene and germanene are most likely to lie in the graphene route and help us achieve room temperature superconductivity.

5.2 Is Free Standing Silicene Stable?

Beginning with hexasilabenzene, Si_6H_6, a silicon analogue of benzene, chemists have wondered about the existence of stable planar p-π bonded Si based molecules [1, 48]. Their concern is that an increased Si-bond length, a simple consequence of a 60% increase in atomic radius, will weaken the p-π bond, leading to reduced aromaticity and increased chemically reactivity. It is also an experimental fact that free Si_6H_6 molecule has not been synthesized so far.

Silicene is an infinitely extended planar p-π bonded network of Si atoms. Density functional theory (DFT) calculations and beyond argue for a stable free standing graphene. On the other hand, Sheka in 2009 [39, 40] and Hoffmann in 2013 [41], have questioned stability and very existence of free standing silicene because of reduced aromaticity and enhanced radicalization. This, according to them, will make silicene react with any molecular dirt.

Radicalization in quantum chemistry is creation of unpaired lone electron in certain molecular orbitals in the ground state; the loners are generically weakly coupled to other loners, if present. In the context of periodic systems such as a crystalline solid we interpret it to mean an extreme Mott localization and formation of nearly decoupled spins. In this Mott insulating state quantum fluctuations lead to residual (superexchange) couplings among spins and quantum magnetism.

Free standing silicene has not synthesized experimentally so far. A key stabilizing factor, namely metallic substrates, Ag, Ir and ZrB_2 is needed. Further, a strong hybridization between Ag bands and Si orbitals at the fermi level leads to significant modification of electronic properties of free standing silicene.

In what follows, by *stable Mott insulating single layer silicene* we mean the following. We assume that (metallic) substrate stabilizes a single layer silicene and at the same time does not significantly modify the Mott or doped Mott insulator character, that we are after.

5.3 Mott Insulating Silicene: Theoretical Support

In this section, using recent estimates of the Hubbard U, nearest neighbour repulsion V, and Mott's argument for metal insulator transition invoking screened long range Coulomb interaction, we will argue a Mott insulating ground state for silicene.

We begin with a summary of basic quantum chemistry and band theory results for silicene. Silicon, located just below carbon in the periodic table has a larger atomic radius ~1.17 Å, in contrast to a smaller value ~0.77 Å for carbon. This leads to a *significant* $3p$-π *bond stretching:* the Si–Si distance is ~2.3 Å, while the C–C distance is only 1.4 Å in graphene.

Unlike graphene, Si allows a small sp^3 admixture to sp^2 bonding, resulting in puckered σ-bonds. That is, Si atoms of the two triangular sublattices of the honeycomb net undergo small and opposite displacements, leading to a net c-axis stretching ~0.4 a.u. A reduced $3s$-$3p$ level separation in the Si atom, compared to $2s$-$2p$ level separation in the C atom encourages a small sp^3 hybridization.

Electronic structure calculations [8–10] predict a semimetal band structure qualitatively similar to graphene, containing two Dirac cones at the K and K' points. A major difference from graphene is a three-fold reduction in the π-π^* bandwidth.

Following a pioneering work of Sorella and Tosatti [35] a recent estimate [34] gives an accurate value of the critical value $U/t \sim 3.8$ for the metal to Mott transition in the Hubbard model on a honeycomb lattice. In a very recent many-body theory, Schüler et al. [38] estimate a value $U/t \sim 4.1$ for silicene. This puts silicene on the Mott insulator side, close to the phase transition point. However, Schüler et al. argue that inclusion of nearest neighbour repulsion $V/t \sim 2.31$, will reduce U to an effective $\tilde{U} \approx U - V$ and bring silicene back to a semimetallic state like graphene. In what follows we argue, on the contrary, that the presence of a finite V will hasten Mott localization and reinforce a Mott state through a first order phase transition.

We start with Mott's argument for a first order Mott transition. Mott begins by asking whether a *screened Coulomb interaction* present in the metallic state is sufficient to form a quantum mechanical *bound state of an electron and the hole it left behind at its home site,* at the Fermi level. Within a Hubbard model idealization this happens when the bandwidth becomes comparable to the onsite U; it also means that at very large U every site binds a lone daughter electron. In Mott's argument the transition is pre-empted by reduced screening of long range interaction (as we loose free carriers by bound state formation), through a feed back resulting in a first order phase transition. This aspect is not contained in the simple Hubbard model.

Let us assume that in addition to U we have a non-zero nearest neighbour coulomb interaction V, which is a leading term in the long range part of screened coulomb interaction. According to Mott's arguments, V will add to the already existing on site attraction U between an electron and the hole it left behind. This additional attraction (widening of the potential well) decreases kinetic energy of bound state. Resulting increase in binding energy hasten bound state formation. Roughly, the onsite Hubbard U gets enhanced: $\tilde{U} \approx U + \alpha V$, where $\alpha \sim 1$.

Thus we conclude, by what we believe to be a correct use of important estimates of Schüler et al. [38] that silicene is in the Mott insulating side of the metal insulator transition point. We hope to address this point in some detail in a separate paper.

5.4 Mott Insulating Silicene: Phenomenological Support

In this section we briefly review experimental results in silicene which are anomalous, from the point of view of a Dirac metal. But, as we will discuss latter, they seem normal from Mott insulator point of view.

Doped Hole and a Narrow Band

In a work that has received a wide attention, using ARPES, Vogt et al. [13] show the presence of a Dirac cone dispersion over a wide energy range below Fermi level, with a Fermi velocity v_F comparable to that in graphene. There is an ongoing debate [49–51] on whether this is primarily a silicene p-π band or primarily a Ag metallic bands or a strong hybrid. Recent ARPES result shows [52] multiple (more than the expected double) Dirac cones in the silicene on Ag system. This has been attributed to a strong hybridization of silicene and silver states at the Fermi level.

On the other hand, ARPES study of silicene grown on ZrB_2 by Fleurence et al. [15] shows a very different behaviour. ZrB_2 is a low carrier density metal with small Fermi pockets. It is expected to have less electronic influence on silicene. Fleurence et al. find a remarkably narrow band around around K and K' points (marked X_2 in Fig. 5.1b), lying below a finite gap ~0.3 eV. Its spectral weight vanishes in two-third of the Brillouin zone. An extrapolation from the shape of the visible part of the band gives us a total band width ~1 eV. This is to be contrasted with the total π-π^* band width of 6 eV as given by LDA calculations. Thus there is a band width renormalization by a factor of 10. Further, three silicene bands denoted by X_1, X_2 and X_2' all lie in an energy interval of 0.5 eV. This spectrum is reminiscent of hole spectrum in Mott insulating La_2CuO_4.

A Superconducting Gap Anomaly

Chen et al. have observed [17] a superconductor like gap structure in their STS study of silicene on Ag and a T_c ~35 to 40 K. They provide arguments in support of a gap

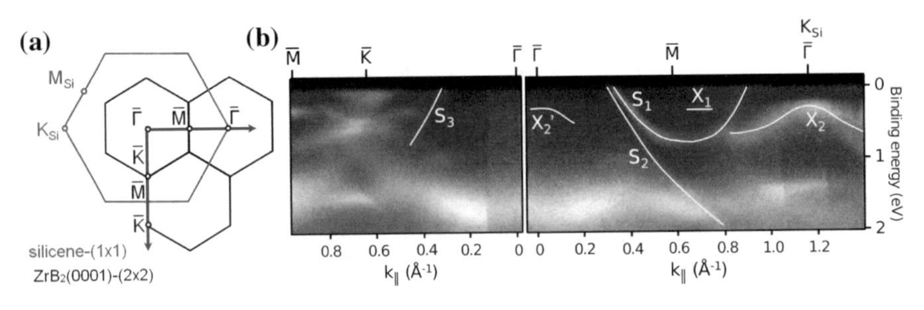

Fig. 5.1 ARPES data of silicene on ZrB_2 (redrawn after [15]). A hole like narrow band of width (marked X_2), below a gap of 0.3 eV at K point is seen in the left panel. This band is only 1 eV wide, compared to 6 eV π-π^* band width given by the local density approximation (LDA). It has appreciable spectral weight only in one third of the Brillouin zone. We interpret this band as a strongly renormalized hole band of a Mott insulator. Interestingly, all silicene bands X_1, X_2 and X_2' lie within an energy interval of 0.5 eV. S's refer to substrate bands

arising from superconductivity. A large $2\Delta / k_B T_c \sim 20$ makes it anomalous. Further experiments are needed to substantiate this important result.

Absence of Landau Level Formation

An STM work [18] in the presence of a magnetic field as large as 7 T does not find an expected Landau level splitting. In our view, a doped hole looses nearly all its quasiparticle weight because of the strong electron-electron interaction. We find that quasiparticle line broadening, as inferred from experiment is larger than the expected Landau level splitting, making it invisible in STS measurements. In graphene, because of quasi particle coherence, one sees Landau level structures for similar magnetic fields in the STS measurements.

Silicene Lattice Reconstructions

Recent experiments on silicene grown on metal surfaces exhibit [53–55] lattice reconstructions, $\sqrt{3} \times \sqrt{3}$, $\sqrt{7} \times \sqrt{7}$, $\sqrt{13} \times \sqrt{13}$, 4×4 etc. They are temperature dependent and some of them exhibit finite temperature phase transitions. Further, reconstructions are accompanied by *space dependent* sp^3 *mixing and consequent bond length modulation*. Even within band theory there is no Fermi surface. So a standard route for density wave instabilities namely Fermi surface nesting is absent. This explains why reconstructions or charge density wave (CDW) order is not seen in graphene. Graphene is stiff, in view of a $\sim$9 eV wide filled π-band and an associated large *aromaticity*. We suggest that p-π electrons in silicene, because of Mott localization, respond strongly to substrate perturbations by making use of the soft sp^3 hybridization option and corresponding c-axis displacements.

Fermi-Arc-Like Electron Pockets

Electron doping in silicene deposited on ZnB_2 by alkali metal leads to appearance of small Fermi-arc- or pockets-like features at the M points [19]. This is not easily explained within LDA band structure result. A doped Mott insulator on the other hand, does not obey Luttinger theorem in the usual fashion and can have unusual emergent Fermi pockets and Fermi arcs—a striking example is the under doped cuprates.

Silicene Transistor and Resistance Oscillations

Using an innovative technique a silicene transistor has been successfully fabricated [20, 21]. It exhibits a desired on off resistance radio as gate voltage is varied. Further, it exhibits an intriguing resistance oscillation at room temperatures, resembling a Fabry–Perot type of quantum interference.

Thus we have some theoretical arguments and a few phenomenological facts in support of a Mott insulator picture for neutral silicene. In the following we will build a model, address the above experimental observations from the Mott insulator point of view.

5.5 Spin Liquid State in Silicene?

Let us first discuss nature of the hypothesized Mott insulating state of neutral silicene. From ARPES experiment [15] we estimate a Mott Hubbard gap of $\sim$0.6 eV. In transition metal oxide and organic Mott insulators, Mott gap is often comparable to bandwidths. As the inferred Mott gap, 0.6 eV is only a tenth of the total π-π^* band width $\sim$6 eV silicene is a *Mott insulator with a small charge gap.*

In a Mott insulator low energy degree of freedom are spins and spin-spin interaction arise from super (kinetic) exchange processes. One can estimate J, using the standard expression, as $J \approx 4t^2/U^2$. The values $t \sim 1.14$ eV and an effective $U \sim 5$ eV, discussed earlier for silicene, gives us a $J \approx 1$ eV. As J is comparable to the Mott Hubbard gap of 0.6 eV, a strong virtual charge density and charge current fluctuations will renormalize J to lower values. Further, higher order cyclic exchanges will be also present.

Ignoring spin-orbit coupling for the moment, our effective spin Hamiltonian in the Mott insulating state is a spin-half Heisenberg Hamiltonian

$$H_s = J \sum_{\langle ij \rangle} \left(\mathbf{S}_i \cdot \mathbf{S}_j - \frac{1}{4} \right) + 4 \text{ and } 6 \text{ spin terms.} \tag{5.1}$$

If the nearest-neighbour J dominates, we will have long range antiferromagnetic order in the ground state. As recent results show, within nearest neighbour repulsive Hubbard model one is unlikely to stabilize a spin liquid phase [36]. In the present case, because of softness of puckering, there may be valence bond order, rather than a spin liquid, without doping. However, as in cuprates, this may not be an important practical issue. This is because we expect that even a small density of dopants through their dynamics will destroy long range antiferromagnetic (AFM) order and valence bond order and stabilize some kind of spin liquid state (pseudogap phase) containing incoherent dopant charges. It is this doping induced spin liquid state that will determine nature of superconductivity over a range of doping. It is likely that a variety of spin liquids are around the corner.

5.6 Doped Mott Insulator

For dopant dynamics in a Mott insulator, up to a certain range of doping, a projective constraint arising from upper and lower Hubbard band formation and surviving superexchange are important. Thus the relevant model for doped Mott insulator is the t-J model:

$$H_{tJ} = -t \sum_{\langle ij \rangle} (c^\dagger_{i\sigma} c_{j\sigma} + \text{H.c.}) + J \sum_{\langle ij \rangle} \left(\mathbf{S}_i \cdot \mathbf{S}_j - \frac{1}{4} n_i n_j \right), \tag{5.2}$$

with the local constraint $n_{i\uparrow} + n_{i\uparrow} \neq 2$ or 0, for hole and electron doping, respectively. The value of $J, t \sim 1$ eV.

It is known from the study of single hole dynamics in Mott insulating cuprates that a free and coherent propagation of a doped hole or electron (carrying its charge and spin) is frustrated by the strongly quantum entangled spin background. This leads to a significant band narrowing and loss of spectral weight over a large part of the Brillouin zone, as seen in ARPES experiments [56] and t-J model calculations [57]. For cuprates band theory gives a width of $8t \approx 3$–4 eV, while ARPES results in the Mott insulating cuprates give a band width of $2J \sim 0.3$ eV (essentially spin wave bandwidth) for holes. This is a substantial, ten fold reduction of hole band width.

As we saw earlier ARPES study on silicene grown on ZrB_2 gives a parabolic hole band at K point, about 0.3 eV below the Fermi level (band X_2 in Fig. 5.1b) with a bandwidth ~1 eV. Comparing results with cuprates mentioned above, we get a $J \sim 0.5$ eV. This is in the right ball park, as our estimate of $J \sim 1$ eV.

5.7 Doped Mott Insulator and Superconductivity

We start with the t-J model for our doped Mott insulator in a honeycomb lattice and discuss superconductivity. Interestingly, the above model, Eq. (5.2), was studied in a mean field approach first in [58] for graphite. Later the present author independently studied the same model ignoring onsite constraints [44] as a semimicroscopic way of incorporating Pauling's spin singlet (resonating valence bond) correlations in planar *graphitic systems*, within a band theory approach. It leads to a prediction of very high-T_c superconductivity around an optimal doping. Our mean-field theory result of high-T_c superconductivity for graphite was reanalyzed and confirmed by Black-Schaffer and Doniach [59]. They further found a remarkable spin singlet chiral superconducting state, namely a state with $d + id$ symmetry as the most stable mean field solution for the same range of doping. A more conservative repulsive Hubbard model has been studied from superconductivity point of view using variational Monte Carlo method in reference [46], by us and collaborators for graphene. There are other important works addressing the same issue [60–65] by different methods. They all support the possibility of $d + id$ chiral spin singlet superconductivity.

A very recent work [66] studies the t-J model on a honeycomb lattice using a new and powerful variational approach using Grassman Tensor Network states. It confirms $d + id$ superconductivity for the t-J model on a honeycomb lattice for a range of doping.

Using available theory and insights we suggest a phase diagram (Fig. 5.2), qualitatively similar to the cuprates, including the *pseudogap phase*. A new aspect for the honeycomb lattice is presence of a van Hove singularity and an associated nesting (Fig. 5.2) at a doping of $x = 0.25$. It has been suggested that this nesting might stabilize a complex magnetic order [67] or charge density wave order. If our variational calculation [46] performed for graphene, a doped semimetal described by

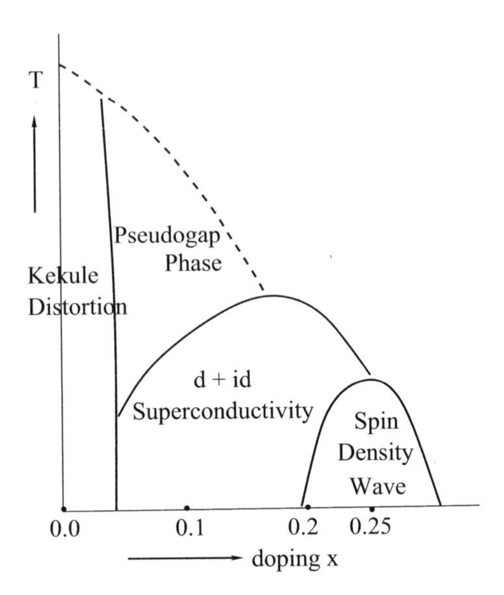

Fig. 5.2 A schematic phase diagram for doped silicene and graphene

intermediate repulsive U Hubbard model, is any guidance, maximum superconducting T_c occurs already around 15%, somewhat similar to cuprates.

The issue of scale of superconducting T_c is very exciting and it shows some promise. As mentioned earlier we have an unusually large $J \sim 1\,\text{eV}$ and $t \sim 1\,\text{eV}$ for our honeycomb lattice. This is to be contrasted with a value of $J \sim 0.15\,\text{eV}$ and $t \sim 0.25\,\text{eV}$ for the square lattice cuprates, known for their record $T_c \sim 90$ K, for single layer cuprates. We have an average four-fold increase in t and J for silicene. Does silicene offer a four-fold increase in T_c? Taking care of lattice structure difference between cuprate and silicene crudely, a scaling gives at least a three-fold increase of T_c. Thus room temperature scales for T_c seems within reach, provided competing orders are taken care of.

As in cuprates, we expect competing orders, charge and spin stripes to challenge the high-T_c superconducting states. Further, as we will see, electron lattice interaction can cause patterns of vertical displacements of Si atoms, arising from sp^3 mixing. It is conceivable that there will be stripes and 2D patterns, corresponding to various valence bond orders. These valence bond localization tendencies will compete and reduce T_c significantly.

The only available, but remarkable indicator for superconductivity in silicene is an STS study [17]. They find a T_c in the range 30–40 K and a large gap of 35 meV; that is, an anomalous value $2\Delta/k_B T_c \approx 20$. If we take the maximal gap value from this experiment and extract a mean field T_c, using weak coupling BCS result, we get a number close to 200 K! We interpret this anomaly in $2\Delta/k_B T_c$, as due to a superconducting state with a large local pairing gap but with T_c reduced by a strong phase fluctuations arising from a static or low frequency local competing orders. This needs to be investigated theoretically and experimentally further.

We have interpreted [68] a quantum oscillation in resistance seen in silicene transistor at room temperatures, as a Fabry–Perot type of interference of preformed bosonic charge $2e$ Cooper pairs, and physics similar to the pseudogap phase of cuprates. Briefly, the gate induced carriers (holons or doublons) enable delocalization and interference of preformed charge $2e$ singlets, even before full phase coherence and superconductivity develops.

There are recent suggestions of superconductivity in silicene based on different mechanisms [69, 70]. We wish to point out that Liu et al. [69] have theoretically studied possibility of spin fluctuation induced chiral $d + id$ superconductivity in *neutral* bilayer silicene. In their mechanism, intralayer hopping between the two Dirac metallic layers produce small electron and hole pockets around K and K' points. In our theory such a bilayer will remain insulating with an additional feature of a spin gap induced by a strong interlayer exchange coupling (spin singlet bond formation between Si atoms along the c-axis). External doping or gate doping by very strong electric fields applied perpendicular to the layers will be needed to create superconductivity in the Mott insulating bilayers.

5.8 Germanene and Stanene

Group IV elements in periodic table C, Si, Ge, Sn and Pb have increasing atomic radii 0.77, 1.17, 1.22, 1.45 and 1.8 Å respectively. The p-π bonds in silicene, germanene and stanene get weaker because of increasing atomic radii, in spite of increasing size of the p orbitals. Ab initio calculations for germanene leads to a few percent enhancement of Ge–Ge bond length, but a substantial increase of puckering and c-axis stretching from ~0.4 Å in silicene to ~0.6 Å in germanene. There is a small reduction in π and π^* bandwidths from 3 to 2.5 eV.

All things being quantitatively nearly equal at the level of model, germanene should be a narrow band Mott insulator. Consequences we have discussed so for, including possibility of quantum spin liquid and high-T_c superconductivity on doping should be anticipated. There are interesting experimental developments with respect to germanene and stanene recently [22–27].

5.9 Summary and Discussion

To summarize, we have hypothesized that silicene and germanene are narrow gap Mott Insulators. This challenges the widely held belief that they are a Dirac metal, like graphene. In making our proposal we have relied heavily on a synthesis of theory and phenomenology. We have used ARPES, STM and recent conductivity results. Mott's arguments, quantum chemical insights and extensive theoretical study of graphene and silicene has helped us in our proposal. Using known results we discussed various solutions for our model, including spectral function of a hole in a Mott insulator, absence of Landau level splitting, superconductivity and so on. Unconventional superconducting order namely $d + id$ and high-T_c's was also discussed.

An important support for Mott localization comes from our translation of the quantum chemical results of Sheka [39, 40] to solid state terminology. Sheka studies nanosilicene clusters and finds an extensive radicalization. Radicalization is extracted using a procedure that uses spin polarized (antiferromagnetic) ab initio solutions. This interesting quantum chemical approximation is less known in solid state context. As indicated earlier we equate *radicalization in finite systems to Mott localization in extended solid state context.* Thus Sheka's finding of extensive radicalization and alleged unstable silicene is a conservative evidence for narrow gap Mott insulator formation and strongly exchange coupled spins.

We have ignored spin-orbit coupling in the present paper, simply to focus on the key physics of strong correlations. Spin orbit coupling will play its own, important unique role, some what different from standard band insulator or semi metal. Within the context of semi metallic graphene interesting effects of spin orbit coupling are being studied [71–73].

Direct and indirect methods should be used to unravel an underlying Mott insulator. Optical conductivity, $\sigma(\omega)$ measurement is an urgent one. It should focus on finding signatures on Hubbard band features. It will be interesting to confirm the existing claim of superconducting gap in STS measurements [17]. Reconstructions should be studied carefully to distinguish valence bond density wave and plaquette resonance density waves. Pseudogap physics needs to be explored using NMR and NQR measurement. Our tantalizing prediction of very high-T_c superconductivity reaching room temperature scales needs to be explored. This needs ways of understanding and overcoming unavoidable competing instabilities such as spin and charge stripes.

Next major experimental and theoretical challenge is to see whether a free standing silicene exists and if it can be synthesized and studied.

I am not aware of any real material for which the t-J model parameters are as large as we have suggested for silicene and germanene, $t, J \sim 1\,\text{eV}$. So we consider silicene and germanene as forefront materials in the race for room temperature superconductivity.

Acknowledgements I thank Professor Yamada-Takamura for giving permission to have a figure redrawn from [15]; Dr. Ayan Datta for a discussion; Dr. Kehui Wu and colleagues for informative discussions at the Silicene meeting in Beijing, 2014. I thank Science and Engineering Research Board (SERB), Government of India for the SERB Distinguished Fellowship. Additional support was provided by the Perimeter Institute for Theoretical Physics. Research at Perimeter Institute is supported by the Government of Canada through the Department of Innovation, Science and Economic Development Canada and by the Province of Ontario through the Ministry of Research, Innovation and Science.

References

1. A. Grassi, G.M. Lombardo, R. Pucci, G.G.N. Angilella, F. Bartha, N.H. March, Chem. Phys. **297**(1), 13 (2004). https://doi.org/10.1016/j.chemphys.2003.10.001
2. N.H. March, A. Rubio, J. Nanomater. **2011**, 932350 (2011). https://doi.org/10.1155/2011/932350
3. G. Forte, A. Grassi, G.M. Lombardo, R. Pucci, G.G.N. Angilella, in *Many-body approaches at different scales: a tribute to Norman H. March on the occasion of his 90th birthday*, ed. by G.G.N. Angilella, C. Amovilli (Springer, New York, 2018), Chap. 19, p. 219. (This volume). https://doi.org/10.1007/978-3-319-72374-7_19
4. G. Baskaran, Silicene and germanene as prospective playgrounds for room temperature superconductivity (2013), arXiv:1309.2242 [cond-mat.str-el]
5. K.S. Novoselov, A.K. Geim, S.V. Morozov, D. Jiang, M.I. Katsnelson, I.V. Grigorieva, S.V. Dubonos, A.A. Firsov, Nature **438**, 197 (2005). https://doi.org/10.1038/nature04233
6. A.K. Geim, K.S. Novoselov, Nat. Mater. **6**(3), 183 (2007). https://doi.org/10.1038/nmat1849
7. A.H. Castro Neto, F. Guinea, N.M.R. Peres, K.S. Novoselov, A.K. Geim, Rev. Mod. Phys. **81**(1), 109 (2009). https://doi.org/10.1103/RevModPhys.81.109
8. K. Takeda, K. Shiraishi, Phys. Rev. B **50**, 14916 (1994). https://doi.org/10.1103/PhysRevB.50.14916
9. G.G. Guzmán-Verri, L.C. Lew Yan Voon, Phys. Rev. B **76**, 075131 (2007). https://doi.org/10.1103/PhysRevB.76.075131
10. S. Cahangirov, M. Topsakal, E. Aktürk, H. Şahin, S. Ciraci, Phys. Rev. Lett. **102**, 236804 (2009). https://doi.org/10.1103/PhysRevLett.102.236804
11. A. Kara, H. Enriquez, A.P. Seitsonen, L.C. Lew Yan Voon, S. Vizzini, B. Aufray, H. Oughaddou, Surf. Sci. Rep. **67**(1) (2012). https://doi.org/10.1016/j.surfrep.2011.10.001
12. P. De Padova, C. Quaresima, C. Ottaviani, P.M. Sheverdyaeva, P. Moras, C. Carbone, D. Topwal, B. Olivieri, A. Kara, H. Oughaddou, B. Aufray, G. Le Lay, Appl. Phys. Lett. **96**(26), 261905 (2010). https://doi.org/10.1063/1.3459143
13. P. Vogt, P. De Padova, C. Quaresima, J. Avila, E. Frantzeskakis, M.C. Asensio, A. Resta, B. Ealet, G. Le Lay, Phys. Rev. Lett. **108**, 155501 (2012). https://doi.org/10.1103/PhysRevLett.108.155501
14. L. Chen, C.C. Liu, B. Feng, X. He, P. Cheng, Z. Ding, S. Meng, Y. Yao, K. Wu, Phys. Rev. Lett. **109**, 056804 (2012). https://doi.org/10.1103/PhysRevLett.109.056804
15. A. Fleurence, R. Friedlein, T. Ozaki, H. Kawai, Y. Wang, Y. Yamada-Takamura, Phys. Rev. Lett. **108**, 245501 (2012). https://doi.org/10.1103/PhysRevLett.108.245501
16. L. Meng, Y. Wang, L. Zhang, S. Du, R. Wu, L. Li, Y. Zhang, G. Li, H. Zhou, W.A. Hofer, H.J. Gao, Nano Lett. **13**(2), 685 (2013). https://doi.org/10.1021/nl304347w
17. L. Chen, B. Feng, K. Wu, Appl. Phys. Lett. **102**(8), 081602 (2013). https://doi.org/10.1063/1.4793998
18. C.L. Lin, R. Arafune, K. Kawahara, M. Kanno, N. Tsukahara, E. Minamitani, Y. Kim, M. Kawai, N. Takagi, Phys. Rev. Lett. **110**, 076801 (2013). https://doi.org/10.1103/PhysRevLett.110.076801
19. R. Friedlein, A. Fleurence, J.T. Sadowski, Y. Yamada-Takamura, Appl. Phys. Lett. **102**(22), 221603 (2013). https://doi.org/10.1063/1.4808214
20. L. Tao, E. Cinquanta, D. Chiappe, C. Grazianetti, M. Fanciulli, M. Dubey, A. Molle, D. Akinwande, Nat. Nanotechnol. **10**(3), 227 (2015). https://doi.org/10.1038/nnano.2014.325
21. G. Le Lay, Nat. Nanotechnol. **10**(3), 202 (2015), News and Views. https://doi.org/10.1038/nnano.2015.10
22. L. Li, S. Lu, J. Pan, Z. Qin, Y.q. Wang, Y. Wang, G. Cao, S. Du, H. Gao, Adv. Mat. **26**(28), 4820 (2014). https://doi.org/10.1002/adma.201400909
23. P. Bampoulis, L. Zhang, A. Safaei, R. van Gastel, B. Poelsema, H. Zandvliet, J. Phys. Cond. Matt. **26**(44), 442001 (2014). https://doi.org/10.1088/0953-8984/26/44/442001
24. M. Derivaz, D. Dentel, R. Stephan, M.C. Hanf, A. Mehdaoui, P. Sonnet, C. Pirri, Nano Lett. **15**(4), 2510 (2015). https://doi.org/10.1021/acs.nanolett.5b00085

25. A. Acun, L. Zhang, P. Bampoulis, M. Farmanbar, A. van Houselt, A.N. Rudenko, M. Lingenfelder, G. Brocks, B. Poelsema, M.I. Katsnelson, H.J.W. Zandvliet, J. Phys. Cond. Matt. **27**(44), 443002 (2015). https://doi.org/10.1088/0953-8984/27/44/443002
26. F. Zhu, W. Chen, Y. Xu, C. Gao, D. Guan, C. Liu, D. Qian, S. Zhang, J. Jia, Nat. Mater. **14**(10), 1020 (2015), Article. https://doi.org/10.1038/nmat4384
27. M.E. Dávila, G. Le Lay, Sci. Rep. **6**, 20714 (2016). https://doi.org/10.1038/srep20714
28. J.G. Bednorz, K.A. Müller, Z. Phys. B **64**, 189 (1986). https://doi.org/10.1007/BF01303701
29. P.W. Anderson, Science **235**, 1196 (1987). https://doi.org/10.1126/science.235.4793.1196
30. G. Baskaran, Z. Zou, P.W. Anderson, Solid State Commun. **63**(11), 973 (1987). https://doi.org/10.1016/0038-1098(87)90642-9
31. G. Baskaran, P.W. Anderson, Phys. Rev. B **37**, 580 (1988). https://doi.org/10.1103/PhysRevB.37.580
32. G. Baskaran, Iran. J. Phys. Res. **6**(3), 234 (2006)
33. X. Wen, *Quantum Field Theory of Many-Body Systems* (Oxford University Press, Oxford, 2007)
34. Z.Y. Meng, T.C. Lang, S. Wessel, F.F. Assaad, A. Muramatsu, Nature **464**(7290), 847 (2010). https://doi.org/10.1038/nature08942
35. S. Sorella, E. Tosatti, Europhys. Lett. **19**(8), 699 (1992). https://doi.org/10.1209/0295-5075/19/8/007
36. S. Sorella, Y. Otsuka, S. Yunoki, Sci. Rep. **2**, 992 (2012). https://doi.org/10.1038/srep00992
37. S.R. Hassan, D. Sénéchal, Phys. Rev. Lett. **110**, 096402 (2013). https://doi.org/10.1103/PhysRevLett.110.096402
38. M. Schüler, M. Rösner, T.O. Wehling, A.I. Lichtenstein, M.I. Katsnelson, Phys. Rev. Lett. **111**, 036601 (2013). https://doi.org/10.1103/PhysRevLett.111.036601
39. E.F. Sheka, May silicene exist? (2009), arXiv:0901.3663 [cond-mat.mtrl-sci]
40. E.F. Sheka, Int. J. Quantum Chem. **113**(4), 612 (2013). https://doi.org/10.1002/qua.24081
41. R. Hoffmann, Ang. Chemie (Int. Ed.) **52**(1), 93 (2013). https://doi.org/10.1002/anie.201206678
42. D. Jose, A. Datta, J. Phys. Chem. C **116**(46), 24639 (2012). https://doi.org/10.1021/jp3084716
43. R. Vidya, G. Baskaran (2017). arXiv:1709.04664
44. G. Baskaran, Phys. Rev. B **65**, 212505 (2002). https://doi.org/10.1103/PhysRevB.65.212505
45. G. Baskaran, S.A. Jafari, Phys. Rev. Lett. **89**, 016402 (2002). https://doi.org/10.1103/PhysRevLett.89.016402
46. S. Pathak, V.B. Shenoy, G. Baskaran, Phys. Rev. B **81**, 085431 (2010). https://doi.org/10.1103/PhysRevB.81.085431
47. G. Baskaran, Pramana **73**(1), 61 (2009). https://doi.org/10.1007/s12043-009-0094-8
48. D.A. Clabo Jr., H.F. Schaefer III, J. Chem. Phys. **84**(3), 1664 (1986). https://doi.org/10.1063/1.450462
49. Y. Wang, H. Cheng, Phys. Rev. B **87**, 245430 (2013). https://doi.org/10.1103/PhysRevB.87.245430
50. Z. Guo, S. Furuya, J. Iwata, A. Oshiyama, J. Phys. Soc. Jpn. **82**(6), 063714 (2013). https://doi.org/10.7566/JPSJ.82.063714
51. S. Cahangirov, M. Audiffred, P. Tang, A. Iacomino, W. Duan, G. Merino, A. Rubio, Phys. Rev. B **88**, 035432 (2013). https://doi.org/10.1103/PhysRevB.88.035432
52. Y. Feng, D. Liu, B. Feng, X. Liu, L. Zhao, Z. Xie, Y. Liu, A. Liang, C. Hu, Y. Hu, S. He, G. Liu, J. Zhang, C. Chen, Z. Xu, L. Chen, K. Wu, Y.T. Liu, H. Lin, Z.Q. Huang, C.H. Hsu, F.C. Chuang, A. Bansil, X.J. Zhou, Proc. Natnl. Acad. Sci. (USA) **113**(51), 14656 (2016). https://doi.org/10.1073/pnas.1613434114, arXiv:1503.06278 [cond-mat.mtrl-sci]
53. C.L. Lin, R. Arafune, K. Kawahara, N. Tsukahara, E. Minamitani, Y. Kim, N. Takagi, M. Kawai, Appl. Phys. Express **5**(4), 045802 (2012). https://doi.org/10.1143/APEX.5.045802
54. Z. Majzik, M. Rachid Tchalala, M. Švec, P. Hapala, H. Enriquez, A. Kara, A.J. Mayne, G. Dujardin, P. Jelínek, H. Oughaddou, J. Phys. Cond. Matt. **25**(22), 225301 (2013). https://doi.org/10.1088/0953-8984/25/22/225301
55. L. Chen, H. Li, B. Feng, Z. Ding, J. Qiu, P. Cheng, K. Wu, S. Meng, Phys. Rev. Lett. **110**, 085504 (2013). https://doi.org/10.1103/PhysRevLett.110.085504

56. T. Tohyama, S. Maekawa, Supercond. Sci. Tech. **13**(4), R17 (2000). https://doi.org/10.1088/0953-2048/13/4/201
57. B.O. Wells, Z.X. Shen, A. Matsuura, D.M. King, M.A. Kastner, M. Greven, R.J. Birgeneau, Phys. Rev. Lett. **74**, 964 (1995). https://doi.org/10.1103/PhysRevLett.74.964
58. T.C. Choy, B.A. McKinnon, Phys. Rev. B **52**, 14539 (1995). https://doi.org/10.1103/PhysRevB.52.14539
59. A.M. Black-Schaffer, S. Doniach, Phys. Rev. B **75**, 134512 (2007). https://doi.org/10.1103/PhysRevB.75.134512
60. C. Honerkamp, Phys. Rev. Lett. **100**, 146404 (2008). https://doi.org/10.1103/PhysRevLett.100.146404
61. W.S. Wang, Y.Y. Xiang, Q.H. Wang, F. Wang, F. Yang, D.H. Lee, Phys. Rev. B **85**, 035414 (2012). https://doi.org/10.1103/PhysRevB.85.035414
62. M.L. Kiesel, C. Platt, W. Hanke, D.A. Abanin, R. Thomale, Phys. Rev. B **86**, 020507 (2012). https://doi.org/10.1103/PhysRevB.86.020507
63. R. Nandkishore, L.S. Levitov, A.V. Chubukov, Nat. Phys. **8**(2), 158 (2012). https://doi.org/10.1038/nphys2208
64. M.M. Scherer, S. Uebelacker, C. Honerkamp, Phys. Rev. B **85**, 235408 (2012). https://doi.org/10.1103/PhysRevB.85.235408
65. J. Vučičević, M.O. Goerbig, M.V. Milovanović, Phys. Rev. B **86**, 214505 (2012). https://doi.org/10.1103/PhysRevB.86.214505
66. Z.C. Gu, H.C. Jiang, D.N. Sheng, H. Yao, L. Balents, X.G. Wen, Phys. Rev. B **88**, 155112 (2013). https://doi.org/10.1103/PhysRevB.88.155112
67. T. Li, Europhys. Lett. **97**(3), 37001 (2012). https://doi.org/10.1209/0295-5075/97/37001
68. G. Baskaran (2017). (Unpublished)
69. F. Liu, C.C. Liu, K. Wu, F. Yang, Y. Yao, Phys. Rev. Lett. **111**, 066804 (2013). https://doi.org/10.1103/PhysRevLett.111.066804
70. W. Wan, Y. Ge, F. Yang, Y. Yao, Europhys. Lett. **104**(3), 36001 (2013). https://doi.org/10.1209/0295-5075/104/36001
71. C.C. Liu, H. Jiang, Y. Yao, Phys. Rev. B **84**, 195430 (2011). https://doi.org/10.1103/PhysRevB.84.195430
72. M. Ezawa, New J. Phys. **14**(3), 033003 (2012). https://doi.org/10.1088/1367-2630/14/3/033003
73. M. Ezawa, Y. Tanaka, N. Nagaosa, Sci. Rep. **3**, 2790 (2013). https://doi.org/10.1038/srep02790

Chapter 6
Molecular Ordering in Covalent Solids: A Simple Lattice Model

F. Siringo

Abstract Some aspects of molecular orientation in covalently-bonded molecular solids are discussed by reviewing a simple model for the molecular ordering of frustrated lattices. The model describes a peculiar phase transition from an isotropic high temperature phase to a low-dimensional anisotropic low-temperature state. The model was studied in the past by several methods ranging from mean-field up to more sophisticated variational Migdal–Kadanoff real space renormalization group and numerical Monte Carlo simulations.

6.1 Introduction

In 1988, Professor Norman March drew my attention to the fascinating problem of metal-insulator transition in molecular solids under pressure (see Ref. [1] for a review). As we discussed in a series of papers [2–4] the electron structure of many covalently bonded molecular solids, like halogens, changes under pressure because of the ordering that takes place in the planes and might give rise to low dimensional chains of molecules. For instance, zig-zag chains are observed in solid iodine [5, 6], hydrogen halides [7, 8], oxygen [9], nitrogen [10] and other molecular solids. The orientational order of the molecules may change according to the thermodynamic conditions giving rise to quite rich phase diagrams: solid-solid transitions may occur where the orientational ordering of molecules plays a special role.

The most studied models of molecular ordering are $O(3)$ symmetric vectorial models, describing weak molecular interactions which arise from dipole fluctuations and give rise to the observed three-dimensional ordering of most molecular Van der Waals solids. In many covalently bonded solids, the simple diatomic molecule has not many directions allowed for its covalent bond and a large degree of frustration is expected, since the coordination number of the covalent bond is quite low. The lower is the allowed coordination number, the higher the frustration which gives rise

F. Siringo (✉)
Dipartimento di Fisica e Astronomia, Università di Catania,
and INFN Sezione di Catania, Via S.Sofia 64, 95123 Catania, Italy
e-mail: fabio.siringo@ct.infn.it

G. G. N. Angilella and C. Amovilli (eds.), *Many-body Approaches at Different Scales*, https://doi.org/10.1007/978-3-319-72374-7_6

to the low-dimensional structures observed in iodine [1] and hydrogen halides [7, 8]. The usual rotational invariant $O(3)$ models cannot describe the existence of such low-dimensional structures and a specific model must be introduced.

A well studied model for polymers and low dimensional ordering is the self-avoiding random walk on a lattice. A more specific and less known model is the so called *molecular* model that was first proposed in Ref. [11] as a very simple d-dimensional lattice model which incorporates some degree of frustration and thus describes some aspects of molecular orientation in covalently bonded molecular solids.

At variance with the more studied self-avoiding walk, the molecular model can only give rise to linear self-avoiding chains of molecules. It consists of a d-dimensional hypercubic lattice with a randomly oriented linear molecule at each site [11–13]. In its simplest version each molecule is only allowed to be oriented towards one of its nearest neighbours. There is an energy gain for any pair of neighbours which are oriented along their common bond (a covalent bond). The existence of preferred orientational axes breaks the rotational invariance of the single molecule as it is likely to occur for any real molecular system under pressure.

Similar lattice models were used for describing the diffusion of particles and molecules inside a polymer, and the growth of one-dimensional islands (polymeric chains; see e.g. Ref. [14], and references therein). The molecular model has also stimulated some work on molecular orientation in nitrogen [15, 16] which goes back to the phenomenology laid down by Pople and Karasz [17]. It was argued [18] that the weak intermolecular bonds between two N_2 molecules could be favourable for the formation of an orientationally disordered *plastic crystal* solid phase, leading to freezing into an orientationally ordered phase. Some experimental data on nitrogen [10] confirm the existence of an orientational disordering temperature in the solid below the melting temperature.

However, as far as we know, the molecular systems which are more closely described by the molecular model are the hydrogen halides HX (X = F, Cl, Br, I). Their low-temperature structures are known to consist of planar chains of molecules in the condensed state while a totally disordered structure is observed with increasing temperature at ambient pressure [7]. Moreover the opposite transition, from orientational disorder to an ordered chain structure, was reported by increasing pressure [8].

Besides its physical motivation, the molecular model is by itself interesting because it provides a simple example of dimensional transmutation. The model undergoes a transition from an high-temperature (or weakly interacting) fully isotropic disordered system, to a low-temperature (or strongly interacting) anisotropic low dimensional broken-symmetry phase. As a consequence of frustration the breaking of symmetry is accompanied by a sort of decomposition of the system in low-dimensional almost independent parts, as observed in solid iodine and hydrogen halides. Such remarkable behaviour requires a space dimension $d > 2$, while for $d = 2$ the model is shown to be equivalent to the exactly solvable two-dimensional Ising model (for a review on the Ising model and related standard techniques, see e.g. Ref. [19]).

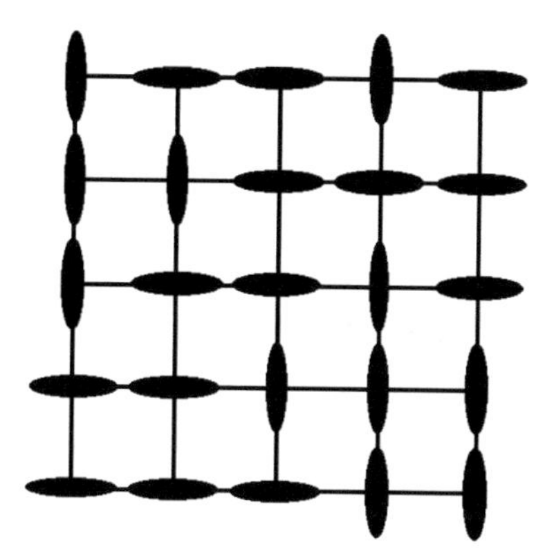

Fig. 6.1 A random configuration of the molecular model for $d = 2$

As shown by Monte Carlo simulations [13], in the broken-symmetry phase the system displays the presence of correlated chains of molecules (polymers) which point towards a common direction inside each two-dimensional sub-set of the lattice (plane). Such planes are weakly correlated in the low-temperature phase, and the system has a two-dimensional behaviour even for $d > 3$. For $d = 3$ the molecular model belongs to a new universality class, since its critical exponent ν turns out to be $\nu = 0.44 \pm 0.02$ by finite size scaling [13]. The universality class of the model describes a broad group of isotropic physical systems characterized by a low-dimensional ordering in their low-temperature phase.

In this contribution, some important features of the molecular model are reviewed and its relevance as a non-trivial extension of the Ising model is discussed.

6.2 The Molecular Model

Exactly solvable models are important for our understanding of more complex systems, and provide a test for approximate techniques. The d-dimensional molecular model shares with the Ising model the $d = 2$ realization, since their equivalence for $d = 2$ can be proven to be exact [11]. While the focus here is on the three-dimensional model, we will take advantage of the existence of an exactly solvable realization for $d = 2$. For $d > 2$, as the frustration increases, the model shows a very different behaviour compared to the Ising or Potts [20] models. These last show a fully d-dimensional broken-symmetry phase while the molecular model is characterized by a low-dimensional ordering inside the planes with negligible correlation among different planes. In fact, for $d = 3$, the model belongs to a different universality class.

Let us consider a d-dimensional hypercubic lattice, with a randomly oriented linear molecule at each site. The molecules are supposed to be symmetric with respect to their centre of mass which is fixed at the lattice site. Only a discrete number of space orientations are allowed for each molecule: we assume that each of them must point towards one of its $2d$ first neighbour sites. This choice can be justified by

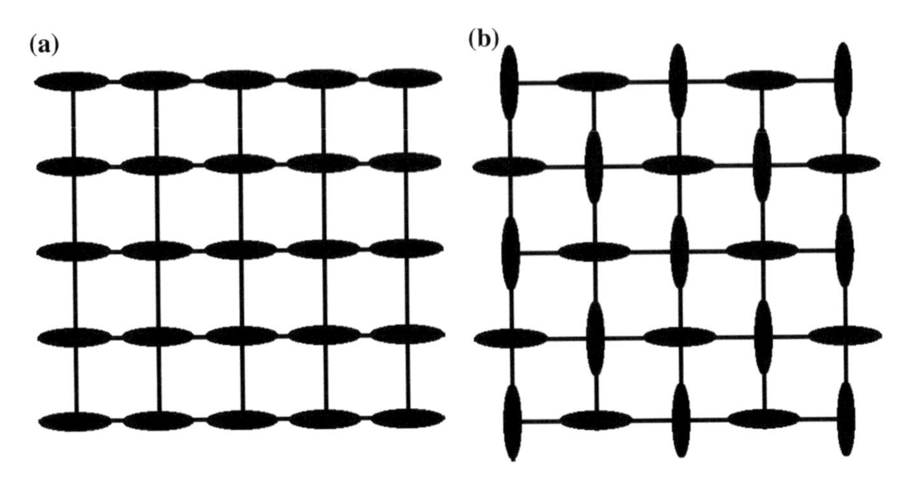

Fig. 6.2 Ground state configurations for the two-dimensional **a** *attractive* and **b** *repulsive* models

the existence of covalent interactions along preferred axes. Then each molecule has d different states corresponding to molecular orientation along the hypercube axes (molecules are symmetric). Finally, each couple of first neighbour molecules, when pointing the one towards the other, are assumed to gain a bonding energy for their directional covalent bond (they touch each other). As shown in Fig. 6.1 for $d = 2$, bonding in a direction excludes any possible bond along the other $(d - 1)$ directions. The coordination number is 2 for any value of d, and the frustration increases with increasing d.

We introduce a versor variable $\hat{w}_{\mathbf{r}}$ for each of the N sites $\mathbf{r}$ of the lattice, with $\hat{w}_{\mathbf{r}} \in \{\hat{x}_1, \hat{x}_2, \ldots \hat{x}_d\}$ pointing towards one of the d hypercube axes x_α. The versors $\hat{x}_\alpha$ are assumed to be orthonormal: $\hat{x}_\alpha \cdot \hat{x}_\gamma = \delta_{\alpha\gamma}$. The partition function follows

$$Z = \sum_{\{\hat{w}\}} e^S = \sum_{\{\hat{w}\}} \exp\left[4\beta \sum_{\mathbf{r},\alpha} (\hat{w}_{\mathbf{r}} \cdot \hat{x}_\alpha)(\hat{w}_{\mathbf{r}+\hat{x}_\alpha} \cdot \hat{x}_\alpha)\right], \tag{6.1}$$

where $\{\hat{w}\}$ indicates a sum over all configurations, α runs from 1 to d, and the lattice spacing is set to unity. The inverse temperature β (in units of binding energy) can be negative for a *repulsive* model, but is assumed positive in the molecular context.

The model may be generalized by introducing an external d-dimensional vectorial field $\mathbf{h}(\alpha)$ at each link. The dependence on α means that the field differs according to the space direction α of the lattice link which joins the sites. The modified partition function reads

$$Z_h = \sum_{\{\hat{w}\}} e^{S_h} = \sum_{\{\hat{w}\}} \exp\left\{4\beta \sum_{\mathbf{r},\alpha} \left[\left(\hat{w}_{\mathbf{r}} \cdot \hat{x}_\alpha\right)\left(\hat{w}_{\mathbf{r}+\hat{x}_\alpha} \cdot \hat{x}_\alpha\right) + \mathbf{h}(\alpha) \cdot \hat{w}_{\mathbf{r}} + \mathbf{h}(\alpha) \cdot \hat{w}_{\mathbf{r}+\hat{x}_\alpha}\right]\right\}. \tag{6.2}$$

It is evident that if the field satisfies the condition

$$\sum_{\alpha} \mathbf{h}(\alpha) = 0, \tag{6.3}$$

then S_h does not depend on $\mathbf{h}$ and $S_h \equiv S$. In such a case the extra degree of freedom provided by $\mathbf{h}$ can be regarded as a sort of internal symmetry of the model. This global symmetry can be made local by allowing the field $\mathbf{h}$ to depend on the site position $\mathbf{r}$. We will only take advantage of the global symmetry in this paper. We notice that such symmetry cannot be seen as a gauge invariance, since in lattice gauge models any gauge change leaves the energy gain unchanged at any link. Here, the field $\mathbf{h}$ changes the energy gain of all the links while the whole action is invariant.

Adopting a more compact notation, the partition function reads

$$Z_h = \sum_{\{\hat{w}\}} e^{S_h} = \sum_{\{\hat{w}\}} e^{\sum_{\mathbf{r},\alpha} \mathscr{L}(\mathbf{r},\alpha)}, \tag{6.4}$$

where the Lagrangian density $\mathscr{L}$ follows as

$$\mathscr{L}(\mathbf{r}, \alpha) = \hat{w}_{\mathbf{r}}^{\dagger} M_{\alpha}(\beta, \mathbf{h}) \hat{w}_{\mathbf{r}+\hat{x}_{\alpha}}. \tag{6.5}$$

Here, the canonical d-dimensional column vector representation of $\mathbb{R}^d$ is employed, with $\hat{x}_1 \equiv (1, 0, 0 \ldots)$, $\hat{x}_2 \equiv (0, 1, 0 \ldots)$, etc. The $d \times d$ matrix M_{α} does not depend on the configurations of the system, and entirely characterizes the model.

The global symmetry of the action provides a simple way to show the equivalence between molecular and Ising models for $d = 2$. For the two-dimensional lattice the condition (6.3) is satisfied by the field $\mathbf{h}(1) = h(\hat{x}_1 - \hat{x}_2)$, $\mathbf{h}(2) = -\mathbf{h}(1)$. The matrices M_{α} follow

$$M_1 = \begin{bmatrix} 4\beta(1+2h) & 0 \\ 0 & -8\beta h \end{bmatrix}, \quad M_2 = \begin{bmatrix} -8\beta h & 0 \\ 0 & 4\beta(1+2h) \end{bmatrix}. \tag{6.6}$$

Then for $h = -1/4$, $M_1 \equiv M_2$, and $\mathscr{L}$ reads

$$L(\mathbf{r}, \alpha) = \beta + \hat{w}_{\mathbf{r}}^{\dagger} \begin{bmatrix} \beta & -\beta \\ -\beta & \beta \end{bmatrix} \hat{w}_{\mathbf{r}+\hat{x}_{\alpha}}. \tag{6.7}$$

Identifying the two-dimensional column versors $\hat{w}$ with spin variables, apart from an inessential factor, Z reduces to the partition function of a two-dimensional Ising model

$$Z = e^{2\beta N} \cdot Z_{\text{Ising}} \tag{6.8}$$

and is exactly solvable. For $\beta \to +\infty$ a ground state is approached with all the molecules oriented along the same direction, and with formation of one-dimensional polymeric chains (Fig. 6.2a); for $\beta \to -\infty$ the repulsive model approaches a zero-energy (no bonds) ground state analogous to the antiferromagnetic configuration of the Ising model (Fig. 6.2b).

For $d \geq 3$ the analogy with the Ising model breaks down, and this is evident from a simple analysis of the ground state configuration. Due to frustration the model has an infinitely degenerate ground state in the thermodynamic limit $N \to \infty$. For instance, in the case $d = 3$, the minimum energy is obtained by orienting all the molecules along a common direction, as for $d = 2$. However the ground state configuration is not unique: the number of molecular bonds does not change if we rotate together all the molecules belonging to an entire layer which is parallel to the original direction of orientation. As a consequence of frustration the total degeneration is $3(2^{(N^{1/3})})$, and the system could even behave like a glass for the large energy barriers which separate each minimum from the other. The phase diagram is expected to be quite rich, with at least a transition point between the high temperature disordered phase and an ordered broken-symmetry low temperature phase.

6.3 Mean Field

For the generic d-dimensional model, some analytical results can be obtained in Mean-Field (MF) approximation: neglecting second order fluctuation terms

$$(\hat{w}_{\mathbf{r}} \cdot \hat{x}_\alpha)(\hat{w}_{\mathbf{r}+\hat{x}_\alpha} \cdot \hat{x}_\alpha) \approx \Delta_\alpha(\hat{w}_{\mathbf{r}} \cdot \hat{x}_\alpha) + \Delta_\alpha(\hat{w}_{\mathbf{r}+\hat{x}_\alpha} \cdot \hat{x}_\alpha) - \Delta_\alpha^2, \tag{6.9}$$

where $\Delta_\alpha = \langle \hat{w}_{\mathbf{r}} \cdot \hat{x}_\alpha \rangle$ is an average over the configurations, and $\sum_\alpha \Delta_\alpha = 1$ (with the obvious bounds $0 \leq \Delta_\alpha \leq 1$). Here, the order parameter Δ_α gives the probability of finding a molecule oriented along the direction of $\hat{x}_\alpha$. The partition function factorizes as

$$Z_{\text{MF}} = \left(\sum_\alpha e^{8\beta\Delta_\alpha}\right)^N \exp\left(-4N\beta \sum_\alpha \Delta_\alpha^2\right) \tag{6.10}$$

and the free energy follows

$$F_{\text{MF}} = -\frac{1}{N\beta} \log Z_{\text{MF}} = 4\sum_\alpha \Delta_\alpha^2 - \frac{1}{\beta} \log\left(\sum_\alpha e^{8\beta\Delta_\alpha}\right). \tag{6.11}$$

The derivative with respect to Δ_μ yields, for the stationary points,

$$\Delta_\mu = \frac{e^{8\beta\Delta_\mu}}{\sum_\alpha e^{8\beta\Delta_\alpha}} \tag{6.12}$$

which satisfies the condition $\sum_\alpha \Delta_\alpha = 1$.

In the high temperature limit $\beta \to 0$, Eq. (6.12) has the unique solution $\Delta_\mu = 1/d$, which reflects the complete random orientation of molecules. In the opposite limit $\beta \to \infty$, apart from such solution, Eq. (6.12) is satisfied by the broken-symmetry field $\Delta_\mu = 1$, $\Delta_\alpha = 0$ for $\alpha \neq \mu$, which obviously corresponds to a minimum for $F_{\rm MF}$. Then at a critical point $\beta = \beta_c$ the high temperature solution must become unstable towards a multivalued minimum configuration. The Hessian matrix is easily evaluated at the stationary points by using Eqs. (6.12) and (6.11):

$$H_{\mu\nu} = \frac{1}{8}\frac{\partial^2 F_{\rm MF}}{\partial\Delta_\mu \partial\Delta_\nu} = \delta_{\mu\nu}\left(1 - 8\beta\Delta_\mu\right) + 8\beta\Delta_\mu\Delta_\nu. \tag{6.13}$$

In the high temperature phase ($\beta < \beta_c$), inserting $\Delta_\mu = 1/d$, the eigenvalue problem

$$\det|H_{\mu\nu} - \lambda\delta_{\mu\nu}| = 0 \tag{6.14}$$

yields

$$\left(1 - \frac{8\beta}{d} - \lambda\right)^{d-1} \cdot (1 - \lambda) = 0. \tag{6.15}$$

Thus the Hessian matrix is positive defined if and only if $\lambda = (1 - 8\beta/d) > 0$. Beyond the critical point $\beta = \beta_c = (d/8)$ the solution $\Delta_\mu = 1/d$ is not a minimum, and a multivalued minimum configuration shows up. Such result obviously agrees with the MF prediction for the Ising model, $\beta_{\rm Ising} = 1/(2d)$, only for the special dimension $d = 2$. For $d > 2$ we observe an increase of β_c with d, to be compared to the opposite trend shown by the Ising model. Such behaviour may be interpreted in terms of the low dimensionality of the ordered phase. Due to frustration the ordering may only occur on a low dimensional scale: for instance in three dimensions each layer has an independent internal ordering. Thus we expect a larger β_c for $d > 2$ since the increasing of d only introduces larger fluctuations, with each molecule having $(d - 2)$ allowed out-of-plane orientations. For $d = 3$ the low temperature phase can be regarded as a quenched disordered superposition of layers which are internally ordered along different in-plane directions. As a consequence of frustration the system shows a two-dimensional character below the critical point while behaving as truly three-dimensional in the high temperature domain. In MF the neglecting of some fluctuations usually leads to a critical temperature which overestimate the exact value (i.e. the critical inverse temperature β_c is underestimated). For $d = 2$, the MF prediction is $\beta_c = 0.25$ to be compared with the exact value $\beta_c = 0.4407$. For $d = 3$ the MF prediction $\beta_c = d/8 = 0.375$ should provide a lower bound to the unknown exact value.

6.4 Migdal–Kadanoff Decimation

A very powerful tool for the study of lattice models is the the Migdal–Kadanoff [19, 21–23] method of link displacement and decimation. A link displacement may be introduced by considering that the configurational average of the Lagrangian density

$\mathscr{L}$ in Eq. (6.5) must be translationally invariant

$$\langle \mathscr{L}(\mathbf{r}, \alpha) \rangle = \langle \mathscr{L}(\mathbf{r}', \alpha) \rangle. \tag{6.16}$$

Then, defining

$$\Gamma_\alpha(\mathbf{r}, \mathbf{r}') = \mathscr{L}(\mathbf{r}, \alpha) - \mathscr{L}(\mathbf{r}', \alpha), \tag{6.17}$$

we can state that $\langle \Gamma_\alpha(\mathbf{r}, \mathbf{r}') \rangle = 0$ and the same holds for any sum Γ over an arbitrary set of such terms

$$\Gamma = \sum \Gamma_\alpha(\mathbf{r}, \mathbf{r}'). \tag{6.18}$$

Replacing the action S_h by the sum $S_h + \Gamma$, and assuming that the condition Eq. (6.3) is verified (so that we can drop the h in S_h and Z_h which are invariant), the modified partition function Z_Γ can be approximated by cumulant expansion as

$$Z_\Gamma = \sum_{\{\hat{w}\}} e^{S+\Gamma} = Z \cdot \langle e^\Gamma \rangle \approx Z \left[e^{\langle \Gamma \rangle} \cdot e^{\frac{1}{2}\left(\langle \Gamma^2 \rangle - \langle \Gamma \rangle^2\right)} \right] \tag{6.19}$$

then, since $\langle \Gamma \rangle = 0$,

$$Z_\Gamma \approx Z \cdot e^{\frac{1}{2}\langle \Gamma^2 \rangle}. \tag{6.20}$$

For instance, the sum in Eq. (6.18) could run over all $\alpha \neq 1$, and for appropriate values of the vectors $\mathbf{r}, \mathbf{r}'$, in order to yield a displacement of links which are orthogonal to $\hat{x}_1$. To second order in Γ, the error introduced by link displacement is controlled by the exponential factor in Eq. (6.20).

Link displacement breaks the internal symmetry of the model, so that Z_Γ is no longer invariant for any field change subject to the condition Eq. (6.3). Then we may improve the approximation by using the extra freedom on the choice of $\mathbf{h}$ for minimizing the difference between the approximate partition function Z_Γ and the exact Z.

If $\mathbf{h}$ satisfies the condition Eq. (6.3), then $a\mathbf{h}$ satisfies such condition as well for any choice of the scalar parameter a. Then a special class of invariance transformations can be described by a change of the strength parameter h, assuming the field $\mathbf{h}$ as proportional to h. The following discussion could be easily generalized to other classes of transformations described by more than one parameter. Since Γ is linear in the field $\mathbf{h}$, then in general

$$\Gamma^2 = [A + hB]^2, \tag{6.21}$$

where A and B depend on the configuration of the system. For the average we have

$$\langle \Gamma^2 \rangle = \langle A^2 \rangle + 2h \langle AB \rangle + h^2 \langle B^2 \rangle. \tag{6.22}$$

This last equation, inserted in Eq. (6.20) leads to the following considerations: (i) the coefficient $\langle B^2\rangle$ is positive defined, then the average $\langle \Gamma^2\rangle$ always has a minimum for an appropriate value of $h = h_0$; (ii) in general $\langle AB\rangle \neq 0$ then $h_0 \neq 0$, and a direct use of the Migdal–Kadanoff method on the original model (with no field considered) would yield a larger error; (iii) to the order of approximation under consideration, Z_Γ is stationary at $h = h_0$, and is symmetric around that point, then all the physical properties described by such partition function must result symmetric with respect to h_0. Moreover, at the same order of approximation, any physical observable f will acquire an unphysical dependence on h, and the symmetry around h_0 requires that $\frac{df}{dh} = 0$ for $h = h_0$. Then we expect that all such observables should be stationary at $h = h_0$, and their best estimate should coincide with the extreme value.

As a consequence of the above statements, the Migdal–Kadanoff method can be improved by taking advantage of the global symmetry of the model. By use of the approximate partition function Z_Γ the critical temperature acquires a non-physical field dependence, but the best estimate of β_c is its stationary value corresponding to $h = h_0$. The method can be seen as a variational method with the best approximation achieved by the minimum in the inverse temperature.

Such stationary condition resembles the principle of *minimal sensitivity* introduced by Stevenson [24] for determining the best renormalization parameters whenever the physical amplitudes depend on them (and they should not). In our context, since the critical temperature should not depend on the choice of the field strength h, the best value for such field is the one which makes the critical temperature less sensitive i.e. the stationary point. However, according to Eqs. (6.20) and (6.22), here we have got a formal proof of the stationary condition up to second order of the cumulant expansion.

The method may be used by performing a displacement of links that are orthogonal to $\hat{x}_1$, and then a one-dimensional decimation along the $\alpha = 1$ axis. According to such program let us define the alternative $d \times d$ matrix $t_\alpha(\beta, h)$

$$e^{\mathscr{L}(\mathbf{r},\alpha)} = \hat{w}_{\mathbf{r}}^\dagger t_\alpha(\beta, h)\hat{w}_{\mathbf{r}+\hat{x}_\alpha}. \tag{6.23}$$

The partition function follows

$$Z_h = \sum_{\{\hat{w}\}} \prod_{\mathbf{r},\alpha} \left[\hat{w}_{\mathbf{r}}^\dagger t_\alpha(\beta, h)\hat{w}_{\mathbf{r}+\hat{x}_\alpha}\right]. \tag{6.24}$$

After link displacement and decimation along the $\alpha = 1$ axis, the modified partition function reads

$$Z_\Gamma = \sum_{\{\hat{w}\}} \prod_{\mathbf{r},\alpha} \left[\hat{w}_{\mathbf{r}}^\dagger \tilde{t}_\alpha(\beta, h)\hat{w}_{\mathbf{r}+\hat{x}_\alpha}\right], \tag{6.25}$$

where the sum and the product run over the configurations and the sites of the new decimated lattice, and

$$\tilde{t}_1(\beta, h) = [t_1(\beta, h)]^\lambda \tag{6.26}$$

$$\tilde{t}_\alpha(\beta, h) = t_\alpha(\lambda\beta, h), \quad \text{for } \alpha \neq 1, \tag{6.27}$$

with λ being the scale factor between the new and the old lattice. A renormalized inverse temperature $\tilde{\beta}_\alpha$ may be defined according to

$$\tilde{t}_1(\beta, h) = t_1(\tilde{\beta}_1, h) \tag{6.28}$$

$$\tilde{\beta}_\alpha = \lambda\beta, \quad \text{for } \alpha \neq 1. \tag{6.29}$$

Eventually, the same scaling operation should be performed consecutively for all the directions in order to obtain an hyper-cubic lattice again. For any finite scaling parameter $\lambda > 1$ the renormalized inverse temperature is anisotropic, but an isotropic fixed-point can be recovered in the limit $\lambda \to 1$. Equations (6.28) and (6.29) define the flow of the renormalized inverse temperature, which changes for any different value of the field strength h. Equation (6.28) has a more explicit aspect in the representation of the common eigenvectors of the matrices t_1 and $\tilde{t}_1 = [t_1]^\lambda$. The rank of such matrices is 2 for any space dimension d, as can be expected from the definition of the model. Then both the matrices can be represented in terms of the two non-vanishing eigenvalues η_1, η_2, which are functions of β and h. Assuming that $\eta_2 \neq 0$, and defining

$$f(\beta, h) = \frac{\eta_1}{\eta_2} \tag{6.30}$$

apart from a regular multiplicative factor for the partition function, the scaling equation (6.28) reads

$$[f(\beta, h)]^\lambda = f(\tilde{\beta}_1, h). \tag{6.31}$$

For any h, the fixed points follow through the standard Migdal–Kadanoff equations

$$\left[f(\lambda^{\alpha-1}\beta_\alpha, h)\right]^\lambda = f(\lambda^{\alpha-d}\beta_\alpha, h). \tag{6.32}$$

When λ is analytically continued up to 1 such equations give the same isotropic fixed point β_c. In fact, the expansion of Eq. (6.32) around $\lambda = 1$ implies (up to first order in $\lambda - 1$)

$$\ln f(\beta_c, h) = -(d-1)\beta_c \left[\frac{1}{f}\frac{df}{d\beta}\right]_{\beta_c}, \tag{6.33}$$

which is an implicit equation for β_c. Such equations yield their best estimate of β_c when the strength of the field h is set to the stationary value h_0.

It is instructive to evaluate the stationary point h_0 for the case $d = 2$ which is equivalent to the two-dimensional Ising model for the choice $h = h_{\mathrm{I}} = -1/4$, as shown in Sect. 6.2. The h invariance of the exact partition function guarantees the equivalence of the two models for any choice of $h \neq h_{\mathrm{I}}$. However, the mere application of the Migdal–Kadanoff equation (6.32) to the simple $h = 0$ molecular model fails to predict even the existence of the fixed point. On the other hand, for $h = h_{\mathrm{I}}$, the very same recurrence Eq. (6.32) are known to predict the exact fixed point in the limit $\lambda \to 1$. That can also be checked by inserting in Eq. (6.33) the exact expression for the fixed point of the two-dimensional Ising model. Such contradictory results are not surprising since, as already discussed, link displacement breaks the h invariance of the model, and the approximate solution is thus dependent on the choice of h. We would like to test the variational method on this exactly solvable model: we look for the stationary point of the function $f(\beta, h)$. The matrices t_α follow from Eq. (6.6)

$$t_1 = \begin{bmatrix} x^2 b & 1 \\ 1 & x^{-2} \end{bmatrix}, \qquad t_2 = \begin{bmatrix} x^{-2} & 1 \\ 1 & xb \end{bmatrix}, \tag{6.34}$$

where $b = \exp(4\beta)$, and $x = \exp(4\beta h)$. Then for the eigenvalues we obtain

$$f(\beta, h) = \frac{\eta_1}{\eta_2} = \frac{(bx^4 + 1) - \sqrt{(bx^4 - 1)^2 + 4x^4}}{(bx^4 + 1) + \sqrt{(bx^4 - 1)^2 + 4x^4}}. \tag{6.35}$$

It can be easily shown that if the derivative of f is zero at a given h independent of β, then the solution β_c of Eq. (6.33) is stationary at that value of h. Differentiating with respect to x, we find that the derivative of f vanishes for $x^4 = 1/b$, which yields $h = -1/4 = h_I$ for any β. As expected, this is the required value in order to recover the Ising model. Thus the Migdal–Kadanoff approximation gives an improving estimate of the critical point as we move from the *molecular* towards the *Ising* representation (where the approximation yields the exact fixed point). We stress that all such representations are equivalent due to the h invariance of the action.

For $d > 2$ no equivalence to standard studied models has been found, and the behaviour seems to be dictated by the strong frustration which does not allow a coordination number higher than two, even for higher dimensions. We will focus on the three-dimensional model in order to compare with the phenomenology.

First of all the fields $\mathbf{h}(\alpha)$ must be defined. An isotropic choice gives a very poor prediction for the transition point, even worse than MF approximation [13]. We cannot use a fully isotropic version of the variational Migdal–Kadanoff method for a system which is not isotropic in its ordered phase. At the transition point the system choices a direction, so that in the ordered phase the correlation length cannot be isotropic: order occurs inside all layers which are orthogonal to the chosen direction, while there is a negligible correlation along such direction. It is more reasonable to describe the ordering inside a single layer, neglecting any correlation among different layers. Inside each layer the correlation length is isotropic, and the two-dimensional variational Migdal–Kadanoff method should give a better description of the transition. The same argument should hold for the generic d-dimensional molecular model. Moreover, the Migdal–Kadanoff method is known to work better for lower dimensions.

Let us take, for the three-dimensional model, the same field we used for $d = 2$, namely $\mathbf{h}(1) = h(\hat{x}_1 - \hat{x}_2)$, $\mathbf{h}(2) = -\mathbf{h}(1)$ and $\mathbf{h}(3) = 0$. The matrix t_1 follows

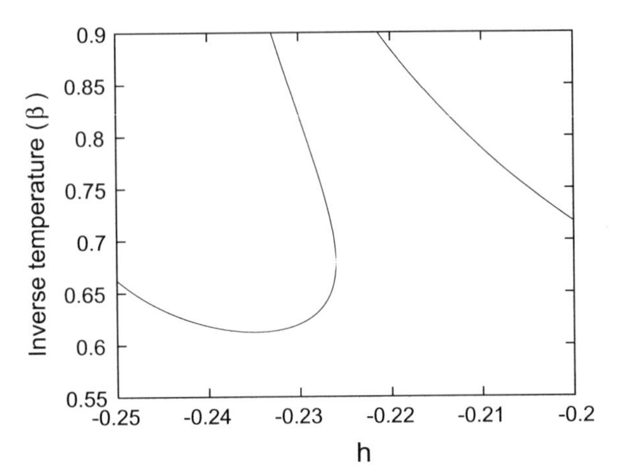

Fig. 6.3 Numerical solutions of the two-dimensional Migdal–Kadanoff equations for a single layer of the three-dimensional molecular model. The critical temperature β is reported as a function of the field strength h. The stationary point is at $h = h_0 = -0.2349$ where $\beta = \beta_c = 0.6122$. For $h > -0.226$ there is no physical solution

$$t_1 = \begin{bmatrix} e^{4\beta+8\beta h} & 1 & e^{4\beta h} \\ 1 & e^{-8\beta h} & e^{-4\beta h} \\ e^{4\beta h} & e^{-4\beta h} & 1 \end{bmatrix} = \begin{bmatrix} bx^2 & 1 & x \\ 1 & 1/x^2 & 1/x \\ x & 1/x & 1 \end{bmatrix}. \tag{6.36}$$

Notice that this is a 3×3 matrix since we are using the two-dimensional method but we are still dealing with the three-dimensional molecular model. The two matrices t_1 and t_2 share the same eigenvalues. Their ratio is

$$f(\beta, h) = \frac{(bx^2 + 1 + 1/x^2) - \sqrt{(bx^2 - 1 - 1/x^2)^2 + 4(1 + x^2)}}{(bx^2 + 1 + 1/x^2) + \sqrt{(bx^2 - 1 - 1/x^2)^2 + 4(1 + x^2)}}. \tag{6.37}$$

Inserting this result in the scaling Eq. (6.33) evaluated at $d = 2$ yields an implicit equation for β_c versus h. The numerical solutions are reported in Fig. 6.3. They share most of the features of the two-dimensional molecular model: (i) There are several solutions but there is no repulsive fixed point for $h = 0$; (ii) the physical solution starts at a negative h which in this case is $h \approx -0.226$; (iii) the physical solution has just one stationary point h_0 where β_c reaches its minimum value. However, in this case the stationary point is at $h_0 = -0.2349$ where $\beta_c = 0.6122$. This estimate of the critical point is not too far from the finite size scaling prediction of numerical simulations $\beta_c = 0.53$ [13]. The result corroborates our understanding of the physics described by the molecular model. Strictly speaking, this two-dimensional variational method describes the transition occurring in a single layer of molecules. However, at variance with the two-dimensional molecular model, each molecule is now allowed to be oriented along three different axes (two in-plane and one out-of-plane orientations). Thus this reasonable prediction for β_c could be regarded as an indirect proof that the correlation between two different layers is negligible, and that in the ordered phase the system behaves as a truly two-dimensional one.

6.5 Concluding Remarks

According to mean-field and finite-size scaling, the three-dimensional molecular model has a second order continuous transition from an isotropic disordered high-temperature phase to an anisotropic two-dimensional ordered low-temperature phase. The three-dimensional realization of the model is the one which more closely describes real molecular systems. For this reason the three-dimensional model has been studied by the variational Migdal–Kadanoff method in some detail.

On the other hand, the two-dimensional model is special by itself for its equivalence to the two-dimensional Ising model, and for the existence of exact analytical results. Thus, for $d = 3$ the model can be seen as a non-trivial extension to higher dimensions of the two-dimensional Ising model. Here 'non-trivial' means that the three-dimensional molecular model does not belong to the universality classes of the standard three-dimensional extensions of the Ising model (three-dimensional Ising and Potts models). The difference is evident from a comparison of the ground state configurations at $T = 0$: highly degenerate and anisotropic in the molecular model (with a two-dimensional character even for higher dimensions); with a small degeneration and fully isotropic in the Potts models (including the Ising one as a special case). By considering the two-dimensional character of the low-temperature phase, the molecular model could be thought to belong to the universality class of the simple two-dimensional Ising or three-states Potts models. However, in the high temperature unbroken-symmetry phase the molecular model is fully isotropic and has a three-dimensional character.

A formal proof of such statements comes from a comparison of the critical exponents. For the three-dimensional molecular model the finite size scaling calculation yields $\nu = 0.44$ [13], to be compared to the two-dimensional two-state (Ising) and three-state Potts models whose exponents are $\nu = 1$ and $\nu = 0.83$, respectively [19], to the three-dimensional Ising model whose exponent is $\nu = 0.64$ [19], and to the three-state three-dimensional Potts model which is known to undergo a first-order transition [25, 26].

The molecular model belongs to a different universality class which is characterized by a sort of dimensional transmutation. In fact order takes place in chains which are arranged in layers, and the disorder-order transition requires a decrease of the effective dimensionality of the system. In the ordered phase the molecules are correlated inside layers, but there is no correlation between molecules which belong to different layers. This understanding of the ordered phase is in agreement with our discussion of the Migdal–Kadanoff variational method. The two-dimensional decimation on a single layer yields a better prediction than the isotropic three-dimensional decimation applied to the whole lattice. On the other hand, the same two-dimensional layer-decimation provides an analytical tool for describing the generic d-dimensional molecular model by a straightforward generalization.

Finally, the universality class of the order-disorder transition which is described by the model deserves some experimental test. Transitions of this kind have been observed in several systems, as discussed in the introduction. Since the critical properties should not depend on the microscopic details of the system we expect that the simple molecular model could predict the correct critical exponent of real orientational transitions occurring in real molecular systems, especially under pressure.

References

1. F. Siringo, R. Pucci, N.H. March, High Press. Res. **2**(2), 109 (1990). https://doi.org/10.1080/08957959008201442
2. F. Siringo, R. Pucci, N.H. March, Phys. Rev. B **37**, 2491 (1988). https://doi.org/10.1103/PhysRevB.37.2491
3. F. Siringo, R. Pucci, N.H. March, Phys. Rev. B **38**, 9567 (1988). https://doi.org/10.1103/PhysRevB.38.9567
4. R. Pucci, F. Siringo, N.H. March, Phys. Rev. B **38**, 9517 (1988). https://doi.org/10.1103/PhysRevB.38.9517
5. M. Pasternak, J.N. Farrell, R.D. Taylor, Phys. Rev. Lett. **58**, 575 (1987). https://doi.org/10.1103/PhysRevLett.58.575
6. M. Pasternak, J.N. Farrell, R.D. Taylor, Solid State Commun. **61**(7), 409 (1987). https://doi.org/10.1016/0038-1098(87)90128-1
7. J. Obriot, F. Fondère, P. Marteau, M. Allavena, J. Chem. Phys. **79**(1), 33 (1983). https://doi.org/10.1063/1.445529
8. P.G. Johannsen, W. Helle, W.B. Holzapfel, J. Phys. Colloq. **45**, 199 (1984). https://doi.org/10.1051/jphyscol:1984836
9. F.A. Gorelli, M. Santoro, L. Ulivi, R. Bini, Physica B **265**(1), 49 (1999). https://doi.org/10.1016/S0921-4526(98)01314-3
10. R. Bini, M. Jordan, L. Ulivi, H.J. Jodl, J. Chem. Phys. **108**(16), 6849 (1998). https://doi.org/10.1063/1.476098
11. F. Siringo, Phys. Lett. A **226**(6), 378 (1997). https://doi.org/10.1016/S0375-9601(96)00959-0
12. F. Siringo, Int. J. Mod. Phys. B **11**(18), 2183 (1997). https://doi.org/10.1142/S021797929700112X
13. F. Siringo, Phys. Rev. E **62**, 6026 (2000). https://doi.org/10.1103/PhysRevE.62.6026
14. K. Mazzitello, J.L. Iguain, C.M. Aldao, H.O. Mártin, Phys. Rev. E **61**, 2954 (2000). https://doi.org/10.1103/PhysRevE.61.2954
15. V. Tozzini, N.H. March, M.P. Tosi, Phys. Chem. Liq. **37**(2), 185 (1999). https://doi.org/10.1080/00319109908045125
16. N.H. March, Physica B **265**(1), 24 (1999). https://doi.org/10.1016/S0921-4526(98)01311-8
17. J.A. Pople, F.E. Karasz, J. Phys. Chem. Solids **18**(1), 28 (1961). https://doi.org/10.1016/0022-3697(61)90080-4
18. C. Vega, E.P.A. Paras, P.A. Monson, J. Chem. Phys. **97**(11), 8543 (1992). https://doi.org/10.1063/1.463372
19. C. Itzykson, J.M. Drouffe, *Statistical Field Theory* (Cambridge University Press, Cambridge, 1989)
20. R.B. Potts, Math. Proc. Camb. Philos. Soc. **48**(1), 106 (1952). https://doi.org/10.1017/S0305004100027419
21. A.A. Migdal, Zh. Eksp. Teor. Fiz. **69**(3), 810 (1975); J. Exp. Theor. Phys. (JETP) **42**(3), 413 (1975)
22. A.A. Migdal, Zh. Eksp. Teor. Fiz. **69**(4), 1457 (1975); J. Exp. Theor. Phys. (JETP) **42**(4), 743 (1975)
23. L.P. Kadanoff, Ann. Phys. **100**(1), 359 (1976). https://doi.org/10.1016/0003-4916(76)90066-X

24. P.M. Stevenson, Phys. Rev. D **23**, 2916 (1981). https://doi.org/10.1103/PhysRevD.23.2916
25. F.Y. Wu, Rev. Mod. Phys. **54**, 235 (1982). https://doi.org/10.1103/RevModPhys.54.235
26. W. Janke, R. Villanova, Nucl. Phys. B **489**(3), 679 (1997). https://doi.org/10.1016/S0550-3213(96)00710-9

Chapter 7
An Ab Initio Evaluation of Mott Properties?

A. Cabo Montes de Oca

Abstract A GW scheme for band calculations is proposed. It rests on the static approximation for the effective potential. A closed system of equations for the determination of the basis of filled and empty states is obtained. The full translational symmetry in the original lattice is allowed to be broken and the wavefunction basis admits non-collinear spin dependences. The results of its planned application to the La_2CuO_4 crystal are expected to reproduce the strong correlation properties which emerged from a previously studied closely related model. A positive result of this study could show an example of the derivation of the Mott properties of a model without the need of introducing auxiliary phenomenological conditions. Thus, a path for derive the properties of strongly correlated electron systems (SCES) from ab initio calculations is suggested.

7.1 Introduction

The development of procedures for band structure calculations is a theme to which an intense research activity has been devoted in modern condensed matter physics. This area of research has a long history due to the existence of important unsolved relevant questions concerning the structure of solids [1–30]. One of those central open problems is the connection between the so called ab initio band evaluation schemes with the Mott phenomenological band calculation approaches. This procedure furnishes a successful description for a wide class of band structures of solids, which are not naturally explained by the assumed fundamental, *first principles* methods [1, 2, 5, 7, 12]. A class of materials which had been in the central point of attention within the existing debate between the two conceptions for band structure calculations are the transition metal oxides (TMO) [7, 10, 12, 16, 17, 20, 21, 25]. A particular compound which is closely related with the TMO is the first high-T_c superconductor material, La_2CuO_4. For this material, the first band structure calculations predicted metal and

A. Cabo Montes de Oca (✉)
Departamento de Física Teórica, Instituto de Cibernética, Matemática y Física,
Calle E, No. 309, e/ 13 y 15, Vedado, Havana, Cuba
e-mail: cabo@icimaf.cu

G. G. N. Angilella and C. Amovilli (eds.), *Many-body Approaches at Different Scales*, https://doi.org/10.1007/978-3-319-72374-7_7

paramagnetic characters, which are largely at variance with its known insulator and antiferromagnetic nature [29]. These two properties are believed to be strong correlation ones, which are not derivable from an ab initio calculational evaluation [16, 25, 31].

In a series of works [32–37], a model was constructed for the CuO planes in La_2CuO_4, which was able not only to describe the Mott strong correlated electron system (SCES) properties of the compound, but also predict the elusive pseudogap states and the quantum phase transition occurring beneath the superconducting dome [38, 39]. That model consisted of a Coulomb interacting electron gas which moves in a square 2D lattice potential with the periodicity of the Cu atoms pertaining to the CuO planes. The main phenomenological parameter was a dielectric constant fixed to screen the interacting Coulomb potential between the electrons. Its value was chosen to equalize the bandwidth of the single tight-binding-like band received after solving the problem in the mean field approximation, with the known bandwidth of the unique band crossing the Fermi level in the original band evaluations of Mattheiss et al. for La_2CuO_4 [29]. This procedure allowed to fix the dielectric constant to a value close to 10. The resulting mean field band was derived by assuming that the orbitals were Bloch functions under translations in the full lattice of the Cu planes, and that the spin projection of the orbitals were well defined values ± 1 along a fixed spatial axis. Further, we sought for iterative solutions of the Hartree–Fock (HF) problem, but starting from an initial set of non collinear single electron wavefunctions, which were also Bloch waves, but only in the sublattice of the Cu atoms in which the known antiferromagnetic (AF) structure of the material shows translational symmetry. That means, on one hand, that we allowed for the solution to break the translation symmetry in the same way as it is done by the AF order, and also to allow for the single particle orbitals to show a non-collinear spin structure. After this, surprisingly, a fully non collinear solution emerged from the iterative process, which not only showed an AF order similar to the one in La_2CuO_4, but which also showed an insulator gap between 1 and 2 eV. Therefore, the main SCES properties of the La_2CuO_4, and the explanations of the pseudogap and the quantum phase transition beneath the dome, were all predicted by the mentioned model.

Therefore, the present work intends to explore the indications following from the previous remarks about the existence of a possible path allowing the derivation of the SCES properties for the cuprate materials and the transition metals oxides within a *first principles* scheme of calculation. Those remarks strongly support the idea about that if the same non-collinear structure of the natural orbitals are assumed and only a reduced translation invariance consistent with the AF order is imposed, under a GW scheme kind of calculation (which is known that generalizes the HF method by including also screening) the possibility exists for evaluating for the La_2CuO_4 crystal, the Mott SCES properties as they emerged for the model. A central element in this possibility, is the fact the GW procedure embodies screening, and precisely in a way, that generalizes the HF method by substituting the simple Coulomb potential by the screened potential. This fact strongly indicates that the effects of the basic phenomenological dielectric constant employed in the model, most be expected to be 'explained' within a more fundamental discussion based in the GW evaluation scheme. We consider that a result like the one suspected will have a relevant meaning: it will show an example of the derivation of the Mott properties of this SCES material from a *first principles* scheme. Thus, in this first work we start by constructing the basic elements of a band calculation method based in the static approximation of the GW procedure, just generalizing the model discussions in references [32–37]. The construction opens the possibility to apply the procedure to real crystals like La_2CuO_4, a study that is expected to be considered elsewhere.

The presentation proceeds as follows: In Sect. 7.2 the general GW equations are presented. A convenient notation is also defined there. Then, Sect. 7.3 describes the static approximation of the general GW equations, which is employed in order to simplify the application of the scheme. The equations are reduced to a kind of generalized HF equations in which the simple Coulomb potential is substituted by the screened potential. This quantity is expressed as a functional of the self-consistent normal orbitals. Therefore, the screening effects which were central in the model discussed in Refs. [32–37], can be derived in the scheme as a result of the screening effects of the bare Coulomb potential. The following Sect. 7.4 defines a convenient Bloch basis for expanding the natural orbitals of the GW method showing the reduced lattice symmetry of the AF order of the material. In Sect. 7.5, this reduced crystal symmetry is employed to simplify the resulting set of equations. Section 7.6 also uses this crystal symmetry to reduce the set of equations to one matrix equation for the set of Bloch functions associated to each value of the quasi-momentum. This completes the presentation of the basic elements of the proposed scheme. The results are resumed in the Summary.

7.2 The *GW* System of Equations

In the discussion below we will closely follow the presentation of the GW method given in Ref. [40]. The GW approximation can be defined by the following set of coupled equations linking the electron self-energy Σ and the electronic polarization P with the electron propagator G and frequency dependent effective potential W through the formulas

$$\Sigma(1,2) = iG(1,2^+)W(1,2), \tag{7.1a}$$

$$P(1,2) = -iG(1,2)G(2,1^+), \tag{7.1b}$$

$$W(1,2) = v(1,2) + v(1,3)P(3,4)G(4,2), \tag{7.1c}$$

$$G(1,2) = G_0(1,2) + G_0(1,3)\Sigma(3,4)G(4,2). \tag{7.1d}$$

Here, v is the interaction potential between the electrons and G_0 is the electron Green function in the absence of the interaction v. The derivation of the equations from the exact many-body description of the system rests in a unique approximation: to consider the exact vertex function with two fermion legs and one boson one, as approximated by its lower order approximation. In these equations the following compact notation for the spatial coordinates and the time had been used, in which a natural number n as an argument of a fermion kernel or wavefunction indicates the set of space time coordinates and spin projection $(\mathbf{x}_n, t_n, s_n)$, and the coordinates are 3D, that is: $\mathbf{x}_n = (x_n^1, x_n^2, x_n^3)$. On another hand, in a bosonic kernel (like the polarization or effective potential) the argument n will only indicate the space time coordinates $(\mathbf{x}_n, t_n)$. Similarly, in the following discussion, the symbol $\mathbf{n}$, in the argument of a Fermi kernel or wavefunction, will denote the set of spatial coordinates plus the spin projection of an electron, $\mathbf{n} \equiv (\mathbf{x}_n, s_n)$. The Einstein convention will also reduce the size of the equations. The coincidence of two arguments given by natural numbers will imply the integration over the space-time variables and the summation over spin projection if they are both arguments of Fermi kernels or electron wavefunctions. If they are both arguments of boson kernels, or one pertains to a boson kernel and the other to a Fermi one, the coincidence will mean only the space-time integrations and the spin projection will correspond to fixed spin argument. Exactly the same definition will have the coincidence of two indices corresponding to vector like natural number indices as $\mathbf{n}$. The free Hamiltonian of the system is defined by the relations

$$H_0(\mathbf{1}, \mathbf{2}) = \delta^{s_1 s_2} \left(\frac{1}{2}\mathbf{p}_1^2 + v_l(\mathbf{x}_1) + v_H(\mathbf{x}_1, t_1) + \phi(\mathbf{x}_1, t_1) \right) \delta^{(D)}(\mathbf{x}_1 - \mathbf{x}_2), \tag{7.2a}$$

$$V(\mathbf{x}_1, t_1) = v_H(\mathbf{x}_1) + \phi(\mathbf{x}_1, t_1), \tag{7.2b}$$

$$\mathbf{p}_1 = -i\frac{\partial}{\partial \mathbf{x}_1}, \tag{7.2c}$$

$$v_l(\mathbf{x}_1) = \sum_{\mathbf{R}\in l} v(\mathbf{x}_1 - \mathbf{R}), \mathbf{R} = p(n_1\mathbf{i} + n_2\mathbf{j}), \quad n_1, n_2 \in Z, \tag{7.2d}$$

$$v(\mathbf{x}_1) = \frac{1}{|\mathbf{x}_1|}, \tag{7.2e}$$

$$v_H(\mathbf{x}_1, t_1) = \int d\mathbf{x}_2 v(\mathbf{x}_1, \mathbf{x}_2)\rho(\mathbf{x}_2, t_1), \tag{7.2f}$$

$$\rho(\mathbf{x}_1, t_1) = -i \sum_{s_1} G(\mathbf{x}_1, s_1, \mathbf{x}_1, s_1, t_1 - t_2)|_{t_2 \to t_1^+}, \tag{7.2g}$$

in which an auxiliary space and time dependent potential field $\phi(\mathbf{x}_1, t_1)$ has been added to the Hartree potential $v_H(\mathbf{x}_1)$ which is generated by the dynamically defined electron density $\rho(\mathbf{x}_1, t_1)$. The function $\delta^{(D)}(\mathbf{x})$ is the Dirac's delta-function in three dimensions, $\delta^{(3)}(\mathbf{x})$. Note that all the quantization procedure depends on the total field V given by the sum of the Hartree field plus the auxiliary one. Note also that the interaction potential v has been defined as equal to the Coulomb one. We will employ the simplifying units

$$\hbar = m = e = 1.$$

The GW equations can also be represented as a system of equations for the determination of a basis set of functions in a similar form as it is done for the Hartree Fock system of equations. For this purpose, the equations for the electron Green function can be substituted by a set of equations for the basis functions. For this purpose, first take into account that the system becomes time-translation invariant after making the auxiliary space and time dependent scalar field ϕ to vanish. Therefore, the temporal Fourier transform of exact electron Green function can be written as

$$G(\mathbf{1}, \mathbf{2}, w) = \int d(t_1 - t_2) G(\mathbf{1}, \mathbf{2}, t_1 - t_2) \exp(iw(t_1 - t_2)). \tag{7.3}$$

Then, following the discussion in [40] $G(\mathbf{1}, \mathbf{2}, w)$ can be written in the form

$$G(\mathbf{1}, \mathbf{2}, w) = \sum_i \frac{\Psi_i(\mathbf{1}, w)\overline{\Psi}_i(\mathbf{2}, w)}{w - E_i(w) + i\sigma \operatorname{sgn}(E_i(w) - \mu)}, \tag{7.4}$$

where the entering wavefunctions satisfy the following eigenvalue equations for each value of the frequency

$$(H_0(\mathbf{1}, \mathbf{2}) + \Sigma(\mathbf{1}, \mathbf{2}, w))\, \Psi_i(\mathbf{2}, w) = E_i(w)\Psi_i(\mathbf{1}, w), \tag{7.5a}$$

$$H_0(\mathbf{1}, \mathbf{2}) = \delta^{s_1 s_2} \left(\frac{1}{2}\mathbf{p}_1^2 + v_l(\mathbf{x}_1) + v_H(\mathbf{x}_1) \right) \delta^{(3)}(\mathbf{x}_1 - \mathbf{x}_2), \tag{7.5b}$$

$$\mathbf{p}_1 = -i\frac{\partial}{\partial \mathbf{x}_1}, \tag{7.5c}$$

$$v_H(\mathbf{x}_1) = \int d\mathbf{x}_2 v(\mathbf{x}_1, \mathbf{x}_2)\rho(\mathbf{x}_2), \tag{7.5d}$$

$$\rho(\mathbf{x}_1) = -i \sum_{s_1} G(\mathbf{x}_1, s_1, \mathbf{x}_1, s_1, t_1 - t_2)|_{t_2 \to t_1^+}, \tag{7.5e}$$

in which the wavefunctions have the quantum numbers i in a compact notation. Due to the crystal sublattice invariance that will be assumed here, among the set of indices defined by i, there will be a momentum pertaining to the Brillouin cell of the reciprocal lattice associated to the sublattices to be defined afterwards. Therefore, when a more explicit representation will be needed the index i will be substituted as $i \to (\mathbf{k}_i, i)$ where the new symbol i now will represent all the quantum numbers of the wavefunction in addition to $\mathbf{k}_i$. This is a system of eigenvalue equations that can be solved once the kernel of the selfenergy Σ is known. Then, the GW system of equations can be also written in the form

$$E_i(w)\Psi_i(\mathbf{1}, w) = (H_0(\mathbf{1}, \mathbf{2}) + \Sigma(\mathbf{1}, \mathbf{2}, w))\, \Psi_i(\mathbf{2}, w), \tag{7.6a}$$

$$H_0(\mathbf{1}, \mathbf{2}) = \delta^{s_1 s_2} \left(\frac{1}{2}\mathbf{p}_1^2 + v_l(\mathbf{x}_1) + v_H(\mathbf{x}_1) \right) \delta^{(D)}(\mathbf{x}_1 - \mathbf{x}_2), \tag{7.6b}$$

$$\Sigma(1, 2) = iG(1, 2^+)W(1, 2), \tag{7.6c}$$

$$P(1, 2) = -iG(1, 2)G(2, 1^+), \tag{7.6d}$$

$$W(1, 2) = \frac{1}{1 - vP}(1, 3)v(3, 2), \tag{7.6e}$$

where for two kernels $K_1(1,2)$ and $K_2(1,2)$ the symbol K_1K_2 indicates the product of kernels $K_1(1,3)\ K_2(3,2)$ and $\frac{1}{1-vP}$ indicates the usual geometric expansion but in terms of the product kernel vP.

7.3 Static Approximation in Terms of the Effective Potential

One serious obstacle in solving the set of equations (7.6) is the fact that the selfenergy depends on the frequency, a fact that complicates finding of the quasiparticle energies $E_i(w)$ which also will depend of the frequency in this case. Let us then start considering an approximation in which this difficulty disappears: the static limit. In this case the frequency-dependent effective potential for all values of the frequency is assumed to be a constant given by its value at zero frequency. Consider the Fourier transform of the selfenergy by also substituting the temporal Fourier expansion of the electron Green functions and the effective potential, which leads to

$$\begin{aligned}\Sigma(\mathbf{1},\mathbf{2},w) &= i\int d(t_1-t_2)\exp(iw(t_1-t_2))G(\mathbf{1},\mathbf{2},t_1-t_2^+)W(\mathbf{1},\mathbf{2},t_1-t_2)\\ &= i\int\frac{dw'}{2\pi}G(\mathbf{1},\mathbf{2},w')W(\mathbf{1},\mathbf{2},w-w')\exp(iw'\delta).\end{aligned}\tag{7.7}$$

Assuming now that W is frequency independent, that is $W(\mathbf{1},\mathbf{2},w)=W(\mathbf{1},\mathbf{2},0)\equiv\widetilde{W}(\mathbf{1},\mathbf{2})$, it directly follows that the selfenergy also turns to be frequency independent since

$$\begin{aligned}\Sigma(\mathbf{1},\mathbf{2},w) &= i\int\frac{dw'}{2\pi}G(\mathbf{1},\mathbf{2},w')\widetilde{W}(\mathbf{1},\mathbf{2})\exp(iw'\delta)\\ &= i\widetilde{W}(\mathbf{1},\mathbf{2})G(\mathbf{1},\mathbf{2},t_1-t_2)|_{t_2\to t_1^+}\\ &\equiv\widetilde{\Sigma}(\mathbf{1},\mathbf{2}).\end{aligned}\tag{7.8}$$

Evaluating $G(\mathbf{1},\mathbf{2},t_1-t_2)|_{t_2\to t_1^+}$ leads to

$$\begin{aligned}\int\frac{dw}{2\pi}G(\mathbf{1},\mathbf{2},w)\exp(iw\delta) &= \sum_i\Psi_i(\mathbf{1})\overline{\Psi}_i(\mathbf{2})\int\frac{dw}{2\pi}\frac{1}{w-E_i+i\sigma\,\mathrm{sgn}(E_i-\mu)}\\ &= i\sum_i^{\mu\geq E_i}\Psi_i(\mathbf{1})\overline{\Psi}_i(\mathbf{2}),\end{aligned}\tag{7.9}$$

where the sum runs over the filled states with energies lower than the chemical potential μ. Thus, the selfenergy in the static approximation takes the following form

$$\widetilde{\Sigma}(\mathbf{1},\mathbf{2}) = -\widetilde{W}(\mathbf{1},\mathbf{2}) \sum_{i}^{\mu \le E_i} \Psi_i(\mathbf{1})\overline{\Psi}_i(\mathbf{2}), \tag{7.10}$$

after which W, the static effective potential, becomes a functional of the basis wavefunctions Ψ_i. This approximation permits to write the equations for the wavefunctions in a form in which nor the energy or the wavefunctions depend on the frequency

$$E_i\Psi_i(\mathbf{1}) = H_0(\mathbf{1},\mathbf{2})\Psi_i(\mathbf{2}) + \widetilde{\Sigma}(\mathbf{1},\mathbf{2})\Psi_i(\mathbf{2}), \tag{7.11a}$$

$$H_0(\mathbf{1},\mathbf{2}) \equiv \delta^{s_1 s_2}\left(\frac{1}{2}\mathbf{p}_1^2 + v_l(\mathbf{x}_1) + v_H(\mathbf{x}_1)\right)\delta^{(D)}(\mathbf{x}_1 - \mathbf{x}_2). \tag{7.11b}$$

The equations determining $\widetilde{\Sigma}(\mathbf{1},\mathbf{2}) = \Sigma(\mathbf{1},\mathbf{2},w)|_{w=0}$ and $\widetilde{W}(\mathbf{1},\mathbf{2}) = W(\mathbf{1},\mathbf{2},w)|_{w=0}$ can be transformed by means of the relations

$$\begin{aligned} W(\mathbf{1},\mathbf{2},w)|_{w=0} &= \left(\varepsilon^{-1}(\mathbf{1},\mathbf{3},w)v(\mathbf{1},\mathbf{3},w)\right)|_{w=0} \\ &= \varepsilon^{-1}(\mathbf{1},\mathbf{3},w)|_{w=0}\, v(\mathbf{3},\mathbf{2},w)|_{w=0}, \end{aligned} \tag{7.12a}$$

$$\begin{aligned} \varepsilon^{-1}(\mathbf{1},\mathbf{2},w)|_{w=0} &= \frac{1}{1-(vP)}(\mathbf{1},\mathbf{2},w)|_{w=0} \\ &= \frac{1}{1-(vP)|_{w=0}}(\mathbf{1},\mathbf{2}), \end{aligned} \tag{7.12b}$$

$$v(\mathbf{3},\mathbf{2},w)|_{w=0} = v(\mathbf{3},\mathbf{2}), \tag{7.12c}$$

$$(vP)|_{w=0}(\mathbf{1},\mathbf{2}) = v(\mathbf{1},\mathbf{3})P(\mathbf{3},\mathbf{2},w)|_{w=0}, \tag{7.12d}$$

which allow to define the static magnitudes

$$\widetilde{\Sigma}(\mathbf{1},\mathbf{2}) = \Sigma(\mathbf{1},\mathbf{2},w)|_{w=0}, \tag{7.13a}$$

$$\widetilde{W}(\mathbf{1},\mathbf{2}) = W(\mathbf{1},\mathbf{2},w)|_{w=0}, \tag{7.13b}$$

$$\widetilde{\varepsilon}^{-1}(\mathbf{1},\mathbf{2}) = \varepsilon^{-1}(\mathbf{1},\mathbf{2},w)|_{w=0}, \tag{7.13c}$$

$$\begin{aligned} \widetilde{P}(\mathbf{1},\mathbf{2}) &= P(\mathbf{1},\mathbf{2},w)|_{w=0}, \\ &= -i\int d(t_1-t_2)G(\mathbf{1},\mathbf{2},t_1-t_2)G(\mathbf{2},\mathbf{1},t_2-t_1-\delta)|_{\delta\to 0^+}, \end{aligned} \tag{7.13d}$$

and write for the static effective potential

$$\widetilde{W}(\mathbf{1},\mathbf{2}) = \widetilde{\varepsilon}^{-1}(\mathbf{1},\mathbf{3})v(\mathbf{3},2), \tag{7.14a}$$

$$\widetilde{\varepsilon}^{-1}(\mathbf{1},\mathbf{2}) = \frac{1}{1-(v\widetilde{P})}(\mathbf{1},\mathbf{2}). \tag{7.14b}$$

In order to find an expression for the static polarization $\widetilde{P}$, let us consider its definition

$$\begin{aligned}\widetilde{P}(\mathbf{1},\mathbf{2}) &= P(\mathbf{1},\mathbf{2},w)|_{w=0},\\ &= -i\int d(t_1-t_2)G(\mathbf{1},\mathbf{2},t_1-t_2)G(\mathbf{2},\mathbf{1},t_2-t_1-\delta)|_{\delta\to 0^+}\\ &= -i\sum_i\sum_j \Psi_i(\mathbf{1})\overline{\Psi}_i(\mathbf{2})\Psi_j(\mathbf{2})\overline{\Psi}_j(\mathbf{1})\times\\ &\int_{-\infty}^{\infty}\frac{dw}{2\pi}\frac{\exp(iw\delta)}{(w-E_i+i\text{sgn}(E_i-\mu))(w-E_j+i\text{sgn}(E_j-\mu))}. \end{aligned} \tag{7.15}$$

But the frequency integral can be explicitly evaluated by using the relation

$$\int_{-\infty}^{\infty}\frac{dw}{2\pi}\frac{\exp(iw\delta)}{(w-E_i+i\text{sgn}(E_i-\mu))(w-E_j+i\text{sgn}(E_j-\mu))} = \frac{i\left(\theta(E_i-\mu)\theta(\mu-E_j)-\theta(\mu-E_i)\theta(E_j-\mu)\right)}{-(E_i-E_j)}, \tag{7.16}$$

to write

$$\begin{aligned}\widetilde{P}(\mathbf{1},\mathbf{2}) &= -\sum_i\sum_j \Psi_i(\mathbf{1})\overline{\Psi}_i(\mathbf{2})\Psi_j(\mathbf{2})\overline{\Psi}_j(\mathbf{1})\\ &\qquad\times\frac{\theta(E_i-\mu)\theta(\mu-E_j)-\theta(\mu-E_i)\theta(E_j-\mu)}{E_i-E_j}\\ &= -\sum_i^{\mu\le E_i}\sum_j^{\mu\ge E_j}\frac{\Psi_i(\mathbf{1})\overline{\Psi}_i(\mathbf{2})\Psi_j(\mathbf{2})\overline{\Psi}_j(\mathbf{1})}{E_i-E_j}+\sum_i^{\mu\ge E_i}\sum_j^{\mu\le E_j}\frac{\Psi_i(\mathbf{1})\overline{\Psi}_i(\mathbf{2})\Psi_j(\mathbf{2})\overline{\Psi}_j(\mathbf{1})}{E_i-E_j}, \end{aligned} \tag{7.17}$$

which expresses the static polarization as functional of the orbitals Ψ_i. Then, the static effective potential is also a functional of the Ψ_i.

The previous expressions allow to write the GW equations in the static approximation as the following set of equations for the determination of the basis functions Ψ_i and its energy spectrum $\{E_i\}$

$$E_i\Psi_i(\mathbf{1}) = H_0(\mathbf{1},\mathbf{2})\Psi_i(\mathbf{2})+\widetilde{\Sigma}(\mathbf{1},\mathbf{2})\Psi_i(\mathbf{2}), \tag{7.18a}$$

$$H_0(\mathbf{1},\mathbf{2}) \equiv \delta^{s_1s_2}\left(\frac{1}{2}\mathbf{p}_1^2+v_l(\mathbf{x}_1)+v_H(\mathbf{x}_1)\right)\delta^{(D)}(\mathbf{x}_1-\mathbf{x}_2), \tag{7.18b}$$

$$\widetilde{\Sigma}(\mathbf{1},\mathbf{2}) = -\widetilde{W}(\mathbf{1},\mathbf{2})\sum_i^{\mu\ge E_i}\Psi_i(\mathbf{1})\overline{\Psi}_i(\mathbf{2}), \tag{7.18c}$$

$$\widetilde{P}(\mathbf{1},\mathbf{2}) = -\sum_{i}^{\mu\leq E_i}\sum_{j}^{\mu\geq E_j}\frac{\Psi_i(\mathbf{1})\overline{\Psi}_i(\mathbf{2})\Psi_j(\mathbf{2})\overline{\Psi}_j(\mathbf{1})}{(E_i - E_j)}$$

$$+\sum_{i}^{\mu\geq E_i}\sum_{j}^{\mu\leq E_j}\frac{\Psi_i(\mathbf{1})\overline{\Psi}_i(\mathbf{2})\Psi_j(\mathbf{2})\overline{\Psi}_j(\mathbf{1})}{(E_i - E_j)}, \tag{7.18d}$$

$$\widetilde{W}(\mathbf{1},\mathbf{2}) = \widetilde{\varepsilon}^{-1}(\mathbf{1},\mathbf{3})v(\mathbf{3},2), \tag{7.18e}$$

$$\widetilde{\varepsilon}^{-1}(\mathbf{1},\mathbf{2}) = \frac{1}{1-(v\widetilde{P})}(\mathbf{1},\mathbf{2}), \tag{7.18f}$$

$$v_H(\mathbf{x}) = \int d\mathbf{x}_1\frac{1}{|\mathbf{x}-\mathbf{x}_1|}\sum_{i}^{\mu\leq E_i}\sum_{s_1=\pm 1}\Psi_i(\mathbf{x}_1,s_1)\overline{\Psi}_i(\mathbf{x}_1,s_1). \tag{7.18g}$$

7.4 The Bloch Basis for the Natural Orbitals

Therefore, following the ideas exposed in the Introduction, in this section we will implement the same allowances for symmetry breaking effects and spin non collinearity as in Ref. [32, 33]. For this purpose, let us decompose the lattice of the unit cell positions $\mathbf{R}$ of the crystal lattice of La_2CuO_4 (to be called the *original* lattice in what follows) in two sublattices indexed by $r = 1, 2$. The two values of the indices r will be defined as follows. Consider the lattice of atoms of the crystal and a particular CuO plane in it, which exhibit AF order in the planar lattice formed by the Cu atoms. This order breaks the original symmetry for translations connecting two neighboring Cu atoms in the CuO plane. Then, we can decompose the whole lattice formed by the Cu atoms within the considered CuO plane in two planar sublattices, the one associated to a given orientation of the spin in the experimentally observed AF order, will be indexed with the value $r = 1$. The other sublattice associated to the opposite experimental orientation of the spin will be indexed by $r = 2$. Now, we can define the origin of coordinates in the lattice in a particular Cu atom corresponding to the index $r = 1$ and also define the x_1 and x_2 axis to be directed from this reference atom to the two nearest neighboring Cu atom to it. Upon this, the x_3 will point from the reference Cu atom in the perpendicular direction to the considered CuO plane, and forming with the other two directions a direct reference frame. Now, it is possible to define the planar lattice vectors that gives the coordinates of all Cu atoms within each of the two sublattices as follows

$$\mathbf{R}_{\text{CuO}}^{(r)} = \sqrt{2}n_1 p\mathbf{q}_1 + \sqrt{2}n_2 p\mathbf{q}_2 + \mathbf{q}^{(r)}, \quad n_1, n_2 \in \mathbf{Z},$$

$$\mathbf{q}^{(r)} = \begin{cases}\mathbf{0}, & r = 1,\\ p\hat{\mathbf{e}}_{x_1}, & r = 2,\end{cases} \tag{7.19}$$

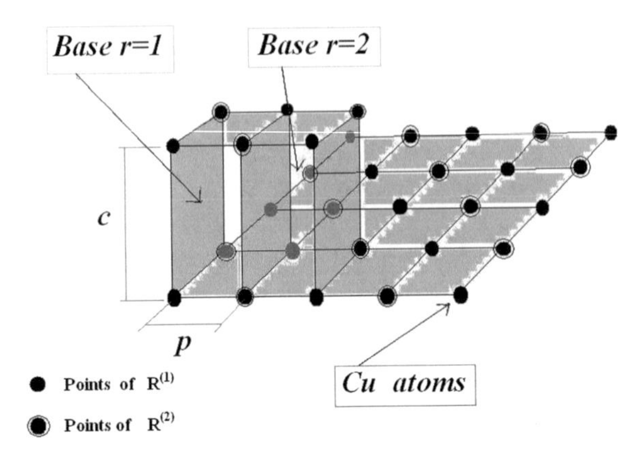

Fig. 7.1 The figure shows the Cu atoms in a CuO plane, whose positions are given by the two kinds of vectors of the sublattices $\mathbf{R}^{(1)}$ and $\mathbf{R}^{(2)}$. Their union defines the whole La_2CuO_4 crystal. This decomposition of the original lattice is done in order to allow the determination of the AF properties of the material through a crystal translational symmetry breakdown. The original basis of the crystal and a copy of it, after translated in the vector $\mathbf{q}^{(2)}$ are also illustrated. The crystal can be constructed after shifting the original basis in all the sublattice vectors $\mathbf{R}^{(1)}$ by also shifting the copy of the basis to all the points of the sublattice $\mathbf{R}^{(2)}$

where $\hat{\mathbf{q}}_1$, $\hat{\mathbf{q}}_2$ are the unit vectors

$$\mathbf{q}_1 = \frac{1}{\sqrt{2}}(\mathbf{e}_{x_1} - \mathbf{e}_{x_2}), \tag{7.20}$$

$$\mathbf{q}_2 = \frac{1}{\sqrt{2}}(\mathbf{e}_{x_1} + \mathbf{e}_{x_2}), \tag{7.21}$$

and $\mathbf{e}_{x_1}$, $\mathbf{e}_{x_2}$ are the unit vectors joining the reference Cu atoms with its two nearest neighbors. Next, we can consider that $p\mathbf{e}_{x_1}$, $p\,\mathbf{e}_{x_2}$ and $c\,\mathbf{e}_{x_3}$ are the unit cell vectors of the La_2CuO_4 lattice, where c is the length of the unit cell along the axis orthogonal to the CuO plane and $\hat{\mathbf{e}}_{x_3}$ is the unit vector along the x_3 axis. Thus the whole lattice defining the La_2CuO_4 crystal can be also decomposed in two sublattices defined by the following set of vectors

$$\begin{aligned} \mathbf{R}^{(r)} &= \mathbf{R}^{(r)}_{\text{CuO}} + n_3\mathbf{q}_3, \qquad\qquad r = 1, 2, \\ &= \sqrt{2}n_1 p\mathbf{q}_1 + \sqrt{2}n_2 p\mathbf{q}_2 + \mathbf{q}^{(r)} + n_3\mathbf{q}_3, \end{aligned} \tag{7.22a}$$

$$\mathbf{q}_3 = c\mathbf{e}_{x_3}. \tag{7.22b}$$

Figure 7.1 shows the defined plane of Cu atoms within a CuO plane and illustrates the points of the two defined sublattices $\mathbf{R}^{(r)}$, $r = 1, 2$ in which the whole crystal is decomposed.

Having already defined the lattice of the band calculation problem on which we will be impose periodic boundary conditions, let us search for basis functions for the self-consistent natural orbitals, as being eigenfunctions of the translations, but only in the reduced group associated to displacements leaving one of the sublattices $\mathbf{R}^{(1)}$, $\mathbf{R}^{(2)}$ invariant (note that the groups of translational invariance of each of the two sublattices are equivalent). In other words, these functions will be assumed to satisfy

$$\hat{T}_{\mathbf{R}^{(1)}}\phi_{\mathbf{k},l} = \exp(i\mathbf{k}\cdot\mathbf{R}^{(1)})\phi_{\mathbf{k},l}, \tag{7.23}$$

in which we have set the label of the orbitals in the form $\eta = (\mathbf{k}, l)$, where the index l, from now on, will indicate an element of the set of atomic orbitals associated to an arbitrary point of the crystal, for a fixed value of the quasimomentum $\mathbf{k}$. Note that translations in the set of vectors $\mathbf{R}^{(1)}$ define the symmetries, either of the same sublattice $\mathbf{R}^{(1)}$ or the one of the other sublattice $\mathbf{R}^{(2)}$. Let us consider now that the system satisfies periodic boundary conditions in order to reduce, as usual, the dimension of the following numerical problem. Therefore, we will impose boundary conditions in order to restrict the problem to the basis of states $\phi_{\mathbf{k},l}$ characterized by momenta $\mathbf{k}$ guaranteeing the periodicity of these states in the boundaries of the region defined by the bounds

$$-\frac{pL}{2} \le x_1 < \frac{pL}{2}, \tag{7.24a}$$

$$-\frac{pL}{2} \le x_2 < \frac{pL}{2}, \tag{7.24b}$$

$$-\frac{cL_3}{2} \le x_3 < \frac{cL_3}{2}, \tag{7.24c}$$

where L is an even integer number defining the size of the periodicity region. These boundary conditions determine the following set of allowed quasimomenta, defining the Brillouin cell $\mathbf{B}$, for the functions satisfying them:

$$\mathbf{k} = \frac{2\pi}{Lp}(n_1\mathbf{e}_{x_1} + n_2\mathbf{e}_{x_2}) + \frac{2\pi}{L_3 c}n_3\mathbf{e}_{x_3},$$
$$n_1, n_2, n_3 \in \mathbf{Z}, \quad -\frac{L}{2} \le (n_1 \pm n_2) < \frac{L}{2}, \quad -\frac{L_3}{2} \le n_3 < \frac{L_3}{2}. \tag{7.25}$$

For the infinite system ($L \to \infty$), $\mathbf{B}$ is continuous as illustrated in the left hand side picture of Fig. 7.2. Note that the number of elements in the reduced translation group of each of both sublattices is a half of the number of elements $N_c = L^2 L_3$ of the translation group, leaving invariant the point lattice of the material after boundary conditions are imposed. The value N_c is the number of cells of the starting La_2CuO_4 lattice, after periodicity has been implemented. Therefore, let us define a starting basis of suitable functions to expand the searched self-consistent GW natural orbitals, in the following form:

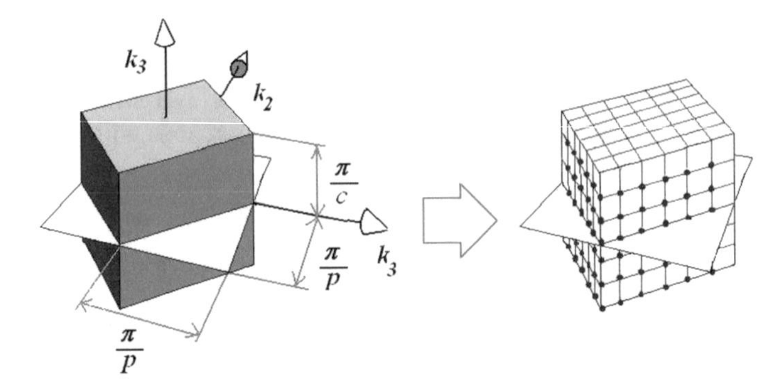

Fig. 7.2 The figure at the left shows the Brillouin cell **B** corresponding to both of the infinite sublattices $\mathbf{R}^{(1)}$ and $\mathbf{R}^{(2)}$. All the continuum of quasimomenta laying inside the box are allowed. The analogous figure on the right shows the point set of inequivalent momenta allowed by imposing the periodic boundary conditions in a box of sizes $2L = 12\,p$, and $2L_3 = 12\,c$. The darker points indicate the 6^3 inequivalent allowed momenta defined by the boundary conditions

$$\varphi_{\mathbf{k}}^{(A,\nu_A,r,\sigma_z)}(\mathbf{x}, s) = \sqrt{\frac{2}{N_c}} u^{\sigma_z}(s) \sum_{\mathbf{R}^{(r)}} \exp(i\mathbf{k}\cdot\mathbf{R}^{(r)})\varphi_{A,\nu_A}(\mathbf{x}-\mathbf{R}_A-\mathbf{R}^{(r)}),$$

$$\hat{\sigma}_z u^{\sigma_z} = \sigma_z u^{\sigma_z}, \quad A = 1, 2, 3, ...n_A, \quad \nu_A = 1, 2, 3, ..., m_A, \tag{7.26}$$

where N_c was defined before and $\hat{\sigma}_z$ is the spin projection operator in the z (x_3) axis. It will be selected as the perpendicular direction to the CuO plane (in the c axis of La_2CuO_4), with eigenvalues $\sigma_z = -1, 1$. The indices r take the values $r = 1, 2$. The states $\varphi_{A,\nu_A}(\mathbf{x}-\mathbf{R}_A)$ are to be chosen as electron wavefunctions (indexed by ν_A) for the atom of kind A in the La_2CuO_4 crystal. The position of this atom in the basis of the *original* crystal is defined by $\mathbf{R}_A$ with respect to the reference frame defined previously, with its origin sitting on a Cu atom of a CuO plane, and axes forming a direct triad, in which the x_3 direction points in the c axis of the crystal. Then, it can be noted that the index r indicates if the atomic Wannier functions $\varphi_{A,\nu_A}(\mathbf{x}-\mathbf{R}_A-\mathbf{R}^{(r)})$ generating the Bloch one, are centered in the points of one or another of the two possible sublattices constructed by shifting the central point of the atomic Wannier orbital. The construction can be interpreted as a one, in which two sets of functions are defined for each atomic orbital: one using this orbital as a Wannier state shifted to the point of the sublattice $r = 1$ and another employing the same state but shifted to the points of the $r = 2$ sublattice.

Note that the basis functions had well defined spin projection on the x_3 axis. It can be noted, that the orthogonality character for states associated to a fixed atom A and state ν_A, is lost only for pairs of wavefunctions in this set, corresponding to different values of the index r. That is, for different sublattices having the same spin projection. However, in general, the basis of states will be not orthonormal. This fact will lead the appearance of an overlapping matrix to enter in the discussion to follow. In order to simplify the notation, let us now define the composite index η in which we will englobe the indices indicating the atom A, the atomic state ν_A, sublattice label r and the spin projection quantum number as the ordered list $\Phi_{\mathbf{k}_i,\eta}$

$$\eta = \{A_\eta, \nu_{A_\eta}, r_\eta, \sigma_{z,\eta}\}. \tag{7.27}$$

With this notation the basis wavefunctions which have been just defined can be written as follows

$$\begin{aligned}\Phi_{\mathbf{k},\eta}(\mathbf{x},s) &\equiv \varphi_{\mathbf{k}}^{\{A_\eta,\nu_{A_\eta},r_\eta,\sigma_{z,\eta}\}}(\mathbf{x},s)\\ &= \sqrt{\frac{2}{N}}u^{\sigma_{z,\eta}}(s)\sum_{\mathbf{R}^{(r_\eta)}}\exp(i\mathbf{k}\cdot\mathbf{R}^{(r_\eta)})\varphi_{A_\eta,\nu_{A_\eta}}(\mathbf{x}-\mathbf{R}_{A_\eta}-\mathbf{R}^{(r)}),\\ &= \sqrt{\frac{2}{N}}u^{\sigma_{z,\eta}}(s)\sum_{\mathbf{R}^{(r_\eta)}}\exp(i\mathbf{k}\cdot\mathbf{R}^{(r_\eta)})\varphi_\eta(\mathbf{x}-\mathbf{R}^{(r_\eta)}) \qquad (7.28a)\end{aligned}$$

$$\varphi_\eta(\mathbf{x}) = \varphi_{A_\eta,\nu_{A_\eta}}(\mathbf{x}-\mathbf{R}_{A_\eta}). \tag{7.28b}$$

Therefore, the orbital function $\Phi_{\mathbf{k},\eta}$ will be considered as a Bloch wavefunction associated to the atom A_η, which is indexed by the symbol η, given by the above defined list $\eta = \{A_\eta, \nu_{A_\eta}, r_\eta, \sigma_{z,\eta}\}$ in which A_η indicates any of the n_{A_η} atoms defining the basis of unit cell, and ν_{A_η} indicates any of the m_{A_η} orbitals describing the same atom A_η. Figure 7.3 illustrates the basis of the La_2CuO_4 crystal and the decomposition of the whole crystal as generated by two similar copies of it after displaced each of them in the two defined sublattices.

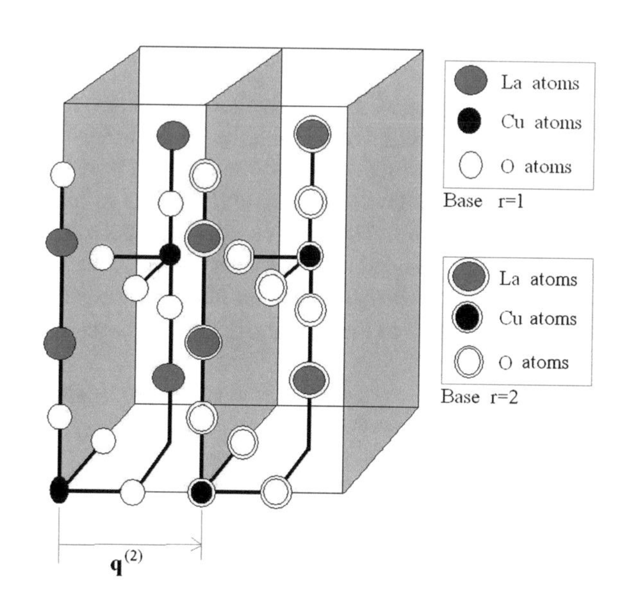

Fig. 7.3 The picture shows the structure of the basis of the La_2CuO_4 crystal. Also illustrated, is the identical basis obtained by shifting the original one in the vector $\mathbf{q}^{(2)}$. These two basis allow to reproduce the whole crystal by shifting the original one to all the vectors in the sublattice $\mathbf{R}^{(1)}$ and additionally displacing the auxiliary basis to all the points of the sublattice $\mathbf{R}^{(2)}$

In ending this section, let us consider the following important properties of any one particle operator $\widehat{O}$ which commutes with all the translations $T_{\mathbf{R}^{(r)}}$

$$\int d\mathbf{x}\overline{\Phi}_{\mathbf{k}',\eta'}(\mathbf{x},s')\widehat{O}^{(s',s)}\Phi_{\mathbf{k},\eta}(\mathbf{x},s) = \int d\mathbf{x}\overline{\Phi}_{\mathbf{k}',\eta'}(\mathbf{x},s')T_{-\mathbf{R}^{(r)}}\widehat{O}^{(s',s)}T_{\mathbf{R}^{(r)}}\Phi_{\mathbf{k},\eta}(\mathbf{x},s),$$
$$= \exp(i(\mathbf{k}-\mathbf{k}')\cdot\mathbf{R}^{(r)})\int d\mathbf{x}\overline{\Phi}_{\mathbf{k}',\eta'}(\mathbf{x},s')\widehat{O}^{(s',s)}\Phi_{\mathbf{k},\eta}(\mathbf{x},s)$$
$$= \frac{2}{N}\sum_{\mathbf{R}^{(r)}}\exp(i(\mathbf{k}-\mathbf{k}')\cdot\mathbf{R}^{(r)})\int d\mathbf{x}\overline{\Phi}_{\mathbf{k}',\eta'}(\mathbf{x},s')\widehat{O}^{(s',s)}\Phi_{\mathbf{k},\eta}(\mathbf{x},s)$$
$$= \delta_{\mathbf{k},\mathbf{k}'}\int d\mathbf{x}\overline{\Phi}_{\mathbf{k}',\eta'}(\mathbf{x},s')\widehat{O}^{(s',s)}\Phi_{\mathbf{k},\eta}(\mathbf{x},s), \tag{7.29}$$

where, thanks to the periodic boundary conditions, the sum over quasimomenta satisfying them obeys

$$\delta_{\mathbf{k},\mathbf{k}'} = \frac{2}{N}\sum_{\mathbf{R}^{(r)}}\exp(i(\mathbf{k}-\mathbf{k}')\cdot\mathbf{R}^{(r)}), \tag{7.30}$$

in which $\delta_{\mathbf{k},\mathbf{k}'}$ is the Kronecker delta function, which is a periodic quantity under shifts in the reciprocal lattice vectors for the sublattice $\mathbf{R}^{(1)}$. In what follows, the complex conjugation of any wave function Φ will be designed alternatively as Φ^* or as $\overline{\Phi}$.

7.4.1 Natural Orbitals

The general objective of this section is to define the functional space in which the GW natural orbitals will be sought for, in order to allow for solutions exhibiting spontaneous breaking of the original crystal translation invariance of the material, and in addition allowing the natural orbitals to having a spin-space non collinear entangled structure. Thus, let us consider the natural orbitals $\Psi^i_{\mathbf{k}}(\mathbf{x},t)$ as indexed by their quasimomentum $\mathbf{k}$ and a set of complementary quantum numbers i. Being Bloch orbitals in the symmetry sublattice of the AF order, they can be expressed as a series expansion in the previously defined basis of orthonormalized functions as

$$\Psi_{\mathbf{k},i}(\mathbf{x},s,t) \equiv \sum_{\eta} B^{i,\eta}_{\mathbf{k}}(t)\Phi_{\mathbf{k},\eta}(\mathbf{x},s). \tag{7.31}$$

As it can be noted, this structure allows for the solutions of the GW system of equations to show both a translational symmetry breaking in accordance with the one in the AF order, and also a non-collinear spin structure. These were the main elements allowed in the model discussed in Refs. [32–37] to describe the Mott properties of

La_2CuO_4 and more surprisingly, the nature of the pseudogap states and the quantum phase transition beneath the dome.

7.5 Using the Translation Invariance of the Sublattices

This section will make use of the invariance of the system under the translations in the sublattices vectors of $\mathbf{R}^{(1)}$. Let us write for definiteness the convention for the spatial Fourier transforms and their inverses associated to a space dependent function $f(\mathbf{x})$ as follows

$$f(\mathbf{x}) = \int \frac{d\mathbf{q}}{(2\pi)^3} f(\mathbf{q}) \exp(i\mathbf{q}\cdot\mathbf{x}), \tag{7.32a}$$

$$f(\mathbf{q}) = \int d\mathbf{x} f(\mathbf{x}) \exp(-i\mathbf{q}\cdot\mathbf{x}). \tag{7.32b}$$

But, the assumed arbitrary value of the momentum $\mathbf{q}$ may always be decomposed as a 'reduced' momentum value $\mathbf{q}_r$ pertaining to the first Brillouin cell of the reciprocal lattice (the one which is associated to the sublattice $\mathbf{R}^{(1)}$), plus a definite vector of the reciprocal lattice $\mathbf{Q}$, in the form

$$\mathbf{q} = \mathbf{q}_r + \mathbf{Q}. \tag{7.33}$$

After this definition, the Fourier expansion can be rewritten as

$$f(\mathbf{x}) = \sum_{\mathbf{Q},\mathbf{Q},'} \int \frac{d\mathbf{q}_r}{(2\pi)^3} f(\mathbf{q}_r, \mathbf{Q}) \exp(i\mathbf{q}_r + \mathbf{Q}).\mathbf{x}), \tag{7.34a}$$

$$f(\mathbf{q}_r, \mathbf{Q}) = \int d\mathbf{x} f(\mathbf{x}) \exp(-i(\mathbf{q}_r + \mathbf{Q}).\mathbf{x}), \tag{7.34b}$$

$$f(\mathbf{q}_r, \mathbf{Q}) \equiv f(\mathbf{q}_r + \mathbf{Q}). \tag{7.34c}$$

Let us now consider a spatial kernel $K(\mathbf{x}_1, \mathbf{x}_2)$ and its double Fourier expansion over its two spatial variables

$$K(\mathbf{x}, \mathbf{x}') = \sum_{\mathbf{Q},\mathbf{Q}'} \int \frac{d\mathbf{q}_r}{(2\pi)^3} \frac{d\mathbf{q}'_r}{(2\pi)^3} K(\mathbf{q}_r, \mathbf{Q}; \mathbf{q}'_r, \mathbf{Q}') \times \exp(i(\mathbf{q}_r.\mathbf{x} + \mathbf{q}'_r.\mathbf{x}' + \mathbf{Q}.\mathbf{x} + \mathbf{Q}'.\mathbf{x}')). \tag{7.35}$$

Now we will assume that the considered kernel represents a correlation function of the system which is assumed to be translationally invariant under spatial shifts in arbitrary vectors pertaining to the sublattice $\mathbf{R}^{(1)}$. Considering that the vectors $\mathbf{Q}$ and $\mathbf{Q}'$ pertain to the reciprocal lattice allows to write

$$K(\mathbf{x},\mathbf{x}') = \frac{1}{N_c}\sum_{\mathbf{R}^{(1)}} K(\mathbf{x}+\mathbf{R}^{(1)},\mathbf{x}'+\mathbf{R}^{(1)})$$
$$= \sum_{\mathbf{Q}}\int \frac{d\mathbf{q}_r}{(2\pi)^3}\frac{d\mathbf{q}'_r}{(2\pi)^3} K(\mathbf{q}_r,\mathbf{Q};\mathbf{q}'_r,\mathbf{Q}')\exp(i(\mathbf{q}_r\cdot\mathbf{x}+\mathbf{q}'_r\cdot\mathbf{x}'+\mathbf{Q}\cdot\mathbf{x}+\mathbf{Q}'\cdot\mathbf{x}'))$$
$$\times\frac{1}{N_c}\sum_{\mathbf{R}^{(1)}}\exp(i(\mathbf{q}_r+\mathbf{q}'_r)\cdot\mathbf{R}^{(1)}). \tag{7.36}$$

But, employing the identities

$$\delta^{(K)}_{\mathbf{q}_r+\mathbf{q}'_r,0} = \frac{1}{N_c}\sum_{\mathbf{R}^{(1)}}\exp(i(\mathbf{q}_r+\mathbf{q}'_r)\cdot\mathbf{R}^{(1)}), \tag{7.37a}$$

$$\int\frac{d\mathbf{q}_r}{(2\pi)^3} \equiv \frac{1}{N_c V_c}\sum_{\mathbf{q}_r}, \tag{7.37b}$$

in which $\delta^{(K)}_{\mathbf{q},\mathbf{q}'}$ is the Kronecker Delta in the indices $\mathbf{q}$, $\mathbf{q}'$, and V_c, N_c is the volume of the unit cell of the sublattice $\mathbf{R}^{(1)}$ and the number of these unit cells which are included in the periodicity region. Then, the kernel can be written in the form

$$K(\mathbf{x},\mathbf{x}') =$$
$$\sum_{\mathbf{Q},\mathbf{Q}'}\int\frac{d\mathbf{q}_r}{(2\pi)^3}\frac{d\mathbf{q}'_r}{(2\pi)^3}\delta^{(K)}_{\mathbf{q}_r+\mathbf{q}'_r,0}K(\mathbf{q}_r,\mathbf{Q};-\mathbf{q}_r,\mathbf{Q}')\exp(i(\mathbf{q}_r\cdot(\mathbf{x}-\mathbf{x}')+\mathbf{Q}\cdot\mathbf{x}+\mathbf{Q}'\cdot\mathbf{x}'))$$
$$= \sum_{\mathbf{Q},\mathbf{Q}'}\int\frac{d\mathbf{q}_r}{(2\pi)^3}\frac{1}{N_c V_c}K(\mathbf{q}_r,\mathbf{Q};-\mathbf{q}_r,\mathbf{Q}')\exp(i(\mathbf{q}_r\cdot(\mathbf{x}-\mathbf{x}')+\mathbf{Q}\cdot\mathbf{x}+\mathbf{Q}'\cdot\mathbf{x}'))$$
$$= \sum_{\mathbf{Q},\mathbf{Q}'}\int\frac{d\mathbf{q}_r}{(2\pi)^3}K(\mathbf{q}_r,\mathbf{Q},\mathbf{Q}')\exp(i(\mathbf{q}_r\cdot(\mathbf{x}-\mathbf{x}')+\mathbf{Q}\cdot\mathbf{x}+\mathbf{Q}'\cdot\mathbf{x}')), \tag{7.38}$$

where the following quantity was defined

$$K(\mathbf{q}_r,\mathbf{Q},\mathbf{Q}') = \frac{1}{N_c V_c}K(\mathbf{q}_r,\mathbf{Q};-\mathbf{q}_r,\mathbf{Q}'). \tag{7.39}$$

Let us now consider the convolution of two kernels

$$K_3(\mathbf{x},\mathbf{x}') = \int d\mathbf{z}K_1(\mathbf{x},\mathbf{z})K_2(\mathbf{z},\mathbf{x}')$$
$$= \sum_{\mathbf{Q}_1,\mathbf{Q}'_1}\int\frac{d\mathbf{q}^1{}_r}{(2\pi)^3}K_1(\mathbf{q}^1{}_r,\mathbf{Q}_1,\mathbf{Q}'_1)\exp(i(\mathbf{q}^1{}_r\cdot\mathbf{x}+\mathbf{Q}_1\cdot\mathbf{x}))$$

$$\times \sum_{\mathbf{Q}_2,\mathbf{Q}_2'} \int \frac{d\mathbf{q}^2{}_r}{(2\pi)^3} K_2(\mathbf{q}^2{}_r, \mathbf{Q}_2, \mathbf{Q}_2') \exp(i(-\mathbf{q}^2{}_r \cdot \mathbf{x}' + \mathbf{Q}_2' \cdot \mathbf{x}'))$$

$$\times \int d\mathbf{z} \exp(i(-\mathbf{q}^1{}_r + \mathbf{q}^2{}_r + \mathbf{Q}_1' + \mathbf{Q}_2) \cdot \mathbf{z})$$

$$= \sum_{\mathbf{Q}_1,\mathbf{Q}_1'} \int \frac{d\mathbf{q}^1{}_r}{(2\pi)^3} K_1(\mathbf{q}^1{}_r, \mathbf{Q}_1, \mathbf{Q}_1') \exp(i(\mathbf{q}^1{}_r \cdot \mathbf{x} + \mathbf{Q}_1 \cdot \mathbf{x}))$$

$$\times \sum_{\mathbf{Q}_2,\mathbf{Q}_2'} \int d\mathbf{q}^2{}_r K_2(\mathbf{q}^2{}_r, \mathbf{Q}_2, \mathbf{Q}_2') \exp(i(-\mathbf{q}^2{}_r \cdot \mathbf{x}' + \mathbf{Q}_2' \cdot \mathbf{x}'))$$

$$\times \delta^{(K)}_{\mathbf{Q}_1'+\mathbf{Q}_2,0} \delta^{(D)}(\mathbf{q}^2{}_r - \mathbf{q}^1{}_r)$$

$$= \sum_{\mathbf{Q}_1,\mathbf{Q}_2'} \int \frac{d\mathbf{q}^1{}_r}{(2\pi)^3} \sum_{\mathbf{Q}_2} K_1(\mathbf{q}^1{}_r, \mathbf{Q}_1, -\mathbf{Q}_2) K_2(\mathbf{q}^1{}_r, \mathbf{Q}_2, \mathbf{Q}_2')$$

$$\times \exp(i(\mathbf{q}^1{}_r \cdot (\mathbf{x} - \mathbf{x}') + \mathbf{Q}_1 \cdot \mathbf{x} + \mathbf{Q}_2' \cdot \mathbf{x}')). \tag{7.40}$$

Thus, the quantities $K_1(\mathbf{q}_r, \mathbf{Q}_1, \mathbf{Q}_2)$ and $K_2(\mathbf{q}_r, \mathbf{Q}_1, \mathbf{Q}_2)$ being associated to the spatio temporal kernels $K_1(\mathbf{x}, \mathbf{x}')$ and $K_2(\mathbf{x}, \mathbf{x}')$ define the quantity $K_3(\mathbf{q}_r, \mathbf{Q}_1, \mathbf{Q}_2)$ related to the product kernel $K_3(\mathbf{x}, \mathbf{x}') = \int d\mathbf{z}\, K_1(\mathbf{x}, \mathbf{z})\, K_2(\mathbf{z}, \mathbf{x}')$ through the special matrix product formula

$$K_3(\mathbf{q}_r, \mathbf{Q}_1, \mathbf{Q}_2) = \sum_{\mathbf{Q}} K_1(\mathbf{q}_r, \mathbf{Q}_1, -\mathbf{Q}) K_2(\mathbf{q}_r, \mathbf{Q}, \mathbf{Q}_2). \tag{7.41}$$

Note that above, in order to simplify, we had to consider that when the kernel under study is a Fermi one, the coordinates in the above formulas, such as $\mathbf{x}_1$ for example, in fact design both the real space coordinates and the spin projection: $(\mathbf{x}_1, s_1)$. Accordingly, the integration $\int d\mathbf{x}_1$ in fact will mean the usual integration in addition to the sum over spin projections as $\sum_{s_1=\pm 1} \int d\mathbf{x}_1$. On the other hand, when the kernel is associated to a boson field, the coordinates $\mathbf{x}_1$ and integrations will mean the usual ones.

7.6 Reducing the Equations to a Matrix Problem for Each Quasi-momentum Value

Finally, in this section the GW equations will be reduced to a set of matrix equations, one for the quasimomentum value $\mathbf{q}$. For this purpose, let us now project the set of Eq. (7.18) on the basis functions defined in Eq. (7.28a) as

$$\overline{\Phi}_{\mathbf{k}_i,\eta}(\mathbf{1}) H_0(\mathbf{1}, \mathbf{2}) \Psi_{\mathbf{k}_i,i}(\mathbf{2}) - \overline{\Phi}_{\mathbf{k}_i,\eta}(\mathbf{1}) \Sigma(\mathbf{1}, \mathbf{2}) \Psi_{\mathbf{k}_i,i}(\mathbf{2}) = E_{\mathbf{k}_i,i} \overline{\Phi}_{\mathbf{k}_i,\eta}(\mathbf{1}) \Psi_{\mathbf{k}_i,i}(\mathbf{2}). \tag{7.42}$$

But, the natural orbitals $\Psi_{k_i,i}$ can be expressed as superpositions of the basis functions defined in (7.28a) accordingly with

$$\Psi_{\mathbf{k}_i,i}(\mathbf{1}) = \sum_{\eta} B_{\mathbf{k}_i}^{i,\eta} \Phi_{\mathbf{k}_i,\eta}(\mathbf{1}), \tag{7.43a}$$

$$\overline{\Psi}_{\mathbf{k}_i,i}(\mathbf{1}) = \sum_{\eta} \overline{\Phi}_{\mathbf{k}_i,\eta}(\mathbf{1}) B_{\mathbf{k}_i}^{\dagger\eta,i}, \tag{7.43b}$$

where for a matrix $A^{k,l}$ its hermitian conjugate is defined as $\mathrm{A}^{\dagger k,l} = \overline{A}^{l,k}$. Substituting these expressions in Eq. (7.42) leads to the following set of matrix equations

$$\left(H_0^{\eta,\eta'}(\mathbf{k}_i) - \Sigma^{\eta,\eta'}(\mathbf{k}_i)\right) B_{\mathbf{k}_i}^{i,\eta'} = E_{\mathbf{k}_i,i} I_{\mathbf{k}_i}^{\eta,\eta'} B_{\mathbf{k}_i}^{i,\eta'}, \tag{7.44}$$

in which the $H_0^{\eta,\eta'}$ and the overlapping $I_{\mathbf{k}_i}^{\eta,\eta'}$ matrices entering are defined as

$$H_0^{\eta,\eta'}(\mathbf{k}_i) = \overline{\Phi}_{\mathbf{k}_i,\eta}(\mathbf{1}) H_0(\mathbf{1},\mathbf{2}) \Phi_{\mathbf{k}_i,\eta'}(\mathbf{2}), \tag{7.45a}$$

$$I_{\mathbf{k}_i}^{\eta,\eta'} = \overline{\Phi}_{\mathbf{k}_i,\eta}(\mathbf{1}) \Phi_{\mathbf{k}_i,\eta'}(\mathbf{1}). \tag{7.45b}$$

Since $\overline{\Phi}_{\mathbf{k}_l,\eta}$ are only approximately orthonormalized, the normalization conditions for the functions Ψ takes the form

$$\sum_{\eta,\eta'} B_{\mathbf{k}_i}^{i,\eta'} I_{\mathbf{k}_i}^{\eta',\eta} B_{\mathbf{k}_i}^{\eta,i} = \delta^{i,i'}. \tag{7.46}$$

The expression for the matrix associated to the selfenergy can be written as follows

$$\begin{aligned}\Sigma^{\eta\eta'}(k_i) &= \overline{\Phi}_{k_i,\eta}(\mathbf{1}) \Sigma(\mathbf{1},\mathbf{2}) \Phi_{k_i,\eta'}(\mathbf{2}) \\ &= -\int\int d\mathbf{x}_1 \mathbf{dx}_2 \overline{\Phi}_{k_i,\eta}(\mathbf{x}_1) \widetilde{W}(\mathbf{x}_1,\mathbf{x}_2) \sum_{i}^{\mu \geq E_i} \Psi_i(\mathbf{x}_1) \overline{\Psi}_i(\mathbf{x}_2) \Phi_{\mathbf{k}_i,\eta}(\mathbf{x}_2).\end{aligned} \tag{7.47}$$

But, using the representation Eq. (7.38) for the static effective potential kernel $\widetilde{W}(\mathbf{x}_1, \mathbf{x}_2)$, the expression for the matrix associated to the selfenergy Σ can rewritten as

$$\begin{aligned}\Sigma^{\eta\eta'}(k_i) = &-\sum_{\mathbf{Q}_1,\mathbf{Q}_2} \int \frac{d\mathbf{q}_r}{(2\pi)^3} \widetilde{W}(\mathbf{q}_r, \mathbf{Q}_1, \mathbf{Q}_2) \sum_{j}^{\mu \geq E_j} B_{\mathbf{k}_j}^{j,\eta''} B_{\mathbf{k}_j}^{\dagger\eta''',j} \\ &\times \int d\mathbf{x}_1 \exp(i\mathbf{x}_1 \cdot (\mathbf{q}_r + \mathbf{Q}_1)) \Phi_{\mathbf{k}_j,\eta''}(\mathbf{x}_1) \overline{\Phi}_{\mathbf{k}_i,\eta}(\mathbf{x}_1) \\ &\times \int \mathbf{dx}_2 \exp(i\mathbf{x}_2 \cdot (-\mathbf{q}_r + \mathbf{Q}_2)) \overline{\Phi}_{\mathbf{k}_j,\eta'''}(\mathbf{x}_2) \Phi_{\mathbf{k}_i,\eta}(\mathbf{x}_2).\end{aligned} \tag{7.48}$$

Now, the Bloch character of the entering wavefunctions and translational invariance,

$$\Phi_{k_i,\eta}(\mathbf{1}+\mathbf{R}^{(1)}) = \exp(i\mathbf{R}^{(1)}\cdot\mathbf{k}_i)\Phi_{k_i,\eta}(\mathbf{1}), \tag{7.49a}$$

$$\widetilde{W}(\mathbf{1},\mathbf{2}) = \widetilde{W}(\mathbf{1}+\mathbf{R}^{(1)},\mathbf{2}+\mathbf{R}^{(1)}), \tag{7.49b}$$

$$\widetilde{\Sigma}(\mathbf{1},\mathbf{2}) = \widetilde{\Sigma}(\mathbf{1}+\mathbf{R}^{(1)},\mathbf{2}+\mathbf{R}^{(1)}), \tag{7.49c}$$

can be used after replacing the integration variables as $\mathbf{x}_1 \rightarrow \mathbf{x}_1 + \mathbf{R}^{(1)}$ (or $\mathbf{x}_2 \rightarrow \mathbf{x}_2 + \mathbf{R}^{(1)}$) and summing over the $\mathbf{R}^{(1)}$ (by also dividing by the number of cells in the periodicity region) to show that $\Sigma^{\eta\eta'}(k_i)$ is proportional to the following factor

$$\frac{1}{N_c}\sum_{\mathbf{R}^{(1)}}\exp(i\mathbf{R}^{(1)}\cdot(\mathbf{q}_r+\mathbf{k}_j-\mathbf{k}_i)) = \delta^{(K)}_{\mathbf{q}+\mathbf{k}_j-\mathbf{k}_i,0}. \tag{7.50}$$

The use of this relation allows to write for the selfenergy matrix

$$\begin{aligned}\Sigma^{\eta\eta'}(k_i) = &-\frac{1}{N_c V_c}\sum_{\mathbf{Q}_1,\mathbf{Q}_2}\sum_{j}^{\mu\geq E_j}\widetilde{W}(\mathbf{k}_i-\mathbf{k}_j,\mathbf{Q}_1,\mathbf{Q}_2)B^{j,\eta''}_{\mathbf{k}_j}B^{\dagger\eta''',j}_{\mathbf{k}_j}\\ &\times\int d\mathbf{x}_1\exp(i\mathbf{x}_1\cdot(\mathbf{k}_i-\mathbf{k}_j+\mathbf{Q}_1))\Phi_{\mathbf{k}_j,\eta''}(\mathbf{x}_1)\overline{\Phi}_{\mathbf{k}_i,\eta}(\mathbf{x}_1)\\ &\times\int \mathbf{dx}_2\exp(i\mathbf{x}_2\cdot(\mathbf{k}_j-\mathbf{k}_i+\mathbf{Q}_2))\overline{\Phi}_{\mathbf{k}_j,\eta'''}(\mathbf{x}_2)\Phi_{\mathbf{k}_i,\eta}(\mathbf{x}_2).\end{aligned} \tag{7.51}$$

A difficult calculational aspect in the above relation is the fact that the effective potential W also depends on the expansion coefficients $B^{j,\eta}_{\mathbf{k}_j}$ to be determined. In order to evaluate W it is necessary to calculate the matrix $\widetilde{W}(\mathbf{k})$ with matrix indices $\mathbf{Q}_1,\mathbf{Q}_2$ defined by

$$\widetilde{W}(\mathbf{k}) \equiv \widetilde{W}^{\mathbf{Q}_1,\mathbf{Q}_2}(\mathbf{k}) = \widetilde{W}(\mathbf{k},\mathbf{Q}_1,\mathbf{Q}_2), \tag{7.52}$$

by using the formula

$$\widetilde{W}(\mathbf{k}) = \frac{1}{1-\widetilde{v}(\mathbf{k})\widetilde{P}(\mathbf{k})}, \tag{7.53}$$

where the matrices $\widetilde{v}$ and $\widetilde{P}$ entering the matrix variant of the geometric series formula are defined by

$$\begin{aligned}\widetilde{P}(\mathbf{k}) &\equiv \widetilde{P}^{\mathbf{Q}_1,\mathbf{Q}_2}(\mathbf{k}) = \widetilde{P}(\mathbf{k},\mathbf{Q}_1,\mathbf{Q}_2),\\ \widetilde{v}(\mathbf{k}) &\equiv \widetilde{v}^{\mathbf{Q}_1,\mathbf{Q}_2}(\mathbf{k}) = v(\mathbf{k}+\mathbf{Q}_1)\delta^{(K)}_{-\mathbf{Q}_2,\mathbf{Q}_1},\\ I &\equiv \delta^{(K)}_{-\mathbf{Q}_2,\mathbf{Q}_1}.\end{aligned}$$

where $v(\mathbf{k})$ is simply the Fourier transform of the Coulomb potential.

Then, after evaluating the double Fourier transform of $\widetilde{P}(\mathbf{x}_1, \mathbf{x}_2)$, Eq. (7.63) in the Appendix provides an expression for the quantity $\widetilde{P}(\mathbf{k}, \mathbf{Q}_1, \mathbf{Q}_2)$ as

$$\begin{aligned}
&\widetilde{P}^{\mathbf{Q}_1,\mathbf{Q}_2}(\mathbf{q}_r) \\
&= -\sum_{(\mathbf{k}_i,i)}\sum_{(\mathbf{k}_j,j)} \frac{\theta(E_{(\mathbf{k}_i,i)}-\mu)\theta(\mu-E_{(\mathbf{k}_i-\mathbf{q}_r,j)})-\theta(\mu-E_{(\mathbf{k}_i,i)})\theta(E_{(\mathbf{k}_i-\mathbf{q}_r,j)}-\mu)}{E_{(\mathbf{k}_i,i)}-E_{(\mathbf{k}_i-\mathbf{q}_r,j)}} \\
&\quad\times N_c^2 \int_{C_0} d\mathbf{x}_1 \int_{C_0} d\mathbf{x}_2 \exp(-i\mathbf{q}_r\cdot(\mathbf{x}_1-\mathbf{x}_2)-i\mathbf{Q}_1\cdot\mathbf{x}_1-i\mathbf{Q}_2\cdot\mathbf{x}_2) \\
&\quad\times \Phi_{\mathbf{k}_i,\eta_1}(\mathbf{x}_1)\overline{\Phi}_{\mathbf{k}_i-\mathbf{q}_r,\eta_2}(\mathbf{x}_1)\overline{\Phi}_{\mathbf{k}_i\eta_1'}(\mathbf{x}_2)\Phi_{\mathbf{k}_i-\mathbf{q}_r,\eta_2'}(\mathbf{x}_2) \\
&\quad\times B_{\mathbf{k}_i}^{i,\eta_1} B_{\mathbf{k}_i-\mathbf{q}_r}^{\dagger\eta_2,j} B_{\mathbf{k}_i}^{\dagger\eta_1',i} B_{\mathbf{k}_i-\mathbf{q}_r}^{j,\eta_2'},
\end{aligned} \tag{7.54}$$

where Eq. (7.31) for the normal states has been used, and in which C_0 indicates a unit cell of the sublattice of points $\mathbf{R}^{(1)}$, say the one having the origin of coordinates as its central point. In reducing the integrals to the small region C_0 it was required to make use of the translational symmetry of the problem.

Therefore, the derived formulas for the selfenergy and polarization as functionals of the coefficients $B_{\mathbf{k}}^{\eta,i}$ allow to determine all quantities required to solve the matrix equations for such matrix of coefficients $B_{\mathbf{k}}^{\eta,i}$ for each fixed momentum value $\mathbf{k}$. Since these coefficients determine the natural orbitals, with this, all the elements required for implementing an iterative evaluation of the band problem are determined. The developing of a numerical implementation of the proposed band calculation method is expected to be considered in coming extensions of the work.

7.7 Summary

We presented the building blocks of a GW kind of band calculation scheme allowing the breaking of translation invariance and the non collinearity of the self-consistent natural orbitals. The starting crystal structure assumed was the one corresponding to La_2CuO_4, in order to prepare the way for its application to this Mott insulator, which was predicted to be a metal in the original Mattheiss band calculations [29]. A reduced crystal symmetry equal to the one exhibited by the known AF order of the material is assumed. This element implements the mentioned possibility for a spatial symmetry breaking. The non collinearity of the resulting natural orbitals is also incorporated by allowing the natural orbitals to have spin projection ± 1 both non vanishing. This freedom made it possible to derive the Mott state and a pseudogap in the model investigated in Refs. [32–37]. The fact that the here proposed static GW expansion directly reduces to the model in question after assuming that the GW effective dielectric response is approximated by the phenomenological constant value of the dielectric constant assumed in the model, strongly suggests that the results of an ab initio band evaluation employing the described GW method, can predict the

Mott properties of La_2CuO_4. The results could furnish a first principles evaluation of the Mott properties for this SCES material. The study of this question will be further considered elsewhere.

Acknowledgements It is for me a real pleasure to participate in this tribute in honor of dear Professor N. H. March, who so much had contributed not only to theoretical condensed matter physics but also to support the formation of Third World physicists. In particular I always remember those visits to the International Center for Theoretical Physics (ICTP, Trieste, Italy) during the eighties, in which the participants coming from developing countries were almost 'forced' to communicate and collaborate among them within seminars and discussion meetings organized by Professors March, Butcher, Garcia-Moliner, etc. In my case, I clearly remember a final meeting with Professors March and Butcher at the end of the first of those visits, in which my motivation was to start applying quantum field theory (QFT) methods in condensed matter problems. In that encounter, they recommended me to consider the investigation of the physics of inhomogeneous electron systems, a line of work that I followed up to nowadays. In particular, this work is closely connected with that initial recommendation.

Appendix

The spatial representation of the static polarization kernel is given by Eq. (7.18d)

$$\widetilde{P}(\mathbf{1},\mathbf{2}) = -\sum_i \sum_j \Psi_i(\mathbf{1})\overline{\Psi}_i(\mathbf{2})\Psi_j(\mathbf{2})\overline{\Psi}_j(\mathbf{1}) \times \frac{\theta(E_i-\mu)\theta(\mu-E_j)-\theta(\mu-E_i)\theta(E_j-\mu)}{E_i-E_j}. \tag{7.55}$$

The expression of its double Fourier transform is

$$\widetilde{P}(\mathbf{q},\mathbf{q}') = \int\int d\mathbf{x}d\mathbf{x}' \exp(-i(\mathbf{q}\cdot\mathbf{x}+\mathbf{q}'\cdot\mathbf{x}'))\widetilde{P}(\mathbf{x},\mathbf{x}'). \tag{7.56}$$

Equation (7.55) thus shows that it is necessary to evaluate the integral

$$I = \int d\mathbf{x}\exp(-i\mathbf{q}\cdot\mathbf{x})\Psi_i(\mathbf{x})\overline{\Psi_j}(\mathbf{x})\int d\mathbf{x}'\exp(-i\mathbf{q}'\cdot\mathbf{x}')\overline{\Psi}_i(\mathbf{x}')\Psi_j(\mathbf{x}'). \tag{7.57}$$

The Bloch property of the functions in the sublattice $\mathbf{R}^{(1)}$ reads

$$\Psi_i(\mathbf{x}+\mathbf{R}^{(1)}) = \exp(i\mathbf{k}_i\cdot\mathbf{R}^{(1)})\Psi_i(\mathbf{x}), \tag{7.58a}$$

$$\overline{\Psi}_i(\mathbf{x}+\mathbf{R}^{(1)}) = \exp(-i\mathbf{k}_i\cdot\mathbf{R}^{(1)})\overline{\Psi}_i(\mathbf{x}). \tag{7.58b}$$

Further, the double integral can be decomposed as a double sum of integrals each taken over a unit cell C_0 of the sublattice as

$$\int\int d\mathbf{x}d\mathbf{x}'F(\mathbf{x},\mathbf{x}') = \sum_{\mathbf{R}^{(1)}}\sum_{\mathbf{R}'^{(1)}}\int_{C_0}\int_{C_0} d\mathbf{x}d\mathbf{x}'F(\mathbf{x}+\mathbf{R}^{(1)},\mathbf{x}'+\mathbf{R}'^{(1)}). \quad (7.59)$$

Now, using the Bloch property and expressing the two arbitrary momenta $\mathbf{q}$ and $\mathbf{q}'$ as decomposed in each part contained in the first Brillouin cell plus a reciprocal lattice vector, allows to write I as follows

$$\begin{aligned}
I &= \sum_{\mathbf{R}^{(1)}}\sum_{\mathbf{R}'^{(1)}}\int_{C_0} d\mathbf{x}\exp(-i\mathbf{q}\cdot(\mathbf{x}+\mathbf{R}^{(1)}))\Psi_i(\mathbf{x}+\mathbf{R}^{(1)})\overline{\Psi_j}(\mathbf{x}+\mathbf{R}^{(1)}) \\
&\quad\times\int_{C_0} d\mathbf{x}'\exp(-i\mathbf{q}'\cdot(\mathbf{x}'+\mathbf{R}'^{(1)}))\overline{\Psi}_i(\mathbf{x}'+\mathbf{R}'^{(1)})\Psi_j(\mathbf{x}'+\mathbf{R}'^{(1)}) \\
&= \sum_{\mathbf{R}^{(1)}}\sum_{\mathbf{R}'^{(1)}}\int_{C_0} d\mathbf{x}\exp(-i\mathbf{q}_r\cdot\mathbf{x}-i\mathbf{Q}\cdot\mathbf{x})\Psi_i(\mathbf{x})\overline{\Psi_j}(\mathbf{x})\exp(i(-\mathbf{q}_r+\mathbf{k}_i-\mathbf{k}_j)\cdot\mathbf{R}^{(1)}) \\
&\quad\times\int_{C_0} d\mathbf{x}'\exp(-i\mathbf{q}_r'\cdot\mathbf{x}'-i\mathbf{Q}'\cdot\mathbf{x}')\overline{\Psi}_i(\mathbf{x}')\Psi_j(\mathbf{x}')\exp(-i(\mathbf{q}_r'+\mathbf{k}_i-\mathbf{k}_j)\cdot\mathbf{R}'^{(1)}) \\
&= N_c^2\delta^{(K)}_{\mathbf{q}_r-\mathbf{k}_i+\mathbf{k}_j,0}\delta^{(K)}_{\mathbf{q}_r'+\mathbf{k}_i-\mathbf{k}_j,0}\int_{C_0} d\mathbf{x}\exp(-i\mathbf{q}_r\cdot\mathbf{x}-i\mathbf{Q}\cdot\mathbf{x})\Psi_i(\mathbf{x})\overline{\Psi_j}(\mathbf{x}) \\
&\quad\times\int_{C_0} d\mathbf{x}'\exp(-i\mathbf{q}_r'\cdot\mathbf{x}'-i\mathbf{Q}'\cdot\mathbf{x}')\overline{\Psi}_i(\mathbf{x}')\Psi_j(\mathbf{x}'). \quad (7.60)
\end{aligned}$$

That is, the integral I vanishes if $\mathbf{q}_r$ and $-\mathbf{q}_r'$ are not equal among them and equal to $\mathbf{k}_i-\mathbf{k}_j$. That is $\mathbf{q}_r$ should be equal to the difference between the quasi-momenta associated to the wavefunctions of the basis. After explicitly writing the momentum quantum numbers by substituting $i\to(\mathbf{k}_i,i)$ and $j\to(\mathbf{k}_j,j)$ in the formula for the double Fourier transform of the static polarization kernel, we can write

$$\begin{aligned}
\widetilde{P}(\mathbf{q},\mathbf{q}') &\equiv \widetilde{P}(\mathbf{q}_r,\mathbf{Q},\mathbf{q}_r',\mathbf{Q}') \\
&= -\sum_{\mathbf{k}_i}\sum_{i}\sum_{j}\frac{\theta(E_{(\mathbf{k}_i i)}-\mu)\theta(\mu-E_{(\mathbf{k}_i-\mathbf{q}_r,j)})-\theta(\mu-E_{\mathbf{k}_i i})\theta(E_{(\mathbf{k}i-\mathbf{q}_r,j)}-\mu)}{E_{(\mathbf{k}_i i)}-E_{(\mathbf{k}i-\mathbf{q}_r,j)}} \\
&\quad\times N_c^2\delta_{\mathbf{q}_r'+\mathbf{q}_r,0}\int_{C_0} d\mathbf{x}\exp(-i\mathbf{q}_r\cdot\mathbf{x}-i\mathbf{Q}\cdot\mathbf{x})\Psi_{(\mathbf{k}_i,i)}(\mathbf{x})\overline{\Psi}_{(\mathbf{k}_i-\mathbf{q}_r,j)}(\mathbf{x}) \\
&\quad\times\int_{C_0} d\mathbf{x}'\exp(i\mathbf{q}_r\cdot\mathbf{x}'-i\mathbf{Q}'\cdot\mathbf{x}')\overline{\Psi}_{(\mathbf{k}_i,i)}(\mathbf{x}')\Psi_{(\mathbf{k}i-\mathbf{q}_r,j)}(\mathbf{x}') \\
&= -N_c^2\delta_{\mathbf{q}_r'+\mathbf{q}_r,0}\sum_{\mathbf{k}_i}\sum_{i}\sum_{j}\frac{\theta(E_{(\mathbf{k}_i i)}-\mu)\theta(\mu-E_{(\mathbf{k}_i-\mathbf{q}_r,j)})-\theta(\mu-E_{\mathbf{k}_i i})\theta(E_{(\mathbf{k}i-\mathbf{q}_r,j)}-\mu)}{E_{(\mathbf{k}_i i)}-E_{(\mathbf{k}_i-\mathbf{q}_r,j)}} \\
&\quad\times\int_{C_0} d\mathbf{x}\exp(-i\mathbf{q}_r\cdot\mathbf{x}-i\mathbf{Q}\cdot\mathbf{x})\Psi_{(\mathbf{k}_i,i)}(\mathbf{x})\overline{\Psi}_{(\mathbf{k}_i-\mathbf{q}_r,j)}(\mathbf{x}) \\
&\quad\times\int_{C_0} d\mathbf{x}'\exp(i\mathbf{q}_r\cdot\mathbf{x}'-i\mathbf{Q}'\cdot\mathbf{x}')\overline{\Psi}_{(\mathbf{k}_i,i)}(\mathbf{x}')\Psi_{(\mathbf{k}_i-\mathbf{q}_r,j)}(\mathbf{x}'). \quad (7.61)
\end{aligned}$$

Now, it is possible to invert the Fourier formula for expressing the static kernel as

$$
\begin{aligned}
\widetilde{P}(\mathbf{x},\mathbf{x}') &= \sum_{\mathbf{Q}}\sum_{\mathbf{Q}'}\int\int \frac{d\mathbf{q}_r}{(2\pi)^3}\frac{d\mathbf{q}'_r}{(2\pi)^3}\exp(i(\mathbf{q}_r\cdot\mathbf{x}+\mathbf{q}'_r\cdot\mathbf{x}'+\mathbf{Q}\cdot\mathbf{x}+\mathbf{Q}'\cdot\mathbf{x}'))\widetilde{P}(\mathbf{q}_r,\mathbf{Q},\mathbf{q}'_r,\mathbf{Q}') \\
&= \sum_{\mathbf{Q}}\sum_{\mathbf{Q}'}\int\int \frac{d\mathbf{q}_r}{(2\pi)^3}\frac{d\mathbf{q}'_r}{(2\pi)^3}\exp(i(\mathbf{q}_r\cdot\mathbf{x}+\mathbf{q}'_r\cdot\mathbf{x}'+\mathbf{Q}\cdot\mathbf{x}+\mathbf{Q}'\cdot\mathbf{x}')) \\
&\times\Bigg(-N_c^2\delta_{\mathbf{q}'_r+\mathbf{q}_r,0}\sum_{\mathbf{k}_i}\sum_i\sum_j\frac{\theta(E_{(\mathbf{k}_i,i)}-\mu)\theta(\mu-E_{\mathbf{k}i-\mathbf{q}_r,j})-\theta(\mu-E_{\mathbf{k}ii})\theta(E_{\mathbf{k}i-\mathbf{q}_r,j}-\mu)}{E_{(\mathbf{k}_i i)}-E_{(\mathbf{k}_i-\mathbf{q}_r,j)}} \\
&\quad\times\int_{C_0}d\mathbf{x}\exp(-i\mathbf{q}_r\cdot\mathbf{x}-i\mathbf{Q}\cdot\mathbf{x})\Psi_{(\mathbf{k}_i,i)}(\mathbf{x})\overline{\Psi}_{(\mathbf{k}_i-\mathbf{q}_r,j)}(\mathbf{x}) \\
&\quad\times\int_{C_0}d\mathbf{x}'\exp(i\mathbf{q}_r\cdot\mathbf{x}'-i\mathbf{Q}'\cdot\mathbf{x}')\overline{\Psi}_{(\mathbf{k}_i,i)}(\mathbf{x}')\Psi_{(\mathbf{k}_i-\mathbf{q}_r,j)}(\mathbf{x}')\Bigg) \\
&= \sum_{\mathbf{Q}}\sum_{\mathbf{Q}'}\int\frac{d\mathbf{q}_r}{(2\pi)^3}\exp(i(\mathbf{q}_r\cdot(\mathbf{x}-\mathbf{x}')+\mathbf{Q}\cdot\mathbf{x}+\mathbf{Q}'\cdot\mathbf{x}'))\widetilde{P}(\mathbf{q}_r,\mathbf{Q},\mathbf{Q}'). \qquad (7.62)
\end{aligned}
$$

The last line defines the associated function $\widetilde{P}(\mathbf{q}_r,\mathbf{Q},\mathbf{Q}')$ and the matrix $\widetilde{P}^{\mathbf{Q},\mathbf{Q}'}(\mathbf{q}_r)$ for the polarization as follows

$$
\begin{aligned}
\widetilde{P}(\mathbf{q}_r,\mathbf{Q},\mathbf{Q}') &= -\frac{1}{N_cV_c}\sum_{\mathbf{k}_i}\sum_i\sum_j\frac{\theta(E_{(\mathbf{k}_i,i)}-\mu)\theta(\mu-E_{(\mathbf{k}_i-\mathbf{q}_r,j)})-\theta(\mu-E_{(\mathbf{k}_i,i)})\theta(E_{(\mathbf{k}_i-\mathbf{q}_r,j)}-\mu)}{E_{(\mathbf{k}_i,i)}-E_{(\mathbf{k}_i-\mathbf{q}_r,j)}} \\
&\times N_c^2\int_{C_0}d\mathbf{x}\exp(-i\mathbf{q}_r\cdot\mathbf{x}-i\mathbf{Q}\cdot\mathbf{x})\Psi_{(\mathbf{k}_i,i)}(\mathbf{x})\overline{\Psi}_{(\mathbf{k}_i-\mathbf{q}_r,j)}(\mathbf{x}) \\
&\times\int_{C_0}d\mathbf{x}'\exp(i\mathbf{q}_r\cdot\mathbf{x}'-i\mathbf{Q}'\cdot\mathbf{x}')\overline{\Psi}_{(\mathbf{k}_i,i)}(\mathbf{x}')\Psi_{(\mathbf{k}_i-\mathbf{q}_r,j)}(\mathbf{x}')) \\
&\equiv \widetilde{P}^{\mathbf{Q},\mathbf{Q}'}(\mathbf{q}_r). \qquad (7.63)
\end{aligned}
$$

References

1. N.F. Mott, Proc. Phys. Soc. A **62**(7), 416 (1949). https://doi.org/10.1088/0370-1298/62/7/303
2. J.C. Slater, Phys. Rev. **81**, 385 (1951). https://doi.org/10.1103/PhysRev.81.385
3. J.C. Slater, *Quantum Theory of Atomic Structure*, vol. 2 (Dover, New York, 1960)
4. J.G. Bednorz, K.A. Müller, Rev. Mod. Phys. **60**, 585 (1988). https://doi.org/10.1103/RevModPhys.60.585
5. E. Dagotto, Rev. Mod. Phys. **66**, 763 (1994). https://doi.org/10.1103/RevModPhys.66.763
6. C. Almasan, M.B. Maple, *Chemistry of High Temperature Superconductors* (World Scientific, Singapore, 1991)
7. Y. Yanase, T. Jujo, T. Nomura, H. Ikeda, T. Hotta, K. Yamada, Phys. Rep. **387**(1–4), 1 (2003). https://doi.org/10.1016/j.physrep.2003.07.002
8. D.J. Van Harlingen, Rev. Mod. Phys. **67**, 515 (1995). https://doi.org/10.1103/RevModPhys.67.515

9. A. Damascelli, Z. Hussain, Z.X. Shen, Rev. Mod. Phys. **75**, 473 (2003). https://doi.org/10.1103/RevModPhys.75.473
10. W.E. Pickett, Rev. Mod. Phys. **61**, 433 (1989). https://doi.org/10.1103/RevModPhys.61.433
11. G. Burns, *High Temperature Superconductivity: An Introduction* (Academic Press, New York, 1992)
12. M. Imada, A. Fujimori, Y. Tokura, Rev. Mod. Phys. **70**, 1039 (1998). https://doi.org/10.1103/RevModPhys.70.1039
13. T. Freltoft, J.P. Remeika, D.E. Moncton, A.S. Cooper, J.E. Fischer, D. Harshman, G. Shirane, S.K. Sinha, D. Vaknin, Phys. Rev. B **36**, 826 (1987). https://doi.org/10.1103/PhysRevB.36.826
14. J.H. de Boer, E.J.W. Verwey, Proc. Phys. Soc. **49**(4S), 59 (1937). https://doi.org/10.1088/0959-5309/49/4S/307
15. N.F. Mott, R. Peierls, Proc. Phys. Soc. **49**(4S), 72 (1937). https://doi.org/10.1088/0959-5309/49/4S/308
16. P.W. Anderson, Phys. Rev. **115**, 2 (1959). https://doi.org/10.1103/PhysRev.115.2
17. J. Hubbard, Proc. R. Soc. Lond. A **276**(1365), 238 (1963). https://doi.org/10.1098/rspa.1963.0204
18. P.W. Anderson, Science **235**, 1196 (1987). https://doi.org/10.1126/science.235.4793.1196
19. E. Fradkin, *Field Theories of Condensed Matter Systems* (Addison-Wesley, Redwood City, 1991)
20. M.C. Gutzwiller, Phys. Rev. **134**, A923 (1964). https://doi.org/10.1103/PhysRev.134.A923
21. M.C. Gutzwiller, Phys. Rev. **137**, A1726 (1965). https://doi.org/10.1103/PhysRev.137.A1726
22. W.F. Brinkman, T.M. Rice, Phys. Rev. B **2**(10), 4302 (1970). https://doi.org/10.1103/PhysRevB.2.4302
23. W. Kohn, Phys. Rev. **133**, A171 (1964). https://doi.org/10.1103/PhysRev.133.A171
24. W. Kohn, L.J. Sham, Phys. Rev. **140**, A1133 (1965). https://doi.org/10.1103/PhysRev.140.A1133
25. K. Terakura, A.R. Williams, T. Oguchi, J. Kübler, Phys. Rev. Lett. **52**, 1830 (1984). https://doi.org/10.1103/PhysRevLett.52.1830
26. D.J. Singh, W.E. Pickett, Phys. Rev. B **44**, 7715 (1991). https://doi.org/10.1103/PhysRevB.44.7715
27. A. Szabo, N.S. Ostlund, *Modern Quantum Chemistry: Introduction to Advanced Electronic Structure Theory* (Dover, New York, 1996)
28. A.L. Fetter, J.D. Walecka, *Quantum Theory of Many-Particle Systems* (Dover, New York, 2004)
29. L.F. Mattheiss, Phys. Rev. Lett. **58**, 1028 (1987). https://doi.org/10.1103/PhysRevLett.58.1028
30. N.F. Mott, *Metal-Insulator Transitions* (Taylor and Francis, London, 1974)
31. B.J. Powell, in *Computational methods for large systems: electronic structure approaches for biotechnology and nanotechnology*, ed. by J.R. Reimers (Wiley, Hoboken, 2011), pp. 309–366. chap. 10, arXiv:0906.1640 [physics.chem-ph]. https://doi.org/10.1002/9780470930779.ch10. ISBN 9780470930779
32. A. Cabo-Bizet, A. Cabo Montes de Oca, Phys. Lett. A **373**(21), 1865 (2009). https://doi.org/10.1016/j.physleta.2009.03.018
33. A. Cabo-Bizet, A. Cabo Montes De Oca, Symmetry **2**(1), 388 (2010). https://doi.org/10.3390/sym2010388
34. V.M. Martinez Alvarez, A. Cabo-Bizet, A. Cabo Montes de Oca, Int. J. Mod. Phys. B **28**(22), 1450146 (2014). https://doi.org/10.1142/S021797921450146X
35. Y. Vielza, A. Cabo Montes de Oca, Revista Cubana de Física **31**(2), 75 (2014)
36. A. Cabo Montes de Oca, N.H. March, A. Cabo-Bizet, Int. J. Mod. Phys. B **28**(04), 1450027 (2014). https://doi.org/10.1142/S0217979214500271
37. A. Cabo-Bizet, A. Cabo Montes de Oca, *Fases de Mott y pseudogap a partir de un modelo simple del* La_2CuO_4*: cómo las fases de Mott y de pseudogap emergen de un modelo simple para las capas* CuO_2 (Editorial Académica Española, Saarbrücken, 2012). ISBN 9783659050541
38. J.L. Tallon, J.W. Loram, Physica C **349**(1–2), 53 (2001). https://doi.org/10.1016/S0921-4534(00)01524-0
39. D.M. Broun, Nat. Phys. **4**(3), 170 (2008). https://doi.org/10.1038/nphys909
40. F. Aryasetiawan, O. Gunnarsson, Rep. Progr. Phys. **61**(3), 237 (1998). https://doi.org/10.1088/0034-4885/61/3/002

Chapter 8
Wavefunctions for Large Electronic Systems

P. Fulde

Dedicated to N. H. March on the occasion of his 90th birthday.

Abstract Wavefunctions for large electron numbers suffer from an exponential growth of the Hilbert space which is required for their description. In fact, as pointed out by W. Kohn, for electron numbers $N > N_0$ where $N_0 \approx 10^3$ they become meaningless (exponential wall problem). Nevertheless, despite of the enormous successes of density functional theory, one would also like to develop electronic structure calculations for solids based on wavefunctions. This is possible if one defines the latter in Liouville space with a cumulant metric instead in Hilbert space. The cluster expansion of the free energy of a classical monatomic gas makes it transparent why cumulants are very well suited also for electronic structure calculations.

8.1 Introduction

Electronic structure calculations, in particular for large systems, are one of the most active and challenging fields in condensed matter physics and quantum chemistry. This remains true despite the fact that the field has almost exploded during the last thirty years. Naively one would think that in an electronic structure calculation one is aiming to determine the many-body wavefunction of the interacting electron system and to derive from it physical properties of interest. This was the path taken when shortly after the rules for dealing with quantum mechanical systems were formulated by Heisenberg [1] and Schrödinger [2] these were applied by Hund, Mullikan, Heitler, London, and others in order to study chemical binding, thereby beginning with the

P. Fulde (✉)
Max-Planck-Institut für Physik Komplexer Systeme, Nöthnitzer Straße 38,
01187 Dresden, Germany
e-mail: fulde@pks.mpg.de

G. G. N. Angilella and C. Amovilli (eds.), *Many-body Approaches at Different Scales*, https://doi.org/10.1007/978-3-319-72374-7_8

H_2 molecule. Since then, the sizes of the quantum chemical systems for which the electronic structures were studied grew continuously. At present, even electrons in molecules with hundreds of atoms have been successfully treated (see, e.g., [3, 4]), as well as in solids with periodic lattices (see, e.g., [5]).

However, with increasing electron number N the dimension of the Hilbert space spanned by the different electronic configurations increases exponentially with N. This led Walter Kohn to the statement [6] that the many-electron wavefunction for a system of more than $N \approx 10^3$ electrons is not a legitimate scientific concept anymore. He referred hereby to wavefunctions expressed in Hilbert space and required that for a legitimate scientific concept two conditions have to be fulfilled: it should be possible to calculate the wavefunction with sufficient accuracy and it should be possible to represent it numerically sufficiently well. Because of the exponential growth of the Hilbert space none of the two conditions can be satisfied when $N \gtrsim 10^3$. Any approximation ψ_{calc} to the exact ground-state wavefunction ψ_0 will have an overlap with the latter of order $|\langle\psi_0|\psi_{\text{calc}}\rangle|^2 = (1-\varepsilon)^N$, which is zero for all purposes, if $N \to \infty$. Here, ε is the minimum error one has to deal with when approximations for the description of an electron are being made. A similar argument applies to the second condition. When it needs $m \geq 2$ bits to describe a single interacting electron, the total number of bits is m^N in order to describe the full electron system and therefore too large in order to be documented.

The exponential growth of Hilbert space is referred to as the exponential wall (EW) problem. The simplest way to get around it is by making use of density-functional theory (DFT), developed by Hohenberg, Kohn and Sham [7, 8] (for extensions see e.g., [9]). Here all degrees of freedom of the electronic system are integrated out, except for the density. No statements are required about the many-electron wavefunctions. The strength of DFT is based on this feature.

Another way of avoiding the EW problem is by reducing the electron Hamiltonian H to its self-consistent field (SCF) part H_{SCF}. This simplifies the problem to a single-electron one with a potential which has to be determined self-consistently. The ground-state wavefunction is given in this case by a single Slater determinant or configuration. Correlation effects are hereby completely neglected and therefore results for various physical quantities are usually of low quality.

Although DFT has revolutionized the field of electronic structure calculations, and can claim fantastic successes, it contains also weaknesses. A general one is that its results depend on the chosen exchange-correlation potential and any approximation to it is essentially uncontrolled. This leads to problems when electronic correlations are strong or when one is dealing with dispersive electron interactions.

For the above reasons it seems worthwhile to develop in parallel to DFT also electronic structure calculations based on wavefunctions. This approach is stimulated by the accuracy of quantum chemical techniques in cases when they have been applied. The question therefore is: does Kohn's correct argument about the inadequacy of wavefunctions for large systems prevent us for doing such calculations on a firm theoretical basis? In short, the answer is no! However, one has to give up characterizing the many-electron wavefunction in Hilbert space. Instead it has to be characterized in Liouville or operator space. The reason is not difficult to see. Consider A very weakly interacting atoms with $N_A \geq 2$ electrons each. The dimension of Hilbert space for the ground state of the electrons on a single atom is of order d^{N_A} with $d \geq 2$. Yet the dimension required for a description of the whole system is $d^{A \cdot N_A}$ despite the fact that in the limit of weak coupling between electrons on different atoms the wavefunction does not contain any additional information compared with the one obtained from a single atom. Thus disconnected correlations are responsible for the EW problem. This suggests to eliminate all disconnected contributions to the wavefunction in order to free oneself from the EW problem. This is easily done with the help of cumulants.

8.2 Use of Cumulants

Cumulants of matrix elements eliminate factorizable contributions to it. In the simplest case the cumulant, denoted by c, of a product of two operators $A_1 A_2$ sandwiched between two vectors ϕ_1 and ϕ_2 in Hilbert space with $\langle \phi_1 \mid \phi_2 \rangle \neq 0$ is

$$\langle \phi_1 | A_1 A_2 | \phi_2 \rangle^c = \frac{\langle \phi_1 | A_1 A_2 | \phi_2 \rangle}{\langle \phi_1 | \phi_2 \rangle} - \frac{\langle \phi_1 | A_1 | \phi_2 \rangle \langle \phi_1 | A_2 | \phi_2 \rangle}{(\langle \phi_1 | \phi_2 \rangle)^2}. \tag{8.1}$$

Note that a replacement of $|\phi_2\rangle$ by $\alpha|\phi_2\rangle$ with $\alpha \neq 0$ leaves the cumulant unchanged. General rules for cumulants are found in the literature, see e.g., [10, 11]. They were first applied in statistical physics [12] when dealing with the classical imperfect gas and later pioneered by Kubo [13] in quantum statistical mechanics. In practice a cumulant implies taking only connected contractions of operators into account when a matrix element or expectation value is evaluated. With this in mind we divide the electronic Hamiltonian H into $H = H_0 + H_1$ so that the ground state of H_0, i.e., $|\Phi_0\rangle$ can be easily calculated. Often H_0 will be H_{SCF}, yet it can also be, e.g., the Kohn-Sham Hamiltonian H_{KS}. The remaining part H_1 contains the residual interactions. We define $|\Phi_0\rangle$ as the vacuum state so that H_1 generates fluctuations, i.e., vacuum fluctuations on it. Next we consider a matrix element of an arbitrary operator A with respect to $|\Phi_0\rangle$, i.e., $\langle \Phi_0 | A | \Phi_0 \rangle = \langle \Phi_0 | A | \Phi_0 \rangle^c$. By a sequence of infinitesimal transformations in Hilbert space, we transform the state $|\Phi_0\rangle$ on the right, into the exact ground state $|\psi_0\rangle$ of H. Then the above expression transforms into

$$\langle \Phi_0 | A | \psi_0 \rangle^c = \langle \Phi_0 | A \Omega | \psi_0 \rangle^c \tag{8.2}$$

with $\Omega = 1 + S$ and S denoting the sum of the infinitesimal increments in the transformation. For a more accurate derivation of Eq. (8.2), see [10]. Note that Ω is not unique since many different paths in Hilbert space may lead from $|\Phi_0\rangle$ to $|\psi_0\rangle$. Yet, the cumulant remains unchanged by these differences.

In quantum mechanics the operator which transforms the ground state of an unperturbed system (here $|\Phi_0\rangle$) into the one of the perturbed system (here $|\psi_0\rangle$) is called Møller or wave operator, i.e., $|\psi_0\rangle = \tilde{\Omega}|\Phi_0\rangle$. Therefore we call Ω in Eq. (8.2) a cumulant wave operator and S in $\Omega = 1 + S$ a cumulant scattering operator. Equation (8.2) suggests to introduce the following metric in Liouville space

$$(A|B) = \langle \Phi_0 | A^+ B | \Phi_0 \rangle^c . \tag{8.3}$$

The metric is not a scalar product since it may vanish or even become negative.

8.3 Cumulant Scattering Matrix

The exponential wall problem does not exist if we define the many-electron wave-function not by a vector in Hilbert space but by the vector $|\Omega)$ in Liouville space. An exponentially small overlap of $\langle \psi_0 | \psi_{\text{cal}} \rangle$ is harmless since the cumulant, e.g., in Eq. (8.1) remains unchanged when ϕ_1 is replaced by $\alpha \phi_1$ with $\alpha \neq 0$. When $|\psi_{\text{cal}}\rangle$ instead of $|\psi_0\rangle$ is considered the only effect is a change of $|S)$ to $|S + \delta S)$. Also a numerical representation of $|\Omega)$ or $|S)$ poses no problems. To see this, we decompose $|S)$ into increments

$$|S) = \left| \sum_I S_I + \sum_{\langle IJ \rangle} \delta S_{IJ} + \sum_{\langle IJK \rangle} \delta S_{IJK} + \ldots \right) \tag{8.4}$$

where I, J, K etc. are site indices. Here, $\delta S_{IJ} = S_{IJ} - S_I - S_J$ etc. When S_I is calculated all electrons in $|\Phi_0\rangle$ are kept frozen except those in orbitals centered at site I. The procedure is similar when S_{IJ}, S_{IJK} etc. are calculated. In each case only a small number of electrons is involved and therefore the different increments of $|S)$ can be documented without problem. Thus the EW problem has been eliminated.

Let us consider $|\Phi_0\rangle$ as the vacuum state of the system. The operators S_I, S_{IJ} etc. can be thought of generating vacuum fluctuations. They modify the energy E_0 of the system according to

$$E_0 = \langle \Phi_0 | H \Omega | \Phi_0 \rangle^c = (H|\Omega). \tag{8.5}$$

When $H_0 = H_{\text{SCF}}$, the correlation energy is simply

$$\begin{aligned} E_{\text{corr}} &= (H_1|\Omega) \\ &= (H_1|S)\,. \end{aligned} \tag{8.6}$$

Because of the cumulant metric, only connected vacuum fluctuations contribute to E_{corr}. This is depicted in Fig. 8.1. Note that I and J need not be nearest neighbors. For a given site I the correlation contributions add up to

$$(H_1|S_I + \frac{1}{2} \sum_{J \neq I} \delta S_{IJ} + \ldots) = E_{\text{corr}}(I) \tag{8.7}$$

and $E_{\text{corr}} = \sum_I E_{\text{corr}}(I)$. This is indicated in Fig. 8.2. The δS_{IJ}, δS_{IJK} etc. decrease rapidly with increasing number of subscripts. They also decrease rapidly with increas-

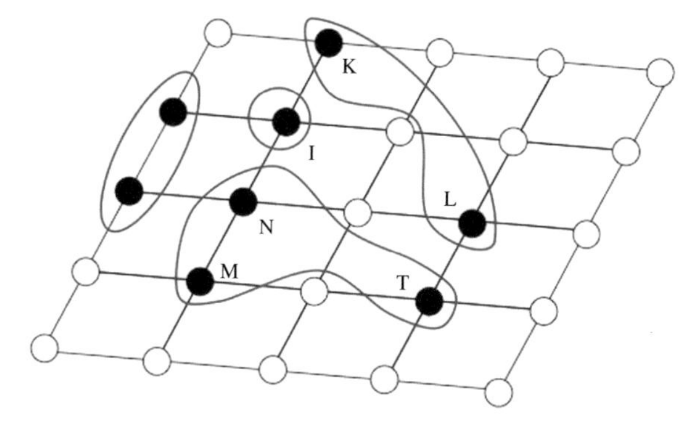

Fig. 8.1 Examples of different vacuum fluctuations S_I, S_{KL}, S_{MNT} contributing to $|S)$.

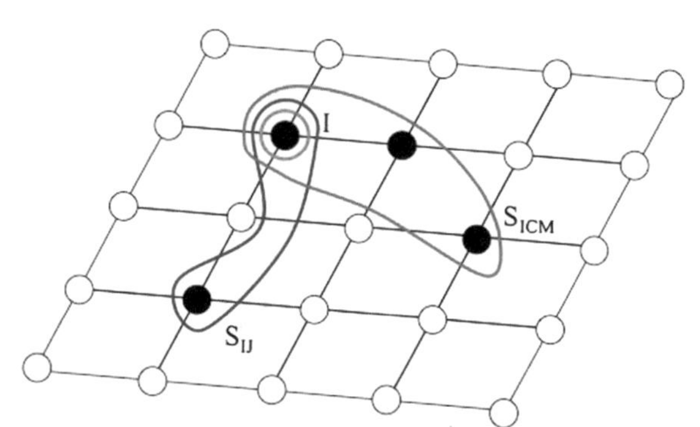

Fig. 8.2 Vacuum fluctuations contributing to $E_{\text{corr}}(I)$. Different colours refer to connected vacuum fluctuations involving electrons on different numbers of sites

ing distances of the sites I, J etc. The slowest decrease when $|\mathbf{R}_I - \mathbf{R}_J|$ increases is taking place for δS_{IJ}. Here $\mathbf{R}_I$ denotes the position of site I. Except in the vicinity of an electronic phase transition the decrease of $(H_1|\delta S_{IJ})$ with $|\mathbf{R}_I - \mathbf{R}_J| \to \infty$ is exponential and defines a characteristic correlation length, e.g., in a solid. The behavior of $(H_1|\delta S_{IJ})$ in this limit tells us whether or not an area law [14] is holding.

We want to point out that the theory presented here has been applied to a number of solids, mainly to semiconductors and insulators. Metallic systems [15, 16] require additional comments since occupied Wannier orbitals are not well localized [17]. Yet, this does not pose a principle problem. A review of the different results obtained is found in [5, 11, 18] although the way is not always obvious in which the computations described there relate to the computational scheme outlined here.

8.4 Classical Imperfect Gas

It is instructive to compare the definition of a wavefunction in Liouville space and the computation of the correlation energy with the one of the temperature T dependent free energy $F(T)$ of a classical imperfect (real) monatomic gas. The limit $F(T = 0)$

yields the classical analogue of the electronic ground-state energy. Let $U = \sum_{i>j} \phi_{ij}$ denote the potential energy of the system where ϕ_{ij} denotes the pair interactions of the gas particles. With $f_{ij} = \exp(-\beta\phi_{ij}) - 1$ and $\beta = (k_B T)^{-1}$ we can write

$$e^{-\beta U} = \prod_{i>j} e^{-\beta\phi_{ij}} = \prod_{i>j} \left(1 + f_{ij}\right). \tag{8.8}$$

The partition function Z is the product $Z = Z_{\mathrm{id}} \cdot Z_U$ of the one for the ideal gas Z_{id} and

$$Z_U = \frac{1}{V^N} \int d\mathbf{r}_1 d\mathbf{r}_2 \ldots dr_N e^{-\beta U(\mathbf{r}_1,\ldots,r_N)}. \tag{8.9}$$

The integration is over the coordinates of the N particles and extends over the full volume V of the system. Thus the free energy $F(T) = F_{\mathrm{id}} + F_U$ can be written in the form

$$\begin{aligned} F &= F_{\mathrm{id}} - k_B T \ln \frac{1}{V^N} \int d\mathbf{r}_1 \ldots d\mathbf{r}_N e^{-\beta U(\mathbf{r}_1,\ldots,r_N)} \\ F_U &= -k_B T \ln \left\langle \prod_{i>j} \left(1 - f_{ij}\right) \right\rangle, \end{aligned} \tag{8.10}$$

where F_{id} is the free energy of the ideal gas.

An often used definition of cumulants is of the form

$$\ln\langle e^{\lambda A}\rangle = \langle e^{\lambda A} - 1\rangle^c \tag{8.11}$$

and demonstrates how cumulants avoid dealing with the logarithm of averages. When applied to F_U we can rewrite the latter as

$$F_U = -k_B T \left\langle \sum_{i<j} f_{ij} + \sum_{i<j} \sum_{\substack{k<l \\ i,j \neq kl}} f_{ij} f_{kl} + \ldots \right\rangle^c. \tag{8.12}$$

Thus to the free energy of a real gas [12] only *linked* pair interactions do contribute. This is sometimes referred to as Mayer's cluster expansion.

This well known classical results suggest strongly that also in quantum mechanics cumulants are a proper tool for calculating energies for large systems of interacting particles. It makes the choice of the metric in Liouville space rather obvious.

The question remains, how the present approach based on a description of wavefunctions in Liouville space with cumulant metric compares with other approaches avoiding the EW problem. A comparison with the Density Matrix Embedding Theory (DMET) [19] was recently worked out [20]. A corresponding one for the Density Matrix Renormalization Group (DMRG) [21] as well as with tensor networks [22] in particular with Matrix Product States [23] is left for the future.

Acknowledgements I would like to thank Hermann Stoll for many fruitful discussions and collaborations on subjects related to this article.

References

1. W. Heisenberg, Z. Phys. **33**, 879 (1925)
2. E. Schrödinger, Ann. de Phys. **79**, 361 (1926)
3. H.J. Werner, P.J. Knowles, G. Knizia, F.R. Manby, M. Schütz, Wiley Interdiscip. Rev. Comput. Mol. Sci. **2**(2), 242 (2012). https://doi.org/10.1002/wcms.82
4. D.G. Liakos, F. Neese, J. Chem, Theory Comput. **11**(9), 4054 (2015). https://doi.org/10.1021/acs.jctc.5b00359
5. B. Paulus, H. Stoll, in *Accurate Condensed-Phase Quantum Chemistry*, ed. by F.R. Manby (CRC Press, Boca Raton, 2011), p. 57
6. W. Kohn, Rev. Mod. Phys. **71**, 1253 (1999). https://doi.org/10.1103/RevModPhys.71.1253
7. P.C. Hohenberg, W. Kohn, Phys. Rev. **136**, B864 (1964). https://doi.org/10.1103/PhysRev.136.B864
8. W. Kohn, L.J. Sham, Phys. Rev. **140**, A1133 (1965). https://doi.org/10.1103/PhysRev.140.A1133
9. N.H. March, in *Electron Correlation in the Solid State*, ed. by N.H. March (Imperial College Press, London, 1999), p. 371. ISBN 1860942008
10. K. Kladko, P. Fulde, Int. J. Quantum Chem. **66**(5), 377 (1998), https://doi.org/10.1002/(SICI)1097-461X(1998)66:5<377::AID-QUA4>3.0.CO;2-S
11. P. Fulde, *Correlated Electrons in Quantum Matter* (World Scientific, Singapore, 2012)
12. J.E. Mayer, M.G. Mayer, *Statistical Mechanics* (Wiley, New York, 1940)
13. R. Kubo, J. Phys. Soc. Jpn. **17**(7), 1100 (1962). https://doi.org/10.1143/JPSJ.17.1100
14. J. Eisert, M. Cramer, M.B. Plenio, Rev. Mod. Phys. **82**, 277 (2010). https://doi.org/10.1103/RevModPhys.82.277
15. B. Paulus, K. Rościszewski, Chem. Phys. Lett. **394**(1–3), 96 (2004). https://doi.org/10.1016/j.cplett.2004.06.118
16. B. Paulus, K. Rościszewski, N. Gaston, P. Schwerdtfeger, H. Stoll, Phys. Rev. B **70**, 165106 (2004). https://doi.org/10.1103/PhysRevB.70.165106
17. W. Kohn, Phys. Rev. **115**, 809 (1959). https://doi.org/10.1103/PhysRev.115.809
18. B. Paulus, Phys. Rep. **428**(1), 1 (2006). https://doi.org/10.1016/j.physrep.2006.01.003
19. G. Knizia, G.K.L. Chan, Phys. Rev. Lett. **109**, 186404 (2012). https://doi.org/10.1103/PhysRevLett.109.186404
20. P. Fulde, H. Stoll, J. Chem. Phys. **146**(19), 194107 (2017). https://doi.org/10.1063/1.4983207
21. U. Schollwöck, Rev. Mod. Phys. **77**(1), 259 (2005). https://doi.org/10.1103/RevModPhys.77.259
22. F. Verstraete, V. Murg, J.I. Cirac, Adv. Phys. **57**(2), 143 (2008). https://doi.org/10.1080/14789940801912366
23. R. Orús, Ann. Phys. **349**, 117 (2014). https://doi.org/10.1016/j.aop.2014.06.013

Chapter 9
Electron Tunneling Excitation of a Coupled Two Impurity System

F. Flores and E. C. Goldberg

Abstract Kondo effects in individual atoms, molecular magnets or quantum dots have received a lot of attention. Systems with two units, like double quantum dots or two atoms, have also attracted some attention due to the interplay between the possible coupling between the spins of the two components and the Kondo effect of each unit. Moreover, the tunneling spectroscopy across one or several magnetic atoms deposited on a metal surface has also been analyzed and shown to present properties associated with spin-flip processes and Kondo resonances. In this paper we analyze the electron tunneling excitations created in a dimer case, assuming that each unit (atom or quantum well) has spin $\frac{1}{2}$. In our approach, the basic Hamiltonian includes the spin-metal hybridization as well as the spin-spin interaction; then, its basic properties are analyzed by means of a Green's function formalism combined with an Equation of Motion method. We present results showing the tunneling differential conductance as a function of the different parameters of the problem and the limits for which spin-flip processes and/or Kondo resonances appear.

9.1 Introduction

Scanning tunneling microscopy (STM) inelastic spin spectroscopy has been used to explore the intrinsic excitations of individual, as well as aggregates of magnetic atoms like Fe or Co on a CuN_2 metal surface [1–6].

F. Flores (✉)
Departamento de Física Teórica de la Materia Condensada and IFIMAC, Universidad Autónoma de Madrid, Cantoblanco, 28049 Madrid, Spain
e-mail: fernando.flores@uam.es

E. C. Goldberg
Instituto de Física del Litoral (CONICET-UNL), Güemes 3450, S3000GLN Santa Fe, Argentina
e-mail: edith.goldberg@santafe-conicet.gov.ar

E. C. Goldberg
Departamento Ingeniería de Materiales, Facultad de Ingeniería Química, Universidad Nacional del Litoral, Santiago del Estero 2829, S3000AOM Santa Fe, Argentina

G. G. N. Angilella and C. Amovilli (eds.), *Many-body Approaches at Different Scales*, https://doi.org/10.1007/978-3-319-72374-7_9

The differential conductance spectra, dI/dV, for a single magnetic atom shows different regimes depending on the atom located below the tip. In the case of Co on Cu_2N, dI/dV shows a few degrees Kondo resonance accompanied by some steps in the spectrum associated with internal excitations of the magnetic atom [5]; for Fe, the tunneling differential conductance only shows equivalent steps but no Kondo resonance [1].

Dimers on Cu_2N, like Co-Co, Fe-Fe or Co-Fe [7–9], have also attracted some attention due to the interplay between the ferro- or antiferromagnetic coupling of the atomic spins and the possible Kondo resonance of each dimer component. The experimental evidence has also shown that in these dimers the ferro- or antiferro- spin-spin interaction depends crucially on the distance between the atoms. The interest in analyzing those dimers is also related to the different properties each independent atom has.

These various experiments have been analyzed by different groups, combining an atomic crystal-field effect with an effective interaction between the tunneling electrons and the atomic spin, described by means of an exchange coupling [10, 11], a spin-assisted Hamiltonian [12–16] or using strong coupling theory [17]. All these approaches are reminiscent of the scattering theory approach used by Kondo [18] to explain experimental results about the resistivity of dilute magnetic impurities in metals.

In previous works [19–21] we have analyzed the STM-differential conductance for individual atoms [19], and for different dimers Fe-Co [20], and Co-Co or Fe-Fe [21], using a new ionic Hamiltonian [22] for describing the charge exchange processes between the metal and the magnetic atoms. In this approach, the tunneling conductance is the result of a co-tunneling process, whereby each atomic spin fluctuates due to the successive jumping between the atom and the contacts.

In this paper we present a theoretical analysis of the simplest model that can be introduced for describing the differential conductance of a dimer, namely, a $\frac{1}{2}$-$\frac{1}{2}$ system. In this model each atom is assumed to have one electron with spin $\frac{1}{2}$, and fluctuate between one and zero electrons exchanging one electron with the metal.

The interest of analyzing this system is related to the convenience of understanding, as a kind of benchmark, the simplest ideal case; this is a way of thinking that has been followed by Professor Norman March along his scientific life, as shown in his study of atoms and molecules [23] and in many other problems [24, 25]. The $\frac{1}{2}$-$\frac{1}{2}$ system is also interesting in itself for its connection with the Co-Co dimer; this dimer presents due to its interaction with the metal a doublet ground state (responsible of the Kondo resonance in the single atom case) and, apparently, this suggests that the Co-Co dimer might be simulated by the $\frac{1}{2}$-$\frac{1}{2}$ case. We will discuss however in this paper the similarities and differences between these two cases.

The paper is organized as follows: in Sect. 9.2 we present our basic Hamiltonians and discuss the differences between the $\frac{1}{2}$-$\frac{1}{2}$ and the Co-Co cases. In Sect. 9.3 we present the Equation of Motion (EOM) method that we have used for solving those Hamiltonians and the way of calculating the tunneling currents. In Sect. 9.4 we present our results for the $\frac{1}{2}$-$\frac{1}{2}$ and the Co-Co dimers, and finally in Sect. 9.5 we present our conclusions.

9.2 Model Hamiltonian For Two Impurities

9.2.1 The $\frac{1}{2}$-$\frac{1}{2}$ Case

Figure 9.1 defines the model we are going to analyze: two atoms with spin $\frac{1}{2}$ are deposited on a metal surface and interact with each other via a Heisenberg term, $J\hat{\mathbf{S}}_1 \cdot \hat{\mathbf{S}}_2$. We assume that each atom donates (or accepts) one electron to (or from) the metal, in such a way that each atomic spin fluctuates between $\frac{1}{2}$ and zero. The atoms and the metal are described by the conventional Hamiltonian:

$$\hat{H}_{\text{atom+metal}} = \sum_{k\sigma} \varepsilon_k \hat{n}_{k\sigma} + \varepsilon_0^A |0,0\rangle_A \langle 0,0|_A + \varepsilon_0^B |0,0\rangle_B \langle 0,0|_B \\ + \varepsilon_1^A \sum_{\sigma} |\tfrac{1}{2},\sigma\rangle_A \langle \tfrac{1}{2},\sigma|_A + \varepsilon_1^B \sum_{\sigma} |\tfrac{1}{2},\sigma\rangle_B \langle \tfrac{1}{2},\sigma|_B. \tag{9.1}$$

In Eq. (9.1), we have used the projection operators based on the following definition of the configurations of atoms A and B: $|0,0\rangle$ for $S=0$, and $|\frac{1}{2},\sigma\rangle$ for $S=\frac{1}{2}$. The k-index denotes the metal states with energy ε_k. The interaction between the atoms and the metal can be described by the following Hamiltonian:

$$\hat{H}_{\text{int}} = \sum_{k\sigma} \left[V_{k\sigma}^{(A)*} c_{k\sigma}^{\dagger} |0,0\rangle_A \langle \tfrac{1}{2},\sigma|_A + \text{H.c.} \right] + \left[V_{k\sigma}^{(B)*} c_{k\sigma}^{\dagger} |0,0\rangle_B \langle \tfrac{1}{2},\sigma|_B + \text{H.c.} \right]. \tag{9.2}$$

We also include the Heisenberg term:

$$\hat{H}_J = J\hat{\mathbf{S}}_1 \cdot \hat{\mathbf{S}}_2. \tag{9.3}$$

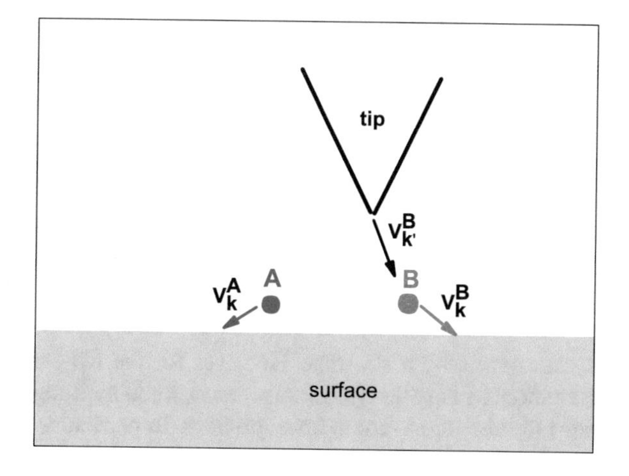

Fig. 9.1 Schematic view of the two atoms (A and B), interacting with the surface and the tip

As we are interested in calculating the tunneling current between a scanning tunneling microscope (STM)-tip and the dimer-$\frac{1}{2}$ (or Co-Co), we have to introduce the tip-dimer interaction by means of a tunneling Hamiltonian, which has the same expression as Hamiltonian (9.2), but where we include now the tip levels, k'; we assume that $V_{k'} \ll V_k$, so that in a first step we will neglect $V_{k'}$ for calculating the dimer electronic properties. In a further step, we will introduce $V_{k'}$ to calculate the tunneling currents.

9.2.2 The Co-Co Case

We will discuss in this section how the Co-Co dimer, for a geometry similar to the one drawn in Fig. 9.1, can be analyzed for a small tunneling bias with a Hamiltonian similar to the one presented for the $\frac{1}{2}$-$\frac{1}{2}$ case, but with some critical differences.

We start this discussion by considering the case of a single Co-atom adsorbed on a metal. We shall assume that, for this case, the Co-atom has spin $\frac{3}{2}$ and that it fluctuates to an atomic state with spin 1, when it exchanges one electron with the metal [19, 22]. Then, by using the notation $|S, M\rangle$ for the atom configurations with total spin S and projection M, the atom and the metal can de described by the following Hamiltonian [22]:

$$\hat{H}_{\text{atom+metal}} = \sum_{k\sigma} \varepsilon_k \hat{n}_{k\sigma} + E^{\frac{3}{2}(A)} \sum_{M} |\tfrac{3}{2}, M\rangle\langle\tfrac{3}{2}, M| + E^{1(A)} \sum_{m} |1, m\rangle\langle 1, m|, \tag{9.4}$$

and the metal/atom interaction by:

$$\hat{H}_{\text{int}} = \sum_{k,M,\sigma} \left[V^{\frac{3}{2}*}_{kM\sigma} \hat{c}^{\dagger}_{k\sigma} |1, M-\sigma\rangle\langle\tfrac{3}{2}, M| + \text{H.c.} \right] \tag{9.5}$$

In Eq. (9.5), $V^S_{kM\sigma} = \sqrt{\frac{S+(-1)^p M}{2S}} V_{kd}$, with p equal to 0 if $\sigma = \frac{1}{2}$, and equal to 1 if $\sigma = -\frac{1}{2}$ [19].

For the Co-atom it is also important to include the anisotropy interaction described by [1]:

$$\hat{H}_{\text{anisotropy}} = D\hat{S}_z^2 + E(\hat{S}_x^2 - \hat{S}_y^2). \tag{9.6}$$

This phenomenological form of the anisotropy term comes from a second order perturbation calculation of the spin-orbit coupling $\lambda\hat{\mathbf{L}} \cdot \hat{\mathbf{S}}$. D and E are parameters we will take from an independent source [1, 3, 5]. Notice that $\hat{H}_{\text{anisotropy}}$ can be written as $D\hat{S}_z^2 + \frac{1}{2}E(\hat{S}_+^2 + \hat{S}_-^2)$, in such a way that for the $S = \frac{1}{2}$ case, it only changes the atomic levels by a constant. However for the Co case, with $S = \frac{3}{2}$ or $S = 1$, the Hamiltonian equation (9.6), introduces some splitting in the atomic levels that we have to take into account appropriately. In particular, for $S = \frac{3}{2}$ we find that the 4 atomic states associated with different values of M are split into 2 doublets: $|\frac{3}{2}, \pm\frac{3}{2}\rangle$ with $E^{\frac{3}{2},\pm\frac{3}{2}} = 9D/4$, and $|\frac{3}{2}, \pm\frac{1}{2}\rangle$ with $E^{\frac{3}{2},\pm\frac{1}{2}} = D/4$ (taking $E = 0$ in Eq. (9.6); see Ref. [7]).

As $E^{\frac{3}{2},\pm\frac{3}{2}} - E^{\frac{3}{2},\pm\frac{1}{2}} = 2D = 10.8$ meV [5], we can neglect the states for energy biases smaller than $2-3$ meV. This approximation leaves the Co-atom with the atomic states $|\frac{3}{2},\pm\frac{1}{2}\rangle$ and, apparently, the problem is similar to the $S=\frac{1}{2}$ case where we only have the doublet $|\frac{1}{2},\pm\frac{1}{2}\rangle$. However, if we consider a dimer and the $J\hat{\mathbf{S}}_1\cdot\hat{\mathbf{S}}_2$ interaction, we find an important difference between the $\frac{1}{2}$-$\frac{1}{2}$ and the Co-Co cases. For the $\frac{1}{2}$-$\frac{1}{2}$ case, the $J\hat{\mathbf{S}}_1\cdot\hat{\mathbf{S}}_2$ interaction breaks the initial quadruplet levels into a triplet (with total spin $S=1$ and $M=1$, 0 and -1) and a singlet (with total spin $S=0$). For the Co-Co case we find, however, that in the reduced space $|\frac{3}{2},\pm\frac{1}{2}\rangle_A\otimes|\frac{3}{2},\pm\frac{1}{2}\rangle_B$ the initial quadruplet levels are split into a doublet and two singlets. The doublet is defined by the states:

$$|\phi_1\rangle = |\tfrac{3}{2},\tfrac{1}{2}\rangle_A\otimes|\tfrac{3}{2},\tfrac{1}{2}\rangle_B, \tag{9.7a}$$

$$|\phi_2\rangle = |\tfrac{3}{2},-\tfrac{1}{2}\rangle_A\otimes|\tfrac{3}{2},-\tfrac{1}{2}\rangle_B, \tag{9.7b}$$

with energies $E_{1,2}=J/4$, and the singlets by:

$$|\phi_3\rangle = \frac{1}{\sqrt{2}}\left(|\tfrac{3}{2},\tfrac{1}{2}\rangle_A\otimes|\tfrac{3}{2},-\tfrac{1}{2}\rangle_B + |\tfrac{3}{2},-\tfrac{1}{2}\rangle_A\otimes|\tfrac{3}{2},\tfrac{1}{2}\rangle_B\right), \tag{9.7c}$$

$$|\phi_4\rangle = \frac{1}{\sqrt{2}}\left(|\tfrac{3}{2},\tfrac{1}{2}\rangle_A\otimes|\tfrac{3}{2},-\tfrac{1}{2}\rangle_B - |\tfrac{3}{2},-\tfrac{1}{2}\rangle_A\otimes|\tfrac{3}{2},\tfrac{1}{2}\rangle_B\right), \tag{9.7d}$$

with energies $E_3=7J/4$ and $E_4=-9J/4$, respectively. Notice that, in the $\frac{1}{2}$-$\frac{1}{2}$ case, the corresponding states $|\phi_1\rangle$, $|\phi_2\rangle$, and $|\phi_3\rangle$ form a triplet with an energy $E_T=J/4$, while $|\phi_4\rangle$ defines the singlet with $E_S=-3J/4$. Regarding the Co-atom with spin $S=1$, consider the states $|1,1\rangle$, $|1,0\rangle$, and $|1,-1\rangle$ together with Hamiltonian equation (9.6). For this case we find a singlet, the $|1,0\rangle$-state, and a doublet, $|1,1\rangle$ and $|1,-1\rangle$, located 2.8 meV above the singlet. In our approximation, consistently with our previous case, we neglect the doublet and keep only the $|1,0\rangle$ state. This means that in the Co-Co case, for the $S=\frac{3}{2}*S=1$ or the $S=1*S=\frac{3}{2}$ states, we are going to work with the functions:

$$S=\frac{3}{2}*S=1\rightarrow|\psi_1\rangle=|\tfrac{3}{2},\tfrac{1}{2}\rangle_A\otimes|1,0\rangle_B \text{ and } |\psi_2\rangle=|\tfrac{3}{2},-\tfrac{1}{2}\rangle_A\otimes|1,0\rangle_B, \tag{9.8a}$$

$$S=1*S=\frac{3}{2}\rightarrow|\chi_1\rangle=|1,0\rangle_A\otimes|\tfrac{3}{2},\tfrac{1}{2}\rangle_B \text{ and } |\chi_1\rangle=|1,0\rangle_A\otimes|\tfrac{3}{2},-\tfrac{1}{2}\rangle_B. \tag{9.8b}$$

Finally, for the Co-Co case with spins $S=1*S=1$, we only include the wave function $\varphi_1\rangle=|1,0\rangle_A\otimes|1,0\rangle_B$.

It is interesting to realize that this approximation for the Co-Co dimer yields a formal problem similar to the one found in the $\frac{1}{2}$-$\frac{1}{2}$ case. In other words, if we introduce the atomic, the Heisenberg and the anisotropy Hamiltonians and make the approximations discussed above, we find for both problems the same Hilbert space but with a critical difference: while for the $S=\frac{1}{2}*S=\frac{1}{2}$ problem we find a triplet, E_T, and a singlet, $E_S=E_T+J$, for the Co-Co case we find that the triplet is split into a doublet of energy E_1, and one singlet with energy E_3, such that $E_3 - E_1 = 3J/2$; the other singlet has an energy E_4, with $E_1 - E_4 = 5J/2$.

We conclude this section mentioning that the Co-Co case is equivalent to the $\frac{1}{2}$-$\frac{1}{2}$ problem if we introduce a fictitious change in the energies of the triplet and singlet states associated with the Heisenberg interaction. Accordingly, we are going to analyze in the next section the $\frac{1}{2}$-$\frac{1}{2}$ case keeping in mind that the energy of the triplet state associated with the $J\hat{\mathbf{S}}_1 \cdot \hat{\mathbf{S}}_2$ interaction has to be split into a doublet and a singlet for the Co-Co case.

9.3 Equation of Motion Solution. Tunneling Currents

9.3.1 A Completely Symmetric Case

For the sake of simplicity, we are going to start our discussion of the EOM (Equations of Motion) method by considering a fully symmetric case for the atoms. This means that we assume to have the same atom/metal interaction for both cases, $V_k^{(A)} = V_k^{(B)}$. It is convenient for this particular case to work with the following molecular states:

Two electrons:

$$|T\uparrow\rangle = |\tfrac{1}{2},\tfrac{1}{2}\rangle_A \otimes |\tfrac{1}{2},\tfrac{1}{2}\rangle_B \equiv |\uparrow,\uparrow\rangle, \tag{9.9a}$$

$$|T\downarrow\rangle = |\tfrac{1}{2},-\tfrac{1}{2}\rangle_A \otimes |\tfrac{1}{2},-\tfrac{1}{2}\rangle_B \equiv |\downarrow,\downarrow\rangle, \tag{9.9b}$$

$$|T0\rangle = \frac{1}{\sqrt{2}}\left(|\tfrac{1}{2},\tfrac{1}{2}\rangle_A \otimes |\tfrac{1}{2},-\tfrac{1}{2}\rangle_B + |\tfrac{1}{2},-\tfrac{1}{2}\rangle_A \otimes |\tfrac{1}{2},\tfrac{1}{2}\rangle_B\right) \equiv \frac{1}{\sqrt{2}}\left(|\uparrow,\downarrow\rangle + |\downarrow,\uparrow\rangle\right), \tag{9.9c}$$

$$|S0\rangle = \frac{1}{\sqrt{2}}\left(|\tfrac{1}{2},\tfrac{1}{2}\rangle_A \otimes |\tfrac{1}{2},-\tfrac{1}{2}\rangle_B - |\tfrac{1}{2},-\tfrac{1}{2}\rangle_A \otimes |\tfrac{1}{2},\tfrac{1}{2}\rangle_B\right) \equiv \frac{1}{\sqrt{2}}\left(|\uparrow,\downarrow\rangle - |\downarrow,\uparrow\rangle\right); \tag{9.9d}$$

One electron:

$$|a_\sigma\rangle = \frac{1}{\sqrt{2}}\left(|\tfrac{1}{2},\sigma\rangle_A \otimes |0,0\rangle_B + |0,0\rangle_A \otimes |\tfrac{1}{2},\sigma\rangle_B\right) \equiv \frac{1}{\sqrt{2}}\left(|\sigma,0\rangle + |0,\sigma\rangle\right), \tag{9.9e}$$

$$|b_\sigma\rangle = \frac{1}{\sqrt{2}}\left(|\tfrac{1}{2},\sigma\rangle_A \otimes |0,0\rangle_B - |0,0\rangle_A \otimes |\tfrac{1}{2},\sigma\rangle_B\right) \equiv \frac{1}{\sqrt{2}}\left(|\sigma,0\rangle - |0,\sigma\rangle\right); \tag{9.9f}$$

Zero electron:

$$|0\rangle = |0,0\rangle_A \otimes |0,0\rangle_B \equiv |0,0\rangle. \tag{9.9g}$$

Using these states, we find that for this symmetric case we have the following molecular interactions: $|\uparrow,\uparrow\rangle$ is coupled to $|b_\uparrow\rangle$ by $-\sqrt{2}V_k$; $(|\uparrow,\downarrow\rangle+|\downarrow,\uparrow\rangle)/\sqrt{2}$ to $|b_\uparrow\rangle$ by $-V_k$, and to $|b_\downarrow\rangle$ by $-V_k$; and finally $|\downarrow,\downarrow\rangle$ to $|b_\downarrow\rangle$ by $-\sqrt{2}V_k$, where $V_k=V_k^{(A)}=V_k^{(B)}$. In other words, in this symmetric case the triplet states are only connected with the $|b_\sigma\rangle$-states. At the same time, $|S0\rangle=(|\uparrow,\downarrow\rangle-|\downarrow,\uparrow\rangle)/\sqrt{2}$ is only connected with the $|a_\sigma\rangle$-states by the interaction $\pm V_k$, while the $|a_\sigma\rangle$-states are also connected to the $|0\rangle$-state by the $\sqrt{2}V_k$ interaction. This indicates that in this problem, we have two groups of wavefunctions connected by the atom-metal interaction. The first group is defined by (i) the triplet states, $|T\uparrow\rangle$, $|T0\rangle$ and $T\downarrow\rangle$, and the $b_\sigma\rangle$-states, and the second one by the states: (ii) $|S0\rangle$ and $|a_\sigma\rangle$ and $|0\rangle$. Now, for an antiferromagnetic interaction ($J>0$) the singlet state, $|S0\rangle$, defines the ground state and we can expect the group of states $|S0\rangle$ and $|a_\sigma\rangle$ to define the contribution to the tunneling current; on the contrary, for $J<0$, the group of states $|T\uparrow\rangle$, $|T0\rangle$, $T\downarrow\rangle$ and $b_\sigma\rangle$ control that tunneling current.

We use the EOM-approach to analyze this problem. Assume first that $J<0$; then, in a first step we introduce the following creation operators:

$$|T\uparrow\rangle\langle b_\uparrow|,\quad |T0\rangle\langle b_\uparrow|,\quad |T0\rangle\langle b_\downarrow|,\quad \text{and}\quad |T\downarrow\rangle\langle b_\downarrow|, \tag{9.10}$$

associated with the creation of one electron in the $|b_\sigma\rangle$-states, and define the different Green's functions:

$$G_{|T\uparrow\rangle\langle b_\uparrow|}\left(|b_\uparrow\rangle\langle T\uparrow|\right)=i\Theta(t'-t)\left\langle\left\{|T\uparrow\rangle\langle b_\uparrow|_{t'};\,|b_\uparrow\rangle\langle T\uparrow|_t\right\}\right\rangle, \tag{9.11a}$$
$$G_{|T0\rangle\langle b_\uparrow|}\left(|b_\uparrow\rangle\langle T0|\right)=i\Theta(t'-t)\left\langle\left\{|T0\rangle\langle b_\uparrow|_{t'};\,|b_\uparrow\rangle\langle T0|_t\right\}\right\rangle, \tag{9.11b}$$
$$G_{|T0\rangle\langle b_\downarrow|}\left(|b_\downarrow\rangle\langle T0|\right)=i\Theta(t'-t)\left\langle\left\{|T0\rangle\langle b_\downarrow|_{t'};\,|b_\downarrow\rangle\langle T0|_t\right\}\right\rangle, \tag{9.11c}$$
$$G_{|T\downarrow\rangle\langle b_\downarrow|}\left(|b_\downarrow\rangle\langle T\downarrow|\right)=i\Theta(t'-t)\left\langle\left\{|T\downarrow\rangle\langle b_\downarrow|_{t'};\,|b_\downarrow\rangle\langle T\downarrow|_t\right\}\right\rangle. \tag{9.11d}$$

Then, we calculate their EOM up to second order in the metal/dimer interaction V_k [19, 20]. In this calculation, off-diagonal components like $G_{|T0\rangle\langle b_\downarrow|}\left(|b_\uparrow\rangle\langle T\uparrow|\right)$ are found to be zero. Fourier-transforming the resulting time-dependent Green functions, we obtain the following equations (with $\omega\equiv\omega-i0^+$, consistently with the advanced character of the Green's functions):

$$\left[\omega-(\varepsilon_0+E_T)-2\alpha-\beta'\right]G_{|T\uparrow\rangle\langle b_\uparrow|}\left(|b_\uparrow\rangle\langle T\uparrow|\right)=C, \tag{9.12a}$$
$$\left[\omega-(\varepsilon_0+E_{T0})-2\alpha+\beta-2\beta''\right]G_{|T0\rangle\langle b_\downarrow|}\left(|b_\downarrow\rangle\langle T0|\right)=D, \tag{9.12b}$$

where

$$\alpha=\sum_k\frac{|V_k|^2}{\omega-\varepsilon_k}, \tag{9.13a}$$

$$\beta = \sum_k \frac{n_k |V_k|^2}{\omega - \varepsilon_k}, \tag{9.13b}$$

$$\beta' = \sum_k \frac{n_k |V_k|^2}{\omega - \varepsilon_k - E_T + E_{T0}}, \tag{9.13c}$$

$$\beta'' = \sum_k \frac{n_k |V_k|^2}{\omega - \varepsilon_k + E_T - E_{T0}}, \tag{9.13d}$$

$$C = \langle |T \uparrow\rangle\langle T \uparrow| + |b_\uparrow\rangle\langle b_\uparrow| \rangle, \tag{9.13e}$$

$$D = \langle |T0\rangle\langle T0| + |b_\uparrow\rangle\langle b_\uparrow| \rangle. \tag{9.13f}$$

For the sake of simplicity, we have neglected on the r.h.s. of Eq. (9.12) contributions of order $|V_k|^2$ as compared with C or D; however, in our numerical calculations shown below these second order terms have been taken into account (see Ref. [20]). Similar equations are obtained for other Green's functions changing the spin sign. In Eq. (9.12) we have taken E_T and E_{T0} as the energies of the triplet states with $M = \pm 1$ and $M = 0$, respectively. For the $\frac{1}{2}$-$\frac{1}{2}$ case, $E_T = E_{T0}$ and $\beta = \beta' = \beta''$; then Eq. (9.12a) is equivalent to Eq. (9.12b), both equations showing a Kondo resonance corresponding to $S = 1$ [22]; our calculations for the tunneling current (see below) confirm this result.

For $J > 0$ we have to consider the $|S0\rangle$, $|a_\sigma\rangle$ and $|0\rangle$-states; then, we build up the following creation operators:

$$|S0\rangle\langle a_\uparrow|, \quad |S0\rangle\langle a_\downarrow|, \quad |a_\uparrow\rangle\langle 0|, \quad \text{and} \quad |a_\downarrow\rangle\langle 0|, \tag{9.14}$$

and their corresponding Green's functions like:

$$G_{|S0\rangle\langle a_\uparrow|}\left(|a_\downarrow\rangle\langle S0|\right), \; G_{|a_\uparrow\rangle\langle 0|}\left(|0\rangle\langle a_\uparrow|\right), \; G_{|a_\uparrow\rangle\langle 0|}\left(|a_\downarrow\rangle\langle S0|\right), \text{ and } G_{|S0\rangle\langle a_\downarrow|}\left(|0\rangle\langle a_\uparrow|\right). \tag{9.15}$$

An EOM calculation up to second order in V_k yields the following equations:

$$\begin{aligned}
&[\omega - (\varepsilon_0 + E_S) - 2\alpha + \beta]\, G_{|S0\rangle\langle a_\downarrow|}\left(|a_\downarrow\rangle\langle S0|\right) \\
&\qquad + \sqrt{2}\beta G_{|S0\rangle\langle a_\downarrow|}\left(|0\rangle\langle a_\uparrow|\right) = \langle |S0\rangle\langle S0| + |a_\downarrow\rangle\langle a_\downarrow| \rangle, && (9.16a)\\
&[\omega - \varepsilon_0 - 2\alpha - 2\beta]\, G_{|S0\rangle\langle a_\downarrow|}\left(|0\rangle\langle a_\uparrow|\right) \\
&\qquad + \sqrt{2}\beta G_{|S0\rangle\langle a_\downarrow|}\left(|a_\downarrow\rangle\langle S0|\right) = 0, && (9.16b)\\
&[\omega - (\varepsilon_0 + E_S) - 2\alpha + \beta]\, G_{|a_\uparrow\rangle\langle 0|}\left(|a_\downarrow\rangle\langle S0|\right) \\
&\qquad + \sqrt{2}\beta G_{|a_\uparrow\rangle\langle 0|}\left(|0\rangle\langle a_\uparrow|\right) = 0, && (9.16c)\\
&[\omega - \varepsilon_0 - 2\alpha - 2\beta]\, G_{|a_\uparrow\rangle\langle 0|}\left(|0\rangle\langle a_\uparrow|\right) \\
&\qquad + \sqrt{2}\beta G_{|a_\uparrow\rangle\langle 0|}\left(|a_\downarrow\rangle\langle S0|\right) = \langle |a_\uparrow\rangle\langle a_\uparrow| + |0\rangle\langle 0| \rangle. && (9.16d)
\end{aligned}$$

These equations provide all the different Green's functions components introduced above. Notice in these equations the mixing of the Green's functions associated with the operators $|a_\uparrow\rangle\langle S0|$ and $|0\rangle\langle a_\uparrow|$; both operators represent the annihilation of one spin up-electron. A similar mixing appears between the Green's functions with the operators $|a_\downarrow\rangle\langle S0|$ and $|0\rangle\langle a_\downarrow|$, associated with the annihilation of one spin down-electron.

The Green's functions, Eqs. (9.12) and (9.16), provide the necessary information for obtaining the different quantities we are interested in. In particular, for calculating the tunneling currents we introduce the tip-dimer interaction $V_{k'}^{(A)} = V_{k'}^{(B)}$ as a perturbation, and use the following equation for the differential conductance G [19], up to a second order in $V_{k'}$:

$$G/G_0 = 4 \sum_{p,q,i,j} \Gamma_{pq,ij} \mathrm{Im} G_{|p\rangle\langle q|} \left(|i\rangle\langle j|\right). \tag{9.17}$$

In Eq. (9.17), G_0 is the quantum conductance $2e^2/h$, and $G_{|p\rangle\langle q|}\left(|i\rangle\langle j|\right)$ represents the different Green's functions defined in Eqs. (9.11) and (9.15), and associated with the set of creation operators: $\{|T\uparrow\rangle\langle b_\uparrow|,\ |T0\rangle\langle b_\downarrow|,\ |S0\rangle\langle a_\downarrow|,\ |a_\uparrow\rangle\langle 0|\}$ and $\{|T\uparrow\rangle\langle a_\uparrow|,\ |T0\rangle\langle a_\downarrow|,\ |S0\rangle\langle b_\downarrow|,\ |b_\uparrow\rangle\langle 0|\}$. The quantities $\Gamma_{pq,ij}$ are defined by the tip/dimer hoppings, $V_{k'iq}^{(A/B)}$ and $V_{k'pj}^{(A/B)*}$ connecting the wavefunctions $|i\rangle$ and $|q\rangle$, or $|p\rangle$ and $|j\rangle$ in $V_{k'iq}\hat{c}_{k'}^\dagger|i\rangle\langle q|$ or $V_{k'pj}^*\hat{c}_{k'}^\dagger|p\rangle\langle j|$:

$$\Gamma_{pq,ij} = \pi \sum_{k'} V_{k'iq} V_{k'iq}^* \delta(\varepsilon - \varepsilon_{k'}). \tag{9.18}$$

In particular, for $J < 0$, we find the following contribution from the $|T\uparrow\rangle$, $|T0\rangle$ and $|T\downarrow\rangle$, and $|b_\sigma\rangle$-states, for spin up electrons:

$$G^{(\uparrow)}/G_0 = 4\Gamma_{b_\uparrow}^{T\uparrow} \mathrm{Im} G_{|T\uparrow\rangle\langle b_\uparrow|}\left(|b_\uparrow\rangle\langle T\uparrow|\right) + 4\Gamma_{b_\downarrow}^{T0} \mathrm{Im} G_{|T0\rangle\langle b_\downarrow|}\left(|b_\downarrow\rangle\langle T0|\right), \tag{9.19}$$

where $\Gamma_{b_\uparrow}^{T\uparrow} = \pi \sum_{k'} |\sqrt{2}V_{k'}|^2 \delta(\omega - \varepsilon_{k'})$ and $\Gamma_{b_\downarrow}^{T0} = \pi \sum_{k'} |V_{k'}|^2 \delta(\omega - \varepsilon_{k'})$ are effective broadenings associated with the atom/tip interaction. In Eq. (9.19), G_0 is the quantum conductance $2e^2/h$.

For $J > 0$ the $|S0\rangle$, $|a_\sigma\rangle$ and $|0\rangle$-states contribute to G, and we find the following differential conductance:

$$\begin{aligned} G^{(\uparrow)}/G_0 =& 4\Gamma_{a_\downarrow}^{S0} \mathrm{Im} G_{|S0\rangle\langle a_\downarrow|}\left(|a_\downarrow\rangle\langle S0|\right) + 4\Gamma_0^{a_\uparrow} \mathrm{Im} G_{|a_\uparrow\rangle\langle 0|}\left(|0\rangle\langle a_\uparrow|\right) \\ &+ 4\Gamma_{a_\downarrow,0}^{S0,a_\uparrow} \mathrm{Im} G_{|a_\uparrow\rangle\langle 0|}\left(|a_\downarrow\rangle\langle S0|\right) + 4\Gamma_{0,a_\downarrow}^{a_\uparrow,S0} \mathrm{Im} G_{|S0\rangle\langle a_\downarrow|}\left(|0\rangle\langle a_\uparrow|\right) \end{aligned} \tag{9.20}$$

where

$$\Gamma_0^{a_\uparrow} = \pi \sum_{k'} |\sqrt{2}V_{k'}|^2 \delta(\omega - \varepsilon_{k'}), \tag{9.21a}$$

$$\Gamma_{a_\downarrow}^{S0} = \pi \sum_{k'} |V_{k'}|^2 \delta(\omega - \varepsilon_{k'}), \tag{9.21b}$$

$$\Gamma_{a_\downarrow,0}^{S0,a_\uparrow} = \Gamma_{0,a_\downarrow}^{a_\uparrow,S0} = \pi \sum_{k'} |V_{k'}|^2 \delta(\omega - \varepsilon_{k'}). \tag{9.21c}$$

Equations (9.12)–(9.20) allow us to calculate the differential conductance in this fully symmetric case. Notice that for a ferromagnetic case with $E_T = E_{T0}$ and $S = 1$, we have to use Eq. (9.12) taking $C = \langle |T \uparrow\rangle\langle T \uparrow| + |b_\uparrow\rangle\langle b_\uparrow| \rangle \simeq \langle |T \uparrow\rangle\langle T \uparrow| \rangle \approx \frac{1}{3}$, and $D = \langle |T0\rangle\langle T0| + |b_\downarrow\rangle\langle b_\downarrow| \rangle \simeq \langle |T0\rangle\langle T0| \rangle \approx \frac{1}{3}$. For an antiferromagnetic case, we have to use Eq. (9.16) and take $\langle |S0\rangle\langle S0| + |a_\downarrow\rangle\langle a_\downarrow| \rangle \simeq \langle |S0\rangle\langle S0| \rangle \approx 1$, as well as $\langle |a_\uparrow\rangle\langle a_\uparrow| + |0\rangle\langle 0| \rangle \simeq 0$.

9.3.2 Dimer with Quasi-independent Atoms

We consider in this section that the atom/metal interactions, $V_k^{(A)}$ and $V_k^{(B)}$, are such that $V_k^{(A)} = V_k^{(B)*} = V_k \exp(i\mathbf{k} \cdot \mathbf{R}/2)$, $\mathbf{R}$ being the vector joining atoms A and B (provisionally, we also assume $V_{k'} = 0$). In general, this atom/metal interaction introduces a mixing of the different groups of states, $\{|T \uparrow\rangle, |T0\rangle, |T \downarrow\rangle, |b_\sigma\rangle\}$, and $\{|S0\rangle, |a_\sigma\rangle, |0\rangle\}$ appearing in the previous discussion. Before considering the EOMs that we find for this case, it is convenient to mention that in those equations, calculated up to second order in V_k, we find terms proportional to $V_k^{(A)} V_k^{(A)*}$, $V_k^{(B)} V_k^{(B)*}$, $V_k^{(A)} V_k^{(B)*}$, and $V_k^{(B)} V_k^{(A)*}$. Notice that the last two terms are proportional to $\sum_k \exp(\pm i\mathbf{k} \cdot \mathbf{R})$, in such a way that the angular integration in k yields the factor $\sin(kR)/(kR)$ (assuming a spherically symmetric metal energy band). For Cu, $k_F = 1.36$ Å^{-1}, so that for the typical atom/atom distances [7], $R > 2$ Å, we find $|\sin(k_F R)/(k_F R)| < 0.2$; this suggests to neglect contributions in the EOMs proportional to $V_k^{(A)} V_k^{(B)*}$ or $V_k^{(B)} V_k^{(A)*}$, as is done in the following discussion [21]. Apparently, this approximation corresponds to having two independent atoms; notice however that we still have the Heisenberg interaction operating between the two atoms, making this case a quasi-independent two atoms problem.

In our EOM approach for this case, we find that the group of creation operators (spin up): $\{|T \uparrow\rangle\langle b_\uparrow|, |T0\rangle\langle b_\downarrow|, |S0\rangle\langle a_\downarrow|, |a_\uparrow\rangle\langle 0|\}$ get mixed by the atom/metal interaction, in such a way that the previous Eqs. (9.12) and (9.16) found for the very symmetric case should be reformulated to include the corresponding off-diagonal Green's function terms (including also the fact that $V_k^{(A)} V_k^{(B)*} = V_k^{(B)} V_k^{(A)*} = 0$). Moreover, with the metal/dimer interactions defined above, $V_k^{(A)} = V_k^{(B)*}$, we also find that the following group of creation operators (spin up): $\{|T \uparrow\rangle\langle a_\uparrow|, |T0\rangle\langle a_\downarrow|, |S0\rangle\langle b_\downarrow|, |b_\uparrow\rangle\langle 0|\}$ yield new Green's functions that we also analyze below.

The EOMs, up to second order in V_k, for the Green's functions associated with the $\{|T \uparrow\rangle\langle b_\uparrow|, |T0\rangle\langle b_\downarrow|, |S0\rangle\langle a_\downarrow|, |a_\uparrow\rangle\langle 0|\}$-operators, yield:

$$\left[\omega-(\varepsilon_0+E_T)-2\alpha+\beta-\tfrac{1}{2}(\beta'+\gamma)\right]G_{|T\uparrow\rangle\langle b_\uparrow|}\left(|b_\uparrow\rangle\langle T\uparrow|\right)=C+\beta G_{|T\uparrow\rangle\langle b_\uparrow|}\left(|0\rangle\langle a_\uparrow|\right), \tag{9.22a}$$

$$\begin{aligned}[\omega-\varepsilon_0-\alpha-3\beta]\,G_{|T\uparrow\rangle\langle b_\uparrow|}\left(|0\rangle\langle a_\uparrow|\right)&=(\beta-\alpha)G_{|T\uparrow\rangle\langle b_\uparrow|}\left(|b_\uparrow\rangle\langle T\uparrow|\right)\\&+\tfrac{1}{\sqrt{2}}(\beta-\alpha)G_{|T\uparrow\rangle\langle b_\uparrow|}\left(|b_\downarrow\rangle\langle T0|\right)-\tfrac{1}{\sqrt{2}}(\beta-\alpha)G_{|T\uparrow\rangle\langle b_\uparrow|}\left(|a_\downarrow\rangle\langle S0|\right),\end{aligned} \tag{9.22b}$$

$$\left[\omega-(\varepsilon_0+E_{T0})-2\alpha+\tfrac{3}{2}\beta-\beta''-\tfrac{1}{2}\gamma_0\right]G_{|T\uparrow\rangle\langle b_\uparrow|}\left(|b_\downarrow\rangle\langle T0|\right)=\tfrac{1}{\sqrt{2}}\beta G_{|T\uparrow\rangle\langle b_\uparrow|}\left(|0\rangle\langle a_\uparrow|\right), \tag{9.22c}$$

$$\left[\omega-(\varepsilon_0+E_{S0})-2\alpha-\tfrac{3}{2}\beta-\gamma'-\tfrac{1}{2}\gamma_0'\right]G_{|T\uparrow\rangle\langle b_\uparrow|}\left(|a_\downarrow\rangle\langle S0|\right)=-\tfrac{1}{\sqrt{2}}\beta G_{|T\uparrow\rangle\langle b_\uparrow|}\left(|0\rangle\langle a_\uparrow|\right), \tag{9.22d}$$

where

$$\gamma=\sum_k\frac{|V_k|^2}{\omega-\varepsilon_k-E_T+E_{S0}}, \tag{9.23a}$$

$$\gamma'=\sum_k\frac{|V_k|^2}{\omega-\varepsilon_k+E_T-E_{S0}}, \tag{9.23b}$$

$$\gamma_0=\sum_k\frac{|V_k|^2}{\omega-\varepsilon_k-E_{T0}+E_{S0}}, \tag{9.23c}$$

$$\gamma_0'=\sum_k\frac{|V_k|^2}{\omega-\varepsilon_k+E_{T0}-E_{S0}}. \tag{9.23d}$$

The energy contributions associated with the Heisenberg interaction, $J\hat{\mathbf{S}}_1\cdot\hat{\mathbf{S}}_2$, have been included in the terms with E_T, E_{T0}, and E_{S0} ($E_T=E_{T0}$ for the $\frac{1}{2}$-$\frac{1}{2}$ case).

Likewise, we also find the following equations associated with the operators $\{|T\uparrow\rangle\langle a_\uparrow|, |T0\rangle\langle a_\downarrow|, |S0\rangle\langle b_\downarrow|, |b_\uparrow\rangle\langle 0|\}$:

$$\left[\omega-(\varepsilon_0+E_T)-2\alpha+\beta-\tfrac{1}{2}(\beta'+\gamma)\right]G_{|T\uparrow\rangle\langle a_\uparrow|}\left(|a_\uparrow\rangle\langle T\uparrow|\right)=F-\beta G_{|T\uparrow\rangle\langle a_\uparrow|}\left(|0\rangle\langle b_\uparrow|\right), \tag{9.24a}$$

$$\begin{aligned}[\omega-\varepsilon_0-\alpha-3\beta]\,G_{|T\uparrow\rangle\langle a_\uparrow|}\left(|0\rangle\langle b_\uparrow|\right)&=-(\beta-\alpha)G_{|T\uparrow\rangle\langle a_\uparrow|}\left(|a_\uparrow\rangle\langle T\uparrow|\right)\\&-\tfrac{1}{\sqrt{2}}(\beta-\alpha)G_{|T\uparrow\rangle\langle a_\uparrow|}\left(|a_\downarrow\rangle\langle T0|\right)+\tfrac{1}{\sqrt{2}}(\beta-\alpha)G_{|T\uparrow\rangle\langle a_\uparrow|}\left(|b_\downarrow\rangle\langle S0|\right),\end{aligned} \tag{9.24b}$$

$$\left[\omega-(\varepsilon_0+E_{T0})-2\alpha+\tfrac{3}{2}\beta-\beta''-\tfrac{1}{2}\gamma_0\right]G_{|T\uparrow\rangle\langle a_\uparrow|}\left(|a_\downarrow\rangle\langle T0|\right)=-\tfrac{1}{\sqrt{2}}\beta G_{|T\uparrow\rangle\langle a_\uparrow|}\left(|0\rangle\langle b_\uparrow|\right), \tag{9.24c}$$

$$\left[\omega-(\varepsilon_0+E_{S0})-2\alpha-\tfrac{3}{2}\beta-\gamma'-\tfrac{1}{2}\gamma_0'\right]G_{|T\uparrow\rangle\langle a_\uparrow|}\left(|a_\downarrow\rangle\langle S0|\right)=\tfrac{1}{\sqrt{2}}\beta G_{|T\uparrow\rangle\langle a_\uparrow|}\left(|0\rangle\langle a_\uparrow|\right), \tag{9.24d}$$

where $F=\left\langle |a_\uparrow\rangle\langle a_\uparrow|+|T\uparrow\rangle\langle T\uparrow|\right\rangle$. Notice that Eqs. (9.22) and (9.24) are similar, differing only in the sign of their r.h.s.; this is related to the fact that changing from the basis set from $\{|a_\sigma\rangle = \frac{1}{\sqrt{2}}(|\sigma,0\rangle + |0,\sigma\rangle), |b_\sigma\rangle = \frac{1}{\sqrt{2}}(|\sigma,0\rangle - |0,\sigma\rangle)\}$ to $\{|\sigma,0\rangle = \frac{1}{\sqrt{2}}(|a_\sigma\rangle + |b_\sigma\rangle), |0,\sigma\rangle = \frac{1}{\sqrt{2}}(|a_\sigma\rangle - |b_\sigma\rangle)\}$ yields equivalent equations for sites A and B, corresponding to the set of operators $\{|T\uparrow\rangle\langle\uparrow,0|, |T0\rangle\langle\downarrow,0|, |S0\rangle\langle\downarrow,0|, |0,\uparrow\rangle\langle 0|\}$ and $\{|T\uparrow\rangle\langle 0,\uparrow|, |T0\rangle\langle 0,\downarrow|, |S0\rangle\langle 0,\downarrow|, |\uparrow,0\rangle\langle 0|\}$, respectively.

Other Green's functions with different arguments, such as $G_{|0\rangle\langle a_\uparrow|}\left(|b_\uparrow\rangle\langle T\uparrow|\right)$ or $G_{|0\rangle\langle a_\uparrow|}\left(|0\rangle\langle a_\uparrow|\right)$, can be obtained from a similar set of equations changing only the independent terms; for example, for the above mentioned Green's functions one should include the independent term $\left\langle |0\rangle\langle 0| + |a_\uparrow\rangle\langle a_\uparrow|\right\rangle$ in the corresponding equation (9.22b).

From these Green's functions one can also calculate the tunneling currents assuming to have some tip/dimer interaction like $V_{k'}^{(A)}$ or $V_{k'}^{(B)}$. For simplicity, we will assume $V_{k'}^{(B)} = 0$ and take $|V_{k'}^{(A)}| = V_{k'}$. Then the tunneling current through atom A is given by:

$$G_A/G_0 = 4\sum \Gamma'_{pq,ij} \mathrm{Im} G_{|p\rangle\langle q|}\left(|i\rangle\langle j|\right). \tag{9.25}$$

Equation (9.25) includes all the Green's functions associated with the sets $\{|T\uparrow\rangle\langle b_\uparrow|, |T0\rangle\langle b_\downarrow|, |S0\rangle\langle a_\downarrow|, |a_\downarrow\rangle\langle 0|\}$ and $\{|T\uparrow\rangle\langle a_\uparrow|, |T0\rangle\langle a_\downarrow|, |S0\rangle\langle b_\downarrow|, |b_\downarrow\rangle\langle 0|\}$. The quantities $\Gamma'_{pq,ij}$ are defined by the hoppings connecting the elements $\langle q|i\rangle$ and $\langle j|p\rangle$ in the way explained above in Eq. (9.18), but for the sake of brevity we will not present more details here.

9.4 Results and Discussion

We start presenting results for the fully symmetric $\frac{1}{2}$-$\frac{1}{2}$ case. In our calculations we use a flat-band approximation for the metal with a half band-width of 10 eV; we also take $\varepsilon_0^A = \varepsilon_0^B = \varepsilon_0 = -1$ eV (the Fermi level is the energy zero), and the following level widths $\Gamma^A = \Gamma^B = \Gamma = \pi\sum_k |V_k|^2\delta(\omega - \varepsilon_k) = 50$ meV. The exchange interaction, J, is taken to be -0.05, -0.1, -0.2 and -0.6 meV for a ferromagnetic interaction, and 0.05, 0.1, 0.2 and 0.6 meV for an antiferromagnetic one; in our calculations we assume, in all the cases, the thermal energy, $k_B T$, to be much smaller than the magnetic interaction between spins ($T = 0.3$ K).

For the $\frac{1}{2}$-$\frac{1}{2}$ case we consider here for the ferromagnetic case ($J < 0$), $E_T = E_{T0}$ and $\beta = \beta' = \beta''$ in Eq. (9.12); moreover, in these equations

$$C = \left\langle |T\uparrow\rangle\langle T\uparrow| + |b_\uparrow\rangle\langle b_\uparrow|\right\rangle \approx \frac{1}{3}, \text{ and } D = \left\langle |T0\rangle\langle T0| + |b_\uparrow\rangle\langle b_\uparrow|\right\rangle \approx \frac{1}{3},$$

so that, as mentioned above, Eqs. (9.12a) and (9.12b) coincide, both equations yielding a Kondo resonance corresponding to $S = 1$ [22]. The results for the differential conductance, see Fig. 9.2, as a function of the applied bias, V, show that resonance around $V = 0$. Notice that our results are practically independent from the value of J; this is easily understood by realizing that in those equations the effect of J is only to shift to $\varepsilon_0 - J/4$ the effective level ε_0.

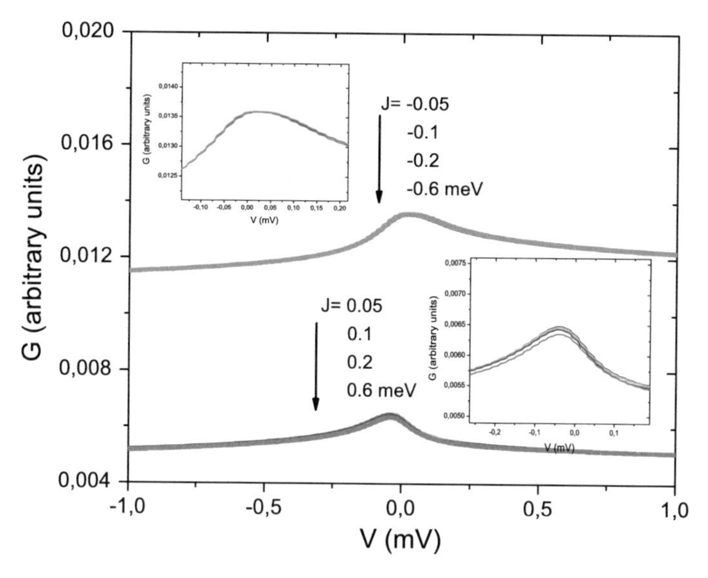

Fig. 9.2 It shows the differential conductance as a function of the applied voltage for the $\frac{1}{2}$-$\frac{1}{2}$ dimer in the fully symmetric case, Eqs. (9.12–9.20). The upper curves correspond to a ferromagnetic interaction, the curves below to an antiferromagnetic one ($|J| = 0.05, 0.1, 0.2, 0.6\,\text{meV}$). The atom energy level is $-1\,\text{eV}$ and the level width $0.05\,\text{eV}$

For the antiferromagnetic case, the singlet $|S0\rangle$ defines the ground state with $E_{S0} = \varepsilon_0 - \frac{3}{4}J$, so that $\langle |S0\rangle\langle S0| \rangle \approx 1$ and $\left\langle |a_\uparrow\rangle\langle a_\uparrow| \right\rangle \approx \left\langle |a_\downarrow\rangle\langle a_\downarrow| \right\rangle \approx \langle |0\rangle\langle 0| \rangle \approx 0$. Then, Eqs. (9.16) and (9.20) yield the differential conductance shown in Fig. 9.2 for different values of J ($J > 0$). These curves also show a very small dependence on J; as in the previous case, this is due to the small change in the energy, from ε_0 to $\varepsilon_0 - \frac{3}{4}J$. On the other hand, it might be unexpected to see the Kondo resonance that also appears in this case, since there is a singlet in the ground state. A careful analysis of Eqs. (9.16a) and (9.16b) shows that that Kondo resonance is a reflection in the differential conductance of the Kondo state associated with the fluctuations between the states $|a_\uparrow\rangle$, $|a_\downarrow\rangle$ and $|0\rangle$, as shown by the off-diagonal Green's function $G_{|S0\rangle\langle a_\downarrow|}\left(|0\rangle\langle a_\uparrow|\right)$ in Eqs. (9.16a) and (9.16b).

Figure 9.3 shows the differential conductance for the antiferromagnetic quasi-independent atoms case, using the same parameters as in the fully symmetric case: $\varepsilon_0^A = \varepsilon_0^B = \varepsilon_0 = -1\,\text{eV}$ and $\Gamma^A = \Gamma^B = \Gamma = \pi \sum_k |V_k|^2 \delta(\omega - \varepsilon_k) = 50\,\text{meV}$; we also assume to have a similar Heisenberg interaction between atoms: $J = 0.05$, 0.10, 0.20 and 0.60 meV. First point to notice is that for $J \to 0$, a Kondo resonance around $V = 0$ starts to develop; this is well understood by considering that in this limit the atoms are completely independent from each other, in such a way that each atom develops a Kondo resonance associated with the $\{|\uparrow, 0\rangle, |\downarrow, 0\rangle, |0, 0\rangle\}$ or the $\{|0, \uparrow\rangle, |0, \downarrow\rangle, |0, 0\rangle\}$ fluctuations. The effect of introducing the Heisenberg interaction is to break that Kondo resonance, by creating a dip around the Fermi energy between the energies $\pm(E_T - E_S) = \pm J$, energies that are associated with the excitation from the singlet to the triplet created by the tunneling electrons.

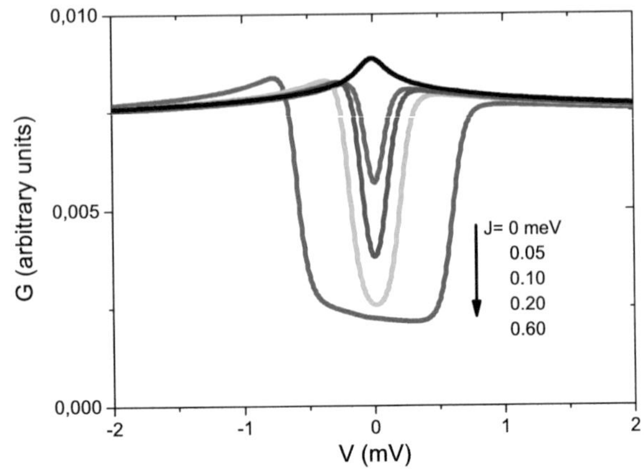

Fig. 9.3 Shows the differential conductance, G, as a function of the applied voltage for the dimer $\frac{1}{2}$-$\frac{1}{2}$ in the quasi-independent atoms case, Eqs. (9.22)–(9.24), for an antiferromagnetic interaction $J = 0.05, 0.1, 0.2, 0.6$ meV. The atom energy level is 1 eV and the level width is 0.05 eV

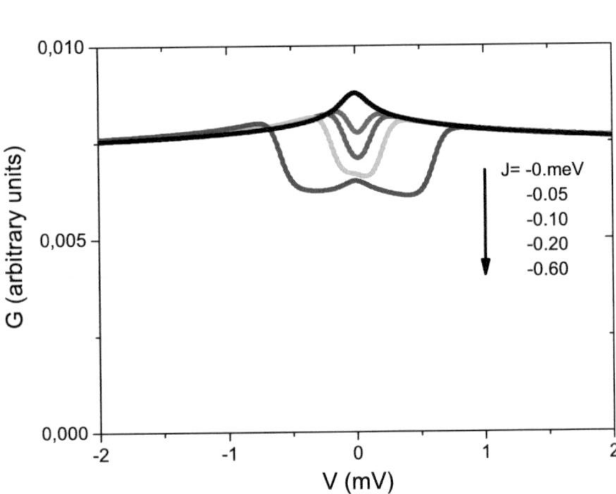

Fig. 9.4 The same as in Fig. 9.2 for a ferromagnetic interaction $J = -0.05$, -0.1, -0.2, -0.6 meV

Figure 9.4 shows similar results for a ferromagnetic interaction. Things are similar to the previous case, with some slight difference. As in the previous case, for $J \to 0$ we recover the Kondo resonance of the isolated atom; the effect of J is also to create a dip around $V = 0$ (the Fermi energy) between $\pm J$. Two differences appear with respect to the previous case: for the ferromagnetic interaction the dip is roughly three times smaller than the one found for the antiferromagnetic case (probably associated with the degeneracy of the triplet state in the ferro case). On the other hand, for very large values of J, we start to see in the differential conductance another resonance around $V = 0$; this is probably associated with the Kondo resonance associated with a triplet state (for J very large, the system should behave like a triplet interacting with the metal density of states).

Finally, Fig. 9.5 shows our results for three different cases (within the quasi-independent atoms model): (a) the $\frac{1}{2}$-$\frac{1}{2}$ dimer; (b) the simplified Co-Co dimer as discussed in this paper; and (c) the Co-Co dimer as analyzed in reference [21], including the lowest energy states for each atom (four considering the $\frac{3}{2} * \frac{3}{2}$ con-

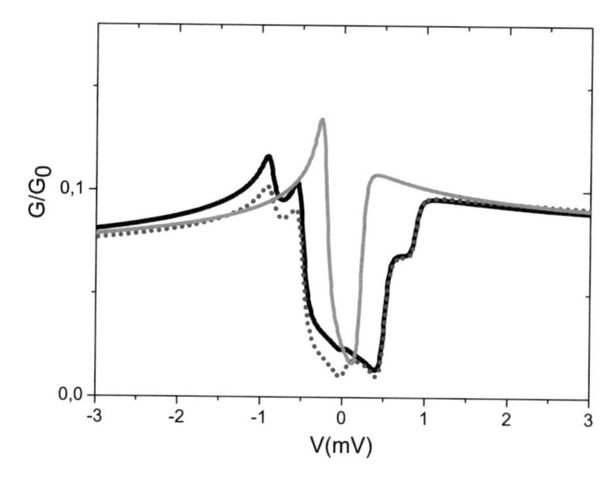

Fig. 9.5 Differential conductance as a function of applied voltage. Black solid line is the simplified Co-Co dimer, gray solid line is the $\frac{1}{2}$-$\frac{1}{2}$ dimer, the dotted line is the Co-Co dimer by considering only the lowest energy states: four in the case of $\frac{3}{2} * \frac{3}{2}$ configurations, two in the case of $\frac{3}{2} * 1$, two in the case of $1 * \frac{3}{2}$ and one in the $1 * 1$ case (see Ref. [21]). The energy (1 eV) and width (0.2 eV) of the Co level are the same used in Ref. [21]

figurations; two in the $1 * \frac{3}{2}$ case; two in the $\frac{3}{2} * 1$ and one in the case $1 * 1$). In these calculations we have taken the following parameters: $\varepsilon_0^A = \varepsilon_0^B = \varepsilon_0 = -1$ eV, $\Gamma^A = \Gamma^B = \Gamma = \pi \sum_k |V_k|^2 \delta(\omega - \varepsilon_k) = 200$ meV, and $J = 0.22$ meV. For the $\frac{1}{2}$-$\frac{1}{2}$ dimer we find results similar to the ones presented in Figs. 9.3 and 9.4: a dip around the Fermi level between energies $\pm J$. For the simplified Co-Co dimer, we find that the dip is surrounded by two steps associated with the excitation energies $\pm|E_T - E_S|$ and $\pm|E_{T0} - E_S|$, namely $5J/2$ and $4J$. For the sake of completeness, we also show similar results calculated for a Co-Co dimer as analyzed in detail including the minimum basis necessary to calculate the differential conductance of the dimer [21]. In Fig. 9.5, we also show the results calculated for this case; this is an indication of the good approximation that our simplified Co-Co dimer represents for a realistic dimer for V smaller than ± 3 meV.

9.5 Summary and Concluding Remarks

In conclusion, we have presented a discussion of the electronic and transport properties of a $\frac{1}{2}$-$\frac{1}{2}$ dimer and have shown how to extend that analysis to a Co-Co dimer. Our results also show that these properties depend crucially on the metal-dimer interaction. In particular, for a $\frac{1}{2}$-$\frac{1}{2}$ fully symmetric case our results indicate that the system develops, for both a ferro- and an antiferromagnetic interaction between the atoms, a Kondo resonance. However, for more realistic interactions we find, for both, $\frac{1}{2}$-$\frac{1}{2}$ and Co-Co cases, that the differential conductance has a typical dip around the Fermi energy that is surrounded by some steps whose energies depend on the triplet and singlet levels. The differences and similarities between the $\frac{1}{2}$-$\frac{1}{2}$ and the Co-Co cases are analyzed and discussed in detail.

Acknowledgements ECG acknowledges financial support by CONICET through Grant No. PIP-201101-00621 and U.N.L. through CAI+D grants. FF acknowledges support from the Spanish Ministerio de Economía y Competitividad (MINECO) under project MAT2014-59966-R, and through the 'María de Maeztu' Programme for Units of Excellence in R&D (MDM-2014-0377).

References

1. C.F. Hirjibehedin, C.P. Lutz, A.J. Heinrich, Science **312**(5776), 1021 (2006). https://doi.org/10.1126/science.1125398
2. C.F. Hirjibehedin, C.Y. Lin, A.F. Otte, M. Ternes, C.P. Lutz, B.A. Jones, A.J. Heinrich, Science **317**(5842), 1199 (2007). https://doi.org/10.1126/science.1146110
3. A.F. Otte, M. Ternes, K. von Bergmann, S. Loth, H. Brune, C.P. Lutz, C.F. Hirjibehedin, A.J. Heinrich, Nat. Phys. **4**(11), 847 (2008). https://doi.org/10.1038/nphys1072
4. R. Bulla, T.A. Costi, T. Pruschke, Rev. Mod. Phys. **80**, 395 (2008). https://doi.org/10.1103/RevModPhys.80.395
5. A.F. Otte, M. Ternes, S. Loth, C.P. Lutz, C.F. Hirjibehedin, A.J. Heinrich, Phys. Rev. Lett. **103**, 107203 (2009). https://doi.org/10.1103/PhysRevLett.103.107203
6. H. Brune, P. Gambardella, Surf. Sci. **603**(10), 1812 (2009). https://doi.org/10.1016/j.susc.2008.11.055. (Special Issue of Surface Science dedicated to Prof. Dr. Dr. h.c. mult. Gerhard Ertl, Nobel-Laureate in Chemistry 2007)
7. A. Spinelli, M. Gerrits, R. Toskovic, B. Bryant, M. Ternes, A.F. Otte, **6**, 10046 (2015). https://doi.org/10.1038/ncomms10046
8. B. Bryant, A. Spinelli, J.J.T. Wagenaar, M. Gerrits, A.F. Otte, Phys. Rev. Lett. **111**, 127203 (2013). https://doi.org/10.1103/PhysRevLett.111.127203
9. B. Bryant, R. Toskovic, A. Ferrón, J.L. Lado, A. Spinelli, J. Fernández-Rossier, A.F. Otte, Nano Lett. **15**(10), 6542 (2015). https://doi.org/10.1021/acs.nanolett.5b02200
10. M. Persson, Phys. Rev. Lett. **103**, 050801 (2009). https://doi.org/10.1103/PhysRevLett.103.050801
11. R. Žitko, T. Pruschke, New J. Phys. **12**(6), 063040 (2010). https://doi.org/10.1088/1367-2630/12/6/063040
12. J. Fernández-Rossier, Phys. Rev. Lett. **102**, 256802 (2009). https://doi.org/10.1103/PhysRevLett.102.256802
13. F. Delgado, J. Fernández-Rossier, Phys. Rev. B **82**, 134414 (2010). https://doi.org/10.1103/PhysRevB.82.134414
14. B. Sothmann, J. König, New J. Phys. **12**(8), 083028 (2010). https://doi.org/10.1088/1367-2630/12/8/083028
15. J. Fransson, O. Eriksson, A.V. Balatsky, Phys. Rev. B **81**, 115454 (2010). https://doi.org/10.1103/PhysRevB.81.115454
16. M. Ternes, New J. Phys. **17**(6), 063016 (2015). https://doi.org/10.1088/1367-2630/17/6/063016
17. N. Lorente, J.P. Gauyacq, Phys. Rev. Lett. **103**, 176601 (2009). https://doi.org/10.1103/PhysRevLett.103.176601
18. J. Kondo, Prog. Theor. Phys. **32**(1), 37 (1964). https://doi.org/10.1143/PTP.32.37
19. E.C. Goldberg, F. Flores, J. Phys.: Cond. Matter **25**(22), 225001 (2013). https://doi.org/10.1088/0953-8984/25/22/225001
20. E.C. Goldberg, F. Flores, Phys. Rev. B **91**, 165408 (2015). https://doi.org/10.1103/PhysRevB.91.165408

21. E.C. Goldberg, F. Flores, Inelastic electron scattering in aggregates of transition metal atoms on metal surfaces. Phys. Rev. B **96**, 115439 (2017)
22. E.C. Goldberg, F. Flores, Phys. Rev. B **77**, 125121 (2008). https://doi.org/10.1103/PhysRevB.77.125121
23. N.H. March, *Electron Density Theory of Atoms and Molecules* (Academic Press, New York, 1992)
24. A. Ayuela, N.H. March, Int. J. Quantum Chem. **110**(15), 2725 (2010). https://doi.org/10.1002/qua.22764
25. F. Flores, N.H. March, I.D. Moore, Surf. Sci. **69**(1), 133 (1977). https://doi.org/10.1016/0039-6028(77)90165-0

Chapter 10
Quantifying the Effect of Point and Line Defect Densities on the melting Temperature in the Transition Metals

C. C. Matthai

Abstract Molecular dynamics simulations of the melting process of bulk copper and gold were performed using Large-scale Atomic/Molecular Massively Parallel Simulator (LAMMPS). The aim of the study was to understand the effects of high pressures and defects on the melting temperature. The simulations were visualised using Visual Molecular Dynamics (VMD). The melting temperature of the perfect crystals were found to be higher than the experimentally observed values. The melting temperature as a function of pressure was determined and found to be in good agreement with experimental results. Vacancies and line defects in the form of dislocations were then introduced into the simulation cell and the melting temperatures recalculated. In both scenarios, we find that the melting temperature decreases as the defect density is increased bringing it closer to the experimentally observed value. Based on the pressure dependence of the melting curve, we conclude that vacancies are not the driving force for the melting transition.

10.1 Introduction

The melting transition has been studied extensively over many decades and over the years many theories of this transition have been expounded. It had been suggested that as the crystal is heated, melting occurs when the atom vibrations become large enough at the melting temperature such that the long range order is lost. Lindemann postulated [1] that the melting temperature T_m could be approximately defined as the point at which the mean interatomic spacing exceeds its equilibrium spacing by 10%. This is the so-called Lindemann criterion for melting. It is now generally agreed that defects, and in particular line defects, play an important role in the melting transition. In dislocation theories of melting, the number of dislocations increases according to some power law at the transition temperature.

More recently, it has been suggested [2] that a wide variety of phase transitions may be formulated in terms of the formation of quasi-particles at the transition. In this

C. C. Matthai (✉)
Department of Physics and Astronomy, Cardiff University, Cardiff, UK
e-mail: Clarence.Matthai@astro.cf.ac.uk

G. G. N. Angilella and C. Amovilli (eds.), *Many-body Approaches at Different Scales*, https://doi.org/10.1007/978-3-319-72374-7_10

phenomenological theory it was proposed that the transition temperature T_t could be written in the form $k_B T_t = E_{char} e^{-\gamma}$ where E_{char} is some characteristic energy associated with the transition and k_B is the usual Boltzmann constant. The quantity γ is related to the energy required to create the quasi-particle appropriate to the transition. It was further proposed that for melting the characteristic energy is related to the bulk modulus, B, through the relation $E_{char} = \Omega B$ where Ω is the atomic volume. The quasi-particles in this formulation are the phonons associated with the shear modulus. The presence of point and line defects allow for the annihilation of these quasi-particles. In this study we have investigated the dependence of the melting temperature on the defect (point and line) density in bulk copper. In addition, we have also investigated the role of pressure, p, in melting as this could be used to test model theories. In particular, the slope of the calculated $T_m(p)$ curve is compared with experimental results.

10.2 Theory and Simulations

10.2.1 Melting Temperature as a Function of Pressure

Simon and Glatzcl (SG) [3] proposed an empirical relation between the melting temperature and the pressure, *viz.*,

$$T_m(p) = T(0)\left(\frac{p}{a} + 1\right)^{1/c} \tag{10.1}$$

where a and c are constants. This relationship only holds for situations where the melting temperature increases with pressure. The slope of the melting curve at zero pressure has also been studied by many researchers. For example, Gilvarry constructed a theory in which the slope of the fusion curve at zero pressure could be expressed through the relation [4]

$$\frac{1}{T_m(p)}\left[\frac{\partial T_m}{\partial p}\right]_{p=0} = (6\gamma_G - 2)\frac{1}{3q B_0} \tag{10.2}$$

where γ_G is the Grüneissen constant and B_0 is the bulk modulus at zero pressure. The quantity q depends on the bulk modulus, the linear expansion coefficient, the melting temperature, the latent heat of melting and the volume change at melting.

In considering theories which relate the melting temperature to point and line defects, Matthai and March [5] used the results of a dislocation-mediated theory of the melting transition [6] to derive a relationship between the slope of the melting curve and the shear modulus, G,

$$\frac{1}{T_m(p)}\left[\frac{\partial T_m}{\partial p}\right]_{p=0} = \frac{1}{G}\frac{dG}{dp} \tag{10.3}$$

They also used the empirical relation between the vacancy formation energy, E_v^f and the melting temperature to give the gradient of the melting temperature curve at zero pressure in terms of the pressure dependence of the vacancy formation energy,

$$\frac{1}{T_m}\left[\frac{\partial T_m}{\partial p}\right]_{p=0} = \frac{1}{E_v^f}\left[\frac{\partial E_v^f}{\partial p}\right]_{p=0} \tag{10.4}$$

In summary, as the slope of the melting curve at zero pressure can be related slope of the shear modulus or vacancy formation energy at zero pressure, this quantity can be used to differentiate between theories of melting.

10.2.2 Simulation Details

A $10 \times 10 \times 10$ periodically repeating face centred cubic crystal with lattice spacing 3.615 Å comprising 4000 atoms was constructed and the LAMMPS package [7] used to carry ou the molecular dynamics simulations. Embedded Atom (EAM) interatomic potentials as formulated by Mei et al. [8] which reproduced the equilibrium properties of Cu were used to characterize the interactions. The time step in the simulations were taken to be 1 fs and the simulations were carried out at constant pressure using a NPT Nose–Hoover ensemble. In the pressure simulations, a hydrostatic pressure was applied to the simulation cell.

10.2.3 Determination of the Melting Temperature

In the molten state the long range order, present in the crystal, disappears and so the liquid state can be identified by examining the pair correlation function (Fig. 10.1), which is in turn related to the structure factor. However, it is not feasible to determine T_m from the structure factor data. In our simulations, the melting temperature was determined using three different approaches. In the first, the Lindemann criterion was applied by analyzing the RMS displacement of the atoms from their initial sites. The other two approaches involved plotting the volume per atom, Ω, and the self-diffusion coefficient, D_s, as a function of temperature. The volume Ω shows a discontinuity at the melting temperature whereas D_s shows a change in the slope at T_m. It may be noted that D_s is zero until T_m as we have considered a perfect crystal with no vacancies and consequently, there is no self-diffusion. From all three approaches, the melting temperature of copper was found to be between 1608 and 1621 K, which is much higher than the experimental value of 1354 K [9].

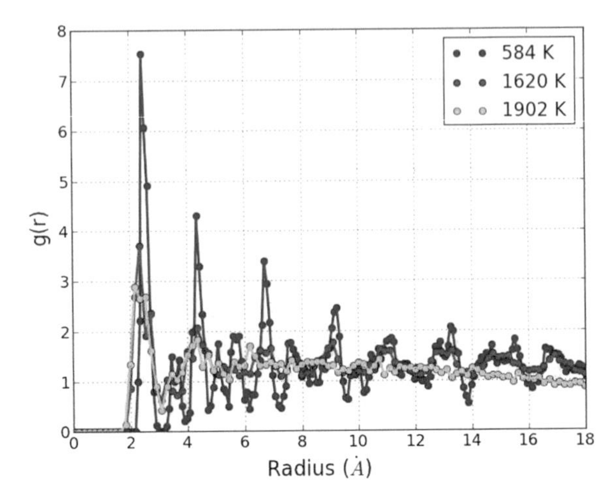

Fig. 10.1 The radial distribution function above, below, and near the melting temperature for copper

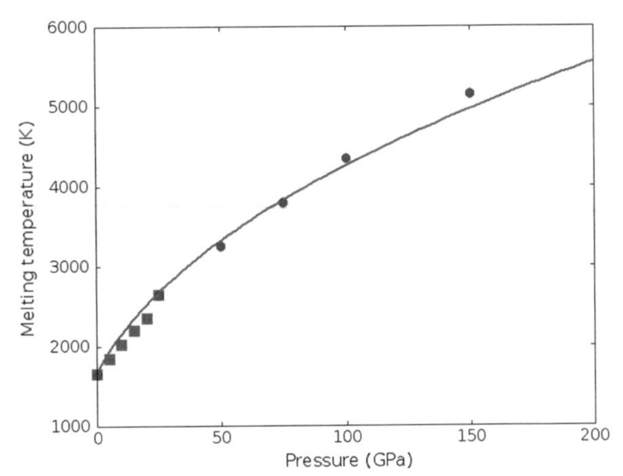

Fig. 10.2 The melting temperature as a function of pressure as found by the Lindemann criterion (filled squares) and the volume change (filled circles) methods. Also shown is the SG fit (full line) from Eq. (10.1), with $a = 11.5$ and $c = 2.4$

10.3 Results

10.3.1 Melting Temperature as a Function of Pressure

The melting temperature as a function of pressure was determined by all the three approaches outlined above. In the case of applying the Lindemann criterion, the pressure was limited to a maximum of 30 GPa. For the other two approaches the maximum pressure was extended to 300 GPa. The results for $T_m(p)$, which are shown in Fig. 10.2, were found to be well described by the SG equation. Additionally, the results for the slope of the melting curve at zero pressure (Table 10.1) are in good agreement with the estimated value of 26.2 K·GPa^{-1} from the work of Japel et al. [10].

Table 10.1 The slope of the melting curve at zero pressure

Melting criteria	T_m (K)	dT_m/dp (K/GPa)
Lindemann	1614 ± 12	46.9
Diffusion	1621 ± 35	28.3
Volume	1608 ± 30	28.0
Experimental [10]	1354 ± 5	26.2

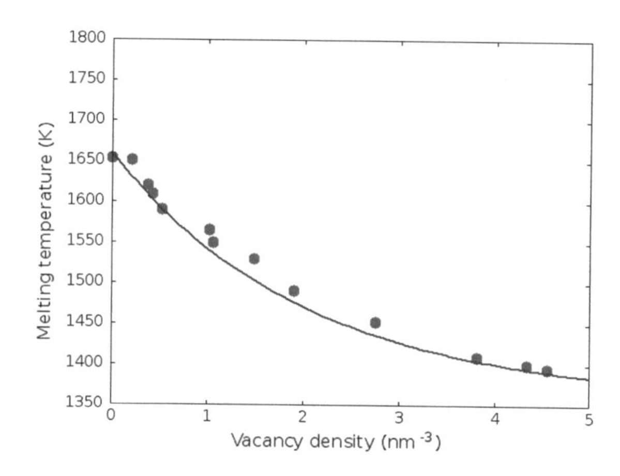

Fig. 10.3 The red circles are the computed melting temperatures for different vacancy densities (ρ_v). Also shown is the line of best fit, $T_m(\rho_v) = (300\, e^{-0.5\rho_v} + 1360)$ K

10.3.2 Effect of Point Defects on the Melting Temperature

As mentioned above, the melting temperature for the perfect crystal was found to be much higher than the experimental value. This is simply a consequence of the simulations not allowing for the creation and evolution of defects in the crystal as the temperature is increased. In order to determine the effect of vacancy density on the melting temperature, vacancies were introduced into the crystal and the constant pressure simulations repeated to determine the melting temperature. These simulations were carried out for densities of up to 5 vacancies per nm^3 and the results are shown in Fig. 10.3. As may be noted, the variation is reasonably well fitted with an exponentially decaying curve.

10.3.3 Effect of Dislocations on the Melting Temperature

It is generally agreed that while there is a relation between the vacancy formation energy and the melting temperature, it is more likely that line defects in the form of dislocations or disclinations are the drivers for the melting transition. We have therefore also carried out simulations aimed at investigating how the dislocation density influences the onset of melting. Since our simulations do not allow for the spon-

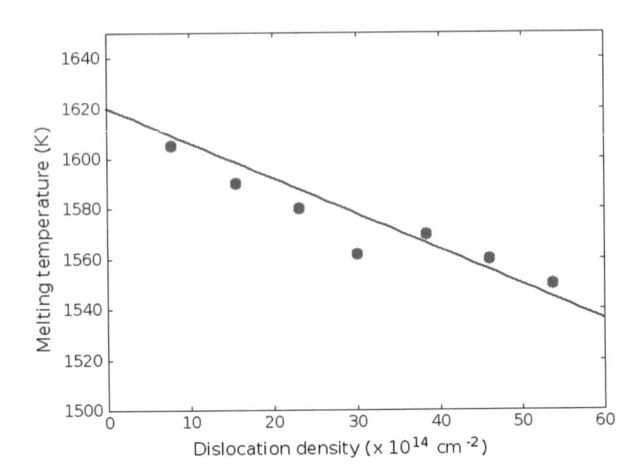

Fig. 10.4 Variation of T_m with dislocation density (filled circles). Also shown is a line of best fit

taneous creation of dislocations, dislocation loops of various lengths and different Burgers vectors were introduced into the crystal and the melting simulations carried out as before. The sites of the dislocation cores were chosen at random. For the sake of consistency, the crystal block was taken to be the same size as for the vacancy simulations. This is not ideal because in order to obtain measurable results, it was necessary to introduce an artificially high dislocation density. However, it should be noted that in dislocation theories of melting, it is expected that as the transition temperature is approached, the dislocation density should diverge. The melting temperature as a function of the number of dislocations is shown in Fig. 10.4. As expected, the melting temperature does decrease with increasing dislocation density but unlike in the case of point defects, the decrease shows a linear dependence.

10.4 Conclusions and Discussion

Computer simulations have been carried out to investigate the influence of point and line defects on the melting temperature. The melting temperature for the perfect crystal is found to follow the SG curve for pressures up to 100 GPa. However, the melting temperatures so found were much higher than that measured experimentally. When defects in the form of vacancies or dislocations are introduced, the computed melting temperature is lowered.

If the relationship between the vacancy formation energy and T_m is as described by Eq. (10.4) holds for high pressures, we would expect E_v^f to have the same variation with pressure as $T_m(p)$. We have therefore also determined the pressure dependence of the mono-vacancy formation energy. This is shown in Fig. 10.5. The data is best fitted by a power law increase over the pressure range investigated. This difference in the pressure dependence of the vacancy formation energy and the melting temperature suggests that it is unlikely that the formation of vacancies is the driver for melting.

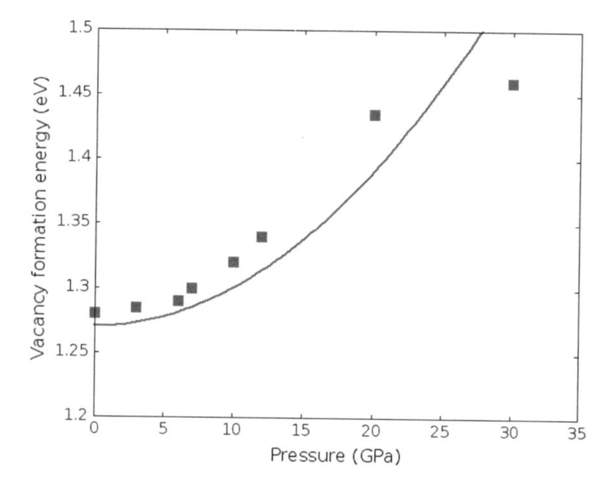

Fig. 10.5 The change in the vacancy formation energy as a function of pressure (filled squares). The solid line is a quadratic fit of the data

The simulations carried out in this study was based on rather small crystal sizes. In order to demonstrate the validity of the results, we are currently carrying out simulations on much large crystal blocks. In addition, calculations to determine the dislocation formation energy as a function of pressure are also under consideration.

Acknowledgements Some of the MD simulations were carried out on the ARCCA computing facilities at Cardiff University. CCM wishes to thank Professor Norman March for the many fruitful discussions on the topic of melting.

References

1. F.A. Lindemann, Phys. Z. **11**, 609 (1910)
2. N.H. March, E.V. Chulkov, P.M. Echenique, C.C. Matthai, Phase Trans. **83**(12), 1085 (2010). https://doi.org/10.1080/01411594.2010.509641
3. F.E. Simon, G. Glatzel, Z. Anorg. (Allg.) Chem. **178**, 309 (1929)
4. F.P. Bundy, H.M. Strong, in *Solid State Physics*, vol. 13, ed. by F. Seitz, D. Turnbull (Academic Press, New York, 1962), pp. 81–146. https://doi.org/10.1016/S0081-1947(08)60456-7
5. C.C. Matthai, N.H. March, Philos. Mag. Lett. **87**(7), 475 (2007). https://doi.org/10.1080/09500830701320307
6. L. Burakovsky, D.L. Preston, R.R. Silbar, J. Appl. Phys. **88**(11), 6294 (2000). https://doi.org/10.1063/1.1323535
7. LAMMPS Molecular Dynamics Simulator (2016), http://lampss.sandia.gov
8. J. Mei, J.W. Davenport, G.W. Fernando, Phys. Rev. B **43**, 4653 (1991). https://doi.org/10.1103/PhysRevB.43.4653
9. H. Brand, D.P. Dobson, L. Vočadlo, I.G. Wood, High Press. Res. **26**(3), 185 (2006). https://doi.org/10.1080/08957950600873089
10. S. Japel, B. Schwager, R. Boehler, M. Ross, Phys. Rev. Lett. **95**, 167801 (2005). https://doi.org/10.1103/PhysRevLett.95.167801

Chapter 11
Application of the Plane-Wave-Based Perturbation Theory to the Density Modulation Induced by a Point Charge in an Electron Gas

I. Nagy and M. L. Glasser

Dedicated to Professor N. H. March on the occasion of his 90th birthday.

Abstract The induced electron density at the position of a single point charge Z embedded in a three-dimensional degenerate electron gas is studied at high densities. The perturbative, plane-wave-based treatment developed within the framework of density matrices by March and Murray (Phys Rev 120: 830, 1960, [1]) is applied here up to second order in Z. Comparison with the result obtained by considering the exact scattering enhancement in a bare Coulomb field is made. The small numerical difference found in the second-order term of the induced density at contact is analyzed following Wigner's (Phys Rev 94: 77, 1954, [2]) similar perturbative treatment of the proton field in the hydrogen atom. The impact of the many-body screening is discussed as well.

I. Nagy
Department of Theoretical Physics, Institute of Physics, Budapest University of Technology and Economics, Budapest 1521, Hungary

I. Nagy · M. L. Glasser
Donostia International Physics Center, P. Manuel de Lardizabal 4, 20018 San Sebastián, Spain

M. L. Glasser (✉)
Department of Physics, Clarkson University, Potsdam, NY 13699-5820, USA
e-mail: lglasser@clarkson.edu

M. L. Glasser
Department of Theoretical Physics, University of Valladolid, 47003 Valladolid, Spain

G. G. N. Angilella and C. Amovilli (eds.), *Many-body Approaches at Different Scales*, https://doi.org/10.1007/978-3-319-72374-7_11

11.1 Introduction

The effect of charged impurities on the properties of a metal is of considerable physical interest, both because of the possibility of deliberately introducing them so as to study the physics of electron-atom interaction in metals, and because most real materials contain impurities which affect their physical properties. By using modern experimental techniques, such as scanning tunneling microscopy (STM) and scanning tunneling spectroscopy (STS), one is able to consider in detail the modulations in the local density of states at the Fermi level. One can probe the induced density in the many-electron system of metals via positron annihilation or Knight-shift measurements with muons. Furthermore, one might consider in calculating the life-time of an added electron in a cold electron gas, the screening of the electron-electron interaction, needed to avoid the divergence in the electron, or intruder charge, self-energy close to the Fermi energy.

There are several instances in which simple approximation methods yield correct results even though the conditions for the applicability of those methods are not fulfilled. The best known example is the calculation of the Rutherford cross section in three dimensions by first-order Born approximation, i.e., by using plane-wave states. Apart from this case, however, one should check the adequacy of this familiar approximation in potential scattering. Such a check is, surprisingly, particularly important in the case of short-range forces [3].

In the present work as a first step the electron gas is assumed to be noninteracting, but perturbed by the potential of an embedded point charge. Thus, the problem is one of quantum mechanical scattering of a *single* electron by the potential of an external charge. The thermodynamics then follows by filling up the new set of energy levels according to Fermi-Dirac statistics corresponding to an *ideal* momentum distribution function, i.e., with unit occupation numbers for one-electron states up to the invariant Fermi level. The continuous spectra of the electron gas Hamiltonian and the perturbed Hamiltonian coincide.

The method applied rests on the well-known paper by March and Murray [1] where the idempotent density matrices, Dirac and canonical, were discussed in reference to imperfections (central fields) in metals. Concretely, we apply March and Murray's second-order prescription, Eq. (6.5) of their paper, to calculate the total electron charge at the position of the embedded point charge in a high-density electron gas. Notice that the corresponding low-density limit was treated earlier [4] by us using their Eq. (4.12), i.e., a third-order differential equation, with a Hulthén-type potential. Since in the present paper we use plane-wave states and find a close similarity with the second-order term based on the exact Coulomb enhancement of scattered waves, we are tempted to refer for analogy to Wigner's second-order perturbational calculation of the energy of the hydrogen atom. Remarkably, in order to provide a physical interpretation, Wigner pointed out that the finite result for the binding energy in his second-order treatment is due to the form (r^{-1}) of the Coulomb potential [2].

11.2 Theory and Results

According to Eq. (6.5) of March and Murray [1], the *second-order* perturbation expression for the total density $n(r=0) = n_0 + n_1(r=0) + n_2(r=0)$ at the position of a charge Z embedded in a paramagnetic electron gas at zero temperature is

$$n_0 = \frac{1}{\pi^2}\int_0^{k_F} dk\, k^2 = \frac{k_F^3}{3\pi^2}, \tag{11.1a}$$

$$n_1(r=0) = \frac{2}{\pi^2}\int_0^{k_F} dk\, k \int_0^\infty ds\, V(s)\, \sin(2ks), \tag{11.1b}$$

$$n_2(r=0) = \frac{4}{\pi^2}\int_0^{k_F} dk \int_0^\infty ds\, V(s)\, \sin(2ks) \int_s^\infty dt\, V(t)\, \sin(2kt), \tag{11.1c}$$

where the ideal momentum distribution function $f_0(0 \le k \le k_F) = 1$ for occupied one-electron momentum eigenstates in a homogeneous degenerate system has been applied. $V(r)$ is the spherical central potential due to the point charge. In writing these equations the expression $x^2 j_0(x)\bar{n}_0(x) = (1/2)\sin(2x)$ is used for the product of the spherical Bessel (j_l) and Neumann ($\bar{n}_l$) functions..

Now, as in Wigner's perturbation theory for binding with an $1/r$ potential, we consider first the unscreened case in our problem, i.e., we take $V(r) = Z/r$. From Eq. (11.1b) we get

$$n_1(r=0) = Z\frac{k_F^2}{2\pi}. \tag{11.2}$$

In dealing with the second-order term in Eq. (11.1c), we change the order of integration and introduce the new variable $t = xs$. Thus, we have

$$n_2(r=0) = Z^2\frac{4}{\pi^2}\int_0^{k_F} dk \int_1^\infty dx\, \frac{1}{x}\int_0^\infty ds\, \frac{\sin(2ks)\sin(2ksx)}{s}. \tag{11.3}$$

The integral with respect to s becomes $I(x) = (1/2)\ln[(x+1)/(x-1)]$ independently of k.

To perform the x-integration in Eq. (11.1c), we employ a convergent ($x \ge 1$) expansion

$$I(x) = \sum_{m=0}^{\infty} \frac{1}{(2m+1)x^{2m+1}} \tag{11.4}$$

by which the remaining integrations are elementary and finally we get

$$n_2(r=0) = Z^2\frac{4}{\pi^2}k_F \sum_{m=0}^{\infty}\frac{1}{(2m+1)^2} = Z^2\frac{k_F}{2} \tag{11.5}$$

since the sum becomes $\lambda(2) = \frac{3}{4}\zeta(2) = \pi^2/8$ in terms of Riemann's zeta-function.

Next, we turn to the exact treatment of scattering states. With Coulomb potential Z/r, the important enhancement factor $E_C(\eta)$ is

$$E_C(\eta) = \frac{2\pi\eta}{1 - e^{-2\pi\eta}} \equiv 1 + \pi\eta + 2\sum_{m=1}^{\infty} \frac{\eta^2}{m^2 + \eta^2} \tag{11.6}$$

in terms of the Sommerfeld parameter $\eta = Z/k$. From this we get

$$\begin{aligned} n(0) &= \frac{1}{\pi^2}\int_0^{k_F} dk\, k^2\, E_C(k) = n_0 + Z\frac{k_F^2}{2\pi} + Z^2\frac{2}{\pi^2}\sum_{m=1}^{\infty}\frac{1}{m^2}\left[k_F - \frac{Z}{m}\arctan\left(\frac{mk_F}{Z}\right)\right] \\ &= n_0 + \frac{Z}{2\pi}k_F^2 + \frac{2Z^2}{\pi^2}k_F\zeta(2) - \frac{Z^3}{\pi}\zeta(3) + \frac{\pi^2 Z^5}{45k_F} + O(k_F^{-3}). \end{aligned} \tag{11.7}$$

In the high-density limit, where $k_F \gg 1$, this expression tends to

$$n(0) = n_0 + Z\frac{k_F^2}{2\pi} + Z^2\frac{k_F}{3} - \frac{Z^3}{\pi}\zeta(3), \tag{11.8}$$

where we used $\zeta(2) = \pi^2/6$. Apart from its sign, the last term corresponds [5, 6] to the total density of the entire spectrum of bound states of a hydrogen-like atom. This k_F-independent term is negligible when $k_F \gg 1$.

By comparing the exact Eq. (11.8) with the perturbative Eq. (11.5), one can see a moderate numerical deviation in the Z^2-order term. This approximate agreement is one of the main results of this work. We can say, following Wigner's early observation, that the second-order perturbation theory developed by March and Murray for charged imperfections in a noninteracting, degenerate electron gas appears to be meaningful.

In the rest of this note, we turn to important subquestions on the role of screening and non-idempotency of the host system. We focus on *changes* in the first-order term $n_1(0)$. By substituting into Eq. (11.1b) a Yukawa-type potential $V_Y(r) = (Z/r)\exp(-\Lambda r)$ we get

$$n_1^{(Y)}(r=0) = \frac{Z}{\pi^2}\left[\left(k_F^2 + \frac{\Lambda^2}{4}\right)\arctan\frac{2k_F}{\Lambda} - \frac{\Lambda k_F}{2}\right] \tag{11.9}$$

which reproduces Eq. (11.2) when $\Lambda = 0$. However, this transparent closed expression allows a deeper analysis of the k_F-dependence of the parameter $\Lambda(k_F)$. One can see that with conventional, Thomas-Fermi scaling of $\Lambda^2 = 4k_F/\pi$ the high density limiting value, $Zk_F^2/2\pi$ will not change. In other words, at that limit the screened potential is penetrable for a very fast electron. One could reduce the numerical factor $(1/2\pi)$ of the limiting form only with an $\Lambda \propto k_F$-scaling. In this manner the high-density pair-correlation function at contact, $g(0) = 1 - |n_1(0)/n_0|$ becomes tunable. Indeed, there are theoretical arguments [7, 8] that such scaling is the only one if the goal is to reproduce the exactly known asymptotic form for this function $g(0) = 1 - (1.4/k_F)$.

We finally come to the challenging problem of the non-idempotency encoded in the reduced one-particle density matrix in momentum space. The diagonal of this matrix is the one-particle momentum distribution function. Until now, we have used the ideal momentum distribution function $f_0(0 \leq k \leq k_F) = 1$. Due to the correlated electron motion, the momentum distribution function $f(k)$ describing the population of plane-wave states of *real particles* (and not Landau's quasiparticles) in the translationally invariant system differs from the ideal momentum distribution function [9]. This is a crucial point since, for instance, in a mean-field Kohn-Sham-like treatments of Density Functional Theory with auxiliary orbitals the population of particles is still the ideal one. Therefore by that method one can not calculate the exact kinetic energy. In fact, all relative-coordinate-dependent many-body complications are hidden in an effective external field.

In order to appreciate the important role of non-idempotency (N) in our problem, we take a simple [10] parametrized $(x \equiv k/k_F)$ expression from the literature

$$f(x) = \begin{cases} a(1)[1 - a(2)x^2], & 0 \leq x \leq 1 \\ a(3)e^{-a(4)(x-1)} + \dfrac{T}{x^8}, & x > 1. \end{cases} \tag{11.10}$$

Here, $a(4) = 4$, $a(3) = (32/13)\delta$, $\delta = (1/3) - a(1)[(1/3) - (1/5)a(2)] - T/5$, $a(2) = 0.048/k_F$, $a(1) = 1 - 0.019/k_F$, and $T = [2/(3\pi k_F)]^2 g(0)$ in terms of the pair-correlation at contact. By performing the integration in Eq. (11.1b) with a Z/r potential, in leading order we get

$$n_1^{(N)}(r = 0) \cong \frac{Zk_F^2}{2\pi}\left[1 - \frac{0.042}{k_F} + O(1/k_F^2)\right]. \tag{11.11}$$

As expected, $n_1^{(N)}(0) < n_1(0)$ due to the nonideality of the momentum distribution function. Relativistic effects [11, 12] which appear at extreme k_F, are beyond the scope of this note.

11.3 Summary

The induced electron density at the position of a single point charge Z embedded in a three-dimensional degenerate electron gas is studied at high densities. The perturbative, plane-wave-based treatment developed within the framework of idempotent density matrices by March and Murray [1] is applied here up to second order in Z. Comparison with the result obtained by considering the exact scattering enhancement in a bare Coulomb field is made. The small *numerical* difference found in the second-order term of the induced density at contact is analyzed following Wigner's [2] similar perturbative treatment of the proton potential in the hydrogen atom.

The impact of many-body screening and non-idempotence in the host system on the perturbative results found are considered as well. Specifically, the impact of a non-ideal momentum distribution function could influence the two-term asymptotic results for the induced density at contact in cases with $V(r) = \pm(1/r)$, i.e., with protons and antiprotons. However, Thomas-Fermi screening modifies the results obtained with bare interactions in a stronger manner. Clearly, further research on the combined effects of these ingredients needed for a physically self-consistent model are still desirable to proceed along the path marked out by the pioneering paper of March on charged imperfections.

Acknowledgements This note is dedicated to Professor Norman H. March. The authors are indebted to him for many useful, enlightening discussions in the past, and hope that the future will allow further fruitful collaborations. One of us (MLG) acknowledges the financial support of MINECO (Project MTM2014-57129-C2-1-P) and Junta de Castilla y Leon (VA057U16).

References

1. N.H. March, A.M. Murray, Phys. Rev. **120**, 830 (1960), Reprinted in Ref. [13]. https://doi.org/10.1103/PhysRev.120.830
2. E.P. Wigner, Phys. Rev. **94**, 77 (1954). https://doi.org/10.1103/PhysRev.94.77
3. R.E. Peierls, *Surprises in Quantum Mechanics* (Princeton University Press, Princeton, 1979)
4. I. Nagy, M.L. Glasser, N.H. March, Phys. Lett. A **373**(35), 3182 (2009). https://doi.org/10.1016/j.physleta.2009.06.051
5. O.J. Heilmann, E.H. Lieb, Phys. Rev. A **52**, 3628 (1995). https://doi.org/10.1103/PhysRevA.52.3628
6. N.H. March, I.A. Howard, I. Nagy, P.M. Echenique, J. Math. Phys. **46**(7), 072104 (2005). https://doi.org/10.1063/1.1947118
7. I. Nagy, J.I. Juaristi, R.D. Muiño, P.M. Echenique, Phys. Rev. B **67**, 073102 (2003). https://doi.org/10.1103/PhysRevB.67.073102
8. M. Corona, P. Gori-Giorgi, J.P. Perdew, Phys. Rev. B **69**, 045108 (2004). https://doi.org/10.1103/PhysRevB.69.045108
9. A.A. Abrikosov, L.P. Gorkov, I.E. Dzyaloshinski, *Methods of Quantum Field Theory in Statistical Physics* (Dover, New York, 1975)
10. B. Barbiellini, A. Bansil, J. Phys. Chem. Solids **62**(12), 2181 (2001). https://doi.org/10.1016/S0022-3697(01)00176-7
11. S.A. Chin, Ann. Phys. (NY) **108**(2), 301 (1977). https://doi.org/10.1016/0003-4916(77)90016-1
12. I. Nagy, J. Phys. B At. Mol. Phys. **19**(11), L421 (1986). https://doi.org/10.1088/0022-3700/19/11/007
13. N.H. March, G.G.N. Angilella (eds.), *Many-Body Theory of Molecules, Clusters, and Condensed Phases* (World Scientific, Singapore, 2009)

Chapter 12
Kovacs Effect and the Relation Between Glasses and Supercooled Liquids

F. Aliotta, R. C. Ponterio, F. Saija and P. V. Giaquinta

Abstract In this note we revisit the Kovacs effect, concerning the way in which the volume of a glass-forming liquid, which has been driven out of equilibrium, changes with time while the system evolves towards a metastable state. The theoretical explanation of this phenomenon has attracted much interest even in recent years, because of its relation with some subtle aspects of the still elusive nature of the glass transition. In fact, even if there is a rather general consensus on the fact that what is experimentally observed on cooling is the dramatic effect produced by the dynamical arrest of slower degrees of freedom over the experimental time scale, it is not yet clear whether this phenomenology can be justified upon assuming the existence of an underlying (possibly, high order) phase transition at lower temperatures.

12.1 Introduction

Understanding the kinetic and thermodynamic routes followed by a supercooled liquid while becoming, through viscous slowdown, a glass well below the freezing point still represents a major challenge in the chemical physics of condensed matter [1, 2]. In fact, in spite of countless efforts on both the theoretical and experimental sides, many significant questions on the very nature of the glassy state still remain unanswered. Even if there is a widespread consensus on the thesis that the experimentally observed glass transition is the macroscopic outcome of relaxation processes which,

F. Aliotta · R. C. Ponterio · F. Saija
CNR-IPCF, Viale F. Stagno d'Alcontres 37, 98158 Messina, Italy
e-mail: aliotta@ipcf.cnr.it

R. C. Ponterio
e-mail: ponterio@ipcf.cnr.it

F. Saija
e-mail: franz.saija@cnr.it

P. V. Giaquinta (✉)
Dipartimento di Scienze Matematiche e Informatiche, Scienze Fisiche e Scienze della Terra, Università degli Studi di Messina, Contrada Papardo, 98166 Messina, Italy
e-mail: paolo.giaquinta@unime.it

G. G. N. Angilella and C. Amovilli (eds.), *Many-body Approaches at Different Scales*, https://doi.org/10.1007/978-3-319-72374-7_12

at low enough temperatures, become much slower than the experimental observation time, it is not clear yet whether such a slowdown of the dynamics of the system can be interpreted, following Kauzmann's original discussion on this point [3], as an indication of an underlying, possibly continuous, phase transition, which would occur below the vitrification point T_g. In addition, it is not clear whether what appears – on the experimental time scale—as a kinetically arrested system may eventually transform into a truly metastable state after a sufficiently long time. However, distinguishing between broken ergodicity and metastability is not an easy task on the operational side and both laboratory and numerical experiments give ambiguous indications on this point [4–10].

Moreover, other relevant questions remain open, one of which concerns the surmised existence of a liquid-liquid phase transition which would be undergone by deeply supercooled metastable water [11–16] and the very possibility of prolonging the associated coexistence line well below the homogeneous nucleation temperature, so as to verify whether it eventually merges with the coexistence line between the experimentally observed low-density and high-density amorphous phases of water [17, 18].

The focus of this note is on a phenomenon, intimately associated with the glass transition, that was originally observed and described by Kovacs [19]. This phenomenon gives useful information on the volumetric time evolution of a glass-forming liquid, which has been originally driven out of equilibrium. The experimental protocol implemented by Kovacs entails three stages: (i) a thermodynamically stable liquid, formerly equilibrated at a temperature T_i, is quenched, at a fixed pressure, to a temperature T_q, not too far below T_g, in such a way that some internal degrees of freedom fall out of equilibrium with the thermal bath; (ii) the system is then left to age for some time, which, however, is not long enough for it to reach a condition of full thermal equilibrium; (iii) the temperature is finally raised to a value T_f, intermediate between T_i and T_q. One can then observe the irreversible evolution of the system as it relaxes from the preset out-of-equilibrium condition at $T = T_q$ to an asymptotic one of metastable equilibrium at $T = T_f$. In particular, the way in which the volume of the sample changes with time exhibits a "memory" effect in that it is found to depend in a non trivial way on the thermal history of the material. In fact, if the final temperature is not too high, the system starts expanding irreversibly to a volume which, after overshooting the equilibrium value at T_f, further increases up to a maximum value that is lower (and reached later) the higher the quenching temperature T_q. Thereafter, the system progressively contracts until it regains its equilibrium volume at $T = T_f$ [20]. The nonmonotic behaviour of the volume and the resulting maximum imply that the values of pressure, temperature, and volume are not sufficient to identify a unique (nonequilibrium) state of the material: in fact, under the conditions outlined above, for assigned values of such three variables, the system can be actually observed in two different "states", corresponding to different stages of the dynamical evolution of the material towards metastable equilibrium.

The Kovacs effect, originally observed in a polymeric substance (polyvinyl acetate), has been observed in a variety of glassy materials. As such, it has been the topic of many experimental and theoretical investigations. In more recent times

Angell and coworkers [21] discussed the phenomenon in the framework of the volumetric behaviour of glass formers in nonergodic regimes, with specific reference to nonlinear relaxation and associated memory effects. Mossa and Sciortino [20] performed molecular dynamics simulations of a model of ortho-terphenyl (OTP) which revealed some fine details of the dynamics of the phenomenon that are not accessible to laboratory experiments; they also performed an analysis of the properties of the potential energy landscapes explored by the system during the relaxation process. A theoretical interpretation of these results was later attempted by Bouchbinder and Langer [22], who resorted to a description of the system based on (separable) configurational and kinetic-vibrational subsystems.

The aim of this note is to illustrate a simple macroscopic model which can be used to describe the dynamic evolution and, correspondingly, the thermodynamic behaviour of a system along the lines traced by Kovacs with its experimental protocol. We shall also discuss the implications of the proposed model as to some aspects of the thermodynamic relation between glasses and metastable liquids.

12.2 Kovacs Effect in Ortho-Terphenyl

As is well known, the glass transition does not occur at a well defined temperature but, rather, over a range of temperatures across which, depending on the time scale of the experiment, a number of internal degrees of freedom of the system become *de facto* arrested. An indirect measure of the effective number of energetically active degrees of freedom over the experimental time scale is given by the isobaric specific heat which, contextually, exhibits a rather sharp drop across T_g. Correspondingly, cusp discontinuities show up in the thermal behaviour of the extensive parameters. However, no latent heat is released which implies that the entropy is continuous across the glass transition.

In the following we shall use OTP as a reference material for our discussion of the Kovacs effect. Liquid OTP can be easily supercooled below its freezing/melting temperature ($T_m = 329.35$ K), down to the glass transition point that is located at a relatively high temperature ($T_g \approx 247$ K). In addition, the glass can be slowly reheated and restored to the metastable liquid phase; this is possible because the homogeneous nucleation temperature (T_H) of this material is lower than T_g. Chang and Bestul used adiabatic calorimetry to measure the heat capacities of liquid, glassy and crystalline OTP at ambient pressure [23]. These data are plotted in Fig. 12.1. The heat capacities of both liquid and crystalline OTP are found to be nearly linear functions of the temperature. Moreover, the heat capacity of glassy OTP ($T < T_g$) is only slightly higher (1.5–2%) than that of the crystal, a difference that is not resolved on the scale of the plot displayed in Fig. 12.1. As noted above, the glass transition is signalled by the abrupt drop of the specific heat; correspondingly, a cusp discontinuity shows up in the molar volume of OTP [24], that is plotted in the upper panel of Fig. 12.2.

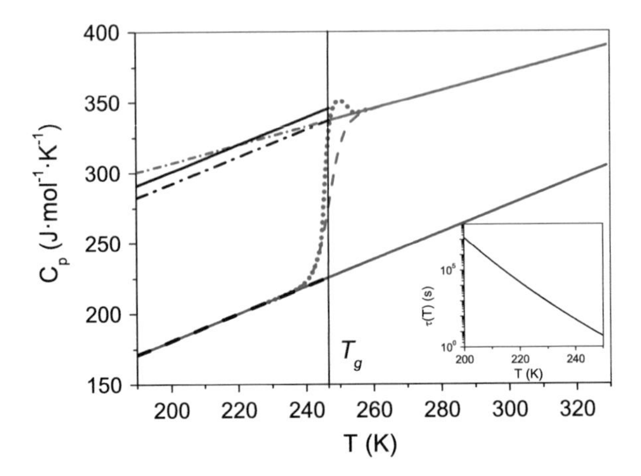

Fig. 12.1 Isobaric molar specific heat of ortho-terphenyl plotted plotted as a function of the temperature below the freezing/melting point at typical modulation angular frequencies ($\approx 10^{-1}$ Hz). Continuous red line: supercooled-liquid branch (linear fit of the experimental data [23]); dash-dotted red line: linear extrapolation of the supercooled-liquid data below the glass transition point; continuous blue line: solid branch (linear fit of the experimental data [23]); dashed red line: crossover between the liquid and glassy branches modelled through Eq. 12.1 and using the relaxation time for slow processes displayed in the inset (see text); dotted red line: typical crossover between the glassy and liquid branches on fast heating the partially aged glass; dash-dotted black line: specific heat of the glass obtained after quenching the metastable fluid at $T = T_g$; continuous black line: effective specific heat of the quenched and partially aged glass with an associated fictive temperature $\widetilde{T}_g < T_g$ (see text). The vertical black line marks the glass transition point ($T_g = 247$ K)

When a system, which was formerly at thermodynamic equilibrium, is cooled, it will start relaxing towards a new equilibrium condition, a process which implies a redistribution of the internal energy among all the degrees of freedom as well as a variety of local and global structural rearrangements over distances and times which can be very long. If the time window of the experimental observation is fixed, only the motions which take place over shorter times will be able to relax. Instead, slower motions will appear somewhat frozen in a state corresponding to that of the system at equilibrium at the initial temperature. Of course, the location of a temporal "boundary" between frozen and active motions depends on the experimental time window. This is the reason why the experimental glass transition temperature cannot be defined in an unambiguous way.

In order to take explicitly into account this crucial aspect, we assumed that the relaxation of the temperature-dependent molar specific heat of our model system, as observed in a typical differential scanning calorimetry experiment, can be represented by the expression [25, 26]:

$$C_P(T;\omega) = \mathrm{Re}\left[C_P^{(c)}(T) + \frac{C_P^{(l)}(T) - C_P^{(c)}(T)}{1 + j\omega\tau(T)} \right], \tag{12.1}$$

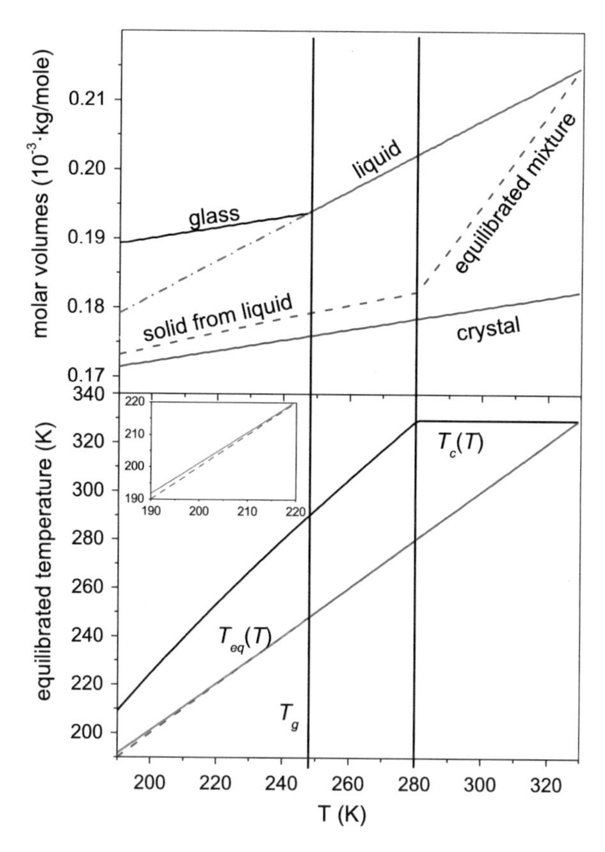

Fig. 12.2 Upper panel: molar volume of ortho-terphenyl plotted as a function of the temperature below the freezing point at ambient pressure. Red continuous line: supercooled liquid; red dash-dotted line: linear extrapolation of the supercooled-liquid data below the glass transition point; black continuous line: glass; blue dashed line: molar volume of the stable phase nucleated by metastable liquid OTP upon spontaneous freezing at adiabatic-isobaric (*i.e.*, isoenthalpic) conditions. The cusp singularity at $T_{\mathrm{solid}} \approx 280$ K marks the boundary between two different outcomes of the irreversible transition eventually undergone by supercooled liquid OTP: for $T > T_{\mathrm{solid}}$ the nucleated phase is a solid-liquid mixture at the freezing/melting temperature T_{m}, whereas for lower temperatures the equilibrium phase is a pure crystalline solid whose temperature decreases with (while still keeping higher than) that of the parent liquid, and whose volume is correspondingly larger than that of the solid (blue continuous line) at T_{m} (for more details see [27]). Lower panel: temperature of the asymptotic metastable phase to which the liquid, originally undercooled to a temperature T, would relax at isoenthalpic conditions (black continuous line); for graphical convenience, we also plot the temperature of the metastable supercooled liquid as a red line—the dashed part being the linear extrapolation below T_{g} of the higher-temperature experimental data—which, by construction, coincides with the first quadrant bisector; the green continuous line (also expanded in the inset) represents the temperature T_{eq} of the metastable state that would be reached asymptotically by the glass as calculated through Eq. (12.4) in the text. The two black vertical lines mark the temperatures T_{g} and T_{solid}, respectively

where $C_P^{(\mathrm{l})}(T)$ and $C_P^{(\mathrm{c})}(T)$ are the isobaric specific heats of the liquid and of the crystal, respectively, $\tau(T)$ is the (temperature dependent) relaxation time of the one single slow process which characterizes the kinetics of the model, $\omega = 2\pi t_{\mathrm{p}}^{-1}$ is the reverse of the experimental sampling time t_{p}, and j is the imaginary unit. The inverse Fourier transformation of Eq. (12.1) leads to the following time dependence of the isobaric specific heat at a given temperature T [25]:

$$C_P(T;t) = C_P^{(\mathrm{c})}(T) + \left[C_P^{(\mathrm{l})}(T) - C_P^{(\mathrm{c})}(T)\right]\left[1 - e^{-\frac{t}{\tau(T)}}\right]. \tag{12.2}$$

The liquid and crystal specific heats show up in the above equations as the long-time (zero-frequency) and short-time (infinite-frequency) values of $C_P(T)$, respectively.

To be more realistic one should actually consider a distribution of relaxation times, whose widths would also depend on the temperature, as well as a more plausible model for the complex susceptibility of the system. However, here we are not as much interested in reproducing the experimental data for OTP in a detailed and quantitative way; in fact, we just want to show that the effect originally observed by Kovacs is the natural outcome of a relaxation process occurring in the system, even in the oversimplified case in which this process is being parametrized with one single relaxation time only. As for its dependence on the temperature, we adopted an Arrhenius-like expression: $\tau(T) = \tau_0 \exp(\Delta E/k_B T)$, and adjusted the values of the two free parameters so as to obtain a rough match with the experimental values reported for OTP in [10]. We also assumed that the relaxation times of the fast processes are much shorter than the observation time in the experiment under consideration. The resulting behaviour of the specific heat across the glass transition region is displayed for a typical angular frequency in Fig. 12.1, whose inset shows the relaxation time that we plugged into Eq. (12.1).

As already noted, when a system, which has previously attained thermodynamic equilibrium at a temperature very close to T_{g}, is rapidly cooled to a lower temperature, fast motions rapidly equilibrate in the new thermal state, whereas slow motions remain substantially "frozen" in the configurational state that the system was in at $T = T_{\mathrm{g}}$. Hence, we can estimate the "effective" specific heat of the quenched fluid as:

$$C_P^{(\mathrm{eff})}(T) \approx C_P^{(\mathrm{c})}(T) + \left[C_P^{(\mathrm{l})}(T_{\mathrm{g}}) - C_P^{(\mathrm{c})}(T_{\mathrm{g}})\right], \tag{12.3}$$

where the term in square brackets on the r.h.s. of Eq. (12.3) is the jump observed in the specific heat of the system at the glass transition point, which approximately quantifies the "hidden" contribution to $C_P^{(\mathrm{eff})}(T)$ that is not resolved by calorimetric measurements on the time scale of the experiment. The statement embodied in Eq. (12.3) conveys an information analogous to that which emerges from the thermal behaviour of the molar volume of the vitrified system; in fact, the experimental data for $v_{\mathrm{g}}(T)$ (see Fig. 12.2) can be reproduced rather accurately by the expression:

$$v_{\mathrm{g}}(T) \approx v_{\mathrm{c}}(T) + \left[v_{\mathrm{l}}(T_{\mathrm{g}}) - v_{\mathrm{c}}(T_{\mathrm{g}})\right], \tag{12.4}$$

where v_l, v_g and v_c are the molar volumes of supercooled liquid, glass, and crystal, respectively. Hence, as already noted before [28], a system which has undergone a dynamical arrest at T_g and which has been then quenched to a lower temperature T_q can be characterized by two temperatures: the actual quenching temperature and an auxiliary temperature (T_g) at which the slow motions have *de facto* arrested over the experimental time scale. Yet, as time goes on even such slow configurational degrees of freedom start to relax, gradually driving the system towards a condition of metastable equilibrium. As far as specific heat and volume are concerned, such an asymptotic state will correspond to a point on the lines traced upon extrapolating (with constant slope, at least for moderate amounts of supercooling) the laboratory data of the liquid branch (see Figs. 12.1 and 12.2). Under adiabatic conditions, this process towards equilibrium will necessarily imply a transfer of energy from the fast motions to the slower ones and, correspondingly, a change of volume. According to this description, the final temperature (T_{eq}) will fall between the glass transition temperature and the quenching temperature. In other words, the out-of-equilibrium glass, obtained through the rapid quenching of the system, will irreversibly relax towards a condition of metastable equilibrium characterized by a higher temperature, and—at least, in the case of OTP—a higher density. We can calculate such intermediate temperature by noting that, under the postulated adiabatic conditions, the enthalpies of the initial and final "states" should be equal, *i.e.*, $H^{(\mathrm{eff})}(T_q) = H^{(l)}(T_{eq})$; hence, it follows that:

$$\int_{T_q}^{T_m} C_P^{(\mathrm{eff})}(T)dT = \int_{T_{eq}}^{T_m} C_P^{(l)}(T)dT, \tag{12.5}$$

where the melting/freezing temperature T_m has been assumed as a common reference temperature for evaluating the enthalpy changes of the system along two paths starting from the unrelaxed and relaxed state, respectively. Obviously, the resulting final equilibrium temperature T_{eq} is a function of the temperature T_q at which the system had been previously quenched. As seen in the lower panel of Fig. 12.2, the system undergoes a moderate heating upon irreversibly relaxing to a metastable condition.

We shall now explore what happens if the system is allowed to exchange energy with an external thermostatic reservoir, which brings us into the conditions of the experiment performed by Kovacs. We consider it useful to carry out our conceptual experiment using two different protocols, the second of which complies more closely with Kovacs' indications.

We first assume that liquid OTP has been equilibrated down to the ordinary vitrification threshold. We then imagine to cool rapidly the system from $T_g = 247$ K to a lower temperature, say $T_q = 198.5$ K. As a result, the system contracts and its molar volume decreases to the value $v_g^{(A)}$, corresponding to point A on the line $v_g(T)$ displayed in Fig. 12.3. Let the system now exchange energy with the thermal reservoir at the quenching temperature for a time (t_{aging}) long enough for its configurational state to change, but nevertheless shorter for the system to reach equilibrium. In order to keep temperatures within a range compatible with the available experimental data for OTP, we chose $t_{aging} = 10^7$ s. While relaxing at constant pressure, the system contracts further. Correspondingly, the isobaric specific heat changes according to Eq. (12.2). After the prescribed time has elapsed, the isobaric specific heat, calculated through Eq. (12.2), has attained the (larger) value $C_P(T_q; t_{aging}) = 295.8\,\mathrm{J\,mol^{-1}\,K^{-1}}$. Following the same line of thought illustrated before on discussing Eq. (12.3), we can infer the "fictive" temperature ($\widetilde{T}_g$) at which the system would have effectively deviated from the metastable-liquid branch, had it been cooled at a *slower* rate than that leading to vitrification at 247 K. Coherently with the assumption underlying Eq. (12.3), we write:

$$C_P(T_q; t_{aging}) \approx C_P^{(c)}(T_q) + \left[C_P^{(l)}(\widetilde{T}_g) - C_P^{(c)}(\widetilde{T}_g)\right]. \tag{12.6}$$

Equation (12.6) allows us to determine the fictive temperature ($\widetilde{T}_g \approx 230$ K) at which the liquid-to-solid jump of the isobaric specific heat is such that the value derived from Eq. (12.6) is equal to that provided by Eq. (12.2) for $T = T_q$ and $t = t_{aging}$. Note that, for temperatures lower than T_g, the specific heat of the metastable liquid used in Eq. (12.6) was estimated through a linear extrapolation of the experimental data. The state of this partially relaxed glass—whose volume, after quenching and aging, has so far dropped to the value $v_g^{(B)}$ (corresponding to point B in Fig. 12.3)—is equivalent to that which would be produced upon rapidly cooling a liquid whose slow dynamics has arrested at 230 K (instead than at 247 K). This latter statement is crucial for explaining the outcome of Kovacs' experiment, whose third and final stage consists in heating the system to a temperature T_f, intermediate between T_q and

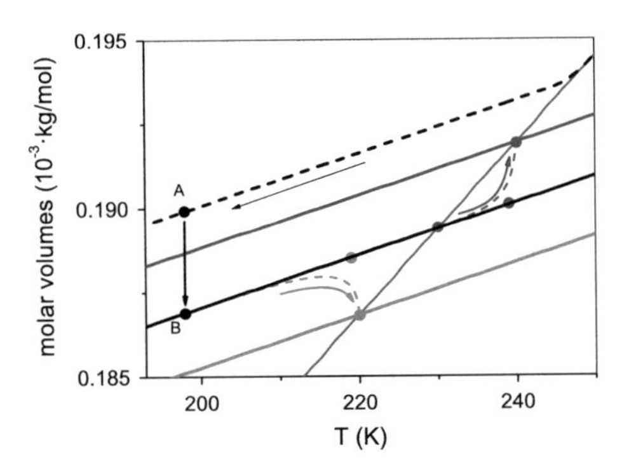

Fig. 12.3 Modified Kovacs' protocol (see text): the sample is equilibrated at T_g and then rapidly cooled down to 198.5 K: the continuous red line is the supercooled liquid line (linearly extrapolated below T_g); the dashed black line represents the molar volume under the above condition and the solid black circle labelled A marks the molar volume attained by the system after quenching; the solid black circle labelled B represents the molar volume achieved by the partially aged system. The continuous black line represents the glass branch that would be followed by the system with a fictive temperature corresponding to the intersection point with the liquid branch. Experiment #1: the sample at B is rapidly heated to 240 K; the blue circles and the blue arrow indicate the time evolution of the molar volume. The blue continuous line represents the glass branch with fictive temperature $\widetilde{T} = 240$ K. Experiment #2: the sample at B is rapidly heated to 220 K; the green circles and the green arrow indicate the time evolution of the molar volume. The green continuous line represents the glass branch with fictive temperature $\widetilde{T} = 220$ K

T_g. In fact, we imagine that, as soon as the time $t = t_{aging}$ has elapsed, the system is immediately coupled with another thermostat whose temperature is T_f and then left free to relax until (metastable) equilibrium has been eventually reached.

We shall now investigate the dynamical behaviour of the system, with specific regard to the way in which the volume changes with time, when partially aged glassy OTP is heated to different temperatures from similarly prepared samples (*i.e.,* samples quenched to the same temperature and aged for the same time). In this way, the outcome of the heating procedure will not be influenced by differing initial conditions of the material. In fact, the states of, say, equally aged samples are generally different at different quenching temperatures because the relaxation times of the system depend on the temperature.

We start inspecting the volumetric behaviour of the system when the final temperature is *higher* than the fictive temperature $\widetilde{T}_g$, while being lower than T_g (see Fig. 12.3). Let us choose $T_f = 240$ K. In the short-time regime, only the fast degrees of freedom of the system react to the modified thermal condition; hence, the volume starts increasing from the value $v_g^{(B)}$, closely following, as the system warms up, the glass line $v_g(T; \widetilde{T}_g)$ that intercepts the supercooled-liquid branch at $T = \widetilde{T}_g$. As soon as the system approaches and eventually surpasses the vitrification threshold, the slower configurational degrees of freedom start relaxing as well. As a result, the volume keeps growing *monotonically*, while departing from the glass line, until the system has eventually equilibrated at the prescribed temperature.

Let us now set the final temperature, at which the system—previously prepared in the same initial state B as in the thought experiment discussed above—is to be heated, to a value *lower* than $\widetilde{T}_g$, say $T_f = 220$ K: in such conditions (see Fig. 12.3), the molar volume of the partially-aged glass at T_f turns out to be larger than the molar volume of the metastable liquid (extrapolated) at $T = 220$ K. Hence, even in this case we would again observe an expansion of the system at short times: the volume would initially increase following the glass line which departs from the metastable branch at the fictive temperature calculated above. However, as soon as the slow configurational degrees of freedom become active, the molar volume would start shrinking, after the initial rise, so as to approach the lower value which corresponds to the asymptotic equilibrium state at the prescribed final temperature.

Hence, in this second thought experiment the molar volume will exhibit a *non-monotonic* time behaviour. The resulting maximum is the distinguishing feature that was originally observed by Kovacs [19]. In the present scheme, as is manifest from Fig. 12.3, on approaching the equilibrium value the volume passes through a maximum only if the final temperature at which the system is heated is lower than the fictive temperature $\widetilde{T}_g$. Moreover, the difference between the maximum value attained by the molar volume and the asymptotic equilibrium value turns out to be larger the larger the difference $(\widetilde{T}_g - T_f)$.

The experimental protocol discussed above partially differs from that originally designed by Kovacs. In fact, this author reported on the time behaviour of the volume of a system which was heated to one single temperature $T_f \leq T_g$ from several lower temperatures at which the system had been previously cooled from the same equilibrium state at a temperature $T_i \geq T_g$. Before heating the system, the quenched liquid was left to age for a variable timespan: the lower the quenching temperature was, the longer the aging time would be.

In order to "simulate" Kovacs' protocol, we proceed as before, assuming that the system has been cooled from T_g to a temperature T_q in the range 190–210 K. The just formed glass is then allowed to relax but for not so long that it may reach equilibrium, say for a time of the order of 10^7 s. Correspondingly, the volume of the aged glass decreases from the value attained soon after quenching, whereas the specific heat increases. Using Eq. (12.2), we calculate the value of the specific heat pertaining to this new state, *i.e.*, $C_P(T_q; t_{aging})$, which, through Eq. (12.6), allows us to infer the fictive temperature $\widetilde{T}_g$. Once we know this datum, we can calculate the value of the molar volume of the partially aged glass which, in our picture, will be equal to the molar volume at $T = T_q$ of a glass whose dynamics has arrested at $T = \widetilde{T}_g$.

We now imagine to put the system in contact with a thermostat at the (higher) temperature $T_f = 230$ K; such a re-heating cycle is repeated a number of times, with the same target temperature but starting from different quenching temperatures in the cited range. In a rather short time (let us say, largely overestimating it, 10 s), fast motions will have fully relaxed whereas the slow degrees of freedom are still frozen. As soon as fast relaxations have occurred, both the specific heat and the molar volume of the glass have contextually increased to the values which correspond to the higher temperature $T = T_f$ on the glass line, departing from the metastable liquid branch at $T = \widetilde{T}_g$. We now use Eq. (12.6), where $C_P^{(c)}$ has been substituted with $C_P(T_q; t_{aging})$, to estimate the value attained by the specific heat once the system has been left free to relax for other 10 s. Following the same procedure outlined above, we then calculate the new fictive temperature corresponding to the updated value of the specific heat at $t = 20$ s, and the resulting value of the molar volume. Upon iterating this procedure, we can trace, with steps of 10 s each, the time evolution of the molar volume of glassy OTP when it has been heated to T_f from different quenching temperatures.

The results predicted by our simplified model are shown in Fig. 12.4 and show that, coherently with what has been observed in both real and numerical experiments

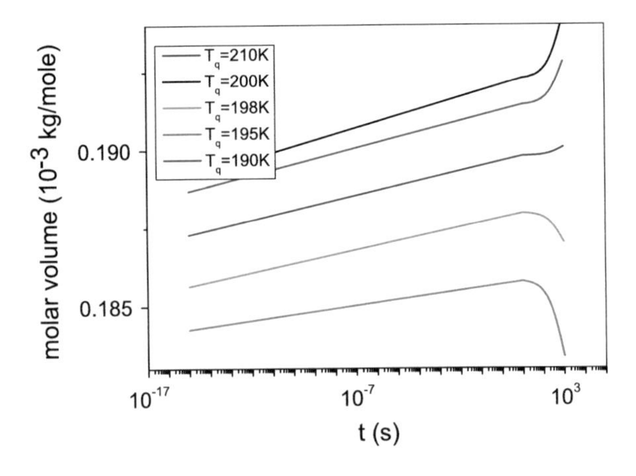

Fig. 12.4 Time evolution of the molar volume according to Kovacs' protocol: the system equilibrated at T_g is rapidly cooled to a temperature T_q in the range 190–210 K and then aged for 10^7 s. After aging, the system is rapidly heated to 230 K and then left free to relax to thermal equilibrium

[19, 20], the molar volume of quenched OTP reaches its asymptotic equilibrium value $v_l(T_f)$ either rising monotonically from the value of the partially aged glass or passing through a maximum at intermediate times, after having initially overshooted $v_l(T_f)$. In the present scheme the occurrence of one or the other alternative behaviour critically depends on whether the final temperature T_f is higher or lower than the temperature T_X at which the extrapolated supercooled-liquid line and the *effective* glass line, onto which the representative state of the glass has "shifted" after the initial aging, cross each other. In both scenarios, the molar volume of the heated glass will initially jump to a value larger than its initial value after quenching and aging. However, if $T_f > T_X$ the specific heat of the rapidly heated glass initially overshoots the value corresponding to the metastable liquid at the same temperature; from there on, the specific heat will decrease with time and this implies a gradual increase of the fictive glass temperature that is calculated at each step and, contextually, of the molar volume of the system. On the other side, if $T_f < T_X$ the specific heat attains a value that is smaller than that corresponding to the metastable liquid at the same final temperature; hence, it will keep on growing with time, which implies that the fictive glass temperature decreases as also does the molar volume.

Following the above discussion we can also interpret other aspects of the glass phenomenology. Imagine, for instance, that we perform a differential scanning calorimetry measurement on a glass which has been previously aged at low temperatures, in such a way that the rate at which the temperature is being changed corresponds to the time scale over which slow motions relax at $T = T_g$. Because of the aging, slow motions are initially equilibrated at a fictive temperature lower than T_g. Hence, when the temperature approaches T_g, an excess of heat from the bath (with respect to the enthalpy reduction originally undergone by the material upon quenching) is required to fully equilibrate the system, thus producing the "endothermic overshoot" that is typically observed in the temperature evolution of the isobaric specific heat.

12.3 Concluding Remarks

In this note we have revisited the Kovacs effect in a simplified picture of the relaxation dynamics underpinning the glass transition. In particular, we have assumed that the time dependence of the isobaric specific heat can be mimicked using just one relaxation time and that fast motions have fully and systematically relaxed over the time step of our calculations. Notwithstanding such rough approximations, the resulting model is found to reproduce correctly, on a qualitative basis, the main features of Kovacs' experiment and to convey some useful insight on the phenomenology of the glass transition.

Our analysis is based on the premise that the glass obtained through the rapid cooling of a supercooled liquid is an out-of-equilibrium system which, however, may asymptotically evolve towards a metastable phase provided it is given enough time to relax. We assume that such a metastable phase cannot be distinguished from that of the "parent" supercooled liquid at a given temperature. The relaxation process undergone by the glass, named "aging", is associated with a gradual change of the molar volume and, correspondingly, of the structural configuration of the material. In this perspective, distinguishing a long-aged glass from a metastable liquid may reduce to a merely semantic question, with obvious consequences on the hypothesis of an underlying (thermodynamic) glass transition. In fact, the apparent differences between the two structural conditions of the material emerge at the crossover temperature at which the experimental time scale becomes shorter than the configurational relaxation time of the system.

However, the postulated equivalence between the asymptotically "equilibrated" glass and the corresponding metastable liquid may be disproved by the existence of a threshold below which the nucleation of the stable crystalline phase can no longer be avoided. In this respect, a candidate threshold would be the Kauzmann temperature [3] at which the entropies of the metastable liquid and of the thermodynamically stable crystalline solid become equal. However, it has been argued that the hypothetical coexistence of a liquid and a solid phase would not be possible at a temperature lower than the equilibrium coexistence temperature [27, 29]. In fact, whenever a supercooled liquid escapes from metastability and freezes, it does so irreversibly and adiabatically with an increase of both entropy and temperature as a consequence of the release of heat. As a result, the transition towards stable equilibrium takes place exothermically and the system warms up while solidifying. Hence, it does not make much sense to compare the entropies of the two phases at the same temperature.

The spontaneous freezing of a a metastable liquid is also associated with a change of volume. The spontaneous formation of a finite solid embryo produces a relatively large density fluctuation which propagates at low frequency, while dissipating, across the whole sample [30]. This also explains why metastable equilibrium is a robust structural condition against fluctuations: even when a thermodynamic fluctuation brings, locally, the system close to the boundary between the metastable and stable equilibrium basins in phase space, dissipative processes are able to back reflect the system trajectory towards the metastable basin.

On cooling, a crossover temperature can be eventually reached at which the volume of the metastable system is equal to the volume of the stable phase that is formed under adiabatic conditions [27]. This is what happens to OTP, as indicated in the upper panel of Fig. 12.2. At such a temperature, the structural re-arrangement which drives the system towards the stable configuration is just a local process which does not need to propagate in order to be completed. In such conditions, local fluctuations are not dissipated away and the local transition, which results in a local increase of the temperature, immediately produces a further fluctuation in the adjacent volumes which can propagate rapidly (over a time scale comparable with the time required for the local rearrangement of a few molecules) across the whole sample. This argument is a different way for saying that, at that temperature, the energetic barrier between metastable and stable equilibrium conditions likely disappears. In this perspective, the observation that in water the volume crossover takes place at a temperature that, at normal pressure, is very close to the widely accepted value of the homogeneous nucleation temperature may not be a mere coincidence [27]. In water, this temperature is definitely higher than the experimental glass-transition temperature and this can explain why metastable liquid water cannot exist at temperatures close to T_g.

Following our argumentation, one would be led to deduce that the existence in water of a crossover temperature at which the transition between the metastable liquid phase and the stable crystalline phase becomes both adiabatic and isochoric strongly supports the idea that any observed amorphous phase observed at very low temperature, which may well appear stable over the observation time scale, has no thermodynamic counterpart and can only be described in kinetic terms. On the contrary, in the case of OTP the volume crossover occurs at a temperature lower than the experimental T_g. Such a difference in the behaviour of glass-forming liquids as far as the relation between T_g and the homogeneous nucleation temperature is concerned, has been already noted several years ago [31].

Summing up, the nature of the glass obtained when a liquid is rapidly cooled to low temperatures, over times shorter than those required for a complete structural rearrangement of the system, depends on the relation existing between the state which has been produced and the metastability basin of the system. For moderate supercoolings, the achieved state is not disconnected from the basin of the metastable liquid phase: hence, the glass, while being out of equilibrium, might still evolve, in principle, towards metastability. However, when quenched at very low temperatures, the system may be driven to a state which is no longer accessible from a metastable disordered phase. In such conditions, should it be able to rearrange itself, its unique, ultimate fate would be that of transforming into a stable solid.

Acknowledgements The authors belonging to the Institute for Chemical-Physical Processes (IPCF) of the National Research Council (CNR) recall with enthusiasm the visit that Professor N. H. March paid to their institute in 2010. PVG expresses his profound gratitude to Professor March who invited him to visit the Imperial College of Science and Technology in London (UK) and later, on repeated occasions, the Theoretical Chemistry Department of the University of Oxford (UK) in the earlier stages of his post-graduation career. Working with and learning from him has always been an influential, unforgettable experience.

References

1. G. Biroli, J.P. Garrahan, J. Chem. Phys. **138**(12), 12A301 (2013). https://doi.org/10.1063/1.4795539
2. M.D. Ediger, P. Harrowell, J. Chem. Phys. **137**(8), 080901 (2012). https://doi.org/10.1063/1.4747326
3. W. Kauzmann, Chem. Rev. **43**(2), 219 (1948). https://doi.org/10.1021/cr60135a002
4. P.N. Pusey, W. van Megen, Nature **320**(6060), 340 (1986). https://doi.org/10.1038/320340a0
5. V. Lubchenko, P.G. Wolynes, Ann. Rev. Phys. Chem. **58**(1), 235 (2007). https://doi.org/10.1146/annurev.physchem.58.032806.104653
6. F.H. Stillinger, J. Chem. Phys. **88**(12), 7818 (1988). https://doi.org/10.1063/1.454295
7. J.P. Eckmann, I. Procaccia, Phys. Rev. E **78**, 011503 (2008). https://doi.org/10.1103/PhysRevE.78.011503
8. W. van Megen, T.C. Mortensen, S.R. Williams, J. Müller, Phys. Rev. E **58**, 6073 (1998). https://doi.org/10.1103/PhysRevE.58.6073

9. L. Boué, H.G.E. Hentschel, V. Ilyin, I. Procaccia, J. Phys. Chem. B **115**(48), 14301 (2011). https://doi.org/10.1021/jp205773c
10. F. Mallamace, C. Corsaro, N. Leone, V. Villari, N. Micali, S. Chen, **4**, 3747 (2014), Scientific Reports. https://doi.org/10.1038/srep03747
11. P.G. Debenedetti, J. Phys.: Condens. Matter **15**(45), R1669 (2003). https://doi.org/10.1088/0953-8984/15/45/R01
12. H.E. Stanley, P. Kumar, L. Xu, Z. Yan, M.G. Mazza, S.V. Buldyrev, S. Chen, F. Mallamace, Physica A **386**(2), 729 (2007), Disorder and Complexity. https://doi.org/10.1016/j.physa.2007.07.044
13. R.J. Speedy, J. Phys. Chem. **86**(6), 982 (1982). https://doi.org/10.1021/j100395a030
14. O. Mishima, L.D. Calvert, E. Whalley, Nature **314**(6006), 76 (1985). https://doi.org/10.1038/314076a0
15. H.E. Stanley, L. Cruz, S.T. Harrington, P.H. Poole, S. Sastry, F. Sciortino, F.W. Starr, R. Zhang, Physica A **236**(1), 19 (1997), *Proceedings of the Workshop on Current Problems in Complex Fluids*. https://doi.org/10.1016/S0378-4371(96)00429-3
16. S. Sastry, P.G. Debenedetti, F. Sciortino, H.E. Stanley, Phys. Rev. E **53**, 6144 (1996). https://doi.org/10.1103/PhysRevE.53.6144
17. O. Mishima, H.E. Stanley, Nature **392**(6672), 164 (1998). https://doi.org/10.1038/32386
18. O. Mishima, Y. Suzuki, Nature **419**(6907), 599 (2002). https://doi.org/10.1038/nature01106
19. A.J. Kovacs, *Transition vitreuse dans les polymères amorphes. Etude phénoménologique* (Springer Berlin Heidelberg, Berlin, Heidelberg), (Fortschritte Der Hochpolymeren-Forschung. Advances in Polymer Science) **3**(3), 394–507 (1964). ISBN 978-3-540-37073-4. https://doi.org/10.1007/BFb0050366
20. S. Mossa, F. Sciortino, Phys. Rev. Lett. **92**, 045504 (2004). https://doi.org/10.1103/PhysRevLett.92.045504
21. C.A. Angell, K.L. Ngai, G.B. McKenna, P.F. McMillan, S.W. Martin, J. Appl. Phys. **88**(6), 3113 (2000). https://doi.org/10.1063/1.1286035
22. E. Bouchbinder, J.S. Langer, Soft Matter **6**, 3065 (2010). https://doi.org/10.1039/C001388A
23. S.S. Chang, A.B. Bestul, J. Chem. Phys. **56**(1), 503 (1972). https://doi.org/10.1063/1.1676895
24. M. Naoki, S. Koeda, J. Phys. Chem. **93**(2), 948 (1989). https://doi.org/10.1021/j100339a078
25. J.E.K. Schawe, Thermochimica Acta **260**, 1 (1995). https://doi.org/10.1016/0040-6031(95)90466-2
26. I. Alig, Thermochimica Acta **304**, 35 (1997), Temperature Modulated Calorimetry. https://doi.org/10.1016/S0040-6031(97)00174-3
27. F. Aliotta, P.V. Giaquinta, M. Pochylski, R.C. Ponterio, S. Prestipino, F. Saija, C. Vasi, J. Chem. Phys. **138**(18), 184504 (2013). https://doi.org/10.1063/1.4803659
28. T.M. Nieuwenhuizen, J. Chem. Phys. **115**(17), 8083 (2001). https://doi.org/10.1063/1.1399036
29. H. Hoffmann, Mat.-wiss. u, Werkstofftech. **43**(6), 528 (2012). https://doi.org/10.1002/mawe.201200673
30. F. Aliotta, P.V. Giaquinta, R.C. Ponterio, S. Prestipino, F. Saija, G. Salvato, C. Vasi, **4**, 7230 (2014), Article, Scientific Reports. https://doi.org/10.1038/srep07230
31. C.A. Angell, E.J. Sare, J. Donnella, D.R. MacFarlane, J. Phys. Chem. **85**(11), 1461 (1981). https://doi.org/10.1021/j150611a001

Chapter 13
Structural Properties of Ionic Aqueous Solutions

P. Gallo, M. Martin Conde, D. Corradini, P. Pugliese and M. Rovere

Abstract On the occasion of the 90th birthday of Norman March, we present here a short review of results on the structural properties of ionic aqueous solutions that we obtained in recent years by computer simulation. In particular we compare structural properties of alkali halides NaCl(aq), KCl(aq), KI(aq) to account for the role of cations and anions of different size. The modifications of the hydration shells and the changes in the water structure induced by the presence of the ions are investigated. It is found that the oxygen–oxygen structure can be strongly distorted at high ionic concentration. The hydrogen bonding however is preserved at all concentrations and temperatures. The relation between the perturbation induced by the ions and the different high density and low density liquid local order of water is also discussed.

13.1 Introduction

> The molecular dynamics experiments of Rahman and Stillinger, carried out on an assembly of 216 molecules over a temperature range, start from the observation that there is a tendency when water molecules interact towards formation of linear hydrogen bonds between neighbours disposed in space in a tetrahedral coordination pattern.
>
> N. H. March and M. P. Tosi, *Atomic Dynamics in Liquids,* (Dover, New York, 1976) [1].

In their book of 1976, Norman March and Mario Tosi describe the first pioneering simulation work on liquid water performed by Rahman and Stillinger [2, 3]. They underline in a short phrase that the main idea of the simulation was to reproduce with a simple potential the hydrogen bond (HB) network of water. The HB network characterizes not only the ice structure but also the short range order in liquid water. A good model for water must reproduce the tetrahedrally coordinated network in order to predict its properties.

P. Gallo · M. Martin Conde · P. Pugliese · M. Rovere (✉)
Dipartimento di Matematica e Fisica, Università “Roma Tre”, Via della Vasca Navale, 84, 00146 Roma, Italy
e-mail: rovere@fis.uniroma3.it

D. Corradini
American Physical Society, 1 Research Road, Ridge, NY 11961, USA

G. G. N. Angilella and C. Amovilli (eds.), *Many-body Approaches at Different Scales*, https://doi.org/10.1007/978-3-319-72374-7_13

Water is involved in almost all the natural phenomena in biology and geology and in a large amount of technological applications. Since the 70s a lot of progress has been done in the study of this system in all its phases. Experimental neutron scattering and various spectroscopic techniques made available a large amount of data on water in a wide range of temperature and pressure. The phase diagram of water is now well known, but there are still a number of open problems related to its anomalous behaviour [4]. The anomalies are more evidenced under extreme conditions of temperature and pressure where however it is more difficult to perform experiments to reach a deep understanding of the phenomena.

Since the Rahman and Stillinger model, the ST2 potential, a number of different potentials for water have been developed and computer simulation has shown to be an essential tool both for the interpretation of experiments and the developing of theoretical approaches.

In many natural phenomena and applications in chemistry, biology and environmental science, it is crucial to understand the role of water as a solvent. In this respect the basic and more common systems are the solutes with dissolved ions. As said at the beginning the phenomena in water are supposed to be dominated by the presence of the HB network so the main issue is to understand how the presence of the ions could perturb the water network and aqueous solutions of salts have been matter of investigation by experiments [5–8] and computer simulations [9–26].

The concept of Hofmeister series has been widely used in the interpretation of the properties of ionic solutions [27, 28], like alkali halides. In this classification scheme anions and cations are ordered according to their properties of enhancing (structure makers) or weakening (structure breakers) the HB network of water. The idea is that structure makers will be strongly hydrated since they break the HB in the surrounding water molecules and the rest of the water molecules can rearrange in an ordered hydration structure. On the contrary, structure breaker ions interact weakly with the water and induce a disordering in the network of water [29].

In recent years, however, this classification scheme in terms of structure makers and structure breakers has been challenged with the idea that this simplified classification scheme of the ions does not take into account other effects like the ion concentration and the different thermodynamic conditions. Recent evidences from experiments and computer simulations indicate that ions perturb the water structure beyond the first hydration shell with an effect similar to the application of pressure on pure water [8, 25, 30].

The research about the anomalies of water has recently increased the interest in the salt solutions in the supercooled region [7, 25, 31]. The anomalous behaviour of water has recently been interpreted by considering water as a mixture of two distinct group of molecules characterized by different arrangement of the HB network [4]. The low density liquid (LDL) component would be characterized locally by a stronger tetrahedral order, while the high density liquid (HDL) component would have a broken, more disordered, HB structure. It is under discussion the effects of the structure making/breaking ions on the HDL/LDL ordering in water.

On the occasion of the 90th birthday of Norman March, we review here the results that we obtained by studying the structural properties of ionic aqueous solutions by means of computer simulation.

We recall in the next section the methodology. In the third section we present the site radial distribution functions of water in the solutions compared with pure water. In the fourth section we discuss the problem of the ionic hydration and the modification that are induced in the structural properties of water. Last section is devoted to the conclusions.

13.2 Model Potentials

As said above, there are a large number of potentials developed for water. In the case of ionic solutions the ion-ion and water-ions interactions must be added. It is obvious that the starting point must be a good model for water. Generally for computer simulation purposes the potential must give enough good results for a number of different properties and it must be simple enough to avoid large computational costs.

The results we present here were obtained using Molecular Dynamics (MD) method. Water is described by the TIP4P site potential [32]. The molecule is represented by a four site rigid model with two hydrogen (H) sites with positive charge of $0.52e$. They are connected to the neutral oxygen (O) site, whose negative charge is attributed to a slightly shifted (0.015 nm) fourth site. The OH bond length is 0.09572 nm, the angle between the two bonds is $\theta = 104.5°$. The oxygen sites interact with a Lennard–Jones potential, the other sites with the Coulomb forces. The TIP4P water potential has been extensively used for studying water at ambient conditions and upon cooling below the freezing point.

The ion-ion and the water sites-ions interactions were also modeled with the combination of the Coulombic and the Lennard–Jones (LJ) potential

$$u_{\alpha\beta}(r) = \frac{q_\alpha q_\beta}{r_{\alpha\beta}} + 4\varepsilon_{\alpha\beta}\left[\left(\frac{\sigma_{\alpha\beta}}{r_{\alpha\beta}}\right)^{12} - \left(\frac{\sigma_{\alpha\beta}}{r_{\alpha\beta}}\right)^{6}\right]. \qquad (13.1)$$

The Jensen and Jorgensen interaction parameters [33] were assumed for the LJ interaction of the ions. The ion-ion and the ion-water LJ parameters were calculated by using the geometrical mixing rules $\varepsilon_{\alpha\beta} = \left(\varepsilon_{\alpha\alpha}\varepsilon_{\beta\beta}\right)^{1/2}$ and $\sigma_{\alpha\beta} = \left(\sigma_{\alpha\alpha}\sigma_{\alpha\beta}\right)^{1/2}$.

Details about the values of the parameters can be found in our previous work [25]. These parameters were found to reproduce very well the structural characteristics and free energies of hydration of the ions.

In the case of KI(aq) we present recent preliminary calculations performed with the use of the force field of Joung and Cheatham (JC) [19] combined with the TIP4P/2005 model for water. The TIP4P/2005 is a modification of the TIP4P potential introduced by Abascal and Vega [34], it works very well in describing the properties of water

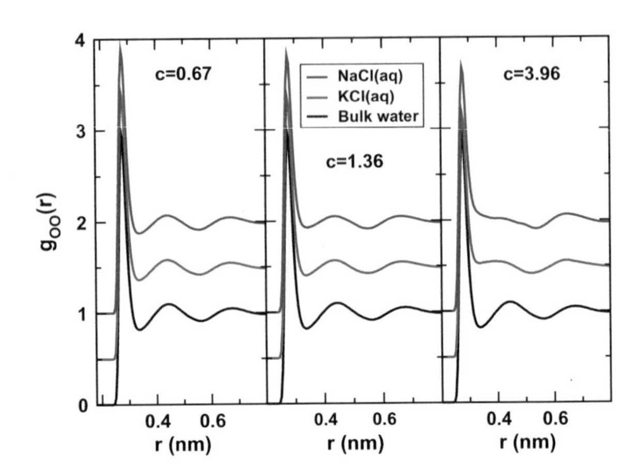

Fig. 13.1 Oxygen-oxygen RDF of bulk water (unshifted) and for NaCl(aq) and KCl(aq) (vertically shifted) at $T = 300$ K at increasing concentrations (in mol/kg) as indicated. Data from Ref. [25]

at ambient temperature and upon cooling. The combination of the JC force field and TIP4P/2005 has been recently successfully tested [26].

The simulations were performed at constant ambient pressure and different temperature with the use of appropriate thermostats. Periodic boundary conditions were applied. The interaction potentials were truncated at $0.9 \div 1.0$ nm. The long range electrostatic interactions were taken into account with the Ewald particle mesh method. The equilibration time was of 20 ns and the averages were calculated on production runs of 30 ns. The parallelized version of the GROMACS package has been used [35].

13.3 Modifications of the Water Structure

We show in Fig. 13.1 the radial distribution functions (RDF) of the water oxygens in solution of NaCl(aq) and KCl(aq) at increasing concentrations at $T = 300$ K compared with the O-O RDF of pure water (TIP4P model).

In the low concentration regime, the presence of ions does not perturb the short range order of liquid water. At intermediate concentration ($c = 1.36$ mol/kg) the first two shells are well defined. At the highest concentration instead there is evidence of a considerable change. In KCl(aq) the second OO shell is not well defined and in particular the first minimum almost disappears. In NaCl(aq) the second shell collapses on the first and all the OO structure appears quite different with respect to pure water, defined also as bulk water.

At variance with the O–O structure the $g_{OH}(r)$ RDF of the solutions, shown in Fig. 13.2, do not present significant changes with respect to bulk water. These results indicate that at high concentration the HB network is still present but it is distorted by the interaction with the ions.

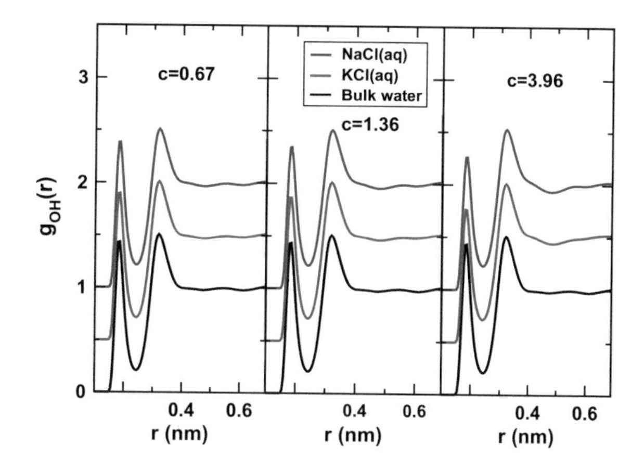

Fig. 13.2 The $g_{OH}(r)$ of water in NaCl(aq) and KCl(aq) (vertically shifted) at $T = 300\,\text{K}$ at increasing concentrations (mol/kg) as indicated, the RDF's are compared with bulk water (not shifted). Data from Ref. [25]

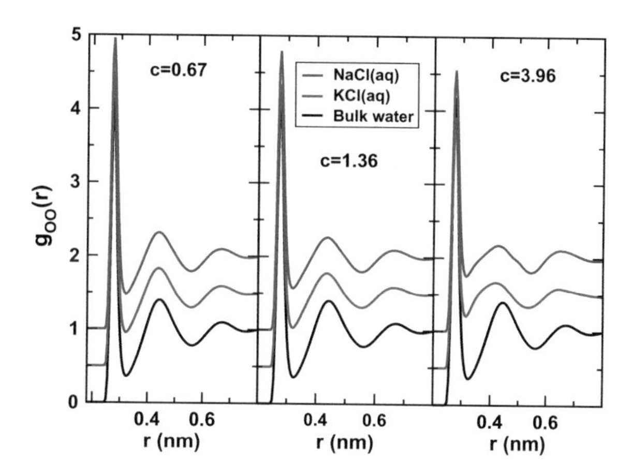

Fig. 13.3 Oxygen-oxygen RDF of bulk water (unshifted) and for NaCl(aq) and KCl(aq) (vertically shifted) at $T = 220\,\text{K}$ at increasing concentrations (mol/kg) as indicated. Data from Ref. [25]

The changes in the second shell of the $g_{OO}(r)$ with unbroken hydrogen bonds are similar to the effects of an high pressure on the liquid water [25].

In the supercooled regime at 220 K, Fig. 13.3, the peaks of the $g_{OO}(r)$ are more well defined for effect of the temperature and they are not changed by the ion interaction at low concentration. At $c = 3.96\,\text{mol/kg}$ the second shell is strongly distorted in particular in the case of NaCl(aq). The $g_{OH}(r)$ (not shown) are not modified by the presence of the ions with respect to the bulk.

The combination of the changes in the second peak of the $g_{OO}(r)$ and the persistence of the HB order is the signature of an incipient HDL phase of water. From this point of view by assuming that the cations are dominating at high concentration, in spite of the different classification of K^+ as structure breaker and Na^+ as structure maker, both of them modify the O–O structure behind the first shell. This is in agreement with recent experiments [7]. We will discuss more this point below.

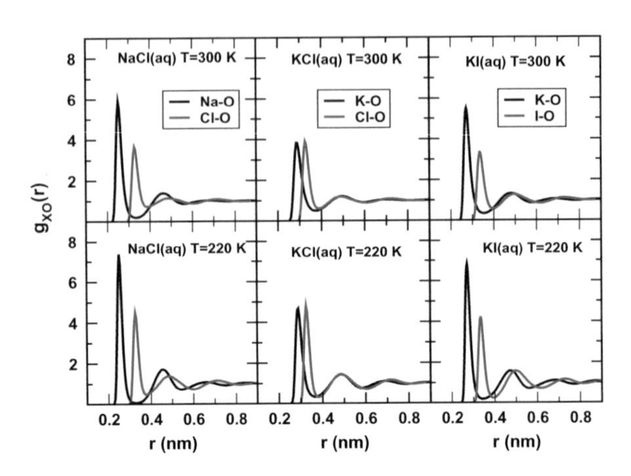

Fig. 13.4 Cation-oxygen and anion-oxygen RDF for NaCl(aq) (left panels), KCl(aq) (central panels), KI(aq) (right panels) at concentration of at $c = 0.67$ mol/kg and temperature $T = 300$ K (top) and $T = 220$ K (bottom). Data on NaCl(aq) and KCl(aq) from Ref. [25]

13.4 Hydration of Ions and Short Range Order

It is expected that the hydration of cations and anions have a strong effect on the solutions. In Fig. 13.4 we present the cation-oxygen and the anion-oxygen RDF in the case of NaCl(aq), KCl(aq) and KI(aq) at the concentration of 0.67 mol/kg at $T = 300$ K and $T = 220$ K. It is evident the formation of well defined shells of oxygens around the ions. The oxygen shells are closer to the cations as expected.

In KCl(aq) the potassium ions show a low hydration. The first K–O peak appears lower and the first minimum broader with respect to the first peak and the first minimum in KI(aq). The two shells of cations and anions in KCl(aq) are penetrating. The possibility of exchanging molecules between the first and the second shell are found in experiments [8] and also ab initio calculations on KCl(aq) [36].

The differences in the hydration shells of cations in KCl(aq) and KI(aq) indicate the interplay between charge and steric effects in the solutions. The role of cations and anions cannot simply been separated. The hydration of K^+ is enhanced in KI(aq) by the presence of I^-. Cations form more defined shells in KI(aq), where the I^- gives a steric contribution in keeping the charge ordering. In KI(aq), as in NaCl(aq) the anion-oxygen first peaks are exactly at the minimum of the cation-oxygen shells indicating a good charge ordering.

Upon cooling similarities and differences between the solution properties are preserved. It is evident that the first shell of cation-oxygen RDF become even more rigid at $T = 220$ K both in NaCl(aq) and KI(aq) in comparison with the KCl(aq).

It is relevant to explore now more in details the effect on water.

We show the three examples for NaCl(aq), KCl(aq) and KI(aq), that are representative of how the ions could be arranged in the water network.

By considering Figs. 13.5 and 13.6, the first Na–O shell appears on the left of the O–O well separated from the Na–H shell. The Cl–O and Cl–H shells appear rigidly shifted with respect to the O–O shells. These results seem indicate that the Na^+ have the tendency to stay out from the HB network, while the Cl^- ions instead could

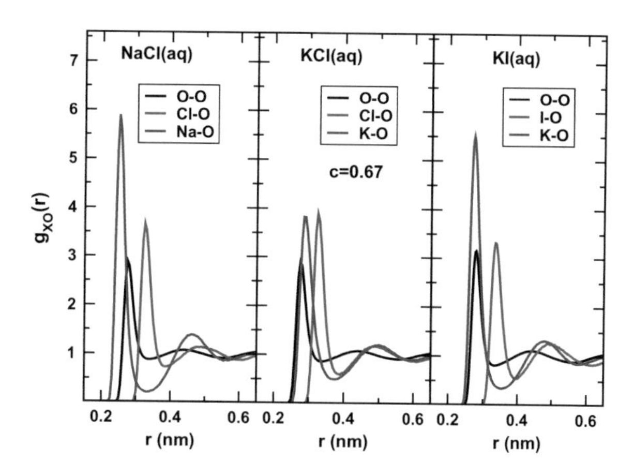

Fig. 13.5 Oxygen-oxygen, anion-oxygen and cation-oxygen RDF at concentration of at $c = 0.67\,\text{mol/kg}$ and temperature $T = 300\,\text{K}$ for NaCl(aq) (left panel), KCl(aq) (central panel) and KI(aq) (right panel). Data on NaCl(aq) and KCl(aq) from Ref. [25]

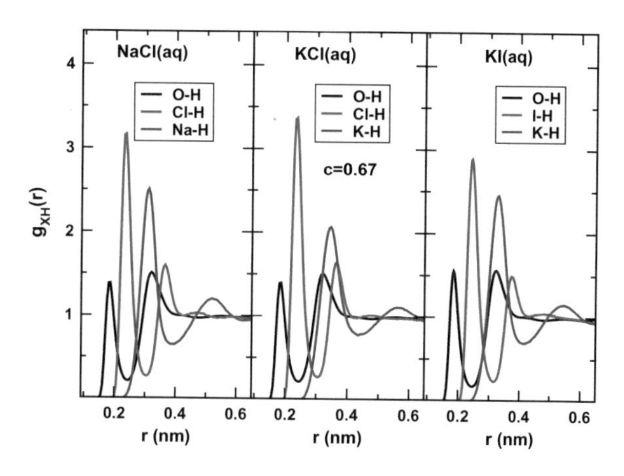

Fig. 13.6 Oxygen-hydrogen, anion-hydrogen and cation-hydrogen RDF at concentration of at $c = 3.96\,\text{mol/kg}$ and temperature $T = 300\,\text{K}$ for NaCl(aq) (left panel), KCl(aq) (central panel) and KI(aq) (right panel). Data on NaCl(aq) and KCl(aq) from Ref. [25]

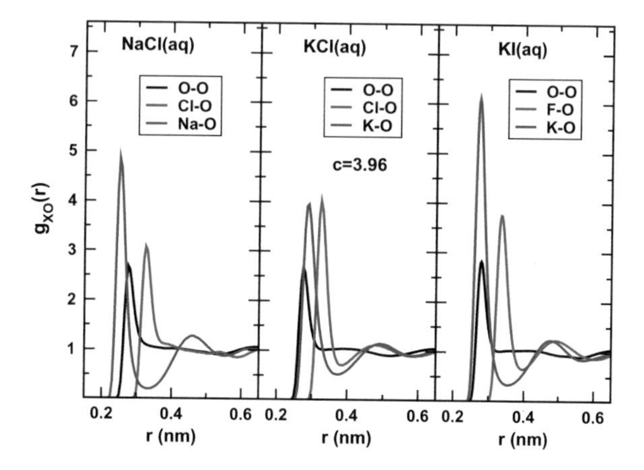

Fig. 13.7 Oxygen-oxygen, anion-oxygen and cation-oxygen RDF at concentration of at $c = 3.96\,\text{mol/kg}$ and temperature $T = 300\,\text{K}$ for NaCl(aq) (left panel), KCl(aq) (central panel) and KI(aq) (right panel). Data on NaCl(aq) and KCl(aq) from Ref. [25]

be inserted in the network of water eventually substituting the oxygens. This is in agreement with recent computer simulation results on Na^+ and Cl^- inclusion in ice [37].

The $g_{ClO}(r)$ and the $g_{ClH}(r)$ are very similar in the three solutions, but in NaCl(aq) and KI(aq) the first peak is more rigid and well separated from the cation-oxygen first peak. In any case the $g_{OO}(r)$ is unchanged as already seen in Fig. 13.1.

At the highest concentration instead we observe modifications of the $g_{OO}(r)$, as can be seen in Fig. 13.7, while the $g_{OH}(r)$, not shown, are very similar to the case $c = 0.67$ mol/kg.

The O–O second shell in the three solutions though the effect is less pronounced in KCl(aq).

In NaCl(aq) the Na^+ ions present at least three well defined shells and it seems that they give the dominating effect. In KI(aq) the ions perturb the network of water and the deformation of the O–O second shell is stronger than in KCl(aq) at the same concentration. It seems that water at this concentration already at $T = 300$ K shows a prevalence of the HDL structure.

13.5 Conclusions

We presented a short review of the results of computer simulations on the structure of ionic aqueous solutions. In particular we considered solutions of potassium halides KCl, KI characterized by anions of different sizes and as consequence different charge densities and compared also with the result for NaCl(aq) to test the effect of changing the cation.

We focused in particular on the hydration of the ions and the effect on the structure of water. In NaCl(aq) and KI(aq) the hydration shells of the ions are well defined. In KI(q) and NaCl(aq) a charge ordering at short range is also present. In KCl(aq) the hydration shells are less well defined. It appears that in KI(aq) the there is an interplay between steric effects due mainly to I^- and charge effects mainly due to K^+.

With increasing concentration the second shell of the oxygen-oxygen RDF is distorted. The short range order of the O-H however is almost unaffected from the presence of ions. So the hydrogen bond network is perturbed in similar way as found in water under pressure.

The changes in the region of the second shell of the oxygen-oxygen RDF are interesting upon cooling or upon increasing concentration since they are related to the presence of a liquid-liquid critical point as a terminal point of a coexistence between LDL and HDL phases of water. Corradini et al. [23], and Corradini and Gallo [31] found evidence of the liquid-liquid critical point in NaCl(aq).

Our study confirms that the traditional classification of ions as structure making/breaking is not able to give a complete prediction of the way in which water structure changes under the effect of ions. On the other hand, taking into account that the ions could perturb water beyond the first shell in the study of ionic solutions is of relevant interest to understand the role of the different ions on the LDL/HDL coexistence in water and the possible effect on the liquid-liquid transition.

References

1. N.H. March, M.P. Tosi, *Atomic Dynamics in Liquids* (Dover, New York, 1976)
2. A. Rahman, F.H. Stillinger, J. Chem. Phys. **55**(7), 3336 (1971). https://doi.org/10.1063/1.1676585
3. F.H. Stillinger, A. Rahman, J. Chem. Phys. **57**(3), 1281 (1972). https://doi.org/10.1063/1.1678388
4. P. Gallo, K. Amann-Winkel, C.A. Angell, M.A. Anisimov, F. Caupin, C. Chakravarty, E. Lascaris, T. Loerting, A.Z. Panagiotopoulos, J. Russo, J.A. Sellberg, H.E. Stanley, H. Tanaka, C. Vega, L. Xu, L.G.M. Pettersson, Chem. Rev. **116**(13), 7463 (2016). https://doi.org/10.1021/acs.chemrev.5b00750
5. Y. Marcus, Chem. Rev. **109**(3), 1346 (2009). https://doi.org/10.1021/cr8003828. PMID: 19236019
6. H. Ohtaki, T. Radnai, Chem. Rev. **93**(3), 1157 (1993). https://doi.org/10.1021/cr00019a014
7. I. Waluyo, D. Nordlund, U. Bergmann, D. Schlesinger, L.G.M. Pettersson, A. Nilsson, J. Chem. Phys. **140**(24), 244506 (2014). https://doi.org/10.1063/1.4881600
8. A.K. Soper, K. Weckström, Biophys. Chem. **124**(3), 180 (2006). https://doi.org/10.1016/j.bpc.2006.04.009
9. R.W. Impey, P.A. Madden, I.R. McDonald, J. Phys. Chem. **87**(25), 5071 (1983). https://doi.org/10.1021/j150643a008
10. S.B. Zhu, G.W. Robinson, J. Chem. Phys. **97**(6), 4336 (1992). https://doi.org/10.1063/1.463903
11. R.M. Lynden-Bell, J.C. Rasaiah, J.P. Noworyta, Pure Appl. Chem. **73**(11), 1721 (2001). https://doi.org/10.1351/pac200173111721
12. S. Koneshan, J.C. Rasaiah, R.M. Lynden-Bell, S.H. Lee, J. Phys. Chem. B **102**(21), 4193 (1998). https://doi.org/10.1021/jp980642x
13. S. Koneshan, J.C. Rasaiah, J. Chem. Phys. **113**(18), 8125 (2000). https://doi.org/10.1063/1.1314341
14. J.C. Rasaiah, R.M. Lynden-Bell, Philos. Trans. R. Soc. Lond. A **359**(1785), 1545 (2001). https://doi.org/10.1098/rsta.2001.0865
15. A. Chandra, Phys. Rev. Lett. **85**, 768 (2000). https://doi.org/10.1103/PhysRevLett.85.768
16. S. Chowdhuri, A. Chandra, J. Chem. Phys. **115**(8), 3732 (2001). https://doi.org/10.1063/1.1387447
17. S. Chowdhuri, A. Chandra, J. Chem. Phys. **118**(21), 9719 (2003). https://doi.org/10.1063/1.1570405
18. H. Du, J.C. Rasaiah, J.D. Miller, J. Phys. Chem. B **111**(1), 209 (2007). https://doi.org/10.1021/jp064659o
19. I.S. Joung, T.E. Cheatham III, J. Phys. Chem. B **112**(30), 9020 (2008). https://doi.org/10.1021/jp8001614
20. P. Auffinger, T.E. Cheatham III, A.C. Vaiana, J. Chem. Theory Comput. **3**(5), 1851 (2007). https://doi.org/10.1021/ct700143s
21. P.J. Lenart, A. Jusufi, A.Z. Panagiotopoulos, J. Chem. Phys. **126**(4), 044509 (2007). https://doi.org/10.1063/1.2431169
22. D. Corradini, P. Gallo, M. Rovere, J. Chem. Phys. **128**(24), 244508 (2008). https://doi.org/10.1063/1.2939118
23. D. Corradini, M. Rovere, P. Gallo, J. Chem. Phys. **132**(13), 134508 (2010). https://doi.org/10.1063/1.3376776
24. D. Corradini, M. Rovere, P. Gallo, J. Phys. Chem. B **115**(6), 1461 (2011). https://doi.org/10.1021/jp1101237

25. P. Gallo, D. Corradini, M. Rovere, Phys. Chem. Chem. Phys. **13**, 19814 (2011). https://doi.org/10.1039/C1CP22166C
26. J.L. Aragones, M. Rovere, C. Vega, P. Gallo, J. Phys. Chem. B **118**(28), 7680 (2014). https://doi.org/10.1021/jp500937h
27. F. Hofmeister, Arch. Exp. Pathol. Pharmakol. **24**, 247 (1888)
28. V.A. Parsegian, Nature **378**(6555), 335 (1995). https://doi.org/10.1038/378335a0
29. B. Hribar, N.T. Southall, V. Vlachy, K.A. Dill, J. Am. Chem. Soc. **124**(41), 12302 (2002)
30. R. Leberman, A.K. Soper, Nature **378**(6555), 364 (1995). https://doi.org/10.1038/378364a0
31. D. Corradini, P. Gallo, J. Phys. Chem. B **115**(48), 14161 (2011). https://doi.org/10.1021/jp2045977. PMID: 21851078
32. W.L. Jorgensen, J. Chandrasekhar, J.D. Madura, R.W. Impey, M.L. Klein, J. Chem. Phys. **79**(2), 926 (1983). https://doi.org/10.1063/1.445869
33. K.P. Jensen, W.L. Jorgensen, J. Chem. Theory Comput. **2**(6), 1499 (2006). https://doi.org/10.1021/ct600252r
34. J.L.F. Abascal, C. Vega, J. Chem. Phys. **123**(23), 234505 (2005). https://doi.org/10.1063/1.2121687
35. B. Hess, C. Kutzner, D. van der Spoel, E. Lindahl, J. Chem. Theory Comput. **4**(3), 435 (2008). https://doi.org/10.1021/ct700301q
36. L.M. Ramaniah, M. Bernasconi, M. Parrinello, J. Chem. Phys. **111**(4), 1587 (1999). https://doi.org/10.1063/1.479418
37. M.M. Conde, M. Rovere, P. Gallo, Phys. Chem. Chem. Phys. **19**, 9566 (2017). https://doi.org/10.1039/C7CP00665A

Chapter 14
Atomic Spectra Calculations for Fusion Plasma Engineering Using a Solvable Model Potential

M. E. Charro and L. M. Nieto

Dedicated to Professor N. H. March on the occasion of his 90th birthday.

Abstract The analysis of the atomic spectra emitted by highly ionized atoms is a field of extraordinary richness and a part of atomic physics with applications in astrophysics, engineering, fusion plasma and materials research. Certain elements have attracted considerable attention because they are useful for spectroscopic diagnostics in fusion plasmas, where a prediction of the experimental spectra is required. Taking into account this fact, the Relativistic Quantum Defect Orbital (RQDO) method has been applied to calculate relevant atomic data, as transition rates for emission lines, in a high number of atoms and ions. This formalism, unlike sophisticated and costly self-consistent-field procedures, is a simple but reliable analytical method based on exactly solvable model potentials, a type of problems that always attracted Professor March's attention. The method has the great advantage of a low computational cost, which is not increased as the atomic system becomes heavier. In this work, a highlight of this method is presented, together with an overview of the main atomic data obtained using it, which are useful in engineering for fusion plasma diagnostic.

M. E. Charro
Faculty of Education, University of Valladolid, 47011 Valladolid, Spain
e-mail: echarro@dce.uva.es

L. M. Nieto (✉)
Departamento de Física Teórica, Atómica y Óptica, Universidad de Valladolid, 47011 Valladolid, Spain
e-mail: luismiguel.nieto.calzada@uva.es

L. M. Nieto
Instituto de Investigación en Matemáticas (IMUVA), Facultad de Ciencias, Universidad de Valladolid, 47011 Valladolid, Spain

G. G. N. Angilella and C. Amovilli (eds.), *Many-body Approaches at Different Scales*, https://doi.org/10.1007/978-3-319-72374-7_14

14.1 Introduction

Present-day magnetic fusion devices, especially tokamaks, can generate plasmas with electron temperatures near 10 keV, and future machines may even reach temperatures of 25 keV or more. This means that they can produce ions with very high charge even from heavy elements. From the beginning, highly charged ions have been an important component of magnetically confined plasmas, and their presence has been highly advantageous for both plasma diagnostics and basic atomic physics studies, and harmful for the operation of a given device, if present in large quantities. The good and bad properties of highly charged ions derive from the fact that they radiate when embedded in a sea of electrons. Partially ionized heavy elements radiate profusely, mostly in the extreme ultraviolet wavelength range, while ions stripped to a few electrons within a closed shell, radiate predominantly in the X-ray range. This radiation can be used to diagnose the plasma conditions, such as the electron temperature, electron density, ion temperature, ion transport and diffusion, and bulk plasma motion. On the other hand, the radiation from highly charged ions contributes to the overall power loss of the plasma: if the plasma contains too many heavy ions, the associated radiative power loss can be severe and prevent ignition and burn. The studies of plasma physics are also among the wide range of research interests of Professor N. H. March, as can be seen in Refs. [1–9].

Highly charged ions play a crucial role in magnetic fusion plasmas given that they are used for many diagnostic purposes in magnetic fusion research [10]. These plasmas are excellent sources for producing highly charged ions and plenty of radiation for learning their atomic properties. These studies include calibration of density diagnostics, X-ray production by charge exchange, line identifications and accurate wavelength measurements, and benchmark data for ionization balance calculations. Studies of magnetic fusion plasmas also consume a large amount of atomic data, especially in order to develop new spectral diagnostics. In this way, line identification has been a diagnostic necessity in fusion research in order to identify the impurities that are inadvertently released into the plasma as different heating scenarios are explored. Early work focused on the transition metals, such as titanium, chromium, iron, and nickel, as well as on noble gases that could be admixed to the plasma. Once the possibility was created to inject any type of material into the plasma via laser injection, spectra of other metals were also investigated [11].

However, it is now clear that under conditions which prevail in low-density laboratory tokamak plasmas (where collisional deexcitation of metastable states is rather slow, leading to buildup of population of metastable levels), forbidden transitions, i.e., electric quadrupole and magnetic dipole transitions, gain in intensity and can be used to infer information about plasma temperature and dynamics. Forbidden lines due to magnetic dipole and electric quadrupole transitions between fine-structure levels of the ground and lower lying excited configurations of highly ionized atoms have been used for diagnostic of laboratory plasmas related to fusion devices. For those lines, fusion specialists need accurate atomic data, such as radiative transition probabilities. To respond to this need, atomic physicists have put a great deal of effort into computing (and sometimes measuring) the required data.

To carry out those calculations, a number of ab initio codes are available, some of them have proved to yield rather accurate transition probability data. However, most of them are highly time-consuming and often infested with convergence problems. As an alternative, semiempirical methods are widely recognized that, for this type of studies, they are of a clear convenience given that combine reliability and simplicity. In most of the semiempirical methods currently employed, which are quite often derived from a modification of a hydrogen-like wave equation, the various interactions are given a different weighting. Deviations from the Coulomb potential, often classified as 'penetration' and 'polarization' (depending on the degree of overlap between the active and passive electrons, in the context of electronic transitions) may, at least to a reasonable extent, be accounted for by a model Hamiltonian that contains a parameter related to the quantum defect. More specifically, the two above effects seem to be adequately described in the relativistic quantum defect orbital (RQDO) method [12, 13], which has been reformulated for the calculation of atomic data for forbidden transitions [14].

The chapter is organized as follows. In the next section we will give a brief overview of the main aspects of the RQDO method, followed by the extension of this formalism to $E2$ transitions in Sect. 14.3. In Sect. 14.4, we illustrate how the use of the systematic trend of the atomic data along an isoelectronic sequence allows to predict data for new ions and to analyze the influence of the relativistic effect for highly ionized atoms. Section 14.5 is dedicated to show the calculated spectra of some of the ions with most interest for plasma diagnostic in fusion devices. Some concluding remarks put the end to the chapter.

14.2 The Relativistic Quantum Defect Orbital Method

The relativistic formulation of the quantum defect orbital formalism, as proposed by Karwowski and Martín [12, 13], was based on the decoupling of the Dirac second-order equation, and the interpretation of the resulting solutions, that, unlike previous models based on the quantum defect, provides exact eigenfunctions of a model Hamiltonian. The resulting orbitals are also valid in the core region retaining approximate core-valence orthogonality.

For a Coulomb potential $V(r) = -Z/r$, after the elimination of the spin and angular variables, the relevant radial part of the Dirac equation reads as follows in atomic units (used throughout the whole chapter)

$$\left[-\frac{d^2}{dr^2} + \frac{s(s \pm 1)}{r^2} - \frac{2Z(1+\alpha^2 E)}{r}\right] \phi_k{}^{\pm} = E(2+\alpha^2 E)\phi_k{}^{\pm}, \qquad (14.1)$$

where $\phi_k{}^{+}$ and $\phi_k{}^{-}$ are related to the *small* and *large* components of the Dirac wave functions, E is the difference between the total and the rest energy of the electron, k is the relativistic angular momentum quantum number

$$k = \pm(j + 1/2), \qquad \text{where} \qquad j = \ell \pm 1/2, \qquad (14.2)$$

with ℓ the orbital quantum number, α is the fine structure constant, and

$$s = k\sqrt{1 - \frac{\alpha^2 Z^2}{k^2}}. \tag{14.3}$$

Two interpretations may be given to the solutions of equation (14.1): first, they give us the two components of the relativistic wave function which is solution of the corresponding Dirac equation; second, they give us two solutions (corresponding to k and $-k$) of a single scalar equation for a scalar (quasirelativistic) wave function ψ_k. A quasirelativistic theory is not only considerably simpler than the relativistic one, but, most important, it closely resembles the Schrödinger formulation. As a consequence, one may rather easily implement the quasirelativistic formalism in the majority of the methods being developed for the nonrelativistic theory. Equation (14.1) is formally very similar to the radial hydrogenic equation and passes into it in a trivial manner in the nonrelativistic limit of $\alpha \to 0$.

The quantum defect orbitals (QDO) are solutions of the Schrödinger equation

$$\left[-\frac{d^2}{dr^2} + \frac{\chi(\chi+1)}{r^2} - \frac{2Z_{net}}{r}\right]{\psi_k}^{QD} = 2E^{QD}{\psi_k}^{QD}, \tag{14.4}$$

where Z_{net} is the nuclear charge on the active electron at large r and

$$\chi = \ell - \delta + c, \tag{14.5}$$

being δ the so-called *quantum defect*, and c an integer chosen to ensure the correct number of nodes and the normalization of the radial wave function. The eigenvalue E^{QD} in Eq. (14.4) depends only on the noninteger part of χ, being independent on c. The quantum defect δ is empirically obtained from the following expression:

$$E^{QD} = E^x = -\frac{Z_{net}^2}{2(n-\delta)^2}, \tag{14.6}$$

where E^x is the experimental energy and n is the principal quantum number of the nonrelativistic theory. A straightforward observation of Eqs. (14.1) and (14.4) clearly prove that the formal mathematical structures of the QDO theory and the scalar relativistic theory are the same. This formal similarity [12, 13] allowed us to reinterpret the QDO theory so that it would account for the major part of the relativistic effects.

Analogously to the nonrelativistic case shown in Eq. (14.4), the relativistic quantum defect orbital (RQDO) equation is written as follows

$$\left[-\frac{d^2}{dr^2} + \frac{\Lambda(\Lambda+1)}{r^2} - \frac{2Z'_{net}}{r}\right]{\psi_k}^{RD} = 2e^{RD}{\psi_k}^{RD}, \tag{14.7}$$

where Z'_{net} depends on quantities already defined

$$Z'_{net} = Z_{net}(1 + \alpha^2 E^x), \tag{14.8}$$

and the parameter Λ is such that

$$\Lambda = \begin{cases} s - 1 - \delta' + c & \text{when } j = \ell + 1/2, \\ -s - \delta' + c & \text{when } j = \ell - 1/2. \end{cases} \tag{14.9}$$

We have just introduced the so-called *relativistic quantum defect* δ', a key element of the present method, which will be determined empirically from the experimental energy E^x using the expression [13]

$$-\frac{Z_{net}^2}{2(\eta - \delta')^2} = E^x \frac{1 + \alpha^2 E^x/2}{(1 + \alpha^2 E^x)^2}. \tag{14.10}$$

In the last equation the parameter η is defined as

$$\eta = n - |k| + |s|, \tag{14.11}$$

in terms of quantities already introduced.

It should be stressed that this formulation is 'exact' in the sense that it is equivalent to a four-component formulation based on the standard first-order form of the Dirac equation. All matrix elements, in particular the transition moments, may be expressed in a simple way, using the solutions of the second-order equation. A set of recurrent formulas which are fulfilled by the radial integrals [15] makes the formalism to be very simple and compact. Karwowski and Martín [12] have remarked that the relativistic density distribution approximates very well the exact one at large values of r. At small distances, the quality of the density deteriorates, as happens when the nonrelativistic QDO densities are compared with the exact nonrelativistic ones. Fortunately, the consequences of this drawback are very seldom reflected in the quality of the QDO and RQDO transition probabilities, because the strongest contribution to radial matrix elements comes, in most cases, from large radial distances.

The most important difference between RQDO and QDO equations is the explicit dependence of the former on the total angular momentum quantum number k. As a consequence, values of the relativistic quantum defect are determined from fine structure energies rather than from their centers of gravity. The corresponding relativistic quantum defect orbitals are different for each component of a multiplet and, if $c = 0$, they retain the nodal structure of the large components of the hydrogenic Dirac wave function.

14.3 Extension of RQDO Method to $E2$ Transitions

Let us consider now electric multipole radiation. The operator responsible for this radiation is the electric 2^{ξ}-pole moment, introduced by the general expression

$$Q^{(\xi)} = e \sum r_k^{\xi} \left(\frac{4\pi}{2\xi + 1} \right)^{1/2} Y_{\xi}^{\mu}. \tag{14.12}$$

Using the Racah tensor $C_m^{(\xi)}$, we have

$$Q^{(\xi)} = e \sum r_k^{\xi} C_m^{(\xi)}. \tag{14.13}$$

In the particular case $\xi = 2$, the transitions take place via electric quadrupole mechanism, $E2$. In this context, the electric quadrupole line strength for a transition between two states within the LSJ-coupling (which is the coupling scheme followed throughout this work), in the notation of the classical book by Condon and Shortley [16], is given by the equation

$$S^{(2)}_{nlj,n'l'j'} = \frac{2}{3} |\langle \alpha J \parallel Q^{(2)} \parallel \alpha' J' \rangle|^2, \tag{14.14}$$

where the matrix-elements have the form

$$\langle \alpha J \parallel Q^{(2)} \parallel \alpha' J' \rangle = \sqrt{(2J+1)(2J'+1)} W(SJL'2; LJ') \langle \alpha L \parallel Q^{(2)} \parallel \alpha' L' \rangle \delta_{SS'}. \tag{14.15}$$

The Kronecker delta $\delta_{SS'}$ appears because the electric n-pole operators do not depend on spin, and $W(SJL^{p}rime2; L'J')$ are the Racah W-coefficients, which can be described in terms of Wigner's 6-j symbols as

$$W(SJL'2; LJ') = (-1)^{S+J+L'} \begin{Bmatrix} S & J & L \\ 2 & L' & J' \end{Bmatrix}. \tag{14.16}$$

Now, we define a line factor R_{line} by

$$R_{line}(SLJ, S'L'J') = (2J+1)^{1/2} (2J'+1)^{1/2} W(SJL'2; LJ'), \tag{14.17}$$

and the line strength then becomes

$$\begin{aligned} S^{(2)}_{nlj,n'l'j'} &= \frac{2}{3} (2J+1)(2J'+1) \left(W(SJL'2; LJ') \right)^2 \left| \langle \alpha L \parallel Q^{(2)} \parallel \alpha' L' \rangle \right|^2 \delta_{SS'} \\ &= \frac{2}{3} \left(R_{line}(SLJ, S'L'J') \right)^2 \left| \langle \alpha L \parallel Q^{(2)} \parallel \alpha' L' \rangle \right|^2 \delta_{SS'}. \end{aligned} \tag{14.18}$$

It is important to stress that spin does not change during the transition because, as it was already mentioned, the relevant operator is spin-independent. The following selection rules apply to $E2$ transitions between LS-coupling states: $\Delta S = 0$ and also $\Delta L = +2, +1, 0, -1, -2$. But they do not apply neither from $L = 0$ to $L = 0$, nor from $L = 1$ to $L = 0$.

Therefore, the reduced matrix element $\langle \alpha L \parallel Q^{(2)} \parallel \alpha' L' \rangle$ provides the relative strength of different multiplets. We can write this reduced matrix element as the product of a single-electron reduced matrix element $\langle nL \parallel Q^{(2)} \parallel n'L' \rangle$, which depends only on the quantum numbers of the jumping electron, and a factor R_{mult}, that we shall call the multiplet factor, as follows:

$$\langle \alpha L \parallel Q^{(2)} \parallel \alpha' L' \rangle = R_{mult}(\alpha L, \alpha' L) \langle nL \parallel Q^{(2)} \parallel n'L' \rangle, \tag{14.19}$$

where

$$\langle nL \parallel Q^{(2)} \parallel n'L' \rangle = \langle L \parallel C^{(2)} \parallel L' \rangle \, \langle R_{nlj} | Q^{(2)} | R_{n'l'j'} \rangle. \tag{14.20}$$

In the last equation $\langle R_{nlj} | Q^{(2)} | R_{n'l'j'} \rangle$ is the transition integral and $\langle L \parallel C^{(2)} \parallel L' \rangle$ is the pertinent reduced matrix element, which can be evaluated using 3-j symbols. The multiplet factor R_{mult} may be expressed as follows:

$$R_{mult}(\alpha L, \alpha' L') = (2L+1)^{1/2}(2L'+1)^{1/2} W(L_c L l' 2; l L'), \tag{14.21}$$

the last symbol in Eq. (14.21) being the Racah W-coefficient

$$W(L_c L l' 2; l L') = (-1)^{L_c + L + l'} \begin{Bmatrix} L_c & L & l \\ 2 & l' & L' \end{Bmatrix}. \tag{14.22}$$

Finally, the line strength Eq. (14.18) takes the form (for $S = S'$):

$$S^{(2)}_{nlj,n'l'j'} = \frac{2}{3} \left| R_{line}(SLJ, S'L'J') \; R_{mult}(\alpha L, \alpha' L') \langle L || C^{(2)} || L' \rangle \; \langle R_{nlj} | Q^{(2)} | R_{n'l'j'} \rangle \right|^2. \tag{14.23}$$

The total line strength for a transition between multiplets is equal to the sum of the line strengths of all the multiplet lines:

$$S(\gamma L, \gamma' L') = \sum S(\gamma J, \gamma' J'). \tag{14.24}$$

Thus, the line strength for a multiplet transition is

$$\begin{aligned} S(\gamma L, \gamma' L') &= (2S+1) \left| \langle \alpha L \parallel Q^{(2)} \parallel \alpha' L' \rangle \right|^2 \\ &= (2S+1) \left| R_{mult}(\alpha L, \alpha' L') \langle nL \parallel Q^{(2)} \parallel n'L' \rangle \right|^2. \end{aligned} \tag{14.25}$$

The relationships between the line strength $S^{(2)}$ (in atomic units, $e^2 a_o^4$), the oscillator strength $f^{(2)}$ (dimensionless), and the transition probability $A^{(2)}$ (in s^{-1}) for $E2$ transitions are given by [17]

$$g' A^{(2)} = (8\pi^2 \hbar \alpha / m\lambda^2) g f^{(2)} = (6.6703 \cdot 10^{15} / \lambda^2) g f^{(2)}, \tag{14.26a}$$

$$g' A^{(2)} = (32\pi^5 \alpha c a_o^4 / 15\lambda^5) S^{(2)} = (1.11995 \cdot 10^{18} / \lambda^5) S^{(2)}, \tag{14.26b}$$

where λ is the transition wavelength (in Å), g and g' are the degeneracies of the lower and upper states, respectively, and α is the fine-structure constant.

14.4 The Z-Expansion Theory for $E2$ Transitions

Regularities in individual oscillator strengths along an isoelectronic sequence as functions of the nuclear charge have been predicted from conventional perturbation theory [18, 19], but these results can be extended to $E2$ transitions. Let us denote the transition integral

$$|\langle R_{nlj}|r^{\xi}|R_{n'l'j'}\rangle| = I^{(\xi)}_{nlj,n'l'j'}. \tag{14.27}$$

Then, for a given operator, the variation of a matrix element as a function of Z may be studied by the Rayleigh–Schrödinger perturbation theory: if we introduce $\rho = Zr$ and $\epsilon = EZ^{-2}$ (in atomic units), the expansions of ψ and E in powers of the perturbation parameter Z^{-1} are the following:

$$\psi_{nlj,n'l'j'} = \psi_0 + \psi_1 Z^{-1} + \psi_2 Z^{-2} + \cdots \tag{14.28a}$$

$$E = Z^2(\epsilon_0 + \epsilon_1 Z^{-1} + \epsilon_2 Z^{-2} + \cdots). \tag{14.28b}$$

In particular, the dipole transition integral ($\xi = 1$) is given by the Z-expansion

$$|\langle \psi_{nlj}|\rho|\psi_{n'l'j'}\rangle| = I^{(1)}_{nlj,n'l'j'} = I^{(1)}_0 Z^{-1} + I^{(1)}_1 Z^{-2} + \cdots, \tag{14.29}$$

where I_0 is the corresponding integral for hydrogen and the superscript (1) refers in all the cases to $E1$ or electric dipole transitions.

The dipole line strength, or squared radial integral equation (14.29), may be written as:

$$S^{(1)}_{nlj,n'l'j'} = S^{(1)}_0 Z^{-2} + S^{(1)}_1 Z^{-3} + S^{(1)}_2 Z^{-4} + \cdots, \tag{14.30}$$

and the expression for the $E1$ oscillator strength will be

$$f^{(1)}_{nlj,n'l'j'} = f^{(1)}_0 + f^{(1)}_1 Z^{-1} + f^{(1)}_2 Z^{-2} + \cdots. \tag{14.31}$$

For the line and oscillator strengths, as well as for the transition probability A, it is possible to perform a paralell nuclear charge expansion representation in the quadrupole case, to study systematic trends of $E2$ S or f-values along an isoelectronic sequence. Here, the transition operator (ρ^2) leads to [20]

$$S^{(2)}_{nlj,n'l'j'} = S^{(2)}_0 Z^{-4} + S^{(2)}_1 Z^{-5} + S^{(2)}_2 Z^{-6} + \cdots \tag{14.32}$$

with

$$S^{(2)}_0 = \left|\langle \psi_{nlj}|\rho^2|\psi_{n'l'j'}\rangle\right|^2. \tag{14.33}$$

For the transition probability A, the corresponding expansion is the following,

$$A^{(2)}_{n'l'j',nlj} = A^{(2)}_0 Z^6 + A^{(2)}_1 Z^5 + A^{(2)}_2 Z^4 + A^{(2)}_3 Z^3 + A^{(2)}_4 Z^2 + \cdots, \tag{14.34}$$

and the quadrupole oscillator strength may be written as

$$f^{(2)}_{nlj,n'l'j'} = f^{(2)}_0 Z^2 + f^{(2)}_1 Z + f^{(2)}_2 + f^{(2)}_3 Z^{-1} + f^{(2)}_4 Z^{-2} + \cdots , \quad (14.35)$$

where $f^{(2)}_0$, $f^{(2)}_1$, $f^{(2)}_2$, ... are proportional to some power of $\Delta\epsilon_0$, the hydrogenic transition energy corresponding to the transition under study. It is also very interesting to analyze the behaviour of $f^{(2)}Z^{-2}$ along the isoelectronic sequence. From Eq. (14.35) it follows that

$$f^{(2)}_{nlj,n'l'j'} Z^{-2} = f^{(2)}_0 + f^{(2)}_1 Z^{-1} + f^{(2)}_2 Z^{-2} + \cdots . \quad (14.36)$$

When no change of the principal quantum number occurs during the transition, that is $\Delta n = 0$, and if we ignore relativistic effects, $\Delta\epsilon_0 = 0$, we have

$$f^{(2)}_{lj,l'j'} = f^{(2)}_3 Z^{-1} + f^{(2)}_4 Z^{-2} + \cdots . \quad (14.37)$$

In other words, the curve of $f^{(2)}$ versus Z^{-1} would tend to an asymptotic zero value in the high-Z side of the sequence. Equation (14.37) is expected to be a good approximation to the behaviour of the oscillator strength with Z^{-1} for $E2$ transitions, at least for the first few ions of the sequence of a light element, e.g., Na I, where relativistic effects still are not very important. However, as soon as these effects set in, deviations from Eq. (14.37), as well as from all the expressions in this section, are expected to occur.

In this way, the dependence of the relativistic effects with the nuclear charge Z in $E2$ transitions can be analysed by the calculation of the contribution of the relativistic effects (RE) in intra-configuration transitions, which can be done by comparing the results from the calculations performed with the non-relativistic (QDO) and the relativistic (RQDO) formulations of the Quantum Defect Orbital method. The weighting of the relativistic effects on the oscillator strengths can be measured in the following manner:

$$RE = \left[\frac{f_{QDO} - f_{RQDO}}{f_{QDO}} \right] \cdot 100. \quad (14.38)$$

Relativistic effects lead to a decrease in the magnitude of the oscillator strengths of the ions under study, that is, f_{RQDO} is generally found to be smaller than f_{QDO} for a given $E2$ transition. We have analysed this effect for the electric quadrupole transitions in the Na sequence as a function of the nuclear charge Z, and fitted the RE-value individually for each of the fine-structure transitions to a polynomic function of Z. The fitting formulae obtained for the $nlj \to nlj'$ $E2$-lines has the following general expression

$$RE_{nlj\to nlj'} = a + bZ + cZ^2. \quad (14.39)$$

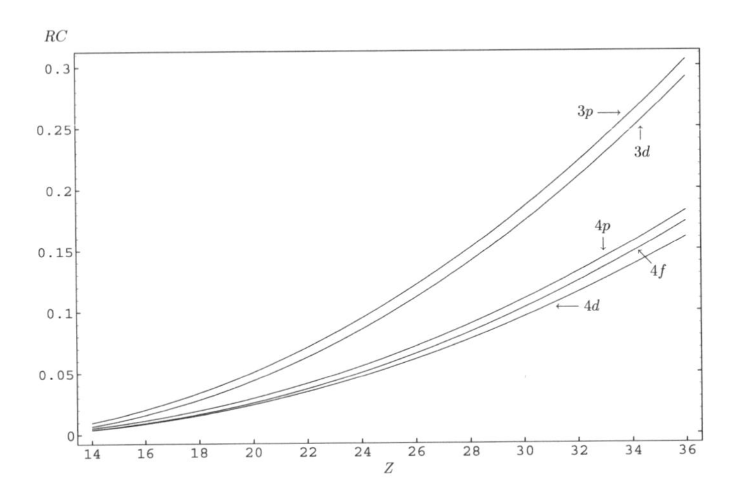

Fig. 14.1 Contribution of the relativistic effects in intraconfigurational fine-structure $E2$ transitions of Na-like ions

These equations can be found in the paper of Charro et al. for Na-like ions [21], and are plotted in Fig. 14.1. It is apparent that the influence of the relativistic effects in the oscillator strengths, for all the fine-structure transitions, decreases as n increases. It is also found that the relativistic effects appear to be generally less important as ℓ increases for a given n. These two features were to be expected in the presently analysed range of ions, where the largest atomic number is $Z = 36$. Hence, the dominant relativistic effects are those of direct character, which are appropriately included in the RQDO procedure.

14.5 Atomic Data and UV and X-ray Atomic Spectra for Ions in Fusion Plasma

Given that the regular behaviour of atomic data as oscillator strengths along isoelectronic sequences has proven to be a useful tool for analysing a large body of f-values, the analysis of the systematic trend of the A-values for $E2$ transitions along isoelectronic sequences was carried out in several previous works using the RQDO method, both for allowed and forbidden transitions. This study may also be exploited to obtain additional oscillator strengths by simple interpolation techniques. From the analysis of systematic trends for $E2$ transitions, several calculations have been performed, in particular the following isoelectronic sequences have been studied:

- B–sequence: $Z = 37 - 82$ (see Ref. [22]).
- K–sequence $Z = 25$ to $Z = 80$ (see Ref. [23]).
- Al–sequence (see Ref. [24]).
- Ga–sequence (see Ref. [25]).
- Na–sequence: $E2$ transitions (see Refs. [21, 26]).

As an example, the behaviour of $E2$ transition probabilities along the boron isoelectronic sequence is graphically analysed in Fig. 14.2. The value of log A has been plotted against the atomic number Z. The available comparative data have also been included [27, 28]. This figure is useful for two purposes: the first one is to show the agreement or deviations among the different sets of data, the second one is to reflect the systematic trends obeyed by the individual RQDO A-values along the isoelectronic sequence, which have long been considered as a qualitative proof of correctness, and can be used for the interpolation or extrapolation of non-calculated data. Inspection of Fig. 14.2 reveals a rather good general agreement between our results and the comparative data.

The interest in line emission from highly-ionized atoms in tokamak devices is, in the first instance, due to the effect of impurities on the overall performance of the tokamak as a fusion device. The most common impurity elements are Ti, Cr, Fe, Ni, the lighter C, N, O, and the rare gases Ne and Ar [29]. Allowed transitions in ions of metals as Zr have been reported using RQDO method [30], but also interesting in tokamaks are the forbidden lines, which are valuable diagnostic monitors of the ion motion, and of the metal impurity concentrations. For $E2$ transitions several calculations using this semiempirical method have been performed in order to predict the UV and X-ray spectra for several metals, in particular highly-charged ions as is the case of Ti XII and Fe XVI (Figs. 14.3 and 14.4). We have performed a simulation of the $E2$ spectrum of these two ions, which include only the intensities of the RQDO $E2$ transitions [31].

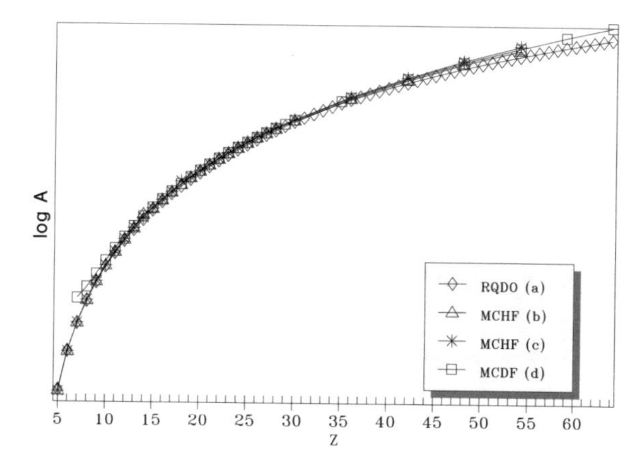

Fig. 14.2 Transition probabilities for $E2$ forbidden lines in boron isoelectronic sequence obtained by Charro et al. [22] in (a), being the comparative data reported by Froese Fischer [27] for (b) and (c), and by Cheng et al. [28] for (d)

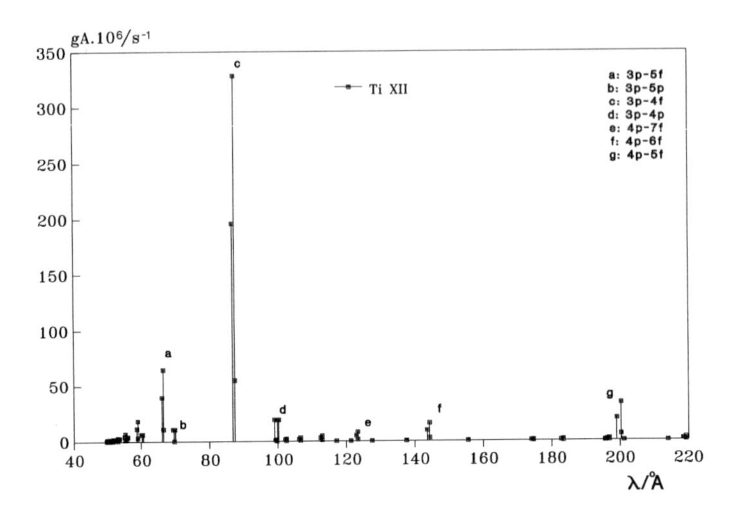

Fig. 14.3 The simulated spectra for Ti XII according to RQDO calculations for forbidden ($E2$) transitions

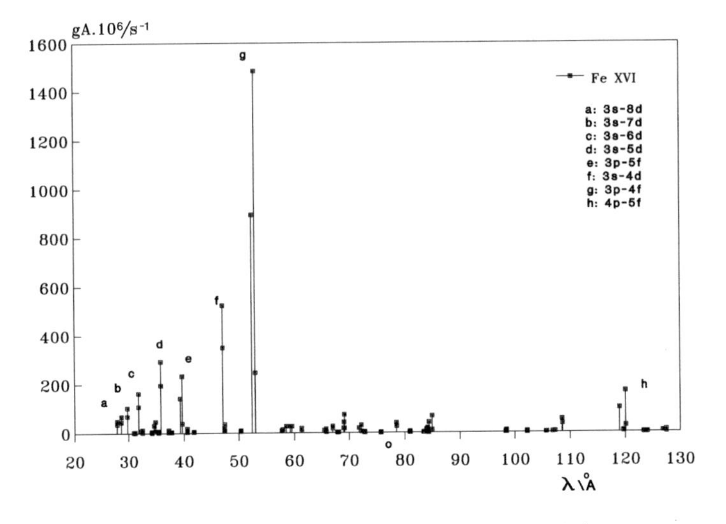

Fig. 14.4 The simulated spectra for Fe XVI according to RQDO calculations

14.6 Concluding Remarks

The RQDO formalism, as opposed to sophisticated and costly self-consistent-field procedures, is a simple but reliable analytical method based on a model Hamiltonian. It has the great advantage of the computational effort not being increased as the atomic system dealt with becomes heavier, and it is also capable of achieving a good balance between computational effort and accuracy of results. The method is a useful tool for calculating UV and X-ray atomic spectra for ions and transitions which are difficult to evaluate, and may play an important role in the future, when fusion becomes a reality.

Acknowledgements Financial support from Spanish MINECO (MTM2014-57129-C2-1-P) and Junta de Castilla y León & FEDER (VA057U16) is acknowledged.

References

1. M.L. Glasser, N.H. March, L.M. Nieto, Phase Trans. **85**(11), 1018 (2012). https://doi.org/10.1080/01411594.2012.656121
2. N.H. March, P. Capuzzi, M.P. Tosi, Phys. Lett. A **327**(2), 226 (2004). https://doi.org/10.1016/j.physleta.2004.05.005
3. N.H. March, M.P. Tosi, Laser Part. Beams **16**(1), 71 (1998), Reprinted in Ref. [32]. https://doi.org/10.1017/S0263034600011782
4. N.H. March, M.P. Tosi, J. Plasma Phys. **57**(1), 121 (1997)
5. M.A. Amato, N.H. March, Laser Part. Beams **14**(4), 685 (1996). https://doi.org/10.1017/S0263034600010405
6. R.E. Robson, B.V. Paranjape, N.H. March, Plasma Phys. Control. Fusion **36**(4), 635 (1994). https://doi.org/10.1088/0741-3335/36/4/005
7. C. Amovilli, N.H. March, S. Pfalzner, Phys. Chem. Liq. **24**(1-2), 79 (1991). https://doi.org/10.1080/00319109108030651
8. J. Mahanty, N.H. March, B.V. Paranjape, Appl. Surf. Sci. **33**, 309 (1988). https://doi.org/10.1016/0169-4332(88)90321-2
9. W.B. Leung, N.H. March, Plasma Phys. **19**(3), 277 (1977). https://doi.org/10.1088/0032-1028/19/3/008
10. B.C. Stratton, M. Bitter, K.W. Hill, D. Hillis, J. Hogan, Fusion Sci. Technol. **53**, 431 (2008)
11. C. Jupén, B. Denne-Hinnov, I. Martinson, L.J. Curtis, Phys. Scr. **68**(4), 230 (2003). https://doi.org/10.1238/Physica.Regular.068a00230
12. J. Karwowski, I. Martín, Phys. Rev. A **43**, 4832 (1991). https://doi.org/10.1103/PhysRevA.43.4832
13. I. Martín, J. Karwowski, J. Phys. B: At. Mol. Opt. Phys. **24**(7), 1539 (1991). https://doi.org/10.1088/0953-4075/24/7/009
14. E. Charro, I. Martín, *Foundations of quantum physics, Anales de Física. Monografías*, vol. 6, ed. by R. Blanco Alcañiz, Á. Mañanez Pérez, S. Marcos Marcos (Real Sociedad Española de Física, Madrid, 2002), pp. 331–349. ISBN 8493215031
15. I. Martín, J. Karwowski, D. Bielinska-Waz, J. Phys. A: Math. Gen. **33**(4), 823 (2000). https://doi.org/10.1088/0305-4470/33/4/315
16. E.U. Condon, G.H. Shortley, *The Theory of Atomic Spectra* (Cambridge University Press, Cambridge, 1935)
17. I.I. Sobelman, *Atomic Spectra and Radiative Transitions* (Springer, Berlin, 1979)
18. M. Cohen, A. Dalgarno, Proc. R. Soc. A **293**(1434), 359 (1966). https://doi.org/10.1098/rspa.1966.0176
19. A. Dalgarno, E.M. Parkinson, Proc. R. Soc. A **301**(1466), 253 (1967). https://doi.org/10.1098/rspa.1967.0206
20. E. Biémont, M. Godefroid, Phys. Scr. **18**(5), 323 (1978). https://doi.org/10.1088/0031-8949/18/5/007
21. E. Charro, I. Martín, J. Phys. B: At. Mol. Opt. Phys. **35**(15), 3227 (2002). https://doi.org/10.1088/0953-4075/35/15/301
22. E. Charro, S. López-Ferrero, I. Martín, J. Phys. B: At. Mol. Opt. Phys. **34**(21), 4243 (2001). https://doi.org/10.1088/0953-4075/34/21/313

23. E. Charro, Z. Curiel, I. Martín, Astron. Astrophys. **387**(3), 1146 (2002). https://doi.org/10.1051/0004-6361:20020288
24. E. Charro, S. López-Ferrero, I. Martín, Astron. Astrophys. **406**(2), 741 (2003). https://doi.org/10.1051/0004-6361:20030660
25. E. Charro, Z. Curiel, I. Martín, Int. J. Quantum Chem. **108**(4), 744 (2008). https://doi.org/10.1002/qua.21552
26. E. Charro, I. Martín, Int. J. Quantum Chem. **90**(1), 403 (2002). https://doi.org/10.1002/qua.10030
27. C. Froese Fischer, J. Phys. B: At. Mol. Phys. **16**(2), 157 (1983). https://doi.org/10.1088/0022-3700/16/2/005
28. K.T. Cheng, Y.K. Kim, J.P. Desclaux, At. Data Nucl. Data Tables **24**(2), 111 (1979). https://doi.org/10.1016/0092-640X(79)90006-8
29. N.J. Peacock, M.F. Stamp, J.D. Silver, Phys. Scr. **1984**(T8), 10 (1984). https://doi.org/10.1088/0031-8949/1984/T8/002
30. E. Charro, J.L. López-Ayuso, I. Martín, J. Phys. B: At. Mol. Opt. Phys. **32**(18), 4555 (1999). https://doi.org/10.1088/0953-4075/32/18/314
31. E. Charro, I. Martín, J. Mol. Struct.: Theochem **621**(1), 75 (2003), 2001 Quitel S.I. https://doi.org/10.1016/S0166-1280(02)00535-3
32. N.H. March, G.G.N. Angilella (eds.), *Many-Body Theory of Molecules, Clusters, and Condensed Phases* (World Scientific, Singapore, 2009). ISBN 9789814271776

Chapter 15
Exceeding the Shockley–Queisser Limit Within the Detailed Balance Framework

M. Bercx, R. Saniz, B. Partoens and D. Lamoen

Abstract The Shockley–Queisser limit is one of the most fundamental results in the field of photovoltaics. Based on the principle of detailed balance, it defines an upper limit for a single junction solar cell that uses an absorber material with a specific band gap. Although methods exist that allow a solar cell to exceed the Shockley–Queisser limit, here we show that it is possible to exceed the Shockley–Queisser limit without considering any of these additions. Merely by introducing an absorptivity that does not assume that every photon with an energy above the band gap is absorbed, efficiencies above the Shockley–Queisser limit are obtained. This is related to the fact that assuming optimal absorption properties also maximizes the recombination current within the detailed balance approach. We conclude that considering a finite thickness for the absorber layer allows the efficiency to exceed the Shockley–Queisser limit, and that this is more likely to occur for materials with small band gaps.

15.1 Introduction

Materials play a central role in the effort to produce cheaper and more efficient solar cells. The discovery of improved absorber materials has the potential to significantly increase the cost-effectiveness of photovoltaic devices, but experimental trial and error methods are often slow and expensive. Here, computational material modeling can provide a valuable assist to the material design process, by screening groups of materials for those that have the best properties.

M. Bercx (✉) · D. Lamoen
EMAT, Department of Physics, University of Antwerp, Groenenborgerlaan 171, 2020 Antwerpen, Belgium
e-mail: marnik.bercx@uantwerpen.be

D. Lamoen
e-mail: dirk.lamoen@uantwerpen.be

R. Saniz · B. Partoens
CMT group, Department of Physics, University of Antwerp, Groenenborgerlaan 171, 2020 Antwerp, Belgium

G. G. N. Angilella and C. Amovilli (eds.), *Many-body Approaches at Different Scales*, https://doi.org/10.1007/978-3-319-72374-7_15

The Shockley–Queisser limit [1] is one of the most well-known metrics to determine the maximum efficiency an absorber material can produce in a single-junction solar cell. It was proposed in 1961 and provides a direct relation between the band gap of a material and its maximum possible efficiency. More recently, Yu and Zunger expanded on the work of Shockley and Queisser by introducing the Spectroscopic Limited Maximum Efficiency [2] (SLME), which takes the absorption coefficient and thickness into consideration for the calculation of the maximum efficiency. The SLME has since been used to investigate the potential of photovoltaic absorber materials such as perovskites [3], direct band gap silicon crystals [4], chalcogenides, and other materials. In our recent work on CuAu-like [5] and Stannite [6] structures, we also used the SLME to study the efficiency of these materials in the context of thin film solar cells. Interestingly, we found several materials with an SLME above the Shockley–Queisser limit, and identified that this is due to the lower recombination current obtained for the material at lower thicknesses.

Since its conception, numerous methods have been proposed to exceed the Shockley–Queisser limiting efficiency [7]. Examples include multi-junction [8, 9] and hot carrier solar cells [10], as well as concepts that use multiple exciton generation [11]. None of these concepts, however, are implemented in the SLME. In this paper, we use a model approach to demonstrate that it is possible to exceed the Shockley–Queisser limit within the detailed balance framework. Simply by dropping the assumption of an infinite absorber layer, i.e. by replacing the Heaviside step function for the absorptivity by a sigmoid function, we obtain efficiencies above the Shockley–Queisser limit. Finally, we analyze for which band gap range a material's efficiency is more likely to exceed the Shockley–Queisser limit.

15.2 Shockley–Queisser Limit

The maximum efficiency η is defined as the maximum output power density P_m divided by the total incoming power density from the solar spectrum P_{in}:

$$\eta = \frac{P_m}{P_{in}}. \tag{15.1}$$

To calculate P_m, the power density $P = JV$ is maximized versus the voltage V, where the current density[1] J is derived from the ideal $J - V$ characteristic of an illuminated solar cell:

$$J = J_{sc} - J_0 \left(e^{eV/k_B T} - 1 \right), \tag{15.2}$$

[1] Note that these current densities are not defined in the conventional way. Rather, they are considered as currents per surface area of the solar cell. This allows us to ignore the surface area of the solar cell in our discussion.

where k_B is Boltzmann's constant, e is the elementary charge, and T is the temperature of the solar cell. The short-circuit current density J_{sc}, also known as the photogenerated current or the illuminated current, is calculated from the number of photons of the solar spectrum that are absorbed by the solar cell:

$$J_{sc} = e \int_0^\infty a(E)\Phi_s(E)dE, \tag{15.3}$$

where $a(E)$ is the absorptivity and $\Phi_s(E)$ is the photon flux density of the solar spectrum. In their original paper, Shockley and Queisser used a blackbody spectrum of $T_s = 6000$ K, but the current convention is to use the AM1.5G solar spectrum [12].

The reverse saturation current density J_0 is calculated by considering the principle of detailed balance, i.e. in equilibrium conditions the rate of photon emission from radiative recombination must be equal to the photon absorption from the surrounding medium. Because the cell is assumed to be attached to an ideal heat sink, the ambient temperature is assumed to be the same as that of the solar cell. Hence, the spectrum of the surrounding medium is that of a black body at cell temperature T:

$$\begin{aligned} J_0 &= e\pi \int_0^\infty a(E)\Phi_{bb}(E)dE \\ &= e\pi \int_0^\infty a(E)\frac{2E^2}{h^3c^2}\frac{dE}{e^{E/k_B T} - 1}, \end{aligned} \tag{15.4}$$

where h is Planck's constant and c is the speed of light. Because of its connection with the recombination of electron-hole pairs at equilibrium, J_0 is also referred to as the recombination current density [13]. This is the convention we will use here.

To obtain the Shockley–Queisser or detailed balance *limit*, Shockley and Queisser made the assumption that the probability of a photon with an energy above the band gap being absorbed by the cell is equal to unity. This corresponds mathematically to setting $a(E)$ to the Heaviside step function, or, from a physical perspective, to considering an infinitely thick absorber layer. Note that in the original expressions, Shockley and Queisser also included a geometrical factor. However, because we assume the solar cell to have a perfect antireflective coating, as well as a reflective back surface, the geometrical factor is equal to unity [14].

15.3 Spectroscopic Limited Maximum Efficiency

Shockley and Queisser's detailed balance limit is considered to be one of the most important results in photovoltaic research. However, as a metric for thin film solar cells, it is somewhat limited in its effectiveness, because it only depends on the band gap of the absorber material in the solar cell. In an attempt to find a more

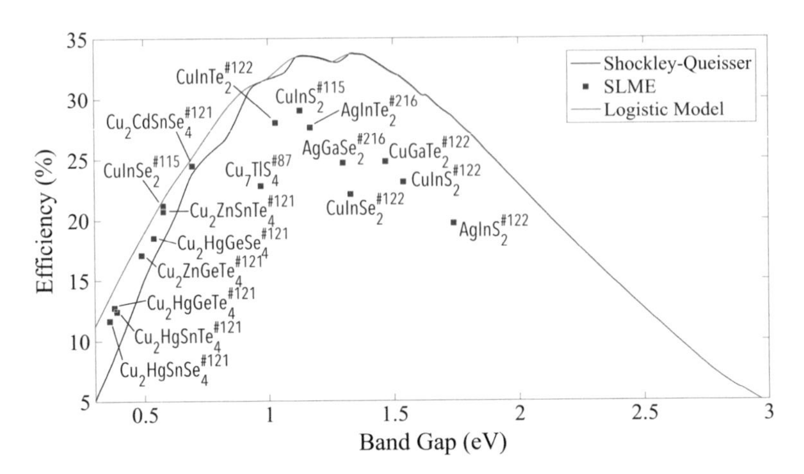

Fig. 15.1 Collection of calculated SLME values from Yu and Zunger [2], as well as our previous work on CuAu-like [5] and Stannite [6] structures. We have added the space group of the material structure as a superscript. The efficiency values were calculated for a thickness of 0.5 μm. The orange curve represents the maximum efficiencies obtained using the logistic model explained in Sect. 15.4

practical screening metric, Yu and Zunger introduced the Spectroscopic Limited Maximum Efficiency [2] (SLME) in 2012. The SLME differs from the detailed balance limit in two ways. First, the absorptivity $a(E)$, taken as a Heaviside step function in the calculation of Shockley and Queisser, is replaced by the absorptivity $a(E) = 1 - e^{-2\alpha(E)L}$, where L is the thickness and $\alpha(E)$ is the absorption coefficient, calculated from first principles. This allows us to use the SLME to study the thickness dependence of the efficiency, an important tool in the study of thin film solar cells.

Second, the SLME also considers the non-radiative recombination in the solar cell by modeling the fraction[2] of radiative recombination as a Boltzmann factor, i.e. $f_r = e^{-\Delta/k_B T}$, with $\Delta = E_g^{da} - E_g$, where E_g and E_g^{da} are the fundamental and direct allowed band gap, respectively. The total recombination current density is then calculated by dividing the radiative recombination current density (Eq. 15.4) by the fraction of radiative recombination. In this work, we only study direct band gap materials (i.e. $E_g = E_g^{da}$), and hence only radiative recombination is considered ($f_r = 1$), just as in the standard calculation of the detailed balance limit.

The SLME has been used to investigate the potential of several classes of photovoltaic absorber materials. In Fig. 15.1, we show a selection of calculated efficiencies of direct band gap materials from previous work [2, 5, 6], compared with the Shockley–Queisser limit. We can see that materials typically used in thin-film photovoltaic cells, e.g. chalcopyrite phase $CuIn(S,Se)_2$, have a high calculated efficiency. We also note other materials that are less studied with high efficiencies,

[2] Actually, Shockley and Queisser also considered the fraction of radiative recombination in their approach. They did not, however, provide a model to calculate it, simply observing that the maximum efficiency is significantly reduced for small fractions f_r.

such as CuAu-like phase $CuInS_2$ and chalcopyrite phase $CuInTe_2$. Most importantly, however, we can see that a significant amount of the presented materials have a calculated efficiency above the Shockley–Queisser limit. Since the calculation of the SLME does not introduce any of the concepts that would typically allow its value to exceed the Shockley–Queisser limit, these results show that for thin-film materials the Shockley–Queisser limit does not necessarily represent an upper limit for the efficiency.

In fact, Shockley and Queisser considered their metric as the detailed balance *limit* because of the assumption that since the step function represents the highest possible absorption spectrum for a material with a specific direct band gap, the resulting efficiency must represent an upper limit. However, as we demonstrated in our previous work [5], this also means the recombination current density J_0 (Eq. 15.4) will be maximal. Since electron-hole recombination results in a loss of electrons contributing to the external current, this has a negative effect on the photovoltaic conversion efficiency. Hence, it is possible that there is an absorptivity function that would result in a higher efficiency than the Shockley–Queisser limit. As we can see in Fig. 15.1, this is exactly what happens for the presented smaller band gap materials.

15.4 Logistic Function Model

The next questions are how far we can exceed the Shockley–Queisser limit, and at which band gaps a material is more likely to do so. Clearly, this will depend on the shape of the absorptivity function. In Fig. 15.2, we show the calculated absorptivity of Cu_2ZnGeS_4 for various thicknesses, derived from the absorption coefficient calculated from first principles (for computational details, we refer the reader to [6]). We can see that the absorptivity has a shape reminiscent of a sigmoid function. In

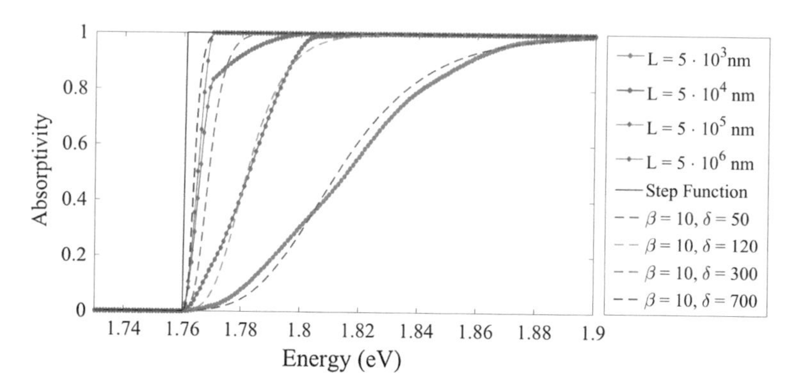

Fig. 15.2 Comparison of the model function with calculated absorptivity spectra for Cu_2ZnGeS_4 at different thicknesses L. We can see that the model function shape matches that of the calculated absorptivity quite well as $L, \delta \to \infty$

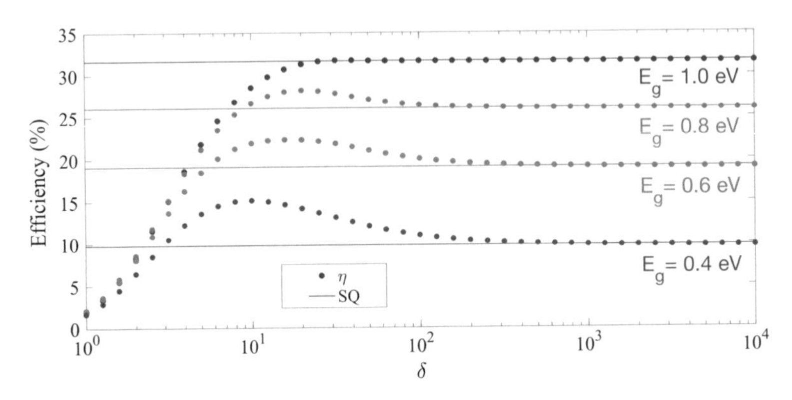

Fig. 15.3 Calculated efficiencies for a range of δ values and a selection of band gaps, compared with the corresponding Shockley–Queisser limit

order to analyze the maximum efficiency for materials with a direct band gap in the range 0.3–3 eV, we model $a(E)$ using a generalized logistic function:

$$a(E) = f(E) = \frac{1}{(1 + e^{-\delta(E-E_g)})^\beta}, \tag{15.5}$$

where E_g is the band gap of the material, and β, δ are parameters that determine the shape of the function. In this model for the absorptivity, the parameter δ is related to the thickness of the material, as for $\delta \to \infty$, $f(E)$ approaches the Heaviside step function (Fig. 15.2). The second parameter (β) is important to make sure that the model function "starts" at the band gap, i.e. that its value for $E < E_g$ is suitably small, so that it can be approximated to zero. Since $f(E_g) = \frac{1}{2^\beta}$, and $f(E) < f(E_g)$ for $E < E_g$, increasing β to a suitably large value gives us this desired function trait. Here, we choose $\beta = 10$ and set $f(E) = 0$ for $E \leq E_g$. As is clear from Fig. 15.2, this model function describes the shape of the calculated absorptivity spectra quite well.

To study the influence of the band gap on the likelihood of the efficiency exceeding the Shockley–Queisser limit, we calculate the efficiency for $\delta \in [1, 10^4]$ and over the band gap range $E_g \in [0.3, 3]$ eV. We show the δ-dependency of the efficiency for a selection of band gap values in Fig. 15.3. We can see that for low band gaps, the calculated efficiency crosses the detailed balance limit of the corresponding band gap, in order to return to the limit value for $\delta \to \infty$. Since δ can be related to the thickness of the material, this implies that for lower band gap materials, there is a thickness that is optimal for the efficiency. Moreover, a clear trend is visible, with the efficiency exceeding the Shockley–Queisser limit more as the band gap is decreased. This is also what we observe when we look at the plot for the maximum efficiency values in Fig. 15.1.

It is interesting to note that the SLME values of the materials that exceed the Shockley–Queisser limit are still below the maximum efficiency for the model absorptivity functions of the corresponding band gap in Fig. 15.1. However, this does not imply that the logistic function maxima curve represents a new upper limit. It is entirely possible that there is another function profile that would allow for higher efficiencies. Using the logistic function approach, we are simply able to observe for which band gap range the Shockley–Queisser limit does not provide a theoretical upper limit.

15.5 Conclusions

In their 1961 paper, Shockley and Queisser characterized their calculated efficiency as an upper limit, because of the assumption that if every photon with an energy above the band gap is absorbed, the obtained efficiency must be maximal. Although this assumption may seem entirely sensible at first glance, it does not consider the fact that it also maximizes the recombination current, which is calculated using the detailed balance principle. Because an increased recombination results in a lower efficiency, this means that lowering the absorptivity can produce higher efficiencies than the Shockley–Queisser limit under the right conditions. By using a model absorptivity function, which closely resembles absorptivity spectra calculated from first principles, we have shown that this can occur for low band gaps. This means that one must take care when dismissing low band gap materials based on their Shockley–Queisser limit, for their actual efficiency at certain thicknesses might still make them suitable for thin film photovoltaic applications.

References

1. W. Shockley, H.J. Queisser, J. Appl. Phys. **32**(3), 510 (1961). https://doi.org/10.1063/1.1736034
2. L. Yu, A. Zunger, Phys. Rev. Lett. **108**, 068701 (2012). https://doi.org/10.1103/PhysRevLett.108.068701
3. W. Meng, B. Saparov, F. Hong, J. Wang, D.B. Mitzi, Y. Yan, Chem. Mater. **28**(3), 821 (2016). https://doi.org/10.1021/acs.chemmater.5b04213
4. I.H. Lee, J. Lee, Y.J. Oh, S. Kim, K.J. Chang, Phys. Rev. B **90**, 115209 (2014). https://doi.org/10.1103/PhysRevB.90.115209
5. M. Bercx, N. Sarmadian, R. Saniz, B. Partoens, D. Lamoen, Phys. Chem. Chem. Phys. **18**, 20542 (2016). https://doi.org/10.1039/C6CP03468C
6. N. Sarmadian, R. Saniz, B. Partoens, D. Lamoen, J. Appl. Phys. **120**(8), 085707 (2016). https://doi.org/10.1063/1.4961562
7. C.A. Nelson, N.R. Monahan, X.Y. Zhu, Energy Environ. Sci. **6**, 3508 (2013). https://doi.org/10.1039/C3EE42098A
8. A.V. Shah, H. Schade, M. Vanecek, J. Meier, E. Vallat-Sauvain, N. Wyrsch, U. Kroll, C. Droz, J. Bailat, Prog. Photovolt. Res. Appl. **12**(2–3), 113 (2004). https://doi.org/10.1002/pip.533
9. P. Heremans, D. Cheyns, B.P. Rand, Acc. Chem. Res. **42**(11), 1740 (2009). https://doi.org/10.1021/ar9000923

10. D. König, K. Casalenuovo, Y. Takeda, G. Conibeer, J.F. Guillemoles, R. Patterson, L.M. Huang, M.A. Green, Phys. E **42**(10), 2862 (2010). https://doi.org/10.1016/j.physe.2009.12.032. 14th International Conference on Modulated Semiconductor Structures
11. M.C. Hanna, A.J. Nozik, J. Appl. Phys. **100**(7), 074510 (2006). https://doi.org/10.1063/1.2356795
12. American Society for Testing and Materials, *ASTM G173-03: Standard tables for reference solar spectral irradiances: direct normal and hemispherical on* 37° *tilted surface* (ASTM International, West Conshohocken, PA, 2012), http://www.astm.org. https://doi.org/10.1520/G0173-03R12
13. A. Cuevas, Energy Procedia **55**, 53 (2014). https://doi.org/10.1016/j.egypro.2014.08.073
14. S. Rühle, Solar Energy **130**, 139 (2016). https://doi.org/10.1016/j.solener.2016.02.015

Part II
Theoretical Chemistry

Chapter 16
Shannon Entropy and Correlation Energy for Electrons in Atoms

C. Amovilli and F. M. Floris

Abstract In this work, we compute Shannon entropy, defined in terms of electron density, for three series of atomic ions including the region of nuclear charges close to the limit at which the ionization potential goes to zero. We use both Hartree–Fock (HF) and quantum Monte Carlo (QMC) densities and we observe a sharp positive deviation of QMC entropy with respect to the HF corresponding value in approaching the limit. We discuss this behaviour taking into account Coulomb correlation, which plays an important role in the weak binding regime.

16.1 Introduction and General Theory

Quantum entropy was formulated by von Neumann [1] by extending to density matrices the classical definition of entropy in terms of probability distributions. A density matrix is the matrix representation of the density operator. For a pure quantum state, the density operator has the dyadic form

$$\hat{\rho} = \frac{|\Psi\rangle\langle\Psi|}{\langle\Psi|\Psi\rangle}, \tag{16.1}$$

and a physical observable associated to an operator $\hat{B}$ has the corresponding expectation value

$$\langle B\rangle = \mathrm{Tr}(\hat{\rho}\hat{B}). \tag{16.2}$$

For the postulates of the quantum theory, if the system is interacting with a reservoir, it cannot be described by a state vector [2]. In this case, the density operator is defined as a weighted sum of single wave function density operators, each weight being the probability of finding the system in a given state. The density operator is thus

---

C. Amovilli (✉) · F. M. Floris
Dipartimento di Chimica e Chimica Industriale, University of Pisa, Via G. Moruzzi 13, 56124 Pisa, Italy
e-mail: claudio.amovilli@unipi.it

G. G. N. Angilella and C. Amovilli (eds.), *Many-body Approaches at Different Scales*, https://doi.org/10.1007/978-3-319-72374-7_16

$$\hat{\rho} = \sum_K w_K |\Psi_K\rangle\langle\Psi_K|, \tag{16.3}$$

where $\sum_K w_K = 1$ and $\langle\Psi_K|\Psi_K\rangle = 1$. For a quantum mechanical system, the von Neumann entropy is then given by the expression

$$S = -\mathrm{Tr}(\hat{\rho}\log\hat{\rho}), \tag{16.4}$$

which reads, by taking the diagonal form, Eq. (16.3), as

$$S = -\sum_K w_K \log w_K. \tag{16.5}$$

Such a quantity is zero for an isolated system in a pure state while is positive for a system 'entangled' with the reservoir as a result of the irreversible process of interaction between the two subsystems. The Von Neumann entropy can also be defined in terms of reduced density matrices. In this case we have the so called entropy of entanglement which refers to a subsystem obtained by a bipartition of the original one. For example, for an isolated system in a pure state partitioned in two subsystems A and B we have (see, e.g., Ref. [3])

$$S_A = -\mathrm{Tr}(\hat{\rho}_A \log\hat{\rho}_A), \tag{16.6}$$

where

$$\hat{\rho}_A = \mathrm{Tr}_B(\hat{\rho}). \tag{16.7}$$

For a system of electrons, the entanglement entropy written in terms of the one-particle density matrix of a pure state has an important physical meaning being related to the interparticle interaction, more precisely to the entanglement of one electron with the others. The electronic one-particle density matrix written in terms of the natural spin orbitals ψ_k is given by

$$\rho_1(x, x') = \sum_k \nu_k \psi_k(x)\psi_k^*(x') \tag{16.8}$$

and leads to

$$S_1 = \sum_k \nu_k \log\frac{1}{\nu_k}, \tag{16.9}$$

in which ν_k are occupation numbers ranging from 0 to 1. For an independent particle system, the occupation numbers are integer and S_1 is zero. This reflects also in the idempotency of ρ_1. Instead, for a real many electron system, where we have two-body interactions, occupation numbers are fractional, the one-particle density matrix is not anymore idempotent and S_1 becomes a positive quantity clearly related to electron correlation.

Collins [4] proposed a relationship between entropy and correlation energy. He conjectured that the entropy given in Eq. (16.9) is proportional to correlation energy. Esquivel et al. [5] supported numerically this conjecture on a series of light atomic ions. Ziesche [6] studied more in detail the relations between entropy and correlation strength. Many other works followed by addressing this point, mainly focussing on the link between the nonidempotency of the one-particle density matrix and complexity measures (see, e.g., Nagy and Romera [7], and references therein). Worth of mentioning is also a recent paper by Grimme and Hansen [8], in which a semiempirical finite temperature free energy functional involving the entanglement entropy is introduced to recover information about static correlation within a density functional theory (DFT) framework.

Since the exact one-particle density matrix for a real system is, in general, unknown and difficult to compute, in this work we focus on a different definition of entropy written in terms of a classical property, namely the electron density. We exploit the fact that the Shannon entropy [9, 10] can be defined also in terms of a continuous probability distribution. Here, we use the following definition

$$S = -\int d^3\mathbf{r}\, \rho(\mathbf{r}) \log \rho(\mathbf{r}), \tag{16.10}$$

where $\rho(\mathbf{r})$ is the one particle electron density. In this work, we perform calculations for atomic ions and we compute the density with Hartree–Fock (HF) and quantum Monte Carlo methods (QMC). This definition of entropy according to Shannon is here used to measure the entanglement of electrons due to the Coulomb correlation. In particular we compare different calculations performed with highly correlated methods with the HF independent particle model. All systems are considered in the ground state so we use the basic result of density functional theory that the electron density brings all information about the N-particle wavefunction. In this regards, the knowledge of Shannon entropy density, namely $-\rho \log \rho$, is sufficient to know any physical observable for a Coulomb system as shown recently by Nagy [11]. Theoretical concepts relating Shannon entropy and many-electron correlation have been discussed by many authors, see for example the work of Delle Site [12]. A further interesting relation between Shannon entropy and correlation energy goes back to the homogeneous electron gas in the high density limit [13]. In this limit, the correlation energy per particle tends to $A + B \log r_s$ where r_s, namely the Wigner–Seitz radius, is explicitly $(\frac{4}{3}\pi\rho)^{-1/3}$ and, within a local density approximation, the relation between the correlation energy and the Shannon entropy is readily obtained. More generally, Shannon entropy is used in the analysis of many physical and chemical phenomena. We mention, for example, the search of relations with other descriptors of Coulomb systems like reactivity indices, in particular hardness and the Fukui function [14]. Moreover, Amovilli and March [15] showed a relation between Shannon and Jaynes entropy [16] (essentially the entanglement entropy of Eq. (16.9)) on a two-particle Moshinsky model atom. Here, we extend their approach to real (Coulomb) atomic systems.

16.2 Calculations

In this work, we limit our test examples to a set of atoms in a spherically symmetric ground state. In order to consider both spin compensated (^1S_g) and spin polarized cases (^4S_u), we show results for He-, N- and Ne-like series of atomic ions. We define the correlation Shannon entropy of this work as

$$\Delta S_{\text{corr}} = S_{\text{QMC}} - S_{\text{HF}}, \tag{16.11}$$

and we analyze the behaviour of this quantity up to the limit of extremely weak binding where the ionization potential goes to zero. Shannon entropy contributions are calculated according to the definition given in Eq. (16.10). Thus, we illustrate results also for fractional nuclear charges. Because we use QMC (for a review of these methods see, e.g., [17, 18]), we concentrate our study only on valence electrons and then we make use of pseudopotentials for N, the number of electrons, greater than 2. The He-like series is treated at the highest level and by employing basis sets of Slater orbitals. For all other systems, we use the Burkatzki, Filippi, and Dolg (BFD) pseudopotentials [19]. In order to perform calculations for nuclei with fractional charges, the corresponding pseudopotentials are derived by interpolation of BFD core potential data of the atoms with the closest nuclear charge. All pseudopotential have the more general form

$$\hat{v}_{\text{PP}} = v_{\text{PP}}^{\text{loc}} + \hat{v}_{\text{PP}}^{\text{nonloc}}, \tag{16.12}$$

with the 'local' part given by the sum

$$v_{\text{PP}}^{\text{loc}} = \frac{1}{r^2} \sum_j a_j r^j e^{-\alpha_j r^2}, \tag{16.13}$$

and the 'non local' part, which accounts for projection over core angular momentum eigenfunctions, namely

$$\hat{v}_{\text{PP}}^{\text{nonloc}} = \sum_{\ell=0}^{L_{\max}} u_\ell(r) |\ell\rangle\langle\ell| \tag{16.14}$$

by the radial functions

$$u_\ell(r) = \frac{1}{r^2} \sum_j b_j^{(\ell)} r^j e^{-\beta_j^{(\ell)} r^2}. \tag{16.15}$$

BFD pseudopotentials are very simple, are not singular and do not have a cusp at the origin, being designed for QMC calculations. These conditions are satisfied by the constraints

$$a_0 = 0, \tag{16.16a}$$

$$a_1 = Z - N_{\text{core}}, \tag{16.16b}$$

$$a_3 = a_1 \alpha_1. \tag{16.16c}$$

These constraints must be preserved in the generation of core potential for fractional nuclear charges. The interpolation used in this study does not give well calibrated effective core potential but we consider this route as a reliable procedure to provide data not available from the literature. At the QMC level, we use a standard Slater–Jastrow (SJ) wavefunction with a Jastrow factor containing electron-nuclear, electron-electron, and electron-electron-nuclear terms [20]. At the variational Monte Carlo level (VMC), all parameters in our SJ wave functions are optimized by using the iterative linear method developed by Umrigar et al. [21]. At the diffusion Monte Carlo (DMC) level, performed in the fixed node approximation, the pseudopotentials are treated beyond the locality approximation using the T-move approach [22]. We used a time step of 0.05 a.u. in all the DMC calculations.

Finally, HF computations have been performed using the GAMESS-US package [23] while for QMC we used the CHAMP suite of programs [24].

16.2.1 He-Like Isoelectronic Series of Atomic Ions

As trial wavefunction for DMC, we have considered a SJ form made of few determinants that includes a Chandrasekar like contribution [25] and double excitations to a shell of p orbitals. For the atomic basis set we used Slater type orbitals.

For a system of electrons confined by a Coulomb potential, the ionization potential goes to zero as the nuclear charge Z approaches a critical value, say Z_{cr}, due to the lack of infinite barriers. Such critical value, for 2 electrons in a singlet state, is known to be 0.911028 [26]. In close proximity of this limit, the electron cloud becomes very diffuse. Amovilli and March [27] studied in detail the behaviour of the ground-state electron density in He-like ions as a function of the nuclear charge. Following their study, by way of example, we plot in Fig. 16.1 the radial density for few ions, calculated at the DMC level, in order to show how the density changes in reducing the confining potential. It is remarkable to know that the atomic sphere containing 99% of the total electronic charge has a radius of about 20 bohr in close proximity of the critical value [27].

In Fig. 16.2, we plot instead the DMC Shannon entropy as a function of the nuclear charge Z compared with that of the hydrogenic ions, namely

$$S_{\text{H-like}}(Z) = 3 - \log \frac{Z^3}{\pi}. \tag{16.17}$$

In the range of Z considered in this plot, the maximum difference occurs in proximity of the critical value. This is the consequence of strong delocalization of electrons near the ionization limit. We remark that in the present study we maintain the normalization

$$N = \int d^3\mathbf{r}\, \rho(\mathbf{r}), \tag{16.18}$$

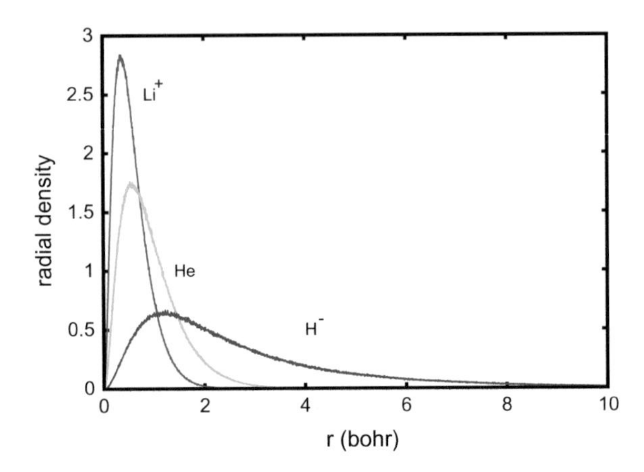

Fig. 16.1 Radial density ($4\pi r^2 \rho$) for three systems of the series of He-like atomic ions considered in this work (^1S_g state)

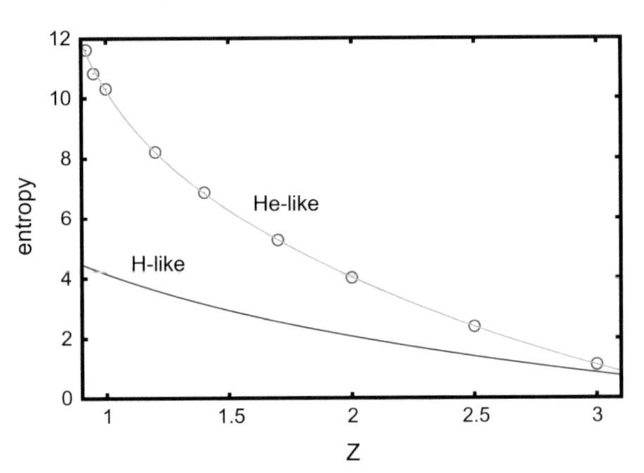

Fig. 16.2 DMC Shannon entropy for He-like and H-like atomic ions as a function of the nuclear charge Z

instead of computing the entropy in terms of the density amplitude $\sigma = \rho/N$. The relation between the two entropies is simply

$$S[\sigma] = \frac{S[\rho]}{N} + \log N. \tag{16.19}$$

For He, the resulting $S[\sigma]$ is 2.700(5) a.u., which is in agreement within statistical error with the very accurate value of 2.7051028 a.u. recently calculated by Ou and Ho [28].

Turning to the electron correlation and according to Eq. (16.11), we compute the correlation Shannon entropy as the difference

$$\Delta S_{\text{corr}} = \int d^3\mathbf{r} \left[\rho_{\text{HF}} \log \rho_{\text{HF}} - \rho_{\text{DMC}} \log \rho_{\text{DMC}}\right]. \tag{16.20}$$

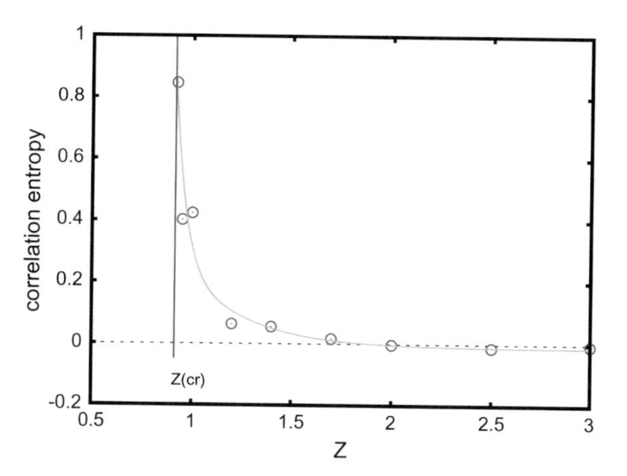

Fig. 16.3 Correlation Shannon entropy as a function of the nuclear charge Z for the He series of atomic ions considered in this work

We note that this quantity contains the Kullback–Leibler (KL) divergence [29] of ρ_{DMC} with respect to ρ_{HF}, this being, by definition,

$$D_{\mathrm{KL}}(\rho_{\mathrm{DMC}}||\rho_{\mathrm{HF}}) = \int d^3\mathbf{r}\, \rho_{\mathrm{DMC}} \log \frac{\rho_{\mathrm{DMC}}}{\rho_{\mathrm{HF}}}. \tag{16.21}$$

However, following Amovilli and March's work [15], which relates directly S_{corr} to Jaynes entropy for the Moshinski atom, we prefer to use the entropy difference, Eq. (16.20), instead of the KL divergence. Correlation Shannon entropy is plotted against Z in Fig. 16.3. This plot shows a sharp increase of the entropy difference in approaching the critical nuclear charge. In such a weak confinement regime, Coulomb correlation tends to be dominant and a measure of this effect seems to be given by $\Delta\, S_{\mathrm{corr}}$.

16.2.2 N-Like Isoelectronic Series of Atomic Ions

For these atomic systems, we have removed the core electrons from the calculations and we have simulated the relevant effect through modified BFD pseudopotentials in order to consider fractional nuclear charges. We used in this case a single determinant SJ wavefunction and a universal basis set made of 12 s and p elementary Gaussian type orbitals.

The ground state is spin polarized with spherical symmetry (^4S_u). The critical binding occurs at a nuclear charge of about 5.85 [30].

In Fig. 16.4 we report the DMC Shannon entropy, and in Fig. 16.5 the difference with the corresponding HF value. As in the previous case, we note also here a significant increase of correlation entropy in approaching the critical Z. Finally, in Fig. 16.6 we show the Z dependence of the deviation from HF of DMC total and kinetic energies given as a percentage of the corresponding DMC values, namely

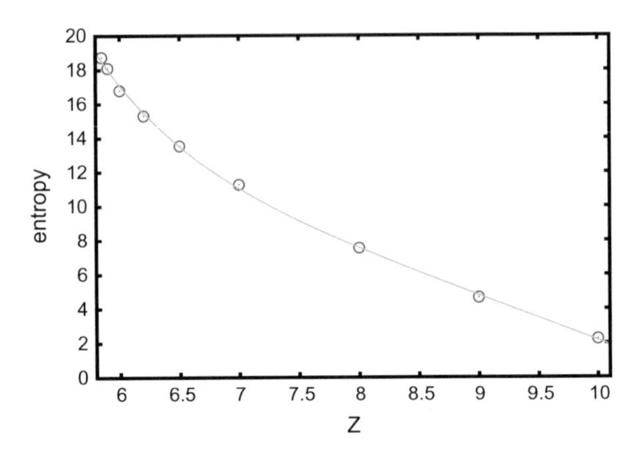

Fig. 16.4 DMC Shannon entropy for N-like atomic ions as a function of the nuclear charge Z

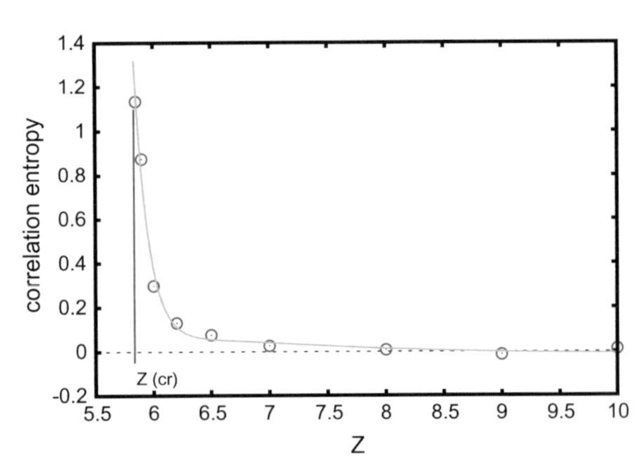

Fig. 16.5 Correlation Shannon entropy as a function of the nuclear charge Z for the N-series of atomic ions considered in this work

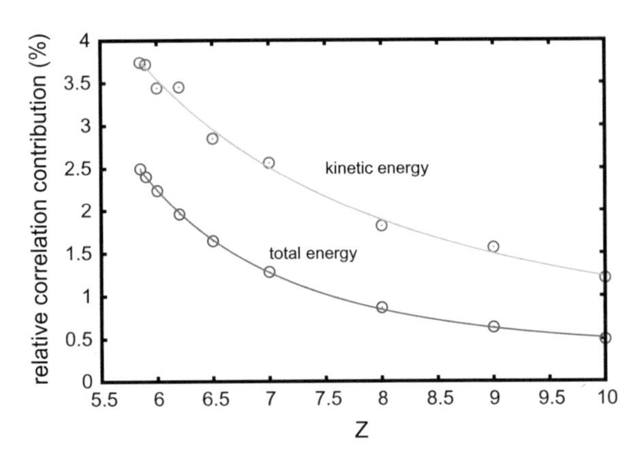

Fig. 16.6 Dependence on nuclear charge Z of correlation energy and correlation kinetic energy, given as percentage of, respectively, total energy and kinetic energy, for the N-like atomic ions considered in this work (^4S_u state)

$$\Delta \mathscr{E}_{r,\mathrm{corr}} = 100 \frac{(E_{\mathrm{DMC}} - E_{\mathrm{HF}})}{E_{\mathrm{DMC}}} \tag{16.22}$$

and

$$\Delta \mathscr{T}_{r,\mathrm{corr}} = 100 \frac{(T_{\mathrm{DMC}} - T_{\mathrm{HF}})}{T_{\mathrm{DMC}}}. \tag{16.23}$$

As expected, the role of correlation becomes more and more important in going to the weak binding regime. From our calculations, generally the effect on kinetic energy seems to be twice that on total energy. We remark that in a weak binding regime, the kinetic energy is lowered due to a greater delocalization of the electron cloud. Results plotted in Figs. 16.5 and 16.6 are consistent with Collins conjecture [4].

16.2.3 Ne-Like Isoelectronic Series of Atomic Ions

Our last test example is that of Ne isoelectronic atomic systems. In this case, we consider eight valence electrons and appropriate BFD modified core potentials. The ground state is a singlet (^1S_g). As for N-like isoelectronic series, we use also here a single determinant SJ wave function and the same universal Gaussian basis set. The critical nuclear charge is about 8.74 [30].

In Fig. 16.7 we plot the DMC Shannon entropy in the range of Z between Z_{cr} and 10.4, in Fig. 16.8 the correlation entropy and, finally, in Fig. 16.9 the fraction of correlation kinetic and total energy. The Shannon entropy takes the maximum value at Z_{cr} as well as all other quantities plotted in the above figures. We can conclude that also for Ne isoelectronic atomic system the behaviour is the same of the previous two cases that differ for the number of electrons and the spin state.

By comparing Figs. 16.3, 16.5, and 16.8, we can see also that the correlation Shannon entropy, as defined in this work, at the critical nuclear charge is very similar, about 1 a.u. We are not able to give an explanation but perhaps this is related to the fact that, in all the three cases, one electron is formally lost in the ionization process.

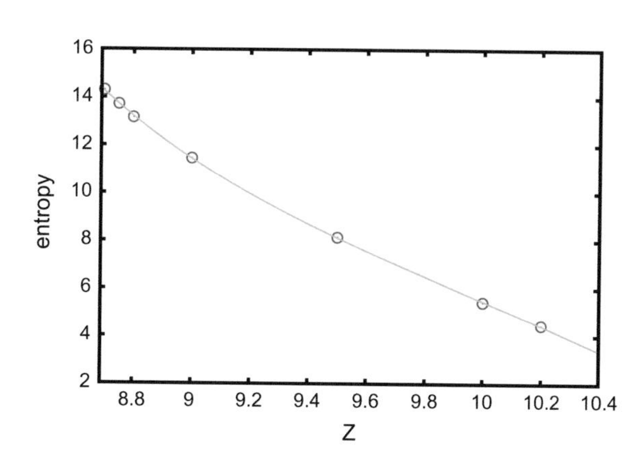

Fig. 16.7 DMC Shannon entropy for Ne-like atomic ions as a function of the nuclear charge Z

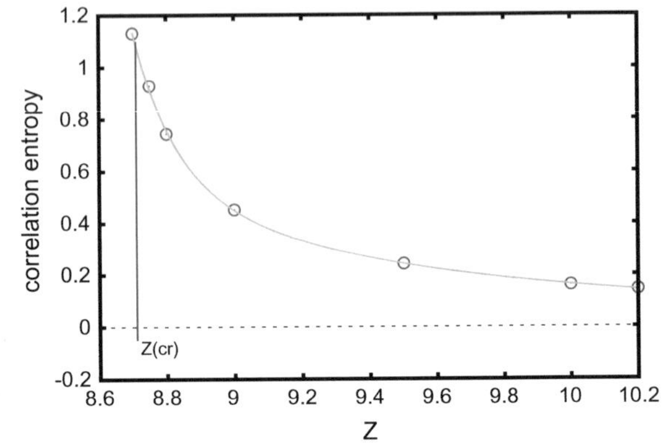

Fig. 16.8 Correlation Shannon entropy as a function of the nuclear charge Z for the Ne series of atomic ions considered in this work

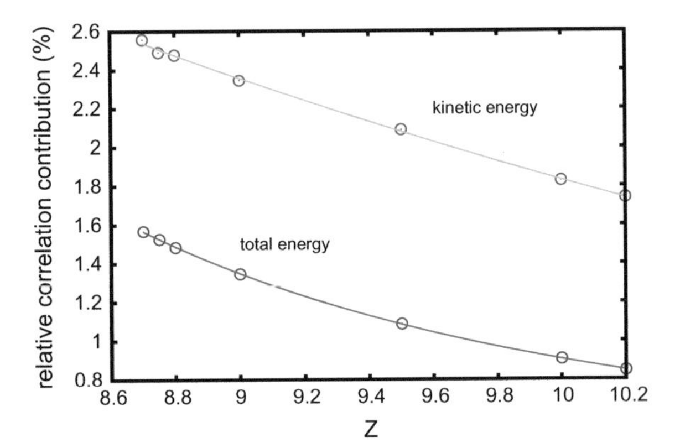

Fig. 16.9 Dependence on nuclear charge Z of correlation energy and correlation kinetic energy, given as percentage of, respectively, total energy and kinetic energy, for the Ne-like atomic ions considered in this work (^1S_g state)

16.3 Conclusions

In this work, we have studied an alternative and more practical way to relate an entropy measure to the correlation energy of atomic systems. We have focussed attention to the case of fractional nuclear charges close to the critical value at which the ionization potential goes to zero. In this regime, the electronic Coulomb correlation plays an important role. Because of the difficulty of computing the one particle density matrix, we avoid the calculation of quantum entanglement entropy and we compute Shannon entropy by means of electron density, a more accessible function for the systems under study. We use densities from DMC and HF calculations. The difference between the two sets of entropy shows a significant positive deviation in approaching the critical nuclear charge. This result suggests that Shannon entropy and the entanglement entropy could be related each other. We have considered He, N and Ne isoelectronic atomic ions and we have found in all cases the same behaviour. For N and Ne series, we have also calculated the fraction of correlation kinetic and total energy. These fractions resulted increased in the region of nuclear charges close to the critical values. This observation seems to be in agreement with Collins conjecture [4]. It is important to remark that, due to the lowering of kinetic energy in weak binding conditions, the fraction of correlation kinetic energy is significant in approaching the limit of zero ionization potential. This point is relevant with DFT in the Kohn–Sham approach where the unknown exchange-correlation functional contains the difference between the exact and the single-particle kinetic energy. If this difference is expected to be high, the method may encounter numerical problems with accuracy. In this regards, it could be interesting to analyze in details the Shannon entropy computed with electron densities obtained by means of the most commonly used functionals.

As for future direction, we intend to extend the present study to polyatomic molecules. In order to pursue this project we need to improve the density estimator from QMC calculations, especially at DMC level of the theory.

A further very interesting aspect is the possible use of Shannon entropy for the fine tuning of existing exchange-correlation functionals in KS-DFT.

Acknowledgements We thank Claudia Filippi for providing the version of CHAMP code under development at the University of Twente (NL).

References

1. J. von Neumann, *Mathematische Grundlagen der Quantenmechanik* (Springer, Berlin, 1932)
2. L. Landau, Z. Phys. **45**(5), 430 (1927). https://doi.org/10.1007/BF01343064
3. D. Janzing, in *Compendium of quantum physics: concepts, experiments, history and philosophy*, ed. by D. Greenberger, K. Hentschel, F. Weinert (Springer, Berlin, 2009), pp. 205–209
4. D.M. Collins, Z. Naturforsch. **48a**, 68 (1993)
5. R.O. Esquivel, A.L. Rodríguez, R.P. Sagar, M. Hô, V.H. Smith, Phys. Rev. A **54**, 259 (1996). https://doi.org/10.1103/PhysRevA.54.259
6. P. Ziesche, Int. J. Quantum Chem. **56**(4), 363 (1995). https://doi.org/10.1002/qua.560560422
7. Á. Nagy, E. Romera, J. Mol. Model. **23**(5), 159 (2017). https://doi.org/10.1007/s00894-017-3331-y
8. S. Grimme, A. Hansen, Angew. Chem. Int. Ed. **54**(42), 12308 (2015). https://doi.org/10.1002/anie.201501887
9. C.E. Shannon, Bell Syst. Tech. J. **27**(3), 379 (1948). https://doi.org/10.1002/j.1538-7305.1948.tb01338.x
10. C.E. Shannon, Bell Syst. Tech. J. **27**(4), 623 (1948). https://doi.org/10.1002/j.1538-7305.1948.tb00917.x
11. Á. Nagy, Chem. Phys. Lett. **556**, 355 (2013). https://doi.org/10.1016/j.cplett.2012.11.065
12. L. Delle Site, Int. J. Quantum Chem. **115**(19), 1396 (2015). https://doi.org/10.1002/qua.24823
13. M. Gell-Mann, K.A. Brueckner, Phys. Rev. **106**, 364 (1957). https://doi.org/10.1103/hysRev.106.364
14. S.P. Fazal, K.D. Sen, G. Gutierrez, P. Fuentealba, Indian J. Chem. A **39**, 48 (2000)
15. C. Amovilli, N.H. March, Phys. Rev. A **69**(5), 054302 (2004). https://doi.org/10.1103/PhysRevA.69.054302
16. E.T. Jaynes, Phys. Rev. **108**, 171 (1957). https://doi.org/10.1103/PhysRev.108.171
17. W.M.C. Foulkes, L. Mitas, R.J. Needs, G. Rajagopal, Rev. Mod. Phys. **73**, 33 (2001). https://doi.org/10.1103/RevModPhys.73.33

18. P.J. Reynolds, D.M. Ceperley, B.J. Alder, W.A. Lester Jr., J. Chem. Phys. **77**(11), 5593 (1982). https://doi.org/10.1063/1.443766
19. M. Burkatzki, C. Filippi, M. Dolg, J. Chem. Phys. **126**(23), 234105 (2007). https://doi.org/10.1063/1.2741534
20. C. Filippi, C.J. Umrigar, J. Chem. Phys. **105**(1), 213 (1996). https://doi.org/10.1063/1.471865
21. C.J. Umrigar, J. Toulouse, C. Filippi, S. Sorella, R.G. Hennig, Phys. Rev. Lett. **98**, 110201 (2007). https://doi.org/10.1103/PhysRevLett.98.110201
22. M. Casula, Phys. Rev. B **74**, 161102 (2006). https://doi.org/10.1103/PhysRevB.74.161102
23. M.W. Schmidt, K.K. Baldridge, J.A. Boatz, S.T. Elbert, M.S. Gordon, J.H. Jensen, S. Koseki, N. Matsunaga, K.A. Nguyen, S. Su, T.L. Windus, M. Dupuis, J.A. Montgomery, J. Comp. Chem. **14**(11), 1347 (1993). https://doi.org/10.1002/jcc.540141112
24. CHAMP is a quantum Monte Carlo program package written by C. J. Umrigar, C. Filippi and collaborators
25. C. Amovilli, N.H. March, I.A. Howard, Á. Nagy, Phys. Lett. A **372**(22), 4053 (2008). https://doi.org/10.1016/j.physleta.2007.11.075
26. G.W.F. Drake, R.A. Swainson, Phys. Rev. A **41**, 1243 (1990). https://doi.org/10.1103/PhysRevA.41.1243
27. C. Amovilli, N.H. March, J. Phys. A Math. Gen. **39**(23), 7349 (2006). https://doi.org/10.1088/0305-4470/39/23/013
28. J.H. Ou, Y.K. Ho, Atoms **5**(2), 15 (2017). https://doi.org/10.3390/atoms5020015
29. S. Kullback, R.A. Leibler, Ann. Math. Stat. **22**(1), 79 (1951). https://doi.org/10.1214/aoms/1177729694
30. H. Hogreve, J. Phys. B At. Mol. Phys. **31**(10), L439 (1998). https://doi.org/10.1088/0953-4075/31/10/001

Chapter 17
Kinetic Energy Density Functionals from Models for the One-Electron Reduced Density Matrix

D. Chakraborty, R. Cuevas-Saavedra and P. W. Ayers

Abstract Orbital-free kinetic energy functionals can be constructed by writing the one-electron reduced density matrix as an approximate functional of the ground-state electron density. In order to utilize this strategy, one needs to impose appropriate N-representability constraints upon the model 1-electron reduced density matrix. We present several constraints of this sort here, the most powerful of which is based upon the March-Santamaria identity for the local kinetic energy.

17.1 Introduction

Practical density-functional theory (DFT) calculations use either the orbital-free method or the Kohn–Sham approach [1]. The orbital-free approach, in which the kinetic energy is directly approximated as an explicit functional of the electron density came first historically, and was a very popular approach through the mid-1980s. However, explicit kinetic energy functionals tend to have poor accuracy, and in particular are subject to variational collapse to chemically absurd solutions with energies that are far too low. This is often attributed to the difficulty of satisfying the Pauli principle using an explicit density functional [2–6], and motivates the idea of using an auxiliary function to evaluate the kinetic energy. Most commonly, one introduces the Kohn–Sham orbitals, which are functionals of the electron density constructed based on the requirement that the energy of the noninteracting Kohn–Sham reference system has the same electron density as the total system. The energy of the noninteracting system is then used a proxy for the energy of the true interacting system, with the remaining correlation-kinetic energy lumped into the exchange-correlation energy functional.

Despite its ubiquity, there is still interest in the orbital-free method, primarily because it is less computationally costly [7–12]. The key obstacle, clearly, is approximating the kinetic energy [7, 9, 10, 13–18] or, alternatively, its functional derivative

D. Chakraborty · R. Cuevas-Saavedra · P. W. Ayers (✉)
Department of Chemistry and Chemical Biology, McMaster University,
Hamilton L8S 4M1, Canada
e-mail: ayers@mcmaster.ca

G. G. N. Angilella and C. Amovilli (eds.), *Many-body Approaches at Different Scales*, https://doi.org/10.1007/978-3-319-72374-7_17

[19–27]. It has been realized that effective calculations are usually nonlocal functionals of the electron density, which motivates introducing new auxiliary quantities like the 1-electron reduced density matrix, the exchange hole, or the linear response function [10–12, 18, 28–42].

Recently, we have proposed a strategy based on writing the 1-electron reduced density matrix as an explicit, nonlocal, functional of the electron density [43]. The exact Levy constrained search functional for the noninteracting (Kohn–Sham) kinetic energy of σ-spin electrons can be expressed in this way [44–46]

$$T_s^\sigma[\rho] = \min_{\{\gamma^\sigma:\ \rho^\sigma(\mathbf{r})=\gamma^\sigma(\mathbf{r},\mathbf{r}),\ \gamma^\sigma=(\gamma^\sigma)^2\}} \int\int \delta(\mathbf{r}-\mathbf{r}')\left(-\frac{1}{2}\nabla_{\mathbf{r}}^2\gamma^\sigma(\mathbf{r},\mathbf{r}')\right) d^3\mathbf{r} d^3\mathbf{r}', \tag{17.1a}$$

$$\gamma_s^\sigma[\rho^\sigma;\mathbf{r},\mathbf{r}'] = \arg \min_{\{\gamma^\sigma:\ \rho^\sigma(\mathbf{r})=\gamma^\sigma(\mathbf{r},\mathbf{r}),\ \gamma^\sigma=(\gamma^\sigma)^2\}} \int\int \delta(\mathbf{r}-\mathbf{r}')\left(-\frac{1}{2}\nabla_{\mathbf{r}}^2\gamma^\sigma(\mathbf{r},\mathbf{r}')\right) d^3\mathbf{r} d^3\mathbf{r}'. \tag{17.1b}$$

Among all idempotent 1-matrices with the correct electron density, the Levy constrained search selects the one with the lowest kinetic energy. This procedure is clearly impractical—it is actually more difficult to construct the Kohn–Sham kinetic energy associated with a specified density than it is to solve the Kohn–Sham equations [47–49]. We therefore proposed an explicit form for the 1-electron reduced matrix,

$$\tilde{\gamma}^\sigma[\rho^\sigma;\mathbf{r},\mathbf{r}'] = \sqrt{\rho^\sigma(\mathbf{r})\rho^\sigma(\mathbf{r}')}\,\tilde{g}\left(k_{\mathrm{F}}^\sigma[\rho^\sigma]\cdot|\mathbf{r}-\mathbf{r}'|\right), \tag{17.2}$$

where the function $\tilde{g}(x)$ must satisfy

$$\tilde{g}(0) = 1, \tag{17.3a}$$

$$\tilde{g}'(0) = 0, \tag{17.3b}$$

$$\tilde{g}''(0) < 1, \tag{17.3c}$$

$$-1 < \tilde{g}(x) \le 1. \tag{17.3d}$$

In general, $\tilde{g}(x) < 0$ for some values of x [50]. Note that Eq. (17.2) is not an approximation: the exact functional is obtained by choosing Fermi wave vector as the 6-dimensional function,

$$k_{\mathrm{F}}^\sigma(\mathbf{r},\mathbf{r}') \equiv \frac{1}{|\mathbf{r}-\mathbf{r}'|}\tilde{g}^{-1}\left(\frac{\tilde{\gamma}^\sigma[\rho^\sigma;\mathbf{r},\mathbf{r}']}{\sqrt{\rho^\sigma(\mathbf{r})\rho^\sigma(\mathbf{r}')}}\right). \tag{17.4}$$

Equation (17.4) is also impractical, of course, so one needs to find practical approximations for the function $\tilde{g}(x)$.

17.2 Model One-Electron Reduced Density Matrices

In our previous work, we chose $\tilde{g}(x)$ based on its form in the uniform electron gas,

$$\tilde{g}^{\sigma\sigma}_{\mathrm{LDA}}(x) = 3\left(\frac{\sin x - x\cos x}{x^3}\right), \tag{17.5}$$

where $x = k^{\sigma}_{\mathrm{F}}(\mathbf{r},\mathbf{r}')|\mathbf{r}-\mathbf{r}'|$. This form, however, is not appropriate for molecules and other insulators, where the 1-electron reduced density matrix decays exponentially with increasing $|\mathbf{r}-\mathbf{r}'|$, with a rate of decay that tends to become faster as the band gap increases [51–56]. This suggests that one should add an exponential damping factor, which leads to a form like

$$\begin{aligned}\tilde{g}^{\sigma\sigma}_{\exp}\left(k^{\sigma}_{\mathrm{F}}(\mathbf{r},\mathbf{r}'), \ell^{\sigma}_{\mathrm{F}}(\mathbf{r},\mathbf{r}'), b\right) &= \tilde{g}^{\sigma\sigma}_{\mathrm{LDA}}\left(k^{\sigma}_{\mathrm{F}}(\mathbf{r},\mathbf{r}')|\mathbf{r}-\mathbf{r}'|\right)\\ &\times \exp\left[\frac{b(\mathbf{r},\mathbf{r}')}{\left|\ell^{\sigma}_{\mathrm{F}}(\mathbf{r},\mathbf{r}')\right|}\left(1-\sqrt{1+\left(\ell^{\sigma}_{\mathrm{F}}(\mathbf{r},\mathbf{r}')|\mathbf{r}-\mathbf{r}'|\right)^2}\right)\right].\end{aligned} \tag{17.6}$$

This form still has an infinite number of nodes, which is not realistic for molecular systems. Choosing an exponential model with a single node is perhaps more appropriate for molecular systems,

$$\begin{aligned}\tilde{g}^{\sigma\sigma}_{\mathrm{IP}}\left(k^{\sigma}_{\mathrm{F}}(\mathbf{r},\mathbf{r}'), a, b\right) &= \left(1 - a k^{\sigma}_{\mathrm{F}}(\mathbf{r},\mathbf{r}')|\mathbf{r}-\mathbf{r}'|^2\right)\\ &\times \exp\left[\frac{b(\mathbf{r},\mathbf{r}')}{\left|\ell^{\sigma}_{\mathrm{F}}(\mathbf{r},\mathbf{r}')\right|}\left(1-\sqrt{1+\left(\ell^{\sigma}_{\mathrm{F}}(\mathbf{r},\mathbf{r}')|\mathbf{r}-\mathbf{r}'|\right)^2}\right)\right].\end{aligned} \tag{17.7}$$

It is reasonable to assume that the functional $b(\mathbf{r},\mathbf{r}')$ is related to the effective ionization potential (IP) at the location of interest, and so a reasonable form is [57–61]

$$b(\mathbf{r},\mathbf{r}') \propto \frac{1}{2}\left(\frac{1}{8}\frac{|\nabla\rho(\mathbf{r})|^2}{\rho^2(\mathbf{r})} + \frac{1}{8}\frac{|\nabla\rho(\mathbf{r}')|^2}{\rho^2(\mathbf{r}')}\right). \tag{17.8}$$

There is enormous flexibility in how one should select the nonlocal component of these functionals, but we choose a form inspired by the weighted density approximation,

$$k^{\sigma}_{\mathrm{F}}(\mathbf{r},\mathbf{r}') = \left[\frac{1}{2}\left(\left(k^{\sigma}_{\mathrm{F}}(\mathbf{r})\right)^p + \left(k^{\sigma}_{\mathrm{F}}(\mathbf{r}')\right)^p\right)\right]^{1/p}, \tag{17.9}$$

where p is a user-specified parameter. This form seems to work relatively well for the exchange and kinetic energies [10, 29, 30, 34–36, 41, 62, 63]. The form in Eq. (17.9) is motivated by the realization that in order to obtain the numerical benefits of orbital-free DFT, the quantities one is considering (here, $k^{\sigma}_{\mathrm{F}}(\mathbf{r},\mathbf{r}')$) must be approximated in terms of three-dimensional functions. Otherwise one could use the six-dimensional Kohn–Sham density matrix, $\gamma^{\sigma}(\mathbf{r}, \mathbf{r}')$, directly.

17.3 Constraints on Model One-Electron Reduced Density Matrices

The free parameters in model one-electron reduced density matrices should be chosen to satisfy exact constraints and, in particular, the Pauli principle. For example, the Kohn–Sham density matrix must be idempotent (cf. Eq. (17.1a)), which leads to the constraint

$$\tilde{\gamma}^{\sigma}(\mathbf{r}, \mathbf{r}'') = \int \tilde{\gamma}^{\sigma}(\mathbf{r}, \mathbf{r}')\tilde{\gamma}^{\sigma}(\mathbf{r}', \mathbf{r}'')d^3\mathbf{r}'. \tag{17.10}$$

As mentioned before, however, it is impractical to consider six-dimensional functions. The simplest way to express Eq. (17.10) as a three-dimensional constraint is to set $\mathbf{r} = \mathbf{r}''$. Then, using the form of the model density matrix (cf. Eq. (17.2)), we have

$$\rho^{\sigma}(\mathbf{r}) = \int \tilde{\gamma}^{\sigma}(\mathbf{r}, \mathbf{r}')\tilde{\gamma}^{\sigma}(\mathbf{r}', \mathbf{r})d^3\mathbf{r}' = \int \left(\sqrt{\rho^{\sigma}(\mathbf{r})}\tilde{g}^{\sigma}(\mathbf{r}, \mathbf{r}')\sqrt{\rho^{\sigma}(\mathbf{r}')}\right)^2 d^3\mathbf{r}', \tag{17.11}$$

which simplifies to the same constraint one uses in the weighted density approximation to the exchange hole [64–68],

$$1 = \int \rho^{\sigma}(\mathbf{r}') \left(\tilde{g}^{\sigma}(\mathbf{r}, \mathbf{r}')\right)^2 d^3\mathbf{r}'. \tag{17.12}$$

Using Eq. (17.12) gives reasonable results, but it is still far from the accuracy we need. This motivates the development of additional constraints.

For example, it is true that for any square-integrable function $\phi(\mathbf{r})$, it must be that

$$\int \tilde{\gamma}^{\sigma}(\mathbf{r}, \mathbf{r}'')\phi(\mathbf{r}'')d^3\mathbf{r}'' = \int\int \tilde{\gamma}^{\sigma}(\mathbf{r}, \mathbf{r}')\tilde{\gamma}^{\sigma}(\mathbf{r}', \mathbf{r}'')\phi(\mathbf{r}'')d^3\mathbf{r}'d^3\mathbf{r}''. \tag{17.13}$$

We clearly cannot force this constraint for all choices of $\phi(\mathbf{r}'')$ (this would be as difficult as enforcing the original idempotency constrain, Eq. (17.10)) but we can use specific functions that might be useful. For example, choosing $\phi(\mathbf{r}'')$ as an atomic $1s$ orbital will prevent the occupation number of that orbital from being too large. An especially simple form, however, is obtained when one uses $\phi(\mathbf{r}'') = \sqrt{\rho^{\sigma}(\mathbf{r}'')}$. Then one has:

$$\int \tilde{g}^{\sigma}(\mathbf{r}, \mathbf{r}'')\rho^{\sigma}(\mathbf{r}'')d^3\mathbf{r}'' = \int\int \tilde{g}^{\sigma}(\mathbf{r}, \mathbf{r}')\tilde{g}^{\sigma}(\mathbf{r}', \mathbf{r}'')\rho^{\sigma}(\mathbf{r}')\rho^{\sigma}(\mathbf{r}'')d^3\mathbf{r}'d^3\mathbf{r}''. \tag{17.14}$$

This constraint is more difficult to apply than Eq. (17.12) because it requires a six-dimensional numerical integration at each point $\mathbf{r}$. However, the same auxiliary basis-set methods that are used in efficient implementations of Eq. (17.12) could be used in Eq. (17.14) also [69]. If one considers $\phi(\mathbf{r})$ to be the gradient operator, then one obtains a constraint with the form of the March-Santamaria expression for the kinetic energy, namely [70–72]

$$\frac{1}{2}\int \left|\nabla_{\mathbf{r}}\tilde{\gamma}^{\sigma}(\mathbf{r},\mathbf{r}')\right|^{2} d^{3}\mathbf{r}' = \frac{1}{2}\left[\nabla_{\mathbf{r}}\cdot\nabla_{\mathbf{r}'}\tilde{\gamma}^{\sigma}(\mathbf{r},\mathbf{r}')\right]_{\mathbf{r}=\mathbf{r}'}. \tag{17.15}$$

This constraint can be implemented with the same computational cost as Eq. (17.12), and helps ensure that the $\mathbf{r}\approx\mathbf{r}'$ portion of the model density matrix is accurate. The right-hand-side of Eq. (17.15) is just the (positive-definite) local kinetic energy [73, 74]. In fact, it is not difficult to see that Eq. (17.15) holds if the idempotency constraint holds near the diagonal,

$$\tilde{\gamma}^{\sigma}(\mathbf{r},\mathbf{r}+\varepsilon\hat{\mathbf{u}}) = \int \tilde{\gamma}^{\sigma}(\mathbf{r},\mathbf{r}')\tilde{\gamma}^{\sigma}(\mathbf{r}',\mathbf{r}+\varepsilon\hat{\mathbf{u}})d^{3}\mathbf{r}', \tag{17.16}$$

where $\hat{\mathbf{u}}$ is a unit vector. Equation (17.16) can be rewritten as

$$\tilde{g}^{\sigma}(\mathbf{r},\mathbf{r}+\varepsilon\hat{\mathbf{u}}) = \int \rho(\mathbf{r}')\tilde{g}^{\sigma}(\mathbf{r}',\mathbf{r}+\varepsilon\hat{\mathbf{u}})d^{3}\mathbf{r}'. \tag{17.17}$$

Expanding both sides in a Taylor series,

$$\begin{aligned}
\tilde{g}^{\sigma}(\mathbf{r},\mathbf{r}) &+ \varepsilon\left[\nabla_{\mathbf{r}''}\tilde{g}^{\sigma}(\mathbf{r},\mathbf{r}'')\right]_{\mathbf{r}''=\mathbf{r}}\cdot\hat{\mathbf{u}} + \varepsilon^{2}\hat{\mathbf{u}}^{\top}\left[\nabla_{\mathbf{r}''}\nabla_{\mathbf{r}''}^{\top}\tilde{g}^{\sigma}(\mathbf{r},\mathbf{r}'')\right]_{\mathbf{r}''=\mathbf{r}}\hat{\mathbf{u}} + \ldots \\
&= \int \rho(\mathbf{r}')\tilde{g}(\mathbf{r},\mathbf{r}')\left\{\tilde{g}^{\sigma}(\mathbf{r}',\mathbf{r}) + \varepsilon\left[\nabla_{\mathbf{r}''}\tilde{g}^{\sigma}(\mathbf{r}',\mathbf{r}'')\right]_{\mathbf{r}''=\mathbf{r}}\cdot\hat{\mathbf{u}}\right. \\
&\qquad \left. + \,\varepsilon^{2}\hat{\mathbf{u}}^{\top}\left[\nabla_{\mathbf{r}''}\nabla_{\mathbf{r}''}^{\top}\tilde{g}^{\sigma}(\mathbf{r}',\mathbf{r}'')\right]_{\mathbf{r}''=\mathbf{r}}\hat{\mathbf{u}} + \ldots\right\} d^{3}\mathbf{r}',
\end{aligned} \tag{17.18}$$

the right-hand-side of this equation simplifies due to Eq. (17.3),

$$\begin{aligned}
\tilde{g}^{\sigma}(\mathbf{r},\mathbf{r}) &+ \varepsilon\left[\nabla_{\mathbf{r}''}\tilde{g}^{\sigma}(\mathbf{r},\mathbf{r}'')\right]_{\mathbf{r}''=\mathbf{r}}\cdot\hat{\mathbf{u}} + \varepsilon^{2}\hat{\mathbf{u}}^{\top}\left[\nabla_{\mathbf{r}''}\nabla_{\mathbf{r}''}^{\top}\tilde{g}^{\sigma}(\mathbf{r},\mathbf{r}'')\right]_{\mathbf{r}''=\mathbf{r}}\hat{\mathbf{u}} + \ldots \\
&= 1 + \varepsilon^{2}\hat{\mathbf{u}}^{\top}\left[\nabla_{\mathbf{r}''}\nabla_{\mathbf{r}''}^{\top}\tilde{g}^{\sigma}(\mathbf{r},\mathbf{r}'')\right]_{\mathbf{r}''=\mathbf{r}}\hat{\mathbf{u}} + \ldots.
\end{aligned} \tag{17.19}$$

Then, equating terms in the expansions order-by-order, one has an infinite set of constraints, of which the lowest-order ones are most important and easiest to apply,

$$1 = \int \rho(\mathbf{r}')\left(\tilde{g}^{\sigma}(\mathbf{r},\mathbf{r}')\right)^{2} d^{3}\mathbf{r}', \tag{17.20a}$$

$$0 = \int \rho(\mathbf{r}')\tilde{g}^{\sigma}(\mathbf{r},\mathbf{r}')\nabla_{\mathbf{r}}\tilde{g}^{\sigma}(\mathbf{r}',\mathbf{r})d^{3}\mathbf{r}'$$

$$= \int \rho(\mathbf{r}')\nabla_{\mathbf{r}}\left(\tilde{g}^{\sigma}(\mathbf{r},\mathbf{r}')\right)^2 d^3\mathbf{r}', \tag{17.20b}$$

$$\left[\nabla_{\mathbf{r}''}\nabla_{\mathbf{r}''}^{\top}\tilde{g}^{\sigma}(\mathbf{r},\mathbf{r}'')\right]_{\mathbf{r}''=\mathbf{r}} = \int \rho(\mathbf{r}')\tilde{g}^{\sigma}(\mathbf{r},\mathbf{r}')\nabla_{\mathbf{r}}\nabla_{\mathbf{r}}^{\top}\tilde{g}^{\sigma}(\mathbf{r}',\mathbf{r})d^3\mathbf{r}', \tag{17.20c}$$

$$\vdots$$

The first of these equations is just the original diagonal condition, Eq. (17.12), which is equivalent to the weighted density approximation equation for the normalization of the exchange hole,

$$-1 = \int \rho(\mathbf{r}')h_x(\mathbf{r},\mathbf{r}')d^3\mathbf{r}'. \tag{17.21}$$

The second equation also can be seen as a constraint on the model density matrix but also on the exchange hole for weighted density approximations,

$$0 = \int \rho(\mathbf{r}')\nabla_{\mathbf{r}}h_x(\mathbf{r},\mathbf{r}')d^3\mathbf{r}'. \tag{17.22}$$

It is just the gradient of both sides of Eq. (17.21), so it should hold automatically when Eq. (17.21) is true. The third constraint in Eq. (17.20) is nontrivial, and cannot be easily written as a constraint on the exchange hole. Overall, Eq. (17.20) are a hierarchy of constraints and, if the full hierarchy is imposed, then the model one-electron reduced density matrix, and the kinetic energy density functional it implies, is guaranteed to be N-representable. Philosophically, then, Eq. (17.20) are similar to other methods for developing density functionals based on hierarchies of constraints [72, 75–83].

17.4 Summary

Approximating the kinetic energy as an explicit functional of the electron density has proven to be a very difficult task, we propose to explore a strategy based on modelling the one-electron density matrix using a weighted density approximation. We propose that the parameters in the weighted density approximation should be determined by constraints associated with its idempotency, and propose Eqs. (17.14) and (17.15) as additions to the usual weighted density approximation condition, Eq. (17.12). In particular, the March-Santamaria identity inspires the hierarchy of derivative constraints in Eq. (17.20); these seem especially promising since they are directly linked to the requirement that the accuracy of the kinetic energy from the model density matrix and, specifically, require that different ways of calculating the kinetic energy from the model density matrix give the same results. In order to simultaneously satisfy constraints Eq. (17.20), one needs a more flexible form for the model density matrix than is provided when one uses the uniform electron gas.

The damped local density approximation model given in Eq. (17.7) is merely one possibility.

The weighted density approximation for the one-electron reduced density matrix, Eq. (17.2), is a very flexible form, and there are an infinite number of potentially useful constraints that can be imposed upon it. We expect that further investigations along these lines will provide more accurate kinetic energy functionals, though innovations are also required so that constraints like Eq. (17.20) can be efficiently imposed. We will address these numerical problems in a follow-up paper [84].

Acknowledgements Support from Sharcnet, NSERC, and the Canada Research Chairs is appreciated. RCS acknowledges financial support from CONACYT, ITESM and DGRI-SEP.

References

1. W. Kohn, L.J. Sham, Phys. Rev. **140**, A1133 (1965). https://doi.org/10.1103/PhysRev.140.A1133
2. F.R. Manby, P.J. Knowles, A.W. Lloyd, Chem. Phys. Lett. **335**(5), 409 (2001). https://doi.org/10.1016/S0009-2614(01)00075-6
3. P.W. Ayers, S. Liu, Phys. Rev. A **75**, 022514 (2007). https://doi.org/10.1103/PhysRevA.75.022514
4. E.S. Kryachko, E.V. Ludea, Phys. Rev. A **43**, 2179 (1991). https://doi.org/10.1103/PhysRevA.43.2179
5. E.V. Ludeña, J. Mol. Struct. Theochem **709**(1), 25 (2004), *A Collection of Papers Presented at the 29th International Congress of Theoretical Chemists of Latin Expression, Marrakech, Morocco, 8–12 September 2003*. https://doi.org/10.1016/j.theochem.2004.03.047
6. E.V. Ludeña, F. Illas, A. Ramirez-Solis, Int. J. Mod. Phys. B **22**(25–26), 4642 (2008). https://doi.org/10.1142/S0217979208050395
7. V.V. Karasiev, R.S. Jones, S.B. Trickey, F.E. Harris, in *New developments in quantum chemistry*, ed. by J.L. Paz, A.J. Hernandez (Transworld Research Network, Kerala, India, 2009)
8. J.D. Chai, J.D. Weeks, Phys. Rev. B **75**, 205122 (2007). https://doi.org/10.1103/PhysRevB.75.205122
9. T.A. Wesolowski, Int. J. Chem. (CHIMIA) **58**(5), 311 (2004). https://doi.org/10.2533/000942904777677885
10. Y.A. Wang, E.A. Carter, S.D. Schwartz, in *Theoretical methods in condensed phase chemistry*, ed. by S.D. Schwartz (Kluwer, Dordrecht, 2000), pp. 117–184. ISBN 9780306469497
11. P. García-González, J.E. Alvarellos, E. Chacón, Phys. Rev. B **53**, 9509 (1996). https://doi.org/10.1103/PhysRevB.53.9509
12. J.M. Dieterich, W.C. Witt, E.A. Carter, J. Comp. Chem. **38**(17), 1552 (2017). https://doi.org/10.1002/jcc.24806
13. H.J. Chen, A.H. Zhou, Numer. Math. Theory Meth. Appl. **1**, 1 (2008)
14. D. García-Aldea, J.E. Alvarellos, in *Advances in computational methods in sciences and engineering, Lecture series on computer and computational sciences*, vol. 4A–4B, ed. by T. Simos, G. Maroulis (Koninklijke Brill NV, Leiden, 2005), pp. 1462–1466. Selected Papers from the International Conference of Computational Methods in Sciences and Engineering (ICCMSE 2005). ISBN 9789067644419
15. S.S. Iyengar, M. Ernzerhof, S.N. Maximoff, G.E. Scuseria, Phys. Rev. A **63**, 052508 (2001). https://doi.org/10.1103/PhysRevA.63.052508
16. G.C. Kin-Lic, N.C. Handy, J. Chem. Phys. **112**(13), 5639 (2000). https://doi.org/10.1063/1.481139

17. A.J. Thakkar, Phys. Rev. A **46**, 6920 (1992). https://doi.org/10.1103/PhysRevA.46.6920
18. D. García-Aldea, J.E. Alvarellos, J. Chem. Phys. **127**(14), 144109 (2007). https://doi.org/10.1063/1.2774974
19. K. Finzel, Int. J. Quantum Chem. **115**(23), 1629 (2015). https://doi.org/10.1002/qua.24986
20. K. Finzel, J. Chem. Phys. **144**(3), 034108 (2016). https://doi.org/10.1063/1.4940035
21. K. Finzel, Theor. Chem. Acc. **135**(4), 87 (2016). https://doi.org/10.1007/s00214-016-1850-8
22. K. Finzel, Int. J. Quantum Chem. **116**(16), 1261 (2016). https://doi.org/10.1002/qua.25169
23. K. Finzel, J. Davidsson, I.A. Abrikosov, Int. J. Quantum Chem. **116**(18), 1337 (2016). https://doi.org/10.1002/qua.25181
24. K. Finzel, P.W. Ayers, Theor. Chem. Acc. **135**(12), 255 (2016). https://doi.org/10.1007/s00214-016-2013-7
25. K. Finzel, Int. J. Quantum Chem. **117**(5), 25329 (2017), e25329. https://doi.org/10.1002/qua.25329
26. K. Finzel, P.W. Ayers, Int. J. Quantum Chem. **117**(10), 25364 (2017), e25364. https://doi.org/10.1002/qua.25364
27. A. Genova, M. Pavanello, (2017), Preprint arXiv:1704.08943 [cond-mat.mtrl-sci]
28. Y.A. Wang, N. Govind, E.A. Carter, Phys. Rev. B **58**, 13465 (1998). https://doi.org/10.1103/PhysRevB.58.13465
29. Y.A. Wang, N. Govind, E.A. Carter, Phys. Rev. B **60**, 16350 (1999). https://doi.org/10.1103/PhysRevB.60.16350
30. B. Zhou, V.L. Ligneres, E.A. Carter, J. Chem. Phys. **122**(4), 044103 (2005). https://doi.org/10.1063/1.1834563
31. L.W. Wang, M.P. Teter, Phys. Rev. B **45**, 13196 (1992). https://doi.org/10.1103/PhysRevB.45.13196
32. E. Smargiassi, P.A. Madden, Phys. Rev. B **49**, 5220 (1994). https://doi.org/10.1103/PhysRevB.49.5220
33. F. Perrot, J. Phys.: Condens. Matt. **6**(2), 431 (1994). https://doi.org/10.1088/0953-8984/6/2/014
34. D. García-Aldea, J.E. Alvarellos, Phys. Rev. A **77**, 022502 (2008). https://doi.org/10.1103/PhysRevA.77.022502
35. D. García-Aldea, J.E. Alvarellos, Phys. Rev. A **76**, 052504 (2007). https://doi.org/10.1103/PhysRevA.76.052504
36. P. García-González, J.E. Alvarellos, E. Chacón, Phys. Rev. A **54**, 1897 (1996). https://doi.org/10.1103/PhysRevA.54.1897
37. C. Huang, E.A. Carter, Phys. Rev. B **81**, 045206 (2010). https://doi.org/10.1103/PhysRevB.81.045206
38. I.V. Ovchinnikov, L.A. Bartell, D. Neuhauser, J. Chem. Phys. **126**(13), 134101 (2007). https://doi.org/10.1063/1.2716667
39. C. Herring, Phys. Rev. A **34**, 2614 (1986). https://doi.org/10.1103/PhysRevA.34.2614
40. E. Chacón, J.E. Alvarellos, P. Tarazona, Phys. Rev. B **32**, 7868 (1985). https://doi.org/10.1103/PhysRevB.32.7868
41. D. García-Aldea, J.E. Alvarellos, J. Chem. Phys. **129**(7), 074103 (2008). https://doi.org/10.1063/1.2968612
42. T. Verstraelen, P.W. Ayers, V. Van Speybroeck, M. Waroquier, J. Chem. Phys. **138**(7), 074108 (2013). https://doi.org/10.1063/1.4791569
43. D. Chakraborty, R. Cuevas-Saavedra, P.W. Ayers, Theor. Chem. Acc. **136**(9), 113 (2017). https://doi.org/10.1007/s00214-017-2149-0
44. M. Levy, Proc. Nat. Acad. Sci. **76**(12), 6062 (1979)
45. M. Levy, J.P. Perdew, in *Density Functional Methods In Physics*, ed. by R.M. Dreizler, J. da Providência (Springer US, Boston, MA, 1985), pp. 11–30. ISBN 978-1-4757-0818-9. https://doi.org/10.1007/978-1-4757-0818-9_2
46. M. Levy, Theor. Comp. Chem. **4**, 3 (1996), Recent Developments and Applications of Modern Density Functional Theory. https://doi.org/10.1016/S1380-7323(96)80083-5
47. Q. Wu, W. Yang, J. Chem. Phys. **118**(6), 2498 (2003). https://doi.org/10.1063/1.1535422

48. Q. Zhao, R.G. Parr, J. Chem. Phys. **98**(1), 543 (1993). https://doi.org/10.1063/1.465093
49. Q. Zhao, R.C. Morrison, R.G. Parr, Phys. Rev. A **50**, 2138 (1994). https://doi.org/10.1103/PhysRevA.50.2138
50. P.W. Ayers, R. Cuevas-Saavedra, D. Chakraborty, Phys. Lett. A **376**(6), 839 (2012). https://doi.org/10.1016/j.physleta.2012.01.028
51. X.P. Li, R.W. Nunes, D. Vanderbilt, Phys. Rev. B **47**, 10891 (1993). https://doi.org/10.1103/PhysRevB.47.10891
52. E. Hernández, M.J. Gillan, C.M. Goringe, Phys. Rev. B **53**, 7147 (1996). https://doi.org/10.1103/PhysRevB.53.7147
53. M. Challacombe, J. Chem. Phys. **110**(5), 2332 (1999). https://doi.org/10.1063/1.477969
54. W. Kohn, Phys. Rev. Lett. **76**, 3168 (1996). https://doi.org/10.1103/PhysRevLett.76.3168
55. R. Baer, M. Head-Gordon, Phys. Rev. Lett. **79**, 3962 (1997). https://doi.org/10.1103/PhysRevLett.79.3962
56. E. Prodan, W. Kohn, Proc. Natl. Acad. Sci. (USA) **102**(33), 11635 (2005). https://doi.org/10.1073/pnas.0505436102
57. G. Sperber, Int. J. Quantum Chem. **5**(2), 189 (1971). https://doi.org/10.1002/qua.560050206
58. P. de Silva, J. Korchowiec, T.A. Wesolowski, J. Chem. Phys. **140**(16), 164301 (2014). https://doi.org/10.1063/1.4871501
59. R.C. Morrison, P.W. Ayers, J. Chem. Phys. **103**(15), 6556 (1995). https://doi.org/10.1063/1.470382
60. R.C. Morrison, C.M. Dixon, J.R. Mizell, Int. J. Quantum Chem. **52**(S28), 309 (1994). https://doi.org/10.1002/qua.560520832
61. W.P. Wang, R.G. Parr, Phys. Rev. A **16**, 891 (1977). https://doi.org/10.1103/PhysRevA.16.891
62. P. García-González, J.E. Alvarellos, E. Chacón, Phys. Rev. A **57**, 4192 (1998). https://doi.org/10.1103/PhysRevA.57.4192
63. P. García-González, J.E. Alvarellos, E. Chacón, Phys. Rev. B **57**, 4857 (1998). https://doi.org/10.1103/PhysRevB.57.4857
64. E. Chacón, P. Tarazona, Phys. Rev. B **37**, 4013 (1988). https://doi.org/10.1103/PhysRevB.37.4013
65. R. Cuevas-Saavedra, D. Chakraborty, P.W. Ayers, Phys. Rev. A **85**, 042519 (2012). https://doi.org/10.1103/PhysRevA.85.042519
66. J.A. Alonso, L.A. Girifalco, Solid State Commun. **24**(2), 135 (1977). https://doi.org/10.1016/0038-1098(77)90591-9
67. J.A. Alonso, L.A. Girifalco, Phys. Rev. B **17**, 3735 (1978). https://doi.org/10.1103/PhysRevB.17.3735
68. O. Gunnarsson, M. Jonson, B.I. Lundqvist, Solid State Commun. **24**(11), 765 (1977). https://doi.org/10.1016/0038-1098(77)91185-1
69. R. Cuevas-Saavedra, P.W. Ayers, Chem. Phys. Lett. **539**, 163 (2012). https://doi.org/10.1016/j.cplett.2012.04.037
70. N.H. March, R. Santamaria, Int. J. Quantum Chem. **39**(4), 585 (1991). https://doi.org/10.1002/qua.560390405
71. D. Chakraborty, P.W. Ayers, J. Math. Chem. **49**(8), 1822 (2011). https://doi.org/10.1007/s10910-011-9861-0
72. P.W. Ayers, J. Math. Phys. **46**(6), 062107 (2005). https://doi.org/10.1063/1.1922071
73. L. Cohen, J. Chem. Phys. **70**(2), 788 (1979). https://doi.org/10.1063/1.437511
74. P.W. Ayers, R.G. Parr, A. Nagy, Int. J. Quantum Chem. **90**(1), 309 (2002). https://doi.org/10.1002/qua.989
75. A.A. Kugler, Phys. Rev. A **41**, 3489 (1990). https://doi.org/10.1103/PhysRevA.41.3489
76. A. Nagy, Phys. Rev. A **47**, 2715 (1993). https://doi.org/10.1103/PhysRevA.47.2715
77. R.G. Parr, S. Liu, A.A. Kugler, A. Nagy, Phys. Rev. A **52**, 969 (1995). https://doi.org/10.1103/PhysRevA.52.969
78. A. Nagy, S. Liu, R.G. Parr, Phys. Rev. A **59**, 3349 (1999). https://doi.org/10.1103/PhysRevA.59.3349
79. A. Nagy, Int. J. Quantum Chem. **106**(5), 1043 (2006). https://doi.org/10.1002/qua.20872

80. P.W. Ayers, Phys. Rev. A **74**, 042502 (2006). https://doi.org/10.1103/PhysRevA.74.042502
81. D. Chakraborty, P.W. Ayers, J. Math. Chem. **49**(8), 1810 (2011). https://doi.org/10.1007/s10910-011-9860-1
82. P.W. Ayers, J. Math. Chem. **44**(2), 311 (2008). https://doi.org/10.1007/s10910-007-9261-7
83. S. Liu, Phys. Rev. A **54**, 1328 (1996). https://doi.org/10.1103/PhysRevA.54.1328
84. M. Chan, R. Cuevas-Saavedra, D. Chakraborty, P.W. Ayers, Computation **5**(4), 42 (2017). https://doi.org/10.3390/computation5040042

Chapter 18
A Gradient Corrected Two-Point Weighted Density Approximation for Exchange Energies

R. Cuevas-Saavedra, D. Chakraborty, M. Chan and P. W. Ayers

Abstract A successful symmetric, two-point, nonlocal weighted density approximation for the exchange energy of atoms and molecules can be constructed using a power mean with constant power p when symmetrizing the exchange-correlation hole [Phys. Rev. A **85**, 042519 (2012)]. In this work, we consider how this parameter depends on the system's charge. Exchange energies for all ions with charge from -1 to $+12$ of the first eighteen atoms of the periodic table are computed and optimized. Appropriate gradient corrections to the current model, based on rational functions, are designed based on the optimal p values we observed for the ionic systems. All of the advantageous features (non-locality, uniform electron gas limit and no self-interaction error) of the original model are preserved.

18.1 Introduction

Density functional theory (DFT) has successfully become the method of choice for computing the electronic structure of large molecules and complex materials [1–5]. However, although density functional theory provides an exact mathematical framework for the electronic structure problem [6–12], its utility in practical calculations is limited by the accuracy of approximate exchange-correlation functionals. This motivates the ongoing research into accurate and feasible approximate exchange-correlation functionals [4, 13–16]. Despite the success of functionals such as the local-density approximation [17–19] (LDA) and generalized gradient approximations [20–23] (GGA) due to their computational efficiency, they are subject to a number of deficiencies. These deficiencies are usually analyzed by determining which exact constraints are (and are not) satisfied by the approximate functionals, though it is also true that no (semi)local functional can ever be exact [24–27].

This has stimulated recent work on nonlocal density functionals, where the exchange-correlation energy is approximated as a six-dimensional integral [28–47],

---

R. Cuevas-Saavedra · D. Chakraborty · M. Chan · P. W. Ayers (✉)
Department of Chemistry and Chemical Biology, McMaster University,
Hamilton, ON L8S 4M1, Canada
e-mail: ayers@mcmaster.ca

G. G. N. Angilella and C. Amovilli (eds.), *Many-body Approaches at Different Scales*, https://doi.org/10.1007/978-3-319-72374-7_18

$$E_{xc}[\rho] = \int\int f[\rho; \mathbf{r}, \mathbf{r}']d^3\mathbf{r}d^3\mathbf{r}'. \tag{18.1}$$

This type of functional is particularly advantageous when the exchange-correlation hole is delocalized [24–26, 48–50] and is essential when for density-functionals that are accurate not only for short-range electron correlations, but also for long-range electron correlations like dispersion [30–37].

Equation (18.1) provides a natural form for the exchange-correlation functional since the exact exchange-correlation energy functional can be written in the form

$$E_{xc}[\rho] = \frac{1}{2}\int\int \frac{\rho(\mathbf{r})\rho(\mathbf{r}')\overline{h_{xc}(\mathbf{r}, \mathbf{r}')}}{|\mathbf{r} - \mathbf{r}'|}d^3\mathbf{r}d^3\mathbf{r}', \tag{18.2}$$

where

$$\overline{h_{xc}(\mathbf{r}, \mathbf{r}')} = \int h_{xc}^{\lambda}(\mathbf{r}, \mathbf{r}')d\lambda \tag{18.3}$$

is the exchange-correlation hole

$$h_{xc}^{\lambda}(\mathbf{r}, \mathbf{r}') = \frac{\rho_2^{\lambda}(\mathbf{r}, \mathbf{r}') - \rho(\mathbf{r})\rho(\mathbf{r}')}{\rho(\mathbf{r})\rho(\mathbf{r}')} \tag{18.4}$$

averaged over the constant-density adiabatic connection path, in which the electron-electron repulsion potential $\lambda/|\mathbf{r} - \mathbf{r}'|$ is increased from the noninteracting limit ($\lambda = 0$) to the physical limit of interest ($\lambda = 1$) [51, 52]. Here,

$$\rho_2^{\lambda}(\mathbf{r}, \mathbf{r}') = \left\langle \Psi^{\lambda} \left| \sum_{j\neq i} \delta(\mathbf{r}_i - \mathbf{r})\delta(\mathbf{r}_j - \mathbf{r}') \right| \Psi^{\lambda} \right\rangle \tag{18.5}$$

is the electron pair density.

There have been several recent attempts to construct nonlocal exchange-correlation functionals using models for the exchange-correlation hole [38–47]. Some of them are based on a variant of the classical Ornstein–Zernike equation [41, 42, 53–57] while some others are two-point weighted density approximations that rely on analytical models of the exchange-correlation hole for the uniform electron gas [58, 59]. The latter approach seems to be promising since they are suggested to be competitive with the best generalized gradient approximation. Moreover, these two-point weighted density approximations are fully nonlocal, have no self-interaction error, approximately fulfill the Pauli principle, and preserve the uniform electron gas limit [38, 39]. In these models the symmetry of the exchange-correlation hole is achieved by means of a generalized mean (power mean).

Motivated by these preliminary results we explore ways to improve the two-point weighted density approximation (2pt-WDA) exchange density functional. Section 18.2 provides a brief overview of the approach, and extends our previous tests to all ions with charges between -1 and $+12$ for the first 36 elements of the periodic table (H–Kr). Based on the dependence of these results on the power used in the generalized mean, we propose a gradient-corrected 2pt-WDA. This model is studied in Sect. 18.3, and our conclusions are summarized in Sect. 18.4.

18.2 The Weighted Density Approximation for Atoms and Ions

The main ingredient of the weighted density approximation discussed here is the exchange hole for the uniform electron gas (UEG)

$$f\left(k_{\mathrm{F},\sigma}^{(x)}|\mathbf{r}-\mathbf{r}'|\right) = -9\left(\frac{\sin\left(k_{\mathrm{F},\sigma}^{(x)}|\mathbf{r}-\mathbf{r}'|\right) - \left(k_{\mathrm{F},\sigma}^{(x)}|\mathbf{r}-\mathbf{r}'|\right)\cos\left(k_{\mathrm{F},\sigma}^{(x)}|\mathbf{r}-\mathbf{r}'|\right)}{\left(k_{\mathrm{F},\sigma}^{(x)}|\mathbf{r}-\mathbf{r}'|\right)^3}\right)^2 \tag{18.6}$$

(cf. Eq. (17.5) in Chap. 17 [60]) where we use the expression for k_{F} from the uniform electron gas,

$$k_{\mathrm{F},\sigma}^{(0)}(\mathbf{r}) = \sqrt{6\pi^2\rho_\sigma(\mathbf{r})}. \tag{18.7}$$

The expression Eq. (18.6) for the exchange hole needs to be symmetric with respect to interchange of $\mathbf{r}$ and $\mathbf{r}'$ because the electron pair density is symmetric [61, 62]. We use the p-mean to symmetrize this formula [38, 39],

$$k_{\mathrm{F},\sigma}^{(x)}(\mathbf{r},\mathbf{r}') = \left[\frac{1}{2}\left(\left(k_{\mathrm{F},\sigma}^{(x)}(\mathbf{r})\right)^p + \left(k_{\mathrm{F},\sigma}^{(x)}(\mathbf{r}')\right)^p\right)\right]^{1/p}. \tag{18.8}$$

Finally, the exchange energy is computed by means of

$$E_x^{(x)}[\rho] \approx \frac{1}{2}\sum_{\sigma=\alpha,\beta}\int\int \frac{\rho_\sigma(\mathbf{r})\rho_\sigma(\mathbf{r}')h_{x,\sigma}^{(x)}(\mathbf{r},\mathbf{r}')}{|\mathbf{r}-\mathbf{r}'|} d^3\mathbf{r}d^3\mathbf{r}', \tag{18.9}$$

where the exchange hole is approximated as

$$h_{x,\sigma}^{(x)}(\mathbf{r},\mathbf{r}') \approx f\left(k_{\mathrm{F},\sigma}^{(x)}(\mathbf{r},\mathbf{r}')|\mathbf{r}-\mathbf{r}'|\right). \tag{18.10}$$

This approach naturally leads to three different types of functional. The 0pt-WDA functional is constructed by the direct use of Eq. (18.7) when computing the exchange hole. This functional is neither symmetric nor normalized. In 1pt-WDA, an effective $k_{\mathrm{F},\sigma}^{(1)}(\mathbf{r})$ is used to enforce the normalization of the hole

$$-1 = \int \rho_\sigma(\mathbf{r}) h_{x;\sigma\sigma}(\mathbf{r}, \mathbf{r}') d^3\mathbf{r}$$
$$= \int \rho_\sigma(\mathbf{r}) f\left(k^{(1)}_{\mathrm{F},\sigma}(\mathbf{r}')|\mathbf{r} - \mathbf{r}'|\right) d^3\mathbf{r}. \quad (18.11)$$

This functional is now normalized but not symmetric. Finally, the 2pts-WDA arises when both symmetry and normalization are imposed.

Following the same computational approach described in our previous work [38, 39], we computed the exchange energy for all ions with charge from -1 to $+12$ of the first 36 atoms in the periodic table (H–Kr). Minimizing the root-mean-square error in the exchange energy over the entire dataset gave a value of $p = 2$ for Eq. (18.8). The weighted density approximation is still competitive with the traditional GGA functionals (B88 [21], PBE [22], and OPTX [23]), and significantly better than the local density approximation [17], but both the average and root-mean-square errors are worse than the results for neutral atoms and small molecules [39]. This finding reinforces our previous observation: p should be density dependent. This is more clearly observed in Tables 18.1 and 18.2, where the average and root-mean-square errors are shown for different charges; the errors increase with increasing charge, indicating that p should be system-dependent.

Table 18.1 Average errors for the atoms and atomic ions with -1 to $+12$ charge, H–Kr, for conventional density functionals (LDA, B88, PBE, OPTX) and the symmetrized weighted density approximations (0pt-WDA, 1pt-WDA, 2pt-WDA) described in Sect. 18.2. The $p = 2$ mean is chosen in Eq. (18.8). The average over all atoms and ions is provided in the bottom row. The rows above that are the average error for species of a given charge; for example, the first row is the average error in the atomic anions

Charge	0pt-WDA	1pt-WDA	2pt-WDA	LDA	B88	PBE	OPT
−1	10.134	3.782	0.014	2.303	−0.049	−1.168	−0.198
0	10.505	3.915	0.007	2.398	−0.034	−1.172	−0.203
1	10.809	3.959	0.018	2.482	−0.010	−1.187	−0.187
2	11.081	3.967	0.013	2.549	−0.006	−1.204	−0.190
3	11.332	3.955	0.003	2.603	−0.011	−1.230	−0.201
4	11.415	3.893	0.085	2.647	0.015	−1.205	−0.175
5	11.794	3.960	0.092	2.757	0.045	−1.206	−0.152
6	12.120	3.994	0.175	2.862	0.078	−1.200	−0.117
7	11.842	3.734	0.226	2.836	0.108	−1.135	−0.067
8	12.067	3.717	0.261	2.905	0.123	−1.139	−0.046
9	12.109	3.605	0.317	2.926	0.123	−1.143	−0.027
10	11.963	3.405	0.376	2.927	0.150	−1.098	0.022
11	12.373	3.570	0.555	3.042	0.170	−1.118	0.048
12	12.520	3.527	0.633	3.096	0.183	−1.118	0.070
all	11.471	3.811	0.167	3.096	0.183	−1.118	0.070

Table 18.2 Root-mean-square errors for the atoms and atomic ions with −1 to +12 charge, H–Kr, for conventional density functionals (LDA, B88, PBE, OPTX) and the symmetrized weighted density approximations (0pt-WDA, 1pt-WDA, 2pt-WDA) described in Sect. 18.2. The $p = 2$ mean is chosen in Eq. (18.8). The root-mean-square error over all atoms and ions is provided in the bottom row

Charge	0pt-WDA	1pt-WDA	2pt-WDA	LDA	B88	PBE	OPT
−1	12.313	5.010	0.323	2.734	0.131	1.426	0.357
0	12.759	5.188	0.351	2.844	0.123	1.433	0.373
1	12.971	5.210	0.352	2.904	0.118	1.447	0.368
2	13.145	5.186	0.368	2.944	0.134	1.460	0.383
3	13.298	5.139	0.493	2.975	0.271	1.505	0.473
4	13.345	5.088	0.378	3.006	0.208	1.464	0.425
5	13.669	5.145	0.405	3.107	0.158	1.444	0.376
6	13.862	5.125	0.394	3.189	0.171	1.417	0.347
7	13.560	4.891	0.417	3.159	0.183	1.346	0.318
8	13.719	4.865	0.444	3.216	0.205	1.343	0.307
9	13.626	4.690	0.475	3.203	0.199	1.345	0.293
10	13.399	4.490	0.518	3.199	0.252	1.287	0.249
11	13.711	4.523	0.683	3.294	0.270	1.304	0.254
12	13.797	4.462	0.751	3.336	0.292	1.302	0.256
all	13.323	4.977	0.751	3.336	0.292	1.302	0.256

18.3 A Preliminary Generalized Gradient Corrected Weighted Density Approximation

From the insight gained in the previous findings, we propose in this section the power p as a rational function of the reduced gradient, $s(\mathbf{r}) = |\nabla\rho(\mathbf{r})|/\rho^{4/3}(\mathbf{r})$,

$$p(s; \mathbf{r}) = \frac{a_0 + a_1 s(\mathbf{r})}{b_0 + s(\mathbf{r})}. \tag{18.12}$$

We opted to use the reduced gradient since p should be a dimensionless quantity. We will not engaged in a detailed optimization of this form here, but defer that to future work. Our goal is merely to explore the possible utility of this form on the performance of the weighted density approximation.

Because the exchange hole must remain symmetric, the power in Eq. (18.8) must be symmetrized. We choose to do this with the form,

$$p(\mathbf{r}, \mathbf{r}') = \left[\frac{1}{2}\left(p^q(s, \mathbf{r}) + p^q(s, \mathbf{r}')\right)\right]^{1/q}. \tag{18.13}$$

The exponent q can be chosen as a free parameter also; however we considered only three possibilities: the arithmetic ($q = 1$), harmonic ($q = -1$) and geometric ($q = 0$) means.

From our previous studies, we have learned that the coefficient a_1 in Eq. (18.12) has to be small to reach convergence, especially in the asymptotic regions of the density (regions where s diverges). For this reason we only allowed a_1 to have the values 0.1 and 0.01. The remaining parameters a_0 and b_0 were assigned values 0.1, 0.5, 1.0, 1.5 and 2.0. This gave a total of 50 functionals (each defined by a specified value of (a_0, b_0, a_1) to test). To speed up the testing, we considered only the neutral atoms and +1, +2, +3, +4, and +5 atomic ions, and only for the first 18 elements of the periodic table (H–Ar). We observed that the results were quite insensitive to the choice of q in Eq. (18.13), typically differing in only the 3rd or 4th decimal. For simplicity we henceforth consider only the results for the simple arithmetic mean ($q = 1$).

Tables 18.3 and 18.4 show the average and rms errors, respectively, for each of the triads considered. For fixed values of a_0, b_0 the errors seem to decrease when

Table 18.3 Average errors for the neutral atoms and atomic ions with +1, +2, +3, +4, and +5 charge, H–Ar, for symmetrized weighted density approximations (0pt-WDA, 1pt-WDA, 2pt-WDA) described in Sect. 18.3. The arithmetic mean ($q = 1$) is chosen

a_0	b_0	a_1	0pt	1pt	2pt
2.00	0.10	0.10	9.612	3.048	0.441
2.00	0.50	0.10	9.485	2.965	0.478
2.00	0.10	0.01	9.503	2.993	0.499
1.50	0.10	0.10	9.388	2.909	0.511
2.00	1.00	0.10	9.364	2.889	0.517
2.00	0.50	0.01	9.386	2.918	0.535
1.50	0.50	0.10	9.282	2.843	0.546
2.00	1.50	0.10	9.269	2.831	0.549
2.00	1.00	0.01	9.274	2.848	0.573
2.00	2.00	0.10	9.193	2.785	0.575
1.50	0.10	0.01	9.270	2.852	0.576
1.50	1.00	0.10	9.181	2.781	0.580
1.00	0.10	0.10	9.147	2.765	0.594
2.00	1.50	0.01	9.188	2.795	0.602
1.50	1.50	0.10	9.104	2.734	0.608
1.50	0.50	0.01	9.175	2.793	0.608
1.00	0.50	0.10	9.066	2.716	0.622
2.00	2.00	0.01	9.119	2.754	0.627
1.50	2.00	0.10	9.041	2.698	0.630
1.50	1.00	0.01	9.086	2.738	0.640
1.00	1.00	0.10	8.989	2.670	0.651
1.50	1.50	0.01	9.018	2.698	0.665
1.00	0.10	0.01	9.020	2.705	0.666
1.00	1.50	0.10	8.930	2.636	0.673
1.50	2.00	0.01	8.963	2.666	0.685
0.50	0.10	0.10	8.889	2.615	0.689
1.00	2.00	0.10	8.883	2.609	0.691
1.00	0.50	0.01	8.952	2.663	0.691
0.50	0.50	0.10	8.836	2.584	0.710
1.00	1.00	0.01	8.888	2.626	0.715
0.50	1.00	0.10	8.786	2.556	0.730
1.00	1.50	0.01	8.839	2.598	0.734
0.50	1.50	0.10	8.748	2.535	0.745
1.00	2.00	0.01	8.801	2.576	0.749
0.50	2.00	0.10	8.718	2.518	0.757
0.50	0.10	0.01	8.751	2.551	0.771
0.10	0.10	0.10	8.667	2.491	0.779
0.50	0.50	0.01	8.713	2.530	0.786
0.10	0.50	0.10	8.641	2.476	0.789
0.10	1.00	0.10	8.615	2.462	0.800
0.50	1.00	0.01	8.678	2.510	0.800
0.10	1.50	0.10	8.596	2.452	0.808
0.50	1.50	0.01	8.652	2.495	0.811
0.10	2.00	0.10	8.581	2.444	0.814
0.50	2.00	0.01	8.631	2.484	0.820
0.10	0.10	0.01	8.521	2.425	0.867
0.10	0.50	0.01	8.511	2.420	0.871
0.10	1.00	0.01	8.502	2.415	0.875
0.10	1.50	0.01	8.495	2.411	0.878
0.10	2.00	0.01	8.490	2.409	0.880

Table 18.4 Root-mean-square for the neutral atoms and atomic ions with +1, +2, +3, +4, and +5 charge, H–Ar, for symmetrized weighted density approximations (0pt-WDA, 1pt-WDA, 2pt-WDA) described in Sect. 18.3. The arithmetic mean ($q = 1$) is chosen

a_0	b_0	a_1	0p-WDA	1p-WDA	2p-WDA
2.00	0.10	0.10	11.382	4.066	0.498
2.00	0.50	0.10	11.220	3.956	0.530
2.00	0.10	0.01	11.230	3.978	0.545
1.50	0.10	0.10	11.101	3.886	0.560
2.00	1.00	0.10	11.065	3.854	0.565
2.00	0.50	0.01	11.080	3.877	0.577
1.50	0.50	0.10	10.966	3.796	0.592
2.00	1.50	0.10	10.946	3.778	0.595
2.00	1.00	0.01	10.938	3.784	0.612
1.50	0.10	0.01	10.939	3.794	0.615
2.00	2.00	0.10	10.850	3.717	0.621
1.50	1.00	0.10	10.838	3.715	0.626
1.00	0.10	0.10	10.800	3.697	0.640
2.00	1.50	0.01	10.829	3.715	0.640
1.50	0.50	0.01	10.818	3.715	0.646
1.50	1.50	0.10	10.740	3.653	0.654
2.00	2.00	0.01	10.742	3.660	0.665
1.00	0.50	0.10	10.696	3.631	0.669
1.50	2.00	0.10	10.662	3.605	0.678
1.50	1.00	0.01	10.705	3.643	0.678
1.00	1.00	0.10	10.599	3.571	0.699
1.50	1.50	0.01	10.618	3.589	0.704
1.00	0.10	0.01	10.627	3.602	0.705
1.00	1.50	0.10	10.525	3.526	0.723
1.50	2.00	0.01	10.549	3.547	0.725
1.00	0.50	0.01	10.540	3.547	0.731
0.50	0.10	0.10	10.476	3.501	0.740
1.00	2.00	0.10	10.466	3.491	0.742
1.00	1.00	0.01	10.458	3.497	0.757
0.50	0.50	0.10	10.409	3.461	0.762
1.00	1.50	0.01	10.397	3.460	0.777
0.50	1.00	0.10	10.347	3.423	0.784
1.00	2.00	0.01	10.349	3.431	0.794
0.50	1.50	0.10	10.300	3.395	0.801
0.50	2.00	0.10	10.262	3.374	0.814
0.50	0.10	0.01	10.291	3.403	0.817
0.50	0.50	0.01	10.243	3.373	0.833
0.10	0.10	0.10	10.200	3.339	0.838
0.50	1.00	0.01	10.199	3.347	0.849
0.10	0.50	0.10	10.166	3.320	0.850
0.50	1.50	0.01	10.165	3.327	0.861
0.10	1.00	0.10	10.135	3.302	0.862
0.50	2.00	0.01	10.139	3.312	0.871
0.10	1.50	0.10	10.111	3.289	0.871
0.10	2.00	0.10	10.092	3.278	0.878
0.10	0.10	0.01	10.003	3.238	0.924
0.10	0.50	0.01	9.991	3.230	0.929
0.10	1.00	0.01	9.980	3.224	0.933
0.10	1.50	0.01	9.971	3.219	0.936
0.10	2.00	0.01	9.965	3.216	0.939

increasing a_1; for fixed values of b_0, a_1 the errors seem to decreasing when increasing a_0. For fixed values of a_0, a_1 the errors seem to decrease when decreasing b_0.

These observations can be understood since s diverges in asymptotic regions. Therefore a small value of a_1 is needed, but if $a_1 > 1$ it becomes practically impossible to satisfy the normalization condition [38, 39]. Near the nucleus a much larger value of p is needed, $p \approx 20$. Therefore it is desirable that the ratio a_0/b_0 should be on the order of 10. Thus, while this is in no sense an optimization of the form in Eq. (18.12), we nonetheless were able to learn something about the underlying principles that must be followed to design a GGA-based weighted density approximation for exchange.

18.4 Summary

We have used a two-points weighted density approximation (2pt-WDA) to compute the exchange energies for all ions with charges from −1 to +12 for the first thirty-six atoms in the periodic table. While the 2pt-WDA is still competitive with popular generalized gradient approximations, its performance worsens for highly charged atomic cations, probably because it is better to use a larger value of the power p in Eq. (18.8) for those systems. This motivated us to build a density functional expression for p by writing p as a rational function of the reduced gradient, $s(\mathbf{r})$. Preliminary tests reveal the order of magnitude for the parameters in the mean, and allow us to suggest the form:

$$p(\mathbf{r}, \mathbf{r}') = \frac{1}{2}\left(p(s; \mathbf{r}) + p(s; \mathbf{r}')\right), \tag{18.14a}$$

$$p(s; \mathbf{r}) = \frac{20 + s(\mathbf{r})}{1 + 10s(\mathbf{r})}. \tag{18.14b}$$

The numerical parameters in Eq. (18.14b) are not optimized, and are merely indicative of important features for the functional form.

Acknowledgements Support from Compute Canada, NSERC, and the Canada Research Chairs is appreciated.

References

1. R.G. Parr, W. Yang, *Density Functional Theory of Atoms and Molecules* (Oxford University Press, Oxford, 1989). ISBN 9780195092769
2. P.W. Ayers, W. Yang, in *ComputatiOnal Medicinal Chemistry for Drug Discovery*, ed. by P. Bultinck, H. de Winter, W. Langenaeker, J.P. Tollenaere (Dekker, New York, 2003), pp. 571–616
3. W. Kohn, A.D. Becke, R.G. Parr, J. Phys. Chem. **100**(31), 12974 (1996). https://doi.org/10.1021/jp960669l
4. A.J. Cohen, P. Mori-Sánchez, W. Yang, Chem. Rev. **112**(1), 289 (2012). https://doi.org/10.1021/cr200107z
5. W. Kohn, Rev. Mod. Phys. **71**, 1253 (1999). https://doi.org/10.1103/RevModPhys.71.1253
6. P.C. Hohenberg, W. Kohn, Phys. Rev. **136**, B864 (1964). https://doi.org/10.1103/PhysRev.136.B864
7. M. Levy, Proc. Nat. Acad. Sci. **76**(12), 6062 (1979)
8. S.M. Valone, J. Chem. Phys. **73**(9), 4653 (1980). https://doi.org/10.1063/1.440656
9. E.H. Lieb, Int. J. Quantum Chem. **24**(3), 243 (1983). https://doi.org/10.1002/qua.560240302
10. W. Yang, P.W. Ayers, Q. Wu, Phys. Rev. Lett. **92**, 146404 (2004). https://doi.org/10.1103/PhysRevLett.92.146404
11. P.W. Ayers, Phys. Rev. A **73**, 012513 (2006). https://doi.org/10.1103/PhysRevA.73.012513
12. H. Eschrig, *The Fundamentals of Density Functional Theory* (Eagle, Leipzig, 2003)
13. A.J. Cohen, P. Mori-Sánchez, W. Yang, Science **321**(5890), 792 (2008). https://doi.org/10.1126/science.1158722
14. J.P. Perdew, A. Ruzsinszky, J. Tao, V.N. Staroverov, G.E. Scuseria, G.I. Csonka, J. Chem. Phys. **123**(6), 062201 (2005). https://doi.org/10.1063/1.1904565
15. M. Ernzerhof, J.P. Perdew, K. Burke, in *Density Functional Theory I: Functionals and Effective Potentials*, ed. by R.F. Nalewajski (Springer, Berlin, Heidelberg, 1996), pp. 1–30. https://doi.org/10.1007/3-540-61091-X_1. ISBN 978-3-540-49945-9
16. J.P. Perdew, A. Ruzsinszky, L.A. Constantin, J. Sun, G.I. Csonka, J. Chem. Theory Comput. **5**(4), 902 (2009). https://doi.org/10.1021/ct800531s

17. W. Kohn, L.J. Sham, Phys. Rev. **140**, A1133 (1965). https://doi.org/10.1103/PhysRev.140.A1133
18. S.H. Vosko, L. Wilk, M. Nusair, Can. J. Phys. **58**(8), 1200 (1980). https://doi.org/10.1139/p80-159
19. J.P. Perdew, Y. Wang, Phys. Rev. B **45**, 13244 (1992). https://doi.org/10.1103/PhysRevB.45.13244
20. J.P. Perdew, W. Yue, Phys. Rev. B **33**, 8800 (1986). https://doi.org/10.1103/PhysRevB.33.8800
21. A.D. Becke, Phys. Rev. A **38**, 3098 (1988). https://doi.org/10.1103/PhysRevA.38.3098
22. J.P. Perdew, K. Burke, M. Ernzerhof, Phys. Rev. Lett. **77**, 3865 (1996), [Erratum Phys. Rev. Lett. **78**, 1396 (1997)]. https://doi.org/10.1103/PhysRevLett.77.3865
23. N.C. Handy, A.J. Cohen, Mol. Phys. **99**(5), 403 (2001). https://doi.org/10.1080/00268970010018431
24. A. Savin, in *Recent Developments of Modern Density Functional Theory*, ed. by J.M. Seminario (Elsevier, New York, 1996), p. 327
25. P.W. Ayers, M. Levy, J. Chem. Phys. **140**(18), 18A537 (2014). https://doi.org/10.1063/1.4871732
26. M. Levy, J.S.M. Anderson, F.H. Zadeh, P.W. Ayers, J. Chem. Phys. **140**(18), 18A538 (2014). https://doi.org/10.1063/1.4871734
27. P. Mori-Sánchez, A.J. Cohen, W. Yang, Phys. Rev. Lett. **102**, 066403 (2009). https://doi.org/10.1103/PhysRevLett.102.066403
28. A. Ruzsinszky, J.P. Perdew, G.I. Csonka, J. Chem. Phys. **134**(11), 114110 (2011). https://doi.org/10.1063/1.3569483
29. A. Ruzsinszky, J.P. Perdew, G.I. Csonka, J. Chem. Theory Comput. **6**(1), 127 (2010). https://doi.org/10.1021/ct900518k
30. A. Puzder, M. Dion, D.C. Langreth, J. Chem. Phys. **124**(16), 164105 (2006). https://doi.org/10.1063/1.2189229
31. T. Thonhauser, A. Puzder, D.C. Langreth, J. Chem. Phys. **124**(16), 164106 (2006). https://doi.org/10.1063/1.2189230
32. M. Dion, H. Rydberg, E. Schröder, D.C. Langreth, B.I. Lundqvist, Phys. Rev. Lett. **92**, 246401 (2004). https://doi.org/10.1103/PhysRevLett.92.246401
33. O.A. Vydrov, T. Van Voorhis, J. Chem. Phys. **133**(24), 244103 (2010). https://doi.org/10.1063/1.3521275
34. O.A. Vydrov, T. Van Voorhis, J. Chem. Phys. **130**(10), 104105 (2009). https://doi.org/10.1063/1.3079684
35. O.A. Vydrov, T. Van Voorhis, Phys. Rev. Lett. **103**, 063004 (2009). https://doi.org/10.1103/PhysRevLett.103.063004
36. O.A. Vydrov, Q. Wu, T. Van Voorhis, J. Chem. Phys. **129**(1), 014106 (2008). https://doi.org/10.1063/1.2948400
37. B.I. Lundqvist, Y. Andersson, H. Shao, S. Chan, D.C. Langreth, Int. J. Quantum Chem. **56**(4), 247 (1995). https://doi.org/10.1002/qua.560560410
38. R. Cuevas-Saavedra, D. Chakraborty, S. Rabi, C. Cárdenas, P.W. Ayers, J. Chem. Theory Comput. **8**(11), 4081 (2012). https://doi.org/10.1021/ct300325t
39. R. Cuevas-Saavedra, D. Chakraborty, P.W. Ayers, Phys. Rev. A **85**, 042519 (2012). https://doi.org/10.1103/PhysRevA.85.042519
40. P.W. Ayers, R. Cuevas-Saavedra, D. Chakraborty, Phys. Lett. A **376**(6), 839 (2012). https://doi.org/10.1016/j.physleta.2012.01.028
41. R. Cuevas-Saavedra, D.C. Thompson, P.W. Ayers, Int. J. Quantum Chem. **116**(11), 852 (2016). https://doi.org/10.1002/qua.25081
42. R. Cuevas-Saavedra, P.W. Ayers, J. Phys. Chem. Solids **73**(5), 670 (2012). https://doi.org/10.1016/j.jpcs.2012.01.004
43. R. Cuevas-Saavedra, P.W. Ayers, Chem. Phys. Lett. **539**, 163 (2012). https://doi.org/10.1016/j.cplett.2012.04.037
44. H. Antaya, Y. Zhou, M. Ernzerhof, Phys. Rev. A **90**, 032513 (2014). https://doi.org/10.1103/PhysRevA.90.032513

45. C.E. Patrick, K.S. Thygesen, J. Chem. Phys. **143**(10), 102802 (2015). https://doi.org/10.1063/1.4919236
46. Y. Zhou, H. Bahmann, M. Ernzerhof, J. Chem. Phys. **143**(12), 124103 (2015). https://doi.org/10.1063/1.4931160
47. J.P. Přecechtělová, H. Bahmann, M. Kaupp, M. Ernzerhof, J. Chem. Phys. **143**(14), 144102 (2015). https://doi.org/10.1063/1.4932074
48. O.V. Gritsenko, B. Ensing, P.R.T. Schipper, E.J. Baerends, J. Phys. Chem. A **104**(37), 8558 (2000). https://doi.org/10.1021/jp001061m
49. Y. Zhang, W. Yang, J. Chem. Phys. **109**(7), 2604 (1998). https://doi.org/10.1063/1.476859
50. A. Savin, Chem. Phys. **356**(1), 91 (2009), Moving Frontiers in Quantum Chemistry. https://doi.org/10.1016/j.chemphys.2008.10.023
51. O. Gunnarsson, B.I. Lundqvist, Phys. Rev. B **13**, 4274 (1976). https://doi.org/10.1103/PhysRevB.13.4274
52. D.C. Langreth, J.P. Perdew, Phys. Rev. B **15**, 2884 (1977). https://doi.org/10.1103/PhysRevB.15.2884
53. M.S. Becker, Phys. Rev. **185**, 168 (1969). https://doi.org/10.1103/PhysRev.185.168
54. N.H. March, Phys. Chem. Liq. **46**(5), 465 (2008). https://doi.org/10.1080/00319100802239503
55. C. Amovilli, N.H. March, Phys. Rev. B **76**, 195104 (2007). https://doi.org/10.1103/PhysRevB.76.195104
56. R. Cuevas-Saavedra, P.W. Ayers, Int. J. Mod. Phys. B **24**(25n26), 5115 (2010). https://doi.org/10.1142/S0217979210057250
57. R. Cuevas-Saavedra, P.W. Ayers, *Exchange-Correlation Functionals from the Identical-Particle Ornstein-Zernike Equation:. Basic Formulation and Numerical Algorithms* (World Scientific, Singapore, 2012), vol. 25, pp. 237–249. https://doi.org/10.1142/9789814340793_0019. ISBN 9789814340793
58. P. Gori-Giorgi, F. Sacchetti, G.B. Bachelet, Phys. Rev. B **61**, 7353 (2000), [**66**, 159901(E) (2002)]. https://doi.org/10.1103/PhysRevB.61.7353
59. P. Gori-Giorgi, J.P. Perdew, Phys. Rev. B **66**, 165118 (2002). https://doi.org/10.1103/PhysRevB.66.165118
60. D. Chakraborty, R. Cuevas-Saavedra, P.W. Ayers, in *Many-Body Approaches at Different Scales: A Tribute to Norman H. March on the Occasion of His 90th Birthday*, ed. by G.G.N. Angilella, C. Amovilli (Springer, New York, 2018), chap. 17, p. 199. (This volume.). https://doi.org/10.1007/978-3-319-72374-7_17
61. J.P.A. Charlesworth, Phys. Rev. B **53**, 12666 (1996). https://doi.org/10.1103/PhysRevB.53.12666
62. P. García-González, J.E. Alvarellos, E. Chacón, P. Tarazona, Phys. Rev. B **62**, 16063 (2000). https://doi.org/10.1103/PhysRevB.62.16063

Chapter 19
From Molecules and Clusters of Atoms to Solid State Properties

G. Forte, A. Grassi, G. M. Lombardo, R. Pucci and G. G. N. Angilella

Abstract Several structural and electronic properties of solid-state systems can be thought of as *emerging* from the correlation of individual molecules in suitable clusters, which may be viewed as precursors of the solid phases. This is reviewed through reference to numerous cases studied by N. H. March and collaborators by quantum chemical methods.

19.1 Introduction

In a seminal paper of 1972, P. W. Anderson stated that qualitatively different properties stem from the aggregation of quantitatively many individual constituents [1]. He concisely phrased such a fact by saying that 'more is different'. Indeed, Bloch states in solid-state crystals, the renormalized, effective properties (such as mass or spin) of Landau quasiparticles in correlated electron liquids, or superconductivity in several materials are probably the best-known examples of many-body effects

G. Forte · A. Grassi · G. M. Lombardo
Dipartimento di Scienze del Farmaco, Università di Catania,
Viale A. Doria, 6, 95125 Catania, Italy
e-mail: gforte@unict.it

A. Grassi
e-mail: agrassi@unict.it

G. M. Lombardo
e-mail: glombardo@dipchi.unict.it

R. Pucci · G. G. N. Angilella (✉)
Dipartimento di Fisica e Astronomia, Università di Catania, and IMM-CNR,
UdR Catania, Via S. Sofia, 64, 95123 Catania, Italy
e-mail: giuseppe.angilella@ct.infn.it

R. Pucci
e-mail: renato.pucci@ct.infn.it

G. G. N. Angilella
Scuola Superiore di Catania, Università di Catania, Via Valdisavoia, 9,
95123 Catania, Italy

G. G. N. Angilella and C. Amovilli (eds.), *Many-body Approaches at Different Scales*, https://doi.org/10.1007/978-3-319-72374-7_19

emerging from the correlation of many (order of the Avogadro number) single particles, that could not be directly ascribed to any individual constituent, or to any small number of them. Actually, the latter example, i.e. superconductivity, is a paradigm for many phase transitions characterized by the spontaneous breaking of some continuous symmetry (gauge invariance, in this case), and one formal requirement from statistical mechanics would actually be the system to be infinite, even though deviations from a true phase transition are hardly visible, experimentally, in finite, but macroscopic samples [2, 3] (see also Ref. [4] for a review).

However, precursor signatures of some properties to be seen in larger complexes can be recognized already in clusters and molecules. Here, we will briefly review some results from quantum chemistry calculations in molecules and clusters which were often motivated by novel materials in the solid state. These results have been obtained by Professor Norman H. March and the present collaboration over the last two decades.

19.2 Periodane

Correlated molecules can be recognized, at least theoretically, as individual units in larger complexes. Such is the case, for example, of periodane. This is a molecular cluster, originally hypothesized by Krüger [5], made by the first atoms in each column of the Periodic Table of the elements (the first 'period', hence its name). Its chemical formula would then be LiBeBCNOF, and its ground-state isomer has been determined by Bera et al. [6] within density functional theory (DFT) applied to the whole cluster. However, much information, especially concerning the geometry of the low-lying isomers of this cluster, could be obtained by applying simpler theories to smaller blocks treated as constituent molecules, such as Hartree–Fock theory to LiOB and coupled cluster singles and doubles (CCSD) theory to FBeCN [7]. In particular, a planar structure for periodane was predicted by Forte et al. [7], much similar to that found by Bera et al. [6].

19.3 C-Based Clusters

Motivated by the very early work by N. H. March on benzene [8], where Thomas-Fermi theory was applied for the first time to study the electronic density in molecules, thereby anticipating density functional theory (DFT) (see Refs. [9, 10] for reviews), Forte et al. [11] followed a 'molecular' approach to study impurity effects in graphene. Graphene, an atomically thick layer of carbon atoms in the honeycomb lattice, was there modeled as the limiting case of an infinitely large cluster of benzene rings. Forte et al. [11] then studied several carbon clusters with the honeycomb geometry, with increasing size, ranging from phenalene, including three benzene rings, up to coronene-61, with 61 benzene rings. In the absence of impurities, Forte et al. [11] found a decreasing value of the difference between the highest occupied molecular orbital (HOMO) and the lowest occupied molecular orbital (LUMO). This was interpreted as evidence of the tendency towards a zero-gap semimetal, as is the case for solid-state graphene. This simplified model enabled the authors to study the effect of impurities, such as that of a chemisorbed H atom, of a vacancy, and a substitutional proton [11]. Their results for the density of states are reminiscent of those available for the local density of states around isolated impurities in graphene (see e.g. Ref. [12]).

This was followed by a study of the orientational properties of a water molecule on coronene at the Hartree–Fock (HF) level [13]. Again, this was motivated by much current experimental interest in the interaction of H_2O with nanographene [14], with possible applications towards water purification by means of graphene filters [15].

Some quite recent theoretical interest in solid lithium carbide under pressure [16] was probably motivated by the more general proposal that several carbon allotropes (including graphite, carbon nanotubes, fullerenes, and graphene) may serve as effective reservoirs for hydrogen storage. Hence, the interest for carbon compounds with the intercalation of alkali metals, such as lithium. Motivated by these studies, Forte et al. [17] then considered the free-space molecule Li_2C_2 plus its dimer and trimer. Albeit at the Hartree–Fock plus Møller–Plesset (HF-MP2) level, their study predicted the $(Li_2C_2)_2$ cluster to have four carbon atoms on a linear chain, whereas the six carbon atoms in the trimer $(Li_2C_2)_3$ were predicted to lie on a hexagon, thus mimicking the precursor structure of graphene. However, the Li atoms prefer each to bond to two carbon atoms, rather than following the geometry of benzene. It was then proposed that it should be possible to study the dependence of these free-space results on pressure by imposing vanishing boundary conditions on the molecular wavefunction on a finite three-dimensional closed surface.

19.4 Si-Based Clusters

A molecular approach was fruitful to study the formation of interstitial $(BO)_n$ impurity complexes in solid-state silicon [18]. To this aim, Forte et al. [19] considered a free space cluster $BOSi_2$, with Si_2 simulating the elementary unit in solid-state silicon. Although the approximation used was Hartree–Fock (HF) theory supplemented by low-order Møller–Plesset (MP2) perturbation theory, the results of Forte et al. [19] were already in good quantitative agreement with the structural details experimentally available for this system in the solid state.

Molecular methods were helpful to gain some insight on the properties of solid-state systems under extreme conditions, in particular under high pressure. One example is provided by the various crystalline forms, which have been predicted to occur for SiO_2 and GeO_2 under pressure [20]. Therefore, Forte et al. [21] considered neutral and cationic free-space oxygen-silicon clusters SiO_n and GeO_n $(1 < n \leq 6)$ at various average internuclear distances.

In addition to silicon oxide, silicon hydride (silane, SiH_4) with intercalated hydrogen under pressure is receiving much attention, both theoretically [22, 23] and experimentally [24], with possible metallization and superconductivity arising at a relatively high temperature [22, 25]. This justified the study within a quantum mechanical approach of a cluster of SiH_4 and two H_2 molecules, together with its dimer and trimer, $(SiH_4(H_2)_2)_n$ ($n = 1, 2, 3$), by Forte et al. [26]. The geometries for the clusters under consideration were found to be in qualitative agreement with those predicted or observed experimentally for the corresponding solid-state phases.

It is worth mentioning at this point the study of Grassi et al. [27], concerning the role of bond-order correlation energy for small molecules containing Si. Also, it was shown that the Löwdin correlation energy E_c per electron, E_c/N, where N is the total number of electrons is nearly constant, with $E_c/N = -0.039 \pm 0.007$ a.u., for some 20 Si-containing molecules in the series SiX_nY_m (where $X, Y =$ H, F, Cl) [28]. A similar result, with a slightly different value of $E_c/N = -0.033 \pm 0.003$ a.u., was found for the closed-shell isoelectronic molecules CH_4, NH_3, H_2O, all having $N = 10$ electrons. These findings supported the conclusion that the Löwdin correlation energy density, $\varepsilon_c(\mathbf{r})$ say, is, albeit approximately, a local functional $\varepsilon_c[\rho]$ of the ground-state electron density $\rho(\mathbf{r})$ at equilibrium. In the lowest degree approximation, somewhat suggested by the behaviour found for the aforementioned molecules, all containing light atomic components, one is led to assume a linear functional relation $\varepsilon_c[\rho] \sim \rho$, which implies in turn that the dominant effect of the Löwdin correlation energy for closed-shell molecules at equilibrium merely consists of a shift in the chemical potential [28].

A related study was directly concerned with Si_6H_6 subjected to bond stretching [29]. This work was motivated by an analysis of the cleavage force in crystalline Si [30], showing that Si–Si bonds in the diamond lattice structure could have comparable elasticity to the bond in the free space H_2 molecule. Of course, Si atoms in such a structure are characterized by sp^3 hybridization, at variance with quasi-sp^2 hybridization in the benzene-like, as yet unsynthesized, free space molecule Si_6H_6. Therefore, Grassi et al. [29] included $3p_z$ orbitals in their unrestricted Hartree–Fock calculations, expecting the formation of molecular π-orbitals. The geometry of the low-lying stable ring was predicted to be slightly buckled, rather than planar, with π-electron localization occurring in the range 1.2–1.5 times the equilibrium distance [29], in good qualitative agreement with earlier variational calculations on H_2 [31]. The latter finding suggested a tendency towards localization of the π-electrons in stretched Si_6H_6, at variance with what happens in the (planar) benzene ring [8]. The cluster Si_6H_6 can also be thought as a precursor to silicene, i.e. a two-dimensional allotrope of silicon, with a hexagonal honeycomb structure similar to that of graphene (see Chap. 5 by Baskaran in the present volume, and references therein, for an account of possible superconductivity in silicene [32]).

19.5 Other Clusters

We conclude by briefly summarizing results for other elemental and molecular clusters.

Lithium clusters Li_n $(1 < n \leq 10)$ were extensively studied by Grassi et al. [33] within Hartree–Fock (HF) theory, supplemented by low-order Møller–Plesset (MP2) correlation corrections. Both geometry, energy, and vibrational frequencies of the low-lying isomers of this class of clusters were determined. Their study pointed out the importance of electron correlation, especially in determining the structure and actual geometry of these clusters. For instance, a comparison of HF + MP2 energies and HF energy plus bond-order correlation energy shows that indeed inclusion of bond-order correlation confirms the hypertetrahedral T_d structure for Li_8, as experimentally observed with Raman spectroscopy [34].

The high-pressure, high-density phase of solid oxygen [35, 36] stimulated Forte et al. [37] to study elemental oxygen clusters O_n $(n = 6, 8, 12)$. In particular, starting from the triplet state of molecular O_2 and in view of the magnetic (and possibly even superconducting [38]) correlations in the solid phases of oxygen, particular attention was devoted to the spin configuration of the stable geometries determined through quantum mechanical calculations, again at the HF + MP2 level (plus further calculations at the CCSD level of approximation). Specifically, while the O_8 cluster in the triplet configuration does not form, one finds a O_8 cluster in the singlet configuration, but unstable. One further result was that the bare electrostatic nuclear-nuclear potential energy U_{nn} correlates closely with the total number of electrons in the equilibrium clusters O_n with increasing $n = 2 - 12$, in accord with the conclusions of Mucci and March [39] on tetrahedral and octahedral molecules.

The analysis of structural and electronic properties of a quite specific metallic ion cluster, viz. $(Li_3Al_4)^-$, enabled Grassi et al. [40] to consider the possible occurrence of 'aromaticity' in a metallic cluster. Indeed, 'aromaticity' is a popular concept, which is usually applied to metallic species [41, 42]. It was originally proposed by Pauling and Wheland [43] in connection with the nature and stability of the chemical bond in benzene and other hydrocarbons, such as naphtalene. An aromatic character is generally thought to enhance the kinetic and thermodynamic stability of such molecules, as well as several other remarkable physical and chemical properties [44, 45]. Therefore, quantum chemistry calculations were performed on $(Li_3Al_4)^-$ to test its possible aromaticity [40]. The main outcome of this study was that the chemical shifts corresponding to the low-lying stable isomers numerically found, were intepreted as clear fingerprints of antiaromaticity in this metal cluster [40].

The issue of long-range orientational ordering in bulk water is a long-standing problem (see e.g. Refs. [46, 47], and references therein), with possible important implications also in biochemistry [48]. This stimulated Howard et al. to study water cages $(H_2O)_n$ and $(D_2O)_n$ $(n = 2$ and $8)$, both structurally and in connection with their interaction with a H_2 molecule [49, 50]. One initial concern of their quantum chemistry calculations was of course that of making contact with available inelastic neutron scattering experiments [51] and Raman spectroscopy results [52, 53] on ice.

However, an important direction for future study relates to larger clusters, whose study may be relevant to understand the solid-liquid-like transition with increasing cluster size [54]. Such studies would provide further insight into the nature of hydrogen bonding, and its relevance for the liquid state, starting from properties at the molecular scale. Clusters larger than the water octamer considered by Howard et al. [49, 50] have been in fact studied by Rousseau et al. [55] within density functional theory. Also, the importance of water molecules and clusters in nanotechnology should be emphasized, with possible applications in wastewater purification [56].

19.6 Conclusions

We have reviewed several quantum chemical studies on molecules and clusters, performed by N.H. March and collaborators over the last few decades. These were mostly motivated by their possible relevance for larger systems, whereof the clusters can be thought as precursors. In several instances, both structural and electronic properties somehow anticipated physical and chemical properties of the corresponding solid-state phases, featuring genuinely many-body effects, such as metallicity, superconductivity, or other phase transitions. The origins of these are then to be thought in the electron correlations, already *emerging* in clusters of a few atoms.

Acknowledgements It is a pleasure to congratulate Professor Norman H. March on his 90th birthday. The University of Catania has been honoured to have been regarded by him as one of his 'fixed points' over the last two decades. The authors are all indebted with him for his scientific direction, for sharing his tremendous physical and chemical insight, as well as for much motivation and invaluable friendship over the years.

References

1. P.W. Anderson, Science **177**(4047), 393 (1972). https://doi.org/10.1126/science.177.4047.393
2. C. Liu, Phil. Sci. **66**, S92 (1999). https://doi.org/10.1086/392718
3. B. Farid, in *Electron Correlation in the Solid State*, Chap. 3 ed. by N.H. March (Imperial College, London, 1999), . ISBN 9781860944079
4. N.H. March, G.G.N. Angilella, *Exactly Solvable Models in Many-Body Theory* (World Scientific, Singapore, 2016)
5. T. Krüger, Int. J. Quantum Chem. **106**(8), 1865 (2006). https://doi.org/10.1002/qua.20948
6. P.P. Bera, P. von R. Schleyer, H.F. Schaefer III, Int. J. Quantum Chem. **107**(12), 2220 (2007). https://doi.org/10.1002/qua.21322
7. G. Forte, A. Grassi, G.M. Lombardo, G.G.N. Angilella, N.H. March, R. Pucci, Phys. Lett. A **372**, 3253 (2008). https://doi.org/10.1016/j.physleta.2008.01.046
8. N.H. March, Acta Cryst. **5**(2), 187 (1952), Reprinted in Ref. [57]. https://doi.org/10.1107/S0365110X52000551
9. N.H. March, Adv. Phys. **6**(21), 1 (1957). https://doi.org/10.1080/00018735700101156
10. N.H. March, in *Theory of the Inhomogeneous Electron Gas*, Chap. 1 ed. by S. Lundqvist, N.H. March (Springer, New York, 1983), p. 1. https://doi.org/10.1007/978-1-4899-0415-7_1

11. G. Forte, A. Grassi, G.M. Lombardo, A. La Magna, G.G.N. Angilella, R. Pucci, R. Vilardi, Phys. Lett. A **372**, 6168 (2008). https://doi.org/10.1016/j.physleta.2008.08.014
12. F.M.D. Pellegrino, G.G.N. Angilella, R. Pucci, Phys. Rev. B **80**, 094203 (2009). https://doi.org/10.1103/PhysRevB.80.094203
13. G. Forte, A. Grassi, G.M. Lombardo, G.G.N. Angilella, N.H. March, R. Pucci, Phys. Chem. Liquids **47**, 599 (2009). https://doi.org/10.1080/00319100903045874
14. F. Schedin, A.K. Geim, S.V. Morozov, E.W. Hill, P. Blake, M.I. Katsnelson, K.S. Novoselov, Nat. Mat. **6**, 652 (2007). https://doi.org/10.1038/nmat1967
15. D. Cohen-Tanugi, J.C. Grossman, Nano Lett. **12**(7), 3602 (2012). https://doi.org/10.1021/nl3012853
16. X.Q. Chen, C.L. Fu, C. Franchini, J. Phys.: Cond. Matter **22**(29), 292201 (2010). https://doi.org/10.1088/0953-8984/22/29/292201
17. G. Forte, G.G.N. Angilella, V. Pittalà, N.H. March, R. Pucci, Phys. Chem. Liq. **50**, 46 (2012). https://doi.org/10.1080/00319104.2010.544019
18. A. Carvalho, R. Jones, M. Sanati, S.K. Estreicher, J. Coutinho, P.R. Briddon, Phys. Rev. B **73**(24), 245210 (2006). https://doi.org/10.1103/PhysRevB.73.245210
19. G. Forte, G.G.N. Angilella, N.H. March, R. Pucci, Chem. Phys. Lett. **608**, 269 (2014). https://doi.org/10.1016/j.cplett.2014.06.020
20. S. Saito, T. Ono, Jpn. J. Appl. Phys. **50**(2R), 021503 (2011). https://doi.org/10.1143/JJAP.50.021503
21. G. Forte, G.G.N. Angilella, V. Pittalà, N.H. March, R. Pucci, Phys. Lett. A **376**(4), 476 (2012). https://doi.org/10.1016/j.physleta.2011.11.049
22. M. Martinez-Canales, A.R. Oganov, Y. Ma, Y. Yan, A.O. Lyakhov, A. Bergara, Phys. Rev. Lett. **102**, 087005 (2009). https://doi.org/10.1103/PhysRevLett.102.087005
23. W. Cui, J. Shi, H. Liu, Y. Yao, H. Wang, T. Iitaka, Y. Ma, **5**, 13039 (2015). https://doi.org/10.1038/srep13039
24. T.A. Strobel, M. Somayazulu, R.J. Hemley, Phys. Rev. Lett. **103**, 065701 (2009). https://doi.org/10.1103/PhysRevLett.103.065701
25. X.J. Chen, V.V. Struzhkin, Y. Song, A.F. Goncharov, M. Ahart, Z. Liu, H.k. Mao, R.J. Hemley, Proc. Natl. Acad. Sci. **105**(1), 20 (2008). https://doi.org/10.1073/pnas.0710473105
26. G. Forte, G.G.N. Angilella, N.H. March, R. Pucci, Phys. Lett. A **374**, 580 (2010). https://doi.org/10.1016/j.physleta.2009.11.039
27. A. Grassi, G.M. Lombardo, G. Forte, G.G.N. Angilella, R. Pucci, N.H. March, Mol. Phys. **104**, 1447 (2006). https://doi.org/10.1080/00268970500509899
28. A. Grassi, G.M. Lombardo, G.G.N. Angilella, G. Forte, N.H. March, C. Van Alsenoy, R. Pucci, Phys. Chem. Liq. **46**(5), 484 (2008). https://doi.org/10.1080/00319100701790069
29. A. Grassi, G.M. Lombardo, R. Pucci, G.G.N. Angilella, F. Bartha, N.H. March, Chem. Phys. **297**(1), 13 (2004). https://doi.org/10.1016/j.chemphys.2003.10.001
30. C.C. Matthai, N.H. March, J. Phys. Chem. Solids **58**(5), 765 (1997). https://doi.org/10.1016/S0022-3697(96)00197-7
31. C.A. Coulson, I. Fischer, Phil. Mag. **40**(303), 386 (1949). https://doi.org/10.1080/14786444908521726
32. G. Baskaran, in Many-Body Approaches at Different Scales: A Tribute to Norman H. March on the Occasion of His 90th Birthday, Chap. 5 ed. by G.G.N. Angilella, C. Amovilli (Springer, New York, 2018), p. 43. (This volume.). https://doi.org/10.1007/978-3-319-72374-7_5
33. A. Grassi, G.M. Lombardo, G.G.N. Angilella, N.H. March, R. Pucci, J. Chem. Phys. **120**(24), 11615 (2004). https://doi.org/10.1063/1.1729954
34. A. Kornath, A. Kaufmann, A. Zoermer, R. Ludwig, J. Chem. Phys. **118**(15), 6957 (2003). https://doi.org/10.1063/1.1555800
35. Y. Freiman, H.J. Jodl, Phys. Rep. **401**(1), 1 (2004). https://doi.org/10.1016/j.physrep.2004.06.002
36. I.N. Goncharenko, O.L. Makarova, L. Ulivi, Phys. Rev. Lett. **93**(5), 055502 (2004). https://doi.org/10.1103/PhysRevLett.93.055502

37. G. Forte, G.G.N. Angilella, N.H. March, R. Pucci, Phys. Lett. A **377**, 801 (2013). https://doi.org/10.1016/j.physleta.2013.01.036
38. K. Shimizu, K. Suhara, M. Ikumo, M.I. Eremets, K. Amaya, Nature **393**, 767 (1998). https://doi.org/10.1038/31656
39. J.F. Mucci, N.H. March, J. Chem. Phys. **71**, 5270 (1979), Reprinted in Ref. [57]. https://doi.org/10.1063/1.438338
40. A. Grassi, G.M. Lombardo, G.G.N. Angilella, N.H. March, R. Pucci, D.J. Klein, A.T. Balaban, Phys. Chem. Liq. **52**, 354 (2014). https://doi.org/10.1080/00319104.2014.862058
41. A.I. Boldyrev, L. Wang, Chem. Rev. **105**(10), 3716 (2005). https://doi.org/10.1021/cr030091t
42. Z. Chen, C.S. Wannere, C. Corminboeuf, R. Puchta, P.v.R. Schleyer, Chem. Rev. **105**(10), 3842 (2005). https://doi.org/10.1021/cr030088+
43. L. Pauling, J. Chem. Phys. **1**(1), 56 (1933). https://doi.org/10.1063/1.1749219
44. P.V.R. Schleyer (ed.). *Chemical Reviews. Special issue on Antiaromaticity*, vol. 101(5) (American Chemical Society, Washington, 2001)
45. M. Randić, Chem. Rev. **103**(9), 3449 (2003). https://doi.org/10.1021/cr9903656
46. D.P. Shelton, J. Chem. Phys. **141**(22), 224506 (2014). https://doi.org/10.1063/1.4903541
47. P. Kumar, G. Franzese, S.V. Buldyrev, H.E. Stanley, Phys. Rev. E **73**, 041505 (2006). https://doi.org/10.1103/PhysRevE.73.041505
48. F. Mallamace, P. Baglioni, C. Corsaro, S.H. Chen, D. Mallamace, C. Vasi, H.E. Stanley, J. Chem. Phys. **141**(16), 165104 (2014). https://doi.org/10.1063/1.4900500
49. I.A. Howard, G.G.N. Angilella, N.H. March, C. Van Alsenoy, Phys. Chem. Liq. **42**(4), 403 (2004). https://doi.org/10.1080/00319100410001697855
50. I.A. Howard, G.G.N. Angilella, N.H. March, C. Van Alsenoy, Phys. Chem. Liq. **43**(5), 441 (2005). https://doi.org/10.1080/00319100500184043
51. J. Li, D. Londono, D.K. Ross, J.L. Finney, S.M. Bennington, A.D. Taylor, J. Phys.: Condens. Matter **4**(9), 2109 (1992). https://doi.org/10.1088/0953-8984/4/9/005
52. J.E. Bertie, E. Whalley, J. Chem. Phys. **40**(6), 1637 (1964). https://doi.org/10.1063/1.1725373
53. J.E. Bertie, E. Whalley, J. Chem. Phys. **40**(6), 1646 (1964). https://doi.org/10.1063/1.1725374
54. C.J. Tsai, K.D. Jordan, J. Chem. Phys. **95**(5), 3850 (1991). https://doi.org/10.1063/1.460788
55. B. Rousseau, C. Van Alsenoy, A. Peeters, F. Bogár, G. Parasi, J. Mol. Structure (Theochem) **666-667**(Supplement C), 41 (2003). https://doi.org/10.1016/j.theochem.2003.08.011
56. X. Qu, P.J.J. Alvarez, Q. Li, Nanotechnology for water and wastewater treatment. Water Res. **47**(12), 3931 (2013). https://doi.org/10.1016/j.watres.2012.09.058
57. N.H. March, G.G.N. Angilella (eds.), *Many-Body Theory of Molecules, Clusters, and Condensed Phases* (World Scientific, Singapore, 2009)

Chapter 20
Alchemical Derivatives of Atoms: A Walk Through the Periodic Table

Robert Balawender, Andrzej Holas, Frank De Proft, Christian Van Alsenoy and Paul Geerlings

Abstract Exploring the Chemical Compound Space is at stake when looking for molecules with optimal properties. In order to guide experimentalists to navigate through this unimaginably huge space, theoreticians should look for efficient and cheap algorithms. One of the strategies put forward some years ago was to look for transmutation of molecular structures, thereby changing their nuclear charge content, for which alchemical derivatives are instrumental. A collection of well tested isolated atom alchemical derivatives would be a basic instrument in a navigation toolbox. In this work, isolated atom alchemical derivatives were evaluated with different techniques, from the more accurate numerical differentiation and Coupled Perturbed Kohn–Sham approaches to the Z^{-1} energy expansion model which upon derivation with respect to Z yields the desired derivatives. For this third approach a systematic, computationally elegant, method is developed to routinely evaluate an optimal set of all expansion coefficients in the energy expansion for a given N. For the lighter elements, $Z = 1 - 18$, the comparison between the three approaches shows that the order of magnitude and sequences in the different approaches are similar paving the way for a walk through the complete Periodic Table by combining the Z^{-1} expansion approach with the National Institute of Standards and Technology (NIST) databank atomic energy values at various levels of LDA. A uniform decrease is retrieved not only for the alchemical potential (the electrostatic potential at the origin) but also for the alchemical hardness, with some minor exceptions. The latter values are relatively strongly influenced by relativistic effects for the heavy elements. The uniform decrease of the first derivative is evidenced and quantified. Periodicity shows up in some exploratory calculations on the third derivative (the hyperhardness) which

R. Balawender · A. Holas
Institute of Physical Chemistry of the Polish Academy of Sciences,
Kasprzaka 44/52, 01-224 Warsaw, Poland

F. De Proft · P. Geerlings
General Chemistry (ALGC), Vrije Universiteit Brussel
(Free University of Brussels, VUB), Pleinlaan 2, 1050 Brussels, Belgium

C. Van Alsenoy (✉)
Department of Chemistry, University of Antwerp, Groenenborgerlaan 171,
2020 Antwerp, Belgium
e-mail: kris.vanalsenoy@uantwerpen.be

G. G. N. Angilella and C. Amovilli (eds.), *Many-body Approaches at Different Scales*, https://doi.org/10.1007/978-3-319-72374-7_20

turn out to be strongly basis set dependent. The Periodic Tables generated could be used in a first step in exploring Chemical Compound Space in a systematic, efficient and cheap way. Some possible refinements (atoms-in-molecules corrections) and extensions (inclusion of mixed Z and N derivatives) are touched upon.

20.1 Introduction

Chemistry and chemists are exploring Chemical or Chemical Compound Space (CCS), the space populated by all imaginable chemicals with natural nuclear charges and real interatomic distances for which chemical interactions exist [1–3]. In their continuous efforts toward Molecular Design, chemists try to 'identify' in that space stable molecules with interesting/optimal properties. Navigating through this space with unimaginable size is costly, certainly for experimentalists in front of a myriad of synthetic problems, but even so for theoreticians. Indeed, exploring Chemical Space by even relatively moderate level ab initio or even semi-empirical brute force techniques may already lead to an astonishing increase of the number and complexity of calculations as the number and complexity of atoms characterizing the subspace investigated is increasing.

This problem urges theoretical and computational chemists to look for efficient ways to drastically shorten the navigation time, as they are indeed supposed to take the lead in this kind of navigations due to the still lower cost of computational experiments as compared to synthetic work, and so to assume their role as a guide for experimentalists. Ingenious alternatives to the brute force approach were presented such as simulated annealing [4], genetic algorithms [5], linear combinations of atomic potentials [6]. The Inverse Molecular Design approach introduced by Beratan and Yang [7, 8] has shown to be very promising. This approach leads to the rational design of new molecules with optimal properties (see for example Refs. [9, 10]).

Another highly promising approach for a more efficient exploration of CCS has been the Alchemical Coupling (AC) concept by von Lilienfeld [11–16]. In this approach, two isoelectronic molecules in CCS can be 'coupled' alchemically through interpolation of their electron-nucleus potentials, an approach on which energy versus nuclear charge Z derivatives appear in a natural way.

Although at first a little bit awkward, these 'alchemical' derivatives [11, 17, 18] find a natural place in the context of Conceptual Density Functional Theory [19–21], where the energy of a molecule is written as a functional of the number of electrons N and the external potential (i.e. due to the nuclei) $v(\mathbf{r})$. Upon perturbing the molecule in N and/or $v(\mathbf{r})$, a perturbation expansion can be written in which in each term a derivative of the type $\partial^n \delta^m E/\partial N^n \delta v(\mathbf{r}_1) \ldots \delta v(\mathbf{r}_m)$ occurs in front of the perturbation, which can be identified as the response function of the system due to the perturbation, independent of the magnitude of the perturbation, and as such an intrinsic property of the system.

Numerous studies appeared involving dN and/or $\delta v(\mathbf{r})$ perturbations. Traditionally, the former perturbation has been at stake in phenomena associated with electron transfer [20]. The density response to the external potential change, the so-called linear response function, is a central example of the latter perturbation. Its physical and mathematical properties were studied [22–24], techniques for its numerical and analytical evaluation were developed [25, 26] and its chemical importance was scrutinized [24, 27–30], up to connections with molecular conductivity [31]. The change in external potential $\delta v(\mathbf{r})$ is typically understood as resulting from a combined change in geometry and charges due to a neighbouring, reacting system; $\delta v(\mathbf{r})$ perturbations due to a change in nuclear charge within the molecule (pure dZ perturbations) have rarely been treated [20]. The interest displayed above to the linear response function then becomes evident as it turns out that in the expression for the second order alchemical derivative the linear response function shows up in complete analogy to the one variable density in the first order ones (vide infra).

In previous work by some of the authors [17, 18] it was shown that the dZ derivatives or alchemical response functions, which have been evaluated up to third order, are of crucial importance/use when exploration of chemical space in a quick and standardized way is necessary. It was indeed shown that on the basis of a single SCF type calculation and the corresponding alchemical derivatives of the reference molecule, the CCS of first neighbours (implying changes in nuclear charge of ± 1) could be fully explored by simple arithmetic operations at negligible cost as compared to the SCF calculations. Transmutation reactions have been studied for the nitrogen molecule, the BN iso-electronic 'alchemical isomers' of benzene and pyrene, and very recently [32] the $(C-C)_n \rightarrow (B-N)_n$ substitution pattern of the archetypical fullerene C_{60}, with n varying from 1 to 30. In all cases, correct sequences of stability of the transmuted products were obtained indicating that the perturbation approach involving alchemical derivatives does have great potential for efficient and accurate screening of Chemical Space.

These considerations encouraged us to consider the following issue. As atomic transmutations are the basis ingredients of a more general transmutation (say of a functional group) it may be interesting to construct a data set of all first- and second-order atomic alchemical derivatives for the Complete Periodic Table so that they might be used as building blocks to have a first idea on the order of magnitude of a given, more complex transmutation. Moreover, they would give us an idea of their periodic behaviour (if any) which could be compared with the well-known periodic behaviour of the corresponding derivatives [20], the electronic chemical potential [33] and the chemical hardness [34].

In this endeavour, we will use three different approaches: (i) a brute force numerical differentiation of E with respect to Z, to the best of our knowledge not used hitherto in this context. Third order derivatives are evaluated 'for the sake of completeness'; note that the corresponding N derivative (the hyperhardness) turned out to be less insightful than its lower order congeners [21, 35]; (ii) an analytical method based on a coupled perturbed Kohn–Sham approach [26, 36] developed by some of the present authors, restricted however to closed shell systems [17, 18]; and (iii) a 'model' approach based on the Z^{-1} expansion approach for atomic and ionic energies [37] studied in detail by Norman March [38] leading, among others, to a beautiful joint paper with the father of conceptual DFT, Robert Parr, on the form of the $E[Z, N]$

function for the total energy of atomic ions [39]. This is one of the very small number of joint papers between these two giants of Density Functional Theory [39, 40].

Comparison of the values for $Z = 1 - 18$ for the three methods will provide a sound basis for the credibility of a fourth road, yielding first and second alchemical derivatives for the complete Periodic Table ($Z = 1 - 91$) on the basis of NIST data [41] on atomic electronic structure calculations (neutral atom and cation) obtained via LDA including relativistic corrections (vide infra), thus permitting 'a walk through the Periodic Table'.

20.2 Theory and Computational Details

As stated above the fundamental quantity in this study is the $E = E[Z, N]$ function expressing the ground-state energy E as a function of the total number of electrons N and the nuclear charge Z. In the context of Conceptual Density Functional Theory, the derivatives $\partial^n E/\partial N^n$, $\partial^m E/\partial Z^m$ can be regarded as response functions to investigate the response of a system when its number of electrons or its nuclear charge is changed. The N-derivatives are widely known and explored, in particular the $n = 1$ case, the electronic chemical potential, identified with minus the electronegativity, and the $n = 2$ case, identified with the chemical hardness. Various computational methods have been proposed for it, starting from a finite difference approach [19] to analytical gradients in a coupled perturbed Hartree–Fock [42–44] or Kohn–Sham approach [45], the larger part being however evaluated numerically using the finite difference Ansatz [19, 20] considering differences between the energy of the considered N-electron system (say a neutral atom or molecule), its cation and anion, at the same geometry (demand for constant external potential).

Let us have some closer look at the $(\partial E/\partial Z)_N$ derivatives. In the case of an isolated atom it is easily seen that the first Z derivative at constant number of electrons, the alchemical potential, can be written as (obviously only the electronic part should be considered here)

$$\mu_{\text{al}} \equiv \left(\frac{\partial E}{\partial Z}\right)_N = \int \left(\frac{\delta \tilde{E}[N, v]}{\delta v(\mathbf{r})}\right)_N \left(\frac{\partial v(\mathbf{r})}{\partial Z}\right)_N d^3\mathbf{r} = -\int \frac{\rho(\mathbf{r})}{|\mathbf{r} - \mathbf{R}|} d^3\mathbf{r}. \tag{20.1}$$

Taking the position $\mathbf{R}$ of the nucleus at the origin further simplifies this expression. The chain rule has been used and $v(\mathbf{r})$ is the external potential in DFT, i.e. the potential felt by the electrons due to the nuclei (in the absence of an external field), which in the case of an atom reduces to $-Z/|\mathbf{r} - \mathbf{R}|$. As on the other hand the response of the energy with respect to the external potential directly follows from the basic equation of DFT and is equal to the electron density function $\rho(\mathbf{r})$, one ends up with a very simple equation stating that the first order alchemical potential is equal to the electrostatic potential due to the electrons and taken at the nucleus, the electronic part of the well-known Molecular Electrostatic Potential [46]. It is also easily seen that this expression corresponds to the electron-nucleus interaction energy V_{en} divided by Z (vide infra)

$$V_{en}[Z, N] = -Z \int \frac{\rho(\mathbf{r})}{|\mathbf{r} - \mathbf{R}|} d^3\mathbf{r} = Z\mu_{\text{al}}[Z, N]. \tag{20.2}$$

Analogously, the alchemical hardness can be written as (again considering the only surviving electronic component)

$$\begin{aligned} \eta_{\text{al}} \equiv \left(\frac{\partial^2 E}{\partial Z^2}\right)_N &= \int\int \left(\frac{\delta^2 \tilde{E}[N, v]}{\delta v(\mathbf{r})\delta v(\mathbf{r}')}\right)_N \left(\frac{\partial v(\mathbf{r})}{\partial Z}\right)_N \left(\frac{\partial v(\mathbf{r}')}{\partial Z}\right)_N d^3\mathbf{r} d^3\mathbf{r}' \\ &= \int\int \chi(\mathbf{r}, \mathbf{r}') \frac{1}{|\mathbf{r} - \mathbf{R}|} \frac{1}{|\mathbf{r}' - \mathbf{R}|} d^3\mathbf{r} d^3\mathbf{r}', \end{aligned} \tag{20.3}$$

where now the linear response function $\chi(\mathbf{r}, \mathbf{r}')$ makes its appearance

$$\chi(\mathbf{r}, \mathbf{r}') = \left(\frac{\delta^2 \tilde{E}[N, v]}{\delta v(\mathbf{r})\delta v(\mathbf{r}')}\right)_N. \tag{20.4}$$

For the evaluation of the alchemical derivatives, Eqs. (20.1) and (20.3), the finite difference approach could be followed again (vide infra). Most, if not all calculations, have been done with a Coupled Perturbed Hartree–Fock or Kohn–Sham approach as described in Refs. [17, 18]. Evaluation of the first order derivative is equivalent with the calculation of the electrostatic potential at the nucleus. Both approaches will be followed and commented below.

In the finite difference approach (for a neutral atom) the energy of an atom with $N = Z$ electrons (fixed) is evaluated at nuclear charge Z, $Z - \delta$, $Z + \delta$, etc. The first-order derivative is then obtained as

$$\mu_{\text{al}}[Z = N, N] = \frac{E[Z + \delta] - E[Z - \delta]}{2\delta}. \tag{20.5}$$

The second one as

$$\eta_{\text{al}}[Z = N, N] = \frac{E[Z + \delta] - 2E[Z] + E[Z - \delta]}{\delta^2}. \tag{20.6}$$

The third-order derivative, denoted by γ_{al}, is obtained as

$$\gamma_{\text{al}}[Z = N, N] = \frac{E[Z + 2\delta] - 2E[Z + \delta] + 2E[Z - \delta] - E[Z - 2\delta]}{2\delta^3}. \tag{20.7}$$

Above, $E[Z', N = Z]$ is abbreviated as $E[Z']$. The δ value used throughout this study was 0.1 as a compromise between numerical accuracy and extent of the perturbation. Test results obtained with a smaller δ value (0.05) invariably gave the same results in all significant figures included in the Tables.

Note that, as an internal check, the second derivatives have also been calculated as

$$\eta_{\text{al}}[Z = N, N] = \frac{\mu_{\text{al}}[Z + \delta, N] - \mu_{\text{al}}[Z - \delta, N]}{2\delta}, \tag{20.8}$$

with μ_{al} taken from Eq. (20.2), leading again to identical results up to all significant figures in Table 20.1 (see Sect. 20.3.1). To the best of our knowledge, the finite difference results with non-integer δ are the first ones communicated in the literature.

All calculations were carried out at DFT level with the standard B3LYP functional [47, 48] and the aug-cc-pCVTZ basis set [49–51], ensuring the presence of tight functions in the basis set and so a proper description of the nuclear region of particular interest in the alchemical transformations as witnessed in our earlier studies [17, 18].

The analytical approach for the derivative calculation, through a Coupled Perturbed Kohn–Sham (CPKS) approach, has been explained in detail in previous reports by some of the present authors [17, 18] where working equations in an atomic basis for the second and third order derivatives of the Hartree–Fock energy and second order for the Kohn–Sham energy were formulated starting from the basic observation that the dependence of the energy (be it Hartree–Fock or Kohn–Sham) on the nuclear charges is two-fold: Z resides in the one electron 'core' energy operator $\hat{h}$ and indirectly in the MOs or, in the present case, the AOs themselves. The CPKS equations are general, i.e. independent of the exchange-correlation operator which has to be plugged in at the moment of the evaluation of the matrix elements at stake. From the numerical examples presented, it was concluded that this technique yields well converged solutions in approaching the energy of a transmuted molecule (say BH_4^-) written as a Taylor expansion for CH_4 changing its C nuclear charge by -1 at constant number of electrons [17]. This CPKS approach was adopted in the present work for the calculation of the second order derivatives with the same exchange-correlation functional and basis as for the numerical differentiation.

The 'third way' to obtain alchemical energy of an atom with nuclear charge Z is based on the Z^{-1} expansion of the energy going back to Hylleraas [37], Layzer [52], Löwdin [53], and March and White [38]. Within this approach, the nonrelativistic atomic ground-state energy is written as

$$E[Z, N] = \sum_{j=0}^{\infty} Z^{2-j} \varepsilon_j^{[N]} = Z^2 \varepsilon_0^{[N]} + Z \varepsilon_1^{[N]} + \varepsilon_2^{[N]} + \sum_{j=3}^{\infty} Z^{2-j} \varepsilon_j^{[N]}, \tag{20.9}$$

where N is the number of electrons. When convenient, arguments are written as superscripts, e.g. $\varepsilon_j[N] \equiv \varepsilon_j^{[N]}$. The expansion coefficients ε_j are independent of Z but depend on N. Although in principle they can be calculated at any j and N, they are in general not known (for an overview see Ref. [54]).

Concentrating on alchemical derivatives using Eq. (20.9), the alchemical potential can be written as

$$\mu_{\mathrm{al}}^{[Z,N]} \equiv \left(\frac{\partial E[Z,N]}{\partial Z}\right)_N = \sum_{j=0}^{\infty}(2-j)Z^{1-j}\varepsilon_j^{[N]}$$

$$= 2Z\varepsilon_0^{[N]} + \varepsilon_1^{[N]} - \sum_{k=1}^{\infty} kZ^{-(k+1)}\varepsilon_{k+2}^{[N]} \qquad (20.10)$$

(note the absence of the $\varepsilon_2^{[N]}$ contribution), and the alchemical hardness as

$$\eta_{\mathrm{al}}^{[Z,N]} \equiv \left(\frac{\partial^2 E[Z,N]}{\partial Z^2}\right)_N = \sum_{j=0}^{\infty}(1-j)(2-j)Z^{-j}\varepsilon_j^{[N]}$$

$$= 2\varepsilon_0^{[N]} + \sum_{k=1}^{\infty} k(k+1)Z^{-(k+2)}\varepsilon_{k+2}^{[N]} \qquad (20.11)$$

(note absence of the $\varepsilon_1^{[N]}$ and $\varepsilon_2^{[N]}$ contributions).

The third order derivative, the alchemical hyperhardness γ can then be written as

$$\gamma_{\mathrm{al}}^{[Z,N]} \equiv \left(\frac{\partial^3 E[Z,N]}{\partial Z^3}\right)_N = -\sum_{j=3}^{\infty} j(j-1)(j-2)Z^{-(j+1)}\varepsilon_j^{[N]}$$

$$= -\sum_{k=1}^{\infty} k(k+1)(k+2)Z^{-(k+3)}\varepsilon_{k+2}^{[N]}. \qquad (20.12)$$

These expressions indicate that once the coefficients $\varepsilon_k^{[N]}$ are known for a given N, the derivatives μ_{al}, η_{al}, and γ_{al} can be obtained for all Z.

March and White [38] expanded the $\varepsilon_k^{[N]}$ coefficients in a power series of $N^{-1/3}$, starting from the conjecture of an asymptotic behaviour of the $\varepsilon_k^{[N]}$, later on confirmed by Tal and Levy [54]:

$$\varepsilon_k^{[N]} = a_k^{[N]}N^{k+1/3} + b_k^{[N]}N^k + c_k^{[N]}N^{k-1/3} + \dots . \qquad (20.13)$$

When inserted into Eq. (20.9), it leads to the following equation which was studied in detail by March and Parrr [39]:

$$E[Z,N] = Z^{7/3}f_1(N/Z) + Z^2 f_2(N/Z) + Z^{5/3}f_3(N/Z) + \dots . \qquad (20.14)$$

The expressions for f_1, f_2, f_3, not explicitly given in their paper, can be retrieved with some algebra and are

$$f_1(N/Z) = a_0^{[N]}(N/Z)^{1/3} + a_1^{[N]}(N/Z)^{4/3} + a_2^{[N]}(N/Z)^{7/3} + \dots , \qquad (20.15a)$$

$$f_2(N/Z) = b_0^{[N]}(N/Z)^0 + b_1^{[N]}(N/Z)^1 + b_2^{[N]}(N/Z)^2 + \dots , \qquad (20.15b)$$

$$f_3(N/Z) = c_0^{[N]}(N/Z)^{-1/3} + c_1^{[N]}(N/Z)^{2/3} + c_2^{[N]}(N/Z)^{5/3} + \dots , \qquad (20.15c)$$

where the regularities in the N/Z exponents are clearly discerned.

Knowledge of the $a_n^{[N]}$, $b_n^{[N]}$, $c_n^{[N]}$ $(n = 1, 2, 3, \ldots)$ yields the $f_k(N/Z)$ and via Eq. (20.14) enables the evaluation of $E[Z, N]$ and its alchemical derivatives. Work along these lines has been done by Gázquez and Vela [55–57] to study the behaviour of the chemical potential of neutral atoms [55]. However, our calculations are not based on this expression because evaluation of $a_n^{[N]}$, $b_n^{[N]}$, $c_n^{[N]}$ is complicated [58–60] and ambiguous to some extent [55].

Asymptotic expansions, like in Eq. (20.9), are often divergent. Nevertheless, when truncated after the Kth term, they may reasonably approximate the expanded function as

$$E[Z, N] = \sum_{j=0}^{\infty} Z^{2-j} \varepsilon_j^{[N]} \approx \sum_{j=0}^{K-1} Z^{2-j} \tilde{\varepsilon}_j^{[N;K]} \equiv \tilde{E}[Z, N; K]. \tag{20.16}$$

The value K_{opt} of K providing the best approximation is specific for each expanded function (here, at fixed N, for the energy, Eq. (20.9), and its derivatives, Eqs. (20.10)–(20.12).

Our proposal here is to calculate $\tilde{\varepsilon}_0$, $\tilde{\varepsilon}_1$, $\ldots$ $\tilde{\varepsilon}_{K-1}$ from the energy expansion $\tilde{E}[Z, N; K]$. The best K and corresponding $\tilde{\varepsilon}_j^{[N;K]}$ are then determined by the requirement that the absolute values of consecutive terms of the expansion are decreasing. So the best value of K is the largest integer K for which the relations

$$\left| Z^2 \tilde{\varepsilon}_0^{[N;K]} \right| > \left| Z \tilde{\varepsilon}_1^{[N;K]} \right| > \left| \tilde{\varepsilon}_2^{[N;K]} \right| > \ldots > \left| Z^{3-K} \tilde{\varepsilon}_{K-1}^{[N;K]} \right| \tag{20.17}$$

are fulfilled. It should be noted that when Eqs. (20.17) are satisfied for $Z = N$, they are also satisfied for $Z > N$.

For the chosen N and K, the values of $\tilde{\varepsilon}_j^{[N;K]}$, $j = 0, 1, \ldots K-1$, in Eq. (20.16) are determined from the given energies $E_{\text{DFT}}[Z, N]$ (i.e. evaluated by means of DFT) of the neutral atom, $Z = N$, and isoelectronic cations, $Z = N+1, N+2, \ldots N+K-1$. It is done by equating these energies to the truncated approximation, Eq. (20.16), i.e., by solving the set of K equations

$$\tilde{E}[Z, N; K] = E_{\text{DFT}}[Z, N], \quad \text{for } Z = N, N+1, N+2, \ldots N+K-1. \tag{20.18}$$

This set of K linear equations written in the matrix form is

$$\begin{pmatrix} N^2 & N & 1 \ldots & N^{3-K} \\ (N+1)^2 & (N+1) & 1 \ldots & (N+1)^{3-K} \\ (N+2)^2 & (N+2) & 1 \ldots & (N+2)^{3-K} \\ \vdots & \vdots & \vdots \ddots & \vdots \\ (N+K-1)^2 & (N+K-1) & 1 \ldots & (N+K-1)^{3-K} \end{pmatrix} \begin{pmatrix} \tilde{\varepsilon}_0^{[N;K]} \\ \tilde{\varepsilon}_1^{[N;K]} \\ \tilde{\varepsilon}_2^{[N;K]} \\ \vdots \\ \tilde{\varepsilon}_{K-1}^{[N;K]} \end{pmatrix} = \begin{pmatrix} E_{\text{DFT}}^{[N,N]} \\ E_{\text{DFT}}^{[N+1,N]} \\ E_{\text{DFT}}^{[N+2,N]} \\ \vdots \\ E_{\text{DFT}}^{[N+K-1,N]} \end{pmatrix}. \tag{20.19}$$

The K value and the corresponding coefficients $\tilde{\varepsilon}_j^{[N;K]}$ were determined for all N values considered, $N = 1 - 18$, all calculations of $E_{\text{DFT}}[Z, N]$ were again done with the B3LYP functional and the aug-cc-pCVTZ basis for reasons of internal consistency.

Note that, as will be seen explicitly in the detailed examples in Sect. 20.3.2, once the values of $\tilde{\varepsilon}_j^{[N;K]}$ for a given, say neutral atom ($Z = N$), have been determined, the alchemical derivatives become available not only for that atom but also for all cations of the isoelectronic series $Z = N + L$, with $L = 1, 2, \ldots K - 1$. In principle, also anions could be considered but in view of their instability only the case of a mono-anion will be considered, though not included in the error analysis.

The final approach, enabling us in an easy way to walk through the Periodic Table, is based on the data in the above mentioned NIST website [41] on the total energies for the ground state neutral configurations and their mono-cations for all atoms with $Z \leq 92$, with three variants of the local density approximation (LDA) [41]. The NIST website tabulates, for both cases and for all LDA versions considered, always the total energy, $E_{\mathrm{NIST}}^{[Z,N]}$, and the electron-nucleus interaction energy, $V_{en,\,\mathrm{NIST}}^{[Z,N]}$, which is equal to the alchemical potential multiplied by Z, Eq. (20.2).

This offers the possibility, for a fixed value of N, $N \leq 91$, to write $K = 4$ linear equations in terms of 4 unknowns $\tilde{\varepsilon}_0^{[N;4]}$, $\tilde{\varepsilon}_1^{[N;4]}$, $\tilde{\varepsilon}_2^{[N;4]}$, $\tilde{\varepsilon}_3^{[N;4]}$, the parameters of the truncated Z^{-1} expansion, namely

$$\begin{cases} \tilde{E}[Z, N; 4] & = E_{\mathrm{NIST}}^{[Z,N]}, \\ \tilde{\mu}_{\mathrm{al}}[Z, N; 4] & = V_{en,\,\mathrm{NIST}}^{[Z,N]}/Z, \quad \text{for } Z = N, N+1. \end{cases} \tag{20.20}$$

Here, $\tilde{\mu}_{\mathrm{al}}$ denotes truncated μ_{al}, Eq. (20.10). This set of 4 linear equations written in the matrix form is given in Appendix 20.A, Eqs. (20.29). The analytical solution of this system of equations in given there too, Eqs. (20.30).

For Part 1, use was made of the BRABO program [61, 62] developed by one of the present authors (CVA), for Part 2 a locally modified version of the GAMESS package [63] was used.

20.3 Results and Discussion

20.3.1 Finite Difference and Analytical Derivatives

In Table 20.1 an overall view of the alchemical derivatives with the three methods described above is given for the atoms with $Z = 1 - 18$. At this moment, we only concentrate on the finite difference and the CPKS method, the latter being used only for closed shell systems.

Concerning the alchemical potential obtained with the numerical differentiation method (the quantity not being calculated as such in CPKS), the results are in line with the B3LYP/6-311G (3df) results by Politzer in his review [46] which can be obtained by dividing his V_{en} values by Z (the small differences can be ascribed to the difference in basis). The values show a monotonous decrease (more negative

Table 20.1 Comparison of numerical, 'Z^{-1}', and analytical (CPKS) results for the first and second alchemical derivatives for $Z = 1 - 18$ at $N = Z$. For CPKS only closed shell atoms are considered. All values are obtained at the B3LYP(G)/aug-cc-pCVTZ level. All data in a.u

	Alchemical derivatives				
	Numerical method		Z^{-1} method		CP KS
	μ_{al}	η_{al}	μ_{al}	η_{al}	η_{al}
H	−0.999	−0.995	−0.625	−1.000	
He	−3.371	−1.964	−3.376	−1.984	−1.945
Li	−5.719	−2.203	−5.715	−2.268	
Be	−8.420	−2.480	−8.419	−2.506	−2.479
B	−11.391	−2.746	−11.390	−2.769	
C	−14.697	−3.001	−14.698	−3.018	
N	−18.337	−3.254	−18.329	−3.309	
O	−22.257	−3.511	−22.266	−3.503	
F	−26.511	−3.765	−26.487	−3.854	
Ne	−31.098	−4.020	−31.131	−4.005	−4.020
Na	−35.418	−3.745	−35.433	−4.181	
Mg	−39.925	−4.098	−39.937	−4.278	−4.103
Al	−44.499	−4.079	−44.515	−4.397	
Si	−49.243	−4.228	−49.258	−4.508	
P	−54.146	−4.363	−54.161	−4.614	
S	−59.181	−4.490	−59.200	−4.710	
Cl	−64.374	−4.618	−64.396	−4.816	
Ar	−69.724	−4.743			−4.743

value) upon increasing Z. This behaviour should be discerned from the corresponding derivative where a pattern of periodicity shows up in the electronegativity which has been identified with this derivative. An interpretation of the monotonous decrease could be found as follows. The change in the alchemical potential when passing from one neutral atom to another, implying $dN = dZ$, can be written as

$$\frac{d\mu_{al}}{dZ} = \left(\frac{\partial \mu_{al}[Z,N]}{\partial Z}\right)_N\Bigg|_{N=Z} + \left(\frac{\partial \mu_{al}[Z,N]}{\partial N}\right)_Z\Bigg|_{N=Z}$$
$$= \int\int \chi(\mathbf{r},\mathbf{r}')\frac{1}{r}\frac{1}{r'}d^3\mathbf{r}d^3\mathbf{r}' - \int \frac{1}{r}\left(\frac{\partial \rho(\mathbf{r})}{\partial N}\right)_Z d^3\mathbf{r}. \tag{20.21}$$

The first term is simply the alchemical hardness calculated for the neutral atom, Eq. (20.3). As the linear response function, $\chi(\mathbf{r},\mathbf{r}')$, is a seminegative definite operator due to the concavity in $v(\mathbf{r})$ of the $E[v, N]$ functional [23], the atomic alchemical hardness, Eq. (20.3), is a non positive number. The $(\partial\rho(\mathbf{r})/\partial N)_Z$ derivative in the atomic case can be identified as the Fukui function, $f(\mathbf{r}) = (\partial\rho(\mathbf{r}; N, v)/\partial N)_v$ [64–66], and the second term is the Fukui potential at the nuclear position [67, 68]. The electrostatic potential of the Fukui function does not show negative value, it exhibits a maximum close to the nuclear position and decays with the distance [68, 69]. Concluding, the change in the alchemical potential for a neutral atom, i.e. when passing from one neutral atom to another, is negative upon increasing Z.

In order to get a more quantitative idea on the this decrease, we use the Z^{-1} expansion for $E[Z, N]$ and put $N = Z$ for a neutral atom case. Equation (20.14) then gives

$$E[Z, Z] = Z^{7/3} f_1(1) + Z^2 f_2(1) + Z^{5/3} f_3(1) + \dots \tag{20.22}$$

In a Thomas–Fermi based Ansatz, March and Parr were able to write [39]

$$V_{ee}[Z, Z] = -\frac{1}{3} Z^{7/3} f_1(1) + \frac{1}{3} Z^{5/3} f_3(1) + \dots, \tag{20.23}$$

where the term in Z^2 vanishes. Using the virial theorem and Eq. (20.2), one gets for $\mu_{\rm al}$

$$\mu_{\rm al}[Z, Z] = \frac{7}{3} Z^{4/3} f_1(1) + 2Z f_2(1) + \frac{5}{3} Z^{2/3} f_3(1) + \dots. \tag{20.24}$$

Based on analytical and numerical results by March and Parr [39], it turns out that $|f_1(1)| > |f_2(1)| > |f_3(1)|$, so that in a first approximation, for not too small Z, Eq. (20.24) can be simplified by retaining only the leading term

$$\mu_{\rm al}[Z, Z] \approx \frac{7}{3} f_1(1) Z^{4/3}. \tag{20.25}$$

This results already shows that $|\mu_{\rm al}|$ is expected to increase as $Z^{4/3}$. Inserting the exact Thomas-Fermi result for $f_1(1)$ (-0.7687) one obtains more precisely $\mu_{\rm al} \approx -1.793 Z^{4/3}$. If instead use is made of the Hartree–Fock results for the noble gas atoms (Table 6.1 in Ref. [19]) via a fitting procedure with $Z^{7/3}$ one obtains $E[Z, Z] \approx -(0.62 \pm 0.05) Z^{7/3}$, yielding $\mu_{\rm al}[Z, Z] \approx -1.45 Z^{4/3}$. When dividing the numerical $\mu_{\rm al}$ values in Table 20.1 by $Z^{4/3}$ for $Z = 3 - 18$, one arrives at an average value of -1.42, decreasing from -1.45 to -1.48 for the large Z sequence. These findings support a $Z^{4/3}$ dependence for $\mu_{\rm al}$ and so for the electrostatic potential at the origin with a proportionality constant of -1.5 except for two light elements.

Passing to the second derivative no results are available in the literature to compare with. It is pleasing to note that for the closed shell systems the results are nearly identical with the two methods, giving confidence in our computational approach. With the exception of Na and Mg at the beginning of the second row, again a uniform decrease (more negative value) is observed, as opposed again to a periodic pattern in the chemical hardness. Note that overall for the $Z = 6 - 18$ values the alchemical hardness is one order of magnitude smaller than the alchemical potential so that even for as large a change in nuclear charge as 1 the second order term in the energy expansion is one order of magnitude smaller than the first one. The same trend persists when going to the higher order term, evaluated only in the finite difference approach (not shown). Its values are not trustworthy yet due to the extreme demand on the basis set close to the nucleus. We therefore did some exploratory calculations with the aug-cc-pCVTZ basis set (Appendix 20.B), where the results show greater diversity between the atoms, are systematically one order of magnitude smaller than the second derivatives and, most interesting, begin to display some periodicity. Periodicity shows up in the sense that for the p-block elements a uniform decrease from left to right is observed in each row, the values for an atom of the lower row being always larger than that of its congener of the higher row suggesting that γ decreases from left to right in a row and increases when going down in a column. The s-block behaviour is less transparent. In other words, one has to eliminate the influence of the two large terms and work at a refined level to discern most probably periodicity in alchemical derivatives. Further research is at stake in this direction.

A final remark concerns the comparison with Balawender's 'atoms in molecules' values of η_{al} for some selected atoms studies on molecular alchemical transformations [18, 19]. Based on some values from Ref. [18] (C in pyrene: −3.192, C in CO: −3.189, N in N_2: −3.314, and O in CO^-: −3.544, values calculated with the same basis set as for the isolated atoms), and the comparison with isolated atom values, Table 20.1, we see that the order of magnitude of the alchemical hardness for the atoms-in-molecules and isolated atoms considered is the same (in fact the values are quite close and the influence of the molecular environment is rather weak) and also their sequence is the same, indicating that the isolated atom values can indeed be used in a preliminary exploration of CCS. For more accurate explorations in CCS a correction term for both μ_{al} and η_{al} (one for each atom type) could be introduced reminiscent of the corrections on the electronic chemical potential or electronegativity and hardness as proposed and fitted by Mortier in his still widely used (cf. force field programs) Electronegativity Equalization Method [70–72] to account for the entrance of an atom in (whatever type of) molecule. For efficient screening of Chemical Compound Space, the isolated atom values may yield first guesses in a very simple way and at very low cost. In this sense it is important to go further down in the Periodic Table, for which the prelude is testing the Z^{-1} approach.

20.3.2 The Z^{-1} Approach

We first demonstrate obtaining of the optimal K and $\tilde{\varepsilon}_j^{[N;K]}$, $j = 0, 1, \ldots K-1$, values for some concrete, neutral atom examples, taking Mg as the first example for which both finite difference and analytical derivatives are available for comparison (cf. Table 20.1). In Table 20.2 we present the different terms in the expansion (Eq. (20.16) with $Z = N = 12$). The criterion of decrease in magnitude of successive terms, Eq. (20.17), is obeyed for $K = 3, 4, 5$ but no longer for $K = 6$, where the third term of the expansion is smaller than the fourth one. Hence, we conclude that in

Table 20.2 Values of the terms in the $\tilde{E}[12, 12, K]$ expansion, Eq. (20.16), for Mg, $N = 12$, as a function of K (see text). Here, $\tilde{\varepsilon}_j$ denotes $\tilde{\varepsilon}_j^{[12,K]}$. The fitting error, Eq. (20.26), is given in the final column. All data in a.u

K	$\lvert\tilde{\varepsilon}_0 N^2\rvert$	$\lvert\tilde{\varepsilon}_1 N\rvert$	$\lvert\tilde{\varepsilon}_2\rvert$	$\lvert\tilde{\varepsilon}_3 N^{-1}\rvert$	$\lvert\tilde{\varepsilon}_4 N^{-2}\rvert$	$\lvert\tilde{\varepsilon}_5 N^{-3}\rvert$	$\lvert\tilde{\varepsilon}_6 N^{-4}\rvert$	$\lvert\tilde{\varepsilon}_7 N^{-5}\rvert$	$\Delta_{en}^{[12,K]}$ (‰)
3	306.5	133.6	27.2						0.48
4	301.8	118.4	10.8	5.9					0.35
5	**306.2**	**138.4**	**44.4**	**19.2**	**7.0**				**0.34**
6	302.1	114.3	11.6	45.9	30.6	8.7			0.34
7	303.2	122.1	11.8	8.4	3.1	7.4	3.2		0.35
8	204.8	607.5	2229.2	3652.4	3293.8	1548.9	278.8	11.7	2.82

this case the optimal K value is 5. With this value, trustworthy $\tilde{\varepsilon}_j^{[N=12;5]}$ values can be obtained. For the sake of comparison, we also tabulate the results for K values lower than the optimal K, starting at $K = 3$ ($K = 2$ would not yield an $\tilde{\varepsilon}_2^{[N=12;K]}$ value essential for $\tilde{E}[Z, N; K]$). The $K > 5$ values will be commented below. The $\tilde{\varepsilon}_0^{[N=12;K]}$ values for the different K values up to 5 are very close to each other, their average (−2.117) being almost identical to the exact 'hydrogenic atom' solution of −2.111 [73]. The other $\tilde{\varepsilon}_j^{[N=12;K]}$ values show larger fluctuations in K with increasing j, but as will be seen below the overall quality of the derivatives increases up to $K = 5$. The values for $K > 5$, given for the sake of comparison, indicate very strong fluctuations.

Since $V_{en,\,\mathrm{DFT}}^{[Z,N]}$ is available from calculated $E_{\mathrm{DFT}}^{[Z,N]}$ and should be equal to $Z\mu_{\mathrm{al}}^{[Z,N]}$, Eq. (20.2), the following error function

$$\Delta_{en}^{[N;K]} = \max_{Z\in\{N,N+1,\ldots N+K-1\}} \left| \frac{\tilde{V}_{en}^{[Z,N;K]} - V_{en,\,\mathrm{DFT}}^{[Z,N]}}{V_{en,\,\mathrm{DFT}}^{[Z,N]}} \right| \tag{20.26}$$

where $\tilde{V}_{en}^{[Z,N,K]} = Z\tilde{\mu}_{al}^{[Z,N,K]}$, Eq. (20.27a), characterizes the accuracy of the Z^{-1} expansion for each K. It is seen in the last column in Table 20.2 that the error decreases from $K = 3$ to the optimal value $K = 5$, and although the errors for $K = 6$ and 7 are only slightly bigger than for $K = 5$, it explodes at $K = 8$. This error analysis confirms $K = 5$ to be optimal, as we found at the beginning of the present Section.

Table 20.3 Alchemical potentials and hardness of the isoelectronic series $N = 12$ with Z varying from 11 to 19 for $K_{\mathrm{opt}} = 5$. The Na anion is given for comparison (see text)

	Na^-	Mg	Al^+	Si^{2+}	P^{3+}	S^{4+}	Cl^{5+}	Ar^{6+}	Ca^{8+}
$\tilde{\mu}_{\mathrm{al}}$	−35.640	−39.937	−44.202	−48.450	−52.689	−56.924	−61.158	−65.392	−73.864
$\Delta\tilde{\mu}_{\mathrm{al}}$	−4.298	−4.265	−4.248	−4.239	−4.235	−4.234	−4.234	−4.236	
$\tilde{\eta}_{\mathrm{al}}$	−4.322	−4.278	−4.255	−4.242	−4.237	−4.234	−4.234	−4.235	−4.238

Table 20.4 Values of the terms in the $\tilde{E}[6, 6, K]$ expansion, Eq. (20.16), for C, as a function of K (see text). Here, $\tilde{\varepsilon}_j$ denotes $\tilde{\varepsilon}_j^{[6,K]}$. The fitting error, Eq. (20.26) is given in the final column. All data in a.u

K	$\lvert\tilde{\varepsilon}_0 N^2\rvert$	$\lvert\tilde{\varepsilon}_1 N\rvert$	$\lvert\tilde{\varepsilon}_2\rvert$	$\lvert\tilde{\varepsilon}_3 N^{-1}\rvert$	$\lvert\tilde{\varepsilon}_4 N^{-2}\rvert$	$\lvert\tilde{\varepsilon}_5 N^{-3}\rvert$	$\lvert\tilde{\varepsilon}_6 N^{-4}\rvert$	$\Delta_{en}^{[6,K]}$ (‰)
3	54.1	20.1	3.8					0.23
4	53.9	19.2	2.8	0.4				0.07
5	**54.0**	**19.7**	**3.7**	**0.3**	**0.2**			**0.04**
6	58.5	49.8	83.3	104.8	68.2	17.6		0.99
7	48.0	855.9	3098.3	5807.1	6058.6	3339.5	759.6	19.63

Having $\tilde{\varepsilon}_j^{[N=12;K]}$ values, the alchemical derivatives can be evaluated. Namely

$$\tilde{\mu}_{\text{al}}^{[Z,N;K]} = \sum_{j=0}^{K-1} (2-j) Z^{1-j} \tilde{\varepsilon}_j^{[N;K]} = 2\tilde{\varepsilon}_0^{[N;K]} Z + \tilde{\varepsilon}_1^{[N;K]} + \ldots + (3-K)\tilde{\varepsilon}_{K-1}^{[N;K]} Z^{2-K}, \tag{20.27a}$$

$$\tilde{\eta}_{\text{al}}^{[Z,N;K]} = \sum_{j=0}^{K-1} (2-j)(1-j) Z^{-j} \tilde{\varepsilon}_j^{[N;K]} = 2\tilde{\varepsilon}_0^{[N;K]} + \ldots + (3-K)(2-K)\tilde{\varepsilon}_{K-1}^{[N;K]} Z^{1-K}. \tag{20.27b}$$

In Table 20.3, we give their values for optimal $K = 5$. The middle line there shows $\Delta\tilde{\mu}_{\text{al}} = \left(\tilde{\mu}_{\text{al}}^{[Z+\Delta Z,N;K]} - \tilde{\mu}_{\text{al}}^{[Z,N;K]}\right)/\Delta Z$ (here $\Delta Z = 1$, except for the Ar cation, for which $\Delta Z = 2$). It can be viewed as the numerical first derivative of μ_{al} with respect to Z, i.e. η_{al}, and consistent with the last line, $\tilde{\eta}_{\text{al}}$. It should be mentioned that the values of $\tilde{\mu}_{\text{al}}^{[Z,N;K]}$ and $\tilde{\eta}_{\text{al}}^{[Z,N;K]}$ obtained by us for the other K (not shown) are almost the same as for $K = 5$, despite a significant dependence of $\tilde{\varepsilon}_j^{[N=12;K]}$ on K. On the other hand, the weak dependence of $\tilde{\eta}_{\text{al}}^{[Z,N;K]}$ on Z is due to the fact that its leading term, $\tilde{\varepsilon}_0$, Eq. (20.27b), is independent of Z.

Note indeed that with the $\tilde{\varepsilon}_j^{[N=12;K]}$ values not only the alchemical derivatives for the neutral system ($Z = N, N$) can be generated, but also those of all its isoelectronic congeners ($Z = N + 1, N$), ($Z = N + 2, N$), etc. Also, the ($Z = N - 1, N$) anion is considered in Table 20.3. More highly charged anions are expected to be unstable. Finally, note that the optimal value for the alchemical potential of neutral Mg (-39.937 a.u.) is close to the one obtained with the numerical method (-39.925 a.u.)

As a second and the last example we consider the carbon atom ($Z = 6$, $N = 6$), an open-shell case where no comparison can be made with analytical derivatives. Table 20.4 indicates that the optimum K value for the energy expansion is also 5 (this is a coincidence with the magnesium case, in the fluorine case for example one gets $K_{\text{opt}} = 4$). The error analysis (last column in Table 20.4) now clearly shows that at higher values than the optimal K, the results start worsening considerably

Table 20.5 Alchemical potential and hardness of the isoelectronic series $N = 6$ with Z varying from 5 to 13 for $K_{\mathrm{opt}} = 5$. The B anion is given for comparison (see text)

	B^-	C	N^+	O^{2+}	F^{3+}	Ne^{4+}	Na^{5+}	Mg^{6+}	Al^{7+}
$\tilde{\mu}_{\mathrm{al}}$	−11.669	−14.698	−17.710	−20.716	−23.719	−26.720	−29.720	−32.720	−35.720
$\Delta\tilde{\mu}_{\mathrm{al}}$	−3.029	−3.012	−3.006	−3.003	−3.001	−3.000	−3.000	−3.000	−3.000
$\tilde{\eta}_{\mathrm{al}}$	−3.044	−3.018	−3.008	−3.004	−3.002	−3.001	−3.000	−3.000	−3.000

at $K = 6$. The values of the alchemical derivatives are shown in Table 20.5. The hardness value for C (−3.018 a.u.) being close to the numerical value (−3.001 a.u.). Note again that the average $\tilde{\varepsilon}_0^{[N=6;K]} = -1.501$ for $K = 3 - 5$ is almost identical to the exact 'hydrogenic atom' solution (−1.500) [73].

In this way, the Z^{-1} method was exploited to evaluate all alchemical potential and hardness values for the neutral atoms ($Z = 1 - 18$), Table 20.1, and their isoelectronic congeners (not shown) by optimizing K and extracting all $\tilde{\varepsilon}_0, \tilde{\varepsilon}_1, \ldots \tilde{\varepsilon}_{K-1}$ for each N value.

Going back to Table 20.1, concentrating on the neutral atoms and comparing with the numerical differentiation values for μ_{al}, an average difference of 0.010 was found for the first-row atoms and 0.016 for second row atoms, the corresponding deviations for the alchemical hardness being 0.037 and 0.269, respectively. The latter value includes an almost systematic lowering of the Z^{-1} method value of −0.2 as compared to the numerical one (the two series show an excellent correlation of $R^2 = 1.00$ for the alchemical potential, and of $R^2 = 0.991$ for the alchemical hardness).

To conclude this section, we depict the difference between the values obtained with $K = 3$, the minimal value considered, and those with optimal K for $\tilde{\varepsilon}_0, \tilde{\varepsilon}_1, \tilde{\varepsilon}_2$ coefficients in Fig. 20.1, for $\tilde{\mu}_{\mathrm{al}}$ and $\tilde{\eta}_{\mathrm{al}}$ in Fig. 20.2. It is seen that despite stronger differences $\left(\tilde{\varepsilon}_j^{[N;K]} - \tilde{\varepsilon}_j^{[N;3]}\right)$ with increasing j (see Fig. 20.1), the two choices of K yield only very small difference in $\tilde{\mu}_{\mathrm{al}}$, almost negligible in $\tilde{\eta}_{\mathrm{al}}$, see Fig. 20.2 (a striking example being that of the fluorine atom). This case indicates that by adding more terms in the series expansion (retaining convergence) individual fluctuations in the $\tilde{\varepsilon}_j$ are compensated (up to the optimal K value). The overall result for $\tilde{\eta}_{\mathrm{al}}$ is already correct at $K = 3$.

We finally rescaled the energy and its derivatives by dividing $\tilde{E}$ by Z^2, $\tilde{\mu}_{\mathrm{al}}$ by $2Z$, and $\tilde{\eta}_{\mathrm{al}}$ by 2, yielding then in all cases $\tilde{\varepsilon}_0^{[N;K_{\mathrm{opt}}]}$ as the first term in the corresponding series expansion

$$\frac{1}{Z^2}\tilde{E}^{[Z,N;K=K_{\mathrm{opt}}]} = \tilde{\varepsilon}_0^{[N;K_{\mathrm{opt}}]} + \frac{1}{Z}\tilde{\varepsilon}_1^{[N;K_{\mathrm{opt}}]} + \frac{1}{Z^2}\tilde{\varepsilon}_2^{[N;K_{\mathrm{opt}}]} + \ldots + Z^{1-K_{\mathrm{opt}}}\tilde{\varepsilon}_{K_{\mathrm{opt}}-1}^{[N;K_{\mathrm{opt}}]}, \tag{20.28a}$$

$$\frac{1}{2Z}\tilde{\mu}_{\mathrm{al}}^{[Z,N;K=K_{\mathrm{opt}}]} = \tilde{\varepsilon}_0^{[N;K_{\mathrm{opt}}]} + \frac{1}{2Z}\tilde{\varepsilon}_1^{[N;K_{\mathrm{opt}}]} + \ldots + \frac{1}{2}(3 - K_{\mathrm{opt}})Z^{1-K_{\mathrm{opt}}}\tilde{\varepsilon}_{K_{\mathrm{opt}}-1}^{[N;K_{\mathrm{opt}}]}, \tag{20.28b}$$

$$\frac{1}{2}\tilde{\eta}_{\mathrm{al}}^{[Z,N;K=K_{\mathrm{opt}}]} = \tilde{\varepsilon}_0^{[N;K_{\mathrm{opt}}]} + \ldots + \ldots + \frac{1}{2}(3 - K_{\mathrm{opt}})(2 - K_{\mathrm{opt}})Z^{1-K_{\mathrm{opt}}}\tilde{\varepsilon}_{K_{\mathrm{opt}}-1}^{[N;K_{\mathrm{opt}}]}. \tag{20.28c}$$

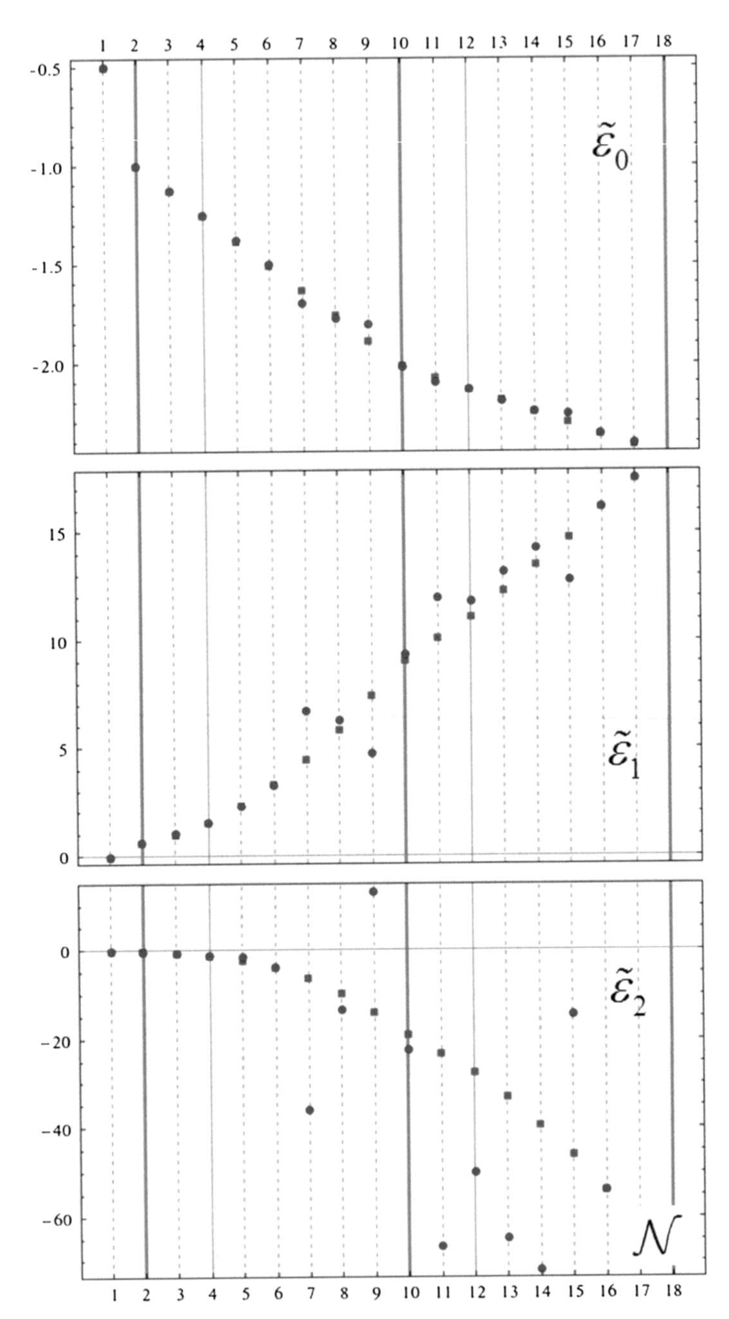

Fig. 20.1 The series expansion coefficients $\tilde{\varepsilon}_0$, $\tilde{\varepsilon}_1$, $\tilde{\varepsilon}_2$ for H to Cl, both for $K = 3$ (red dots) and $K = K_{\text{opt}}$ (blue dots). Values at B3LYP(G)/aug-cc-pCVTZ level

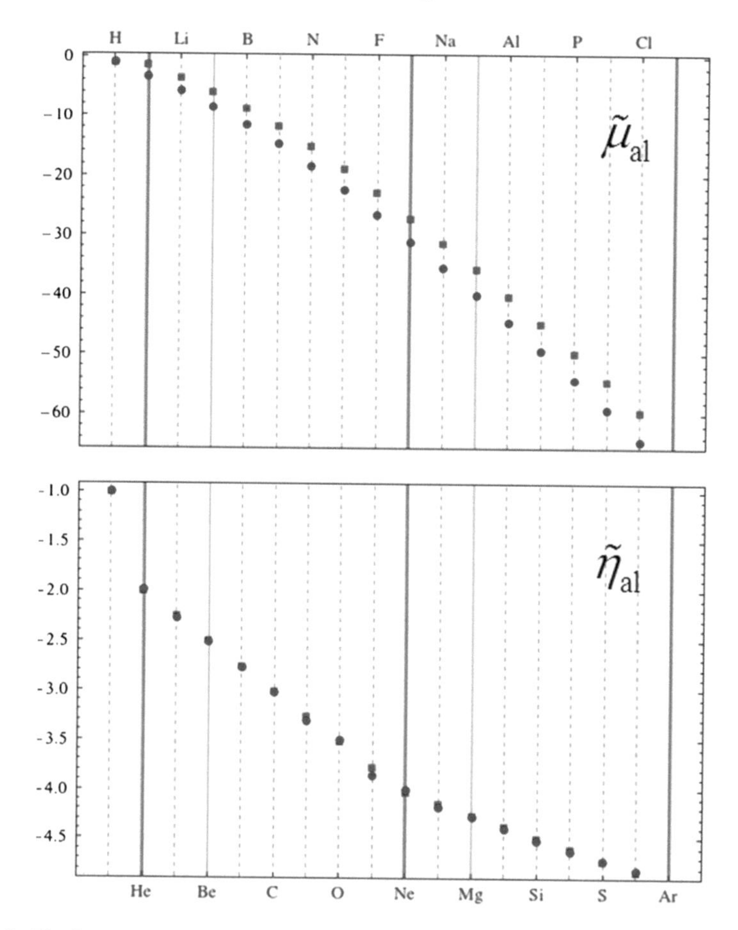

Fig. 20.2 The first and second Z derivative for H to Cl, both for $K = 3$ (red dots) and $K = K_{\text{opt}}$ (blue dots). Values at B3LYP(G)/aug-cc-pCVTZ level

In Fig. 20.3 the three curves illustrate the correction on $\tilde{\varepsilon}_0$, the only low order term in $\tilde{\eta}_{\text{al}}$, by $\tilde{\varepsilon}_1$ (in $\tilde{\mu}_{\text{al}}$), by $\tilde{\varepsilon}_1$ and $\tilde{\varepsilon}_2$ (in $\tilde{E}$), before the omnipresent higher order terms ($j \geq 3$) appear. As $\tilde{\varepsilon}_1$ is positive and $\tilde{\varepsilon}_2$ negative, the orders of magnitude of $\tilde{\varepsilon}_1$ and $\tilde{\varepsilon}_2$ and the $\frac{1}{2}$ coefficient for $\tilde{\varepsilon}_1$ explain the increasing (less negative) values of the scaled $\tilde{\mu}_{\text{al}}$ and $\tilde{E}$ for a given Z. It is interesting that $\tilde{\eta}_{\text{al}}^{[Z,Z;K_{\text{opt}}]}$ shows a linear dependence on Z with different inclinations in two periods.

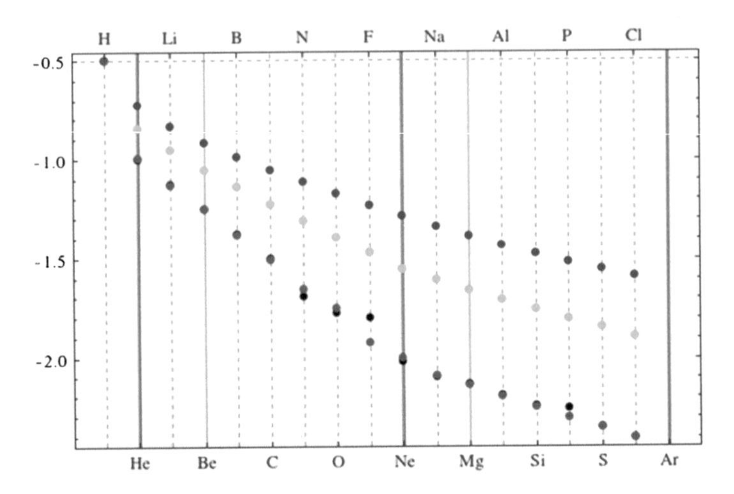

Fig. 20.3 Behaviour of $\tilde{\varepsilon}_0[N, K_{\text{opt}}]$ (black dots), scaled $\tilde{E}$ (blue dots), scaled $\tilde{\mu}_{\text{al}}$ (green dots), and scaled $\tilde{\eta}_{\text{al}}$ (red dots) as a function of $Z = N$

20.3.3 A Walk Through the Periodic Table Using the NIST Data

The system of linear equations, Eq. (20.29) in Appendix 20.A, was solved analytically for every N from 1 to 91, yielding $\tilde{\varepsilon}_0$, $\tilde{\varepsilon}_1$, $\tilde{\varepsilon}_2$, $\tilde{\varepsilon}_3$ for each N. This enabled us to evaluate $\tilde{\mu}_{\text{al}}$ and $\tilde{\eta}_{\text{al}}$ for all neutral atoms from H to Pa. This exercise was performed in the local density approximation (LDA), the local spin density approximation (LSDA), and the relativistic local density approximation (RLDA).

The truncated Z^{-1} expansion of the energy is the basis of the method applied here (a) using E and V_{en} data as input, Eq. (20.20), and the previous method (b) using only E data, Eq. (20.18). To see if (a) and (b) produce similar results, we performed (a) calculations for $N = 1 - 18$ using as input E_{DFT} and $V_{en,\text{DFT}}$ from the same run as in (b) displayed in Sect. 20.3.2. We find the result of method (a) to be very close to method (b) ($R^2 = 0.999$), e.g. for Mg, where $K_{\text{opt}} = 5$ and $\tilde{\eta}_{\text{al}} = -4.278$, the result of method (a) is $\tilde{\eta}_{\text{al}} = -4.265$.

The results for the alchemical hardness on which we will concentrate in view of their more compact range as compared to the alchemical potential (cf. Table 20.1) are depicted under the form of Periodic Tables in Figs. 20.4, 20.5 and 20.6. Starting with LDA, a comparison for H to Cl with the finite difference and, where available, the CPKS results, shows that the values show the correct order of magnitude and sequence for the first-row elements. For the second row elements a systematic error of about 0.03 a.u. should be noticed but again the sequence and the differences within the series are correct. This difference persists in LSDA. In view of the overall agreement in order of magnitude and sequences for the $Z = 1 - 18$ atoms and the two more elaborated approaches in Table 20.1, taking also in account the approximate nature of the LDA ansatz, the NIST data based approach does seem to be trustworthy

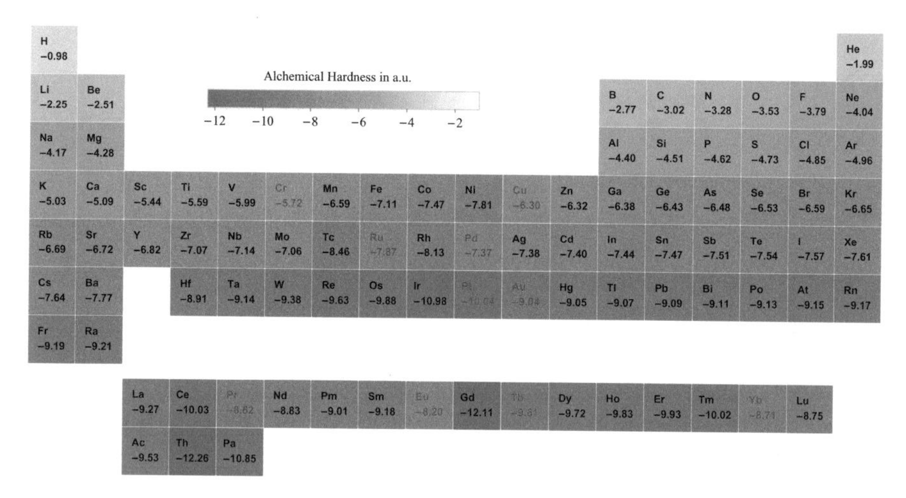

Fig. 20.4 NIST data based Periodic Table of Alchemical hardness: LDA approximation (see text). Colour code: darkening colour indicates more negative hardness. Data in red indicate 'inverted' sequences

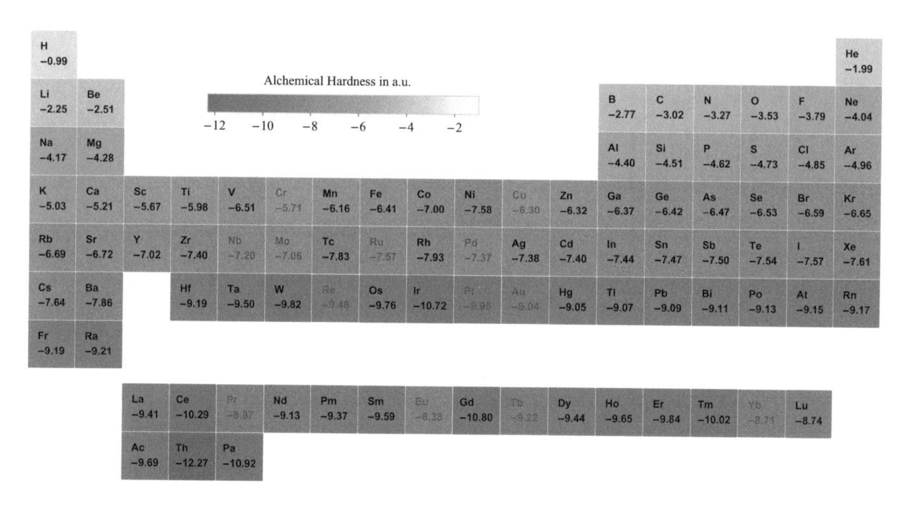

Fig. 20.5 NIST data based Periodic Table of Alchemical hardness: LSDA approximation (see text) Reading the same as in Fig. 20.3

for an at least qualitative walk through the Periodic Table; certainly for the heavier elements, the two approaches described above may become cumbersome, at least due to the basis set issue.

The overall tendencies in the Periodic Table in Fig. 20.4 reveal that the hardness value monotonously decreases (i.e. becomes more negative) along the complete Table, with some exceptions, marked in red, never occurring in main group elements. These exceptions persist in the LSDA approximation. The overall range of hardness

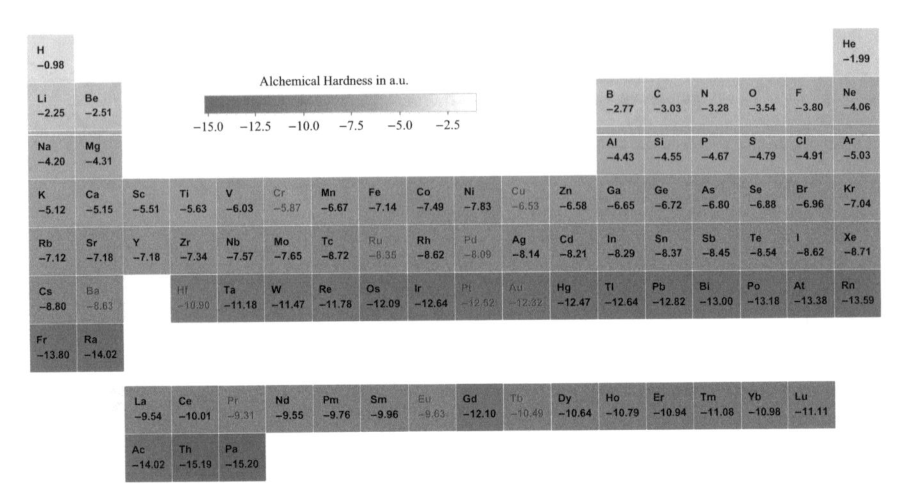

Fig. 20.6 NIST data based Periodic Table of Alchemical hardness: RSDA approximation (see text). Same reading as in Fig. 20.3

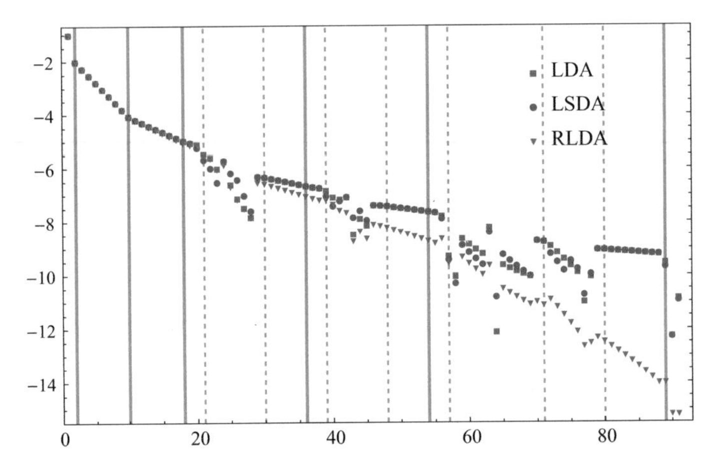

Fig. 20.7 The alchemical hardness as a function of Z for the three LDA variants: LDA, LSDA, RLDA. Vertical lines indicate the periods in the Periodic Table (up to and comprising the sixth period)

is between -1 a.u. for H and -10 a.u. with a few exceptions for the very heavy elements. Along a period, $\tilde{\eta}_{\text{al}}$ decreases between 0.5 and 2.0 units and along a column the decrease is about 1 to 2 units per row. Figure 20.6 shows that inclusion of relativistic effects affects, as expected, the heavy elements with a decrease in hardness by 1 to 4 a.u. The overall tendencies are however unchanged.

In Fig. 20.7 the three LDA-type approaches are compared for all Z values. Overall, the values coincide for the light, main group elements. Deviations between LDA and LSDA values are to be noticed in the Transition Metals and Lanthanides series, be it without a clear pattern, whereas relativistic effects yield systematically, both for main group elements, transition metals and the actinides, more negative $\tilde{\eta}_{\mathrm{al}}$ values, the difference increasing with increasing Z.

20.4 Conclusions

Isolated atom alchemical derivatives were evaluated with different techniques, from the more accurate numerical differentiation and Coupled Perturbed Kohn–Sham approaches to the Z^{-1} expansion model. For this third approach a systematic, extremely-simple method is developed to evaluate systematically an optimal set of all coefficients in the expansion for a given N. For the lighter elements, $Z = 1 - 18$, the comparison between the three approaches shows that the order of magnitude and sequences in the different approaches are similar, paving the way for a walk through the Periodic Table by combining the Z^{-1} expansion approach with the NIST databank. A uniform decrease is retrieved not only for the alchemical potential (the electrostatic potential at the origin) but also for the alchemical hardness, with some minor exceptions. The latter values are relatively strongly influenced by relativistic effects for the heavy elements. Periodicity shows up in some exploratory calculations on the third derivative (the hyperhardness) with turn out to be very basis set dependent.

The Periodic Tables generated could be used in a first step in exploring Chemical Compound Space in a systematic, efficient and cheap way. In a further step 'atoms-in-molecules' corrections could be introduced as in the Electronegativity Equalization method, and the method could be freed from the iso-electronic series limitation by including the 'pure N' $E = E[Z, N]$ derivatives, electronic chemical potential, hardness, etc., and considering mixed N and Z derivatives, the first one being the $\partial^2 E/\partial Z \partial N$ response function which is easily seen to reduce to the Fukui potential [67]. Its relevance in March and Parr's $E = E[Z, N]$ expansion was already touched upon by Cárdenas et al. [68].

Acknowledgements The authors acknowledge financial support by the VUB (Vrije Universiteit Brussel) under the form of a Strategic Research Program (SRP) (PG and FDP), the Interdisciplinary Centre for Mathematical and Computational Modelling computational grant (RB). FDP also acknowledges the Francqui foundation for a position as Francqui research professor. It is both an honour and a pleasure for all of us to dedicate this paper to Professor Norman March, a towering scientist, a true companion and loyal guide on the road to good science, every inch a gentleman. Congratulations, Norman, on the occasion of your 90th birthday!

Appendices

20.A Solution of the System of Equations Using the NIST Data

The system of equations to be considered is

$$N^2\varepsilon_0^{[N]} + N\varepsilon_1^{[N]} + N^{-1}\varepsilon_3^{[N]} = E^{[Z=N,N]}, \tag{20.29a}$$

$$2N\varepsilon_0^{[N]} + \varepsilon_1^{[N]} - N^{-2}\varepsilon_3^{[N]} = V_{en}^{[Z=N,N]}/N, \tag{20.29b}$$

$$(N+1)^2\varepsilon_0^{[N]} + (N+1)\varepsilon_1^{[N]} + \varepsilon_2^{[N]} + (N+1)^{-1}\varepsilon_3^{[N]} = E^{[Z=N+1,N]}, \tag{20.29c}$$

$$2(N+1)\varepsilon_0^{[N]} + \varepsilon_1^{[N]} - (N+1)^{-2}\varepsilon_3^{[N]} = V_{en}^{[Z=N+1,N]}/(N+1), \tag{20.29d}$$

where $E^{[Z=N,N]}$, $V_{en}^{[Z=N,N]}$, and $\varepsilon_j^{[N]}$ mean $E_{\text{NIST}}^{[Z=N,N]}$, $V_{en,\,\text{NIST}}^{[Z=N,N]}$, and $\tilde{\varepsilon}_j^{[N;4]}$.

After some simple algebra, the solutions of Eq. (20.29) can be written as follows

$$\varepsilon_0^{[N]} = V_{en}^{[Z=N+1,N]} + V_{en}^{[Z=N,N]} - (2N+1)\Delta E^+, \tag{20.30a}$$

$$\varepsilon_3^{[N]} = N^2(N+1)^2\left(2\Delta E^+ - \frac{1}{N}V_{en}^{[Z=N,N]} - \frac{1}{N+1}V_{en}^{[Z=N+1,N]}\right), \tag{20.30b}$$

$$\varepsilon_1^{[N]} = \Delta E^+ - (2N+1)\varepsilon_0^{[N]} + \frac{1}{N(N+1)}\varepsilon_3^{[N]}, \tag{20.30c}$$

$$\varepsilon_2^{[N]} = E^{[Z=N,N]} - N^2\varepsilon_0^{[N]} - N\varepsilon_1^{[N]} - N^{-1}\varepsilon_3^{[N]}, \tag{20.30d}$$

with $\Delta E^+ = E^{[Z=N+1,N]} - E^{[Z=N,N]}$.

20.B Alchemical Hyperhardness Values for Li to Cl

The alchemical hyperhardness can be calculated numerically from Eq. (20.7) or as the second derivative of the alchemical potential, Eq. (20.1):

$$\gamma_{\text{al}}[Z=N,N] = \frac{\mu_{\text{al}}[Z+\delta,N] - 2\mu_{\text{al}}[Z,N] + \mu_{\text{al}}[Z-\delta,N]}{\delta^2}. \tag{20.31}$$

The results in Table 20.6 are calculated with aug-cc-pCVTZ basis set.

Table 20.6 Alchemical hyperhardness, γ_{al}, Eq. (20.31)

Li	Be	B	C	N	O	F	Ne
0.044	0.280	0.228	0.190	0.158	0.145	0.131	0.121
Na	Mg	Al	Si	P	S	Cl	Ar
0.103	0.120	0.283	0.242	0.210	0.180	0.139	0.122

References

1. P. Kirkpatrick, C. Ellis, Nature **432**(7019), 823 (2004). https://doi.org/10.1038/432823a
2. C.M. Dobson, Nature **432**(7019), 824 (2004). https://doi.org/10.1038/nature03192
3. O.A. von Lilienfeld, Int. J. Quantum Chem. **113**(12), 1676 (2013). https://doi.org/10.1002/qua.24375
4. A. Franceschetti, A. Zunger, Nature **402**(6757), 60 (1999). https://doi.org/10.1038/46995
5. G.H. Jóhannesson, T. Bligaard, A.V. Ruban, H.L. Skriver, K.W. Jacobsen, J.K. Nørskov, Phys. Rev. Lett. **88**, 255506 (2002). https://doi.org/10.1103/PhysRevLett.88.255506
6. M. Wang, X. Hu, D.N. Beratan, W. Yang, J. Am. Chem. Soc. **128**(10), 3228 (2006). https://doi.org/10.1021/ja0572046
7. C. Kuhn, D.N. Beratan, J. Phys. Chem. **100**(25), 10595 (1996). https://doi.org/10.1021/jp960518i
8. D. Balamurugan, W. Yang, D.N. Beratan, J. Chem. Phys. **129**(17), 174105 (2008). https://doi.org/10.1063/1.2987711
9. F. De Vleeschouwer, A. Chankisjijev, P. Geerlings, F. De Proft, Eur. J. Org. Chem. **2015**(3), 506 (2015). https://doi.org/10.1002/ejoc.201403198
10. F. De Vleeschouwer, P. Geerlings, F. De Proft, Chem. Phys. Chem. **17**(10), 1414 (2016). https://doi.org/10.1002/cphc.201501189
11. O.A. von Lilienfeld, R.D. Lins, U. Rothlisberger, Phys. Rev. Lett. **95**, 153002 (2005). https://doi.org/10.1103/PhysRevLett.95.153002
12. O.A. von Lilienfeld, M.E. Tuckerman, J. Chem. Phys. **125**(15), 154104 (2006). https://doi.org/10.1063/1.2338537
13. O.A. von Lilienfeld, M.E. Tuckerman, J. Chem. Theory Comput. **3**(3), 1083 (2007), PMID: 26627427. https://doi.org/10.1021/ct700002c
14. O.A. von Lilienfeld, J. Chem. Phys. **131**(16), 164102 (2009). https://doi.org/10.1063/1.3249969
15. K.Y.S. Chang, S. Fias, R. Ramakrishnan, O.A. von Lilienfeld, J. Chem. Phys. **144**(17), 174110 (2016). https://doi.org/10.1063/1.4947217
16. M. to Baben, J.O. Achenbach, O.A. von Lilienfeld, J. Chem. Phys. **144**(10), 104103 (2016). https://doi.org/10.1063/1.4943372
17. M. Lesiuk, R. Balawender, J. Zachara, J. Chem. Phys. **136**(3), 034104 (2012). https://doi.org/10.1063/1.3674163
18. R. Balawender, M.A. Welearegay, M. Lesiuk, F. De Proft, P. Geerlings, J. Chem. Theory Comput. **9**(12), 5327 (2013). https://doi.org/10.1021/ct400706g
19. R.G. Parr, W. Yang, *Density Functional Theory of Atoms and Molecules* (Oxford University Press, Oxford, 1989). ISBN 9780195092769
20. P. Geerlings, F. De Proft, W. Langenaeker, Chem. Rev. **103**(5), 1793 (2003). https://doi.org/10.1021/cr990029p
21. P. Geerlings, F. De Proft, Phys. Chem. Chem. Phys. **10**, 3028 (2008). https://doi.org/10.1039/B717671F
22. S. Liu, T. Li, P.W. Ayers, J. Chem. Phys. **131**(11), 114106 (2009). https://doi.org/10.1063/1.3231687
23. P. Geerlings, Z. Boisdenghien, F. De Proft, S. Fias, Theor. Chem. Acc. **135**(9), 213 (2016). https://doi.org/10.1007/s00214-016-1967-9
24. P. Geerlings, S. Fias, Z. Boisdenghien, F. De Proft, Chem. Soc. Rev. **43**, 4989 (2014). https://doi.org/10.1039/C3CS60456J
25. N. Sablon, F. De Proft, P.W. Ayers, P. Geerlings, J. Chem. Phys. **126**(22), 224108 (2007). https://doi.org/10.1063/1.2736698
26. W. Yang, A.J. Cohen, F. De Proft, P. Geerlings, J. Chem. Phys. **136**(14), 144110 (2012). https://doi.org/10.1063/1.3701562
27. N. Sablon, F. De Proft, P. Geerlings, J. Phys. Chem. Lett. **1**(8), 1228 (2010). https://doi.org/10.1021/jz1002132

28. S. Fias, P. Geerlings, P. Ayers, F. De Proft, Phys. Chem. Chem. Phys. **15**, 2882 (2013). https://doi.org/10.1039/C2CP43612D
29. Z. Boisdenghien, S. Fias, F. Da Pieve, F. De Proft, P. Geerlings, Mol. Phys. **113**(13-14), 1890 (2015). https://doi.org/10.1080/00268976.2015.1021110
30. Z. Boisdenghien, C. Van Alsenoy, F. De Proft, P. Geerlings, J. Chem. Theory Comput. **9**(2), 1007 (2013). https://doi.org/10.1021/ct300861r
31. T. Stuyver, S. Fias, F. De Proft, P.W. Fowler, P. Geerlings, J. Chem. Phys. **142**(9), 094103 (2015). https://doi.org/10.1063/1.4913415
32. R. Balawender, M. Lesiuk, F. De Proft, P. Geerlings, J. Chem. Theory Comput. **14**(2), (2018). https://doi.org/10.1021/acs.jctc.7b01114
33. R.G. Parr, R.A. Donnelly, M. Levy, W.E. Palke, J. Chem. Phys. **68**(8), 3801 (1978). https://doi.org/10.1063/1.436185
34. R.G. Parr, R.G. Pearson, J. Am. Chem. Soc. **105**(26), 7512 (1983). https://doi.org/10.1021/ja00364a005
35. P. Fuentealba, R.G. Parr, J. Chem. Phys. **94**(8), 5559 (1991). https://doi.org/10.1063/1.460491
36. W. Kohn, L.J. Sham, Phys. Rev. **137**, A1697 (1965). https://doi.org/10.1103/PhysRev.137.A1697
37. E.A. Hylleraas, Z. Phys. **65**(3), 209 (1930). https://doi.org/10.1007/BF01397032
38. N.H. March, R.J. White, J. Phys. B: At. Mol. Phys. **5**(3), 466 (1972); Reprinted in Ref. [74]. https://doi.org/10.1088/0022-3700/5/3/011
39. N.H. March, R.G. Parr, Proc. Natl. Acad. Sci. (USA) **77**(11), 6285 (1980); Reprinted in Ref. [74]
40. N.H. March, R.G. Parr, J.F. Mucci, Proc. Natl. Acad. Sci. (USA) **78**(10), 5942 (1981)
41. S. Kotochigova, Z.H. Levine, E.L. Shirley, M.D. Stiles, C.W. Clark, Atomic reference data for electronic structure calculations, (National Institute of Standards and Technology, Gaithersburg, MD, 2003), ver. 1.3
42. R. Balawender, L. Komorowski, J. Chem. Phys. **109**(13), 5203 (1998). https://doi.org/10.1063/1.477137
43. R. Balawender, P. Geerlings, J. Chem. Phys. **123**(12), 124102 (2005). https://doi.org/10.1063/1.2012329
44. R. Balawender, P. Geerlings, J. Chem. Phys. **123**(12), 124103 (2005). https://doi.org/10.1063/1.2012330
45. A.J. Cohen, P. Mori-Sánchez, W. Yang, Phys. Rev. B **77**, 115123 (2008). https://doi.org/10.1103/PhysRevB.77.115123
46. P. Politzer, P. Lane, J.S. Murray, in *Reviews of quantum modern chemistry. A celebration of the contributions of Robert G. Parr*, vol. 1, ed. by K.D. Sen (World Scientific, Singapore, 2012), pp. 63–84
47. A.D. Becke, R.M. Dickson, J. Chem. Phys. **89**(5), 2993 (1988). https://doi.org/10.1063/1.455005
48. C. Lee, W. Yang, R.G. Parr, Phys. Rev. B **37**, 785 (1988). https://doi.org/10.1103/PhysRevB.37.785
49. R.A. Kendall, T.H. Dunning Jr., R.J. Harrison, J. Chem. Phys. **96**(9), 6796 (1992). https://doi.org/10.1063/1.462569
50. D.E. Woon, T.H. Dunning Jr., J. Chem. Phys. **98**(2), 1358 (1993). https://doi.org/10.1063/1.464303
51. D.E. Woon, T.H. Dunning Jr., J. Chem. Phys. **100**(4), 2975 (1994). https://doi.org/10.1063/1.466439
52. D. Layzer, Ann. Phys. **8**(2), 271 (1959). https://doi.org/10.1016/0003-4916(59)90023-5
53. P.O. Löwdin, J. Mol. Spectrosc. **3**(1), 46 (1959). https://doi.org/10.1016/0022-2852(59)90006-2
54. Y. Tal, M. Levy, Phys. Rev. A **23**, 408 (1981). https://doi.org/10.1103/PhysRevA.23.408
55. J.L. Gázquez, A. Vela, M. Galván, Phys. Rev. Lett. **56**, 2606 (1986). https://doi.org/10.1103/PhysRevLett.56.2606
56. J.L. Gázquez, A. Vela, Phys. Rev. A **38**, 3264 (1988). https://doi.org/10.1103/PhysRevA.38.3264

57. A. Vela, M. Galván, J.L. Gázquez, Int. J. Quantum Chem. **34**(S22), 329 (1988). https://doi.org/10.1002/qua.560340837
58. B.G. Englert, J. Schwinger, Phys. Rev. A **32**, 26 (1985). https://doi.org/10.1103/PhysRevA.32.26
59. B.G. Englert, J. Schwinger, Phys. Rev. A **32**, 36 (1985). https://doi.org/10.1103/PhysRevA.32.36
60. B.G. Englert, J. Schwinger, Phys. Rev. A **32**, 47 (1985). https://doi.org/10.1103/PhysRevA.32.47
61. C. Van Alsenoy, A. Peeters, J. Mol. Struct. Theochem **286**, 19 (1993). https://doi.org/10.1016/0166-1280(93)87148-7
62. B. Rousseau, C. Van Alsenoy, A. Peeters, F. Bogár, G. Paragi, J. Mol. Struct. Theochem **666**, 41 (2003). The role of chemistry in the evolution of molecular medicine. A Tribute to Professor Albert Szent-Gyorgyi to Celebrate his 110th Birthday. https://doi.org/10.1016/j.theochem.2003.08.011
63. M.W. Schmidt, K.K. Baldridge, J.A. Boatz, S.T. Elbert, M.S. Gordon, J.H. Jensen, S. Koseki, N. Matsunaga, K.A. Nguyen, S. Su, T.L. Windus, M. Dupuis, J.A. Montgomery, J. Comp. Chem. **14**(11), 1347 (1993). https://doi.org/10.1002/jcc.540141112
64. R.G. Parr, W. Yang, J. Am. Chem. Soc. **106**(14), 4049 (1984). https://doi.org/10.1021/ja00326a036
65. W. Yang, R.G. Parr, R. Pucci, J. Chem. Phys. **81**(6), 2862 (1984); Reprinted as chap. 22, p. 303, of Ref. [75]. https://doi.org/10.007/978-3-319-53664-4_22, https://doi.org/10.1063/1.447964
66. E. Echegaray, A. Toro-Labbe, K. Dikmenli, F. Heidar-Zadeh, N. Rabi, S. Rabi, P.W. Ayers, C. Cárdenas, R.G. Parr, J.S.M. Anderson, in Angilella and La Magna [75], chap. 19, pp. 269–288. https://doi.org/10.1007/978-3-319-53664-4_19, ISBN 9783319536637
67. W.P. Ayers, M. Levy, Theor. Chem. Acc. **103**(3), 353 (2000). https://doi.org/10.1007/s002149900093
68. C. Cárdenas, W. Tiznado, P.W. Ayers, P. Fuentealba, J. Phys. Chem. A **115**(11), 2325 (2011). https://doi.org/10.1021/jp109955q
69. P.K. Chattaraj, A. Cedillo, R.G. Parr, J. Chem. Phys. **103**(24), 10621 (1995). https://doi.org/10.1063/1.469847
70. W. Yang, W.J. Mortier, J. Am. Chem. Soc. **108**(19), 5708 (1986). https://doi.org/10.1021/ja00279a008
71. K.A. Van Genechten, W.J. Mortier, P. Geerlings, J. Chem. Phys. **86**(9), 5063 (1987). https://doi.org/10.1063/1.452649
72. W.J. Mortier, in *Electronegativity*, ed. by K.D. Sen, C.K. Jørgensen (Springer, Berlin, 1987), pp. 125–143
73. Y. Tal, L.J. Bartolotti, J. Chem. Phys. **76**(8), 4056 (1982). https://doi.org/10.1063/1.443479
74. N.H. March, G.G.N. Angilella (eds.), *Many-Body Theory of Molecules, Clusters, and Condensed Phases* (World Scientific, Singapore, 2009)
75. G.G.N. Angilella, A. La Magna (eds.), *Correlations in Condensed Matter Under Extreme Conditions: A Tribute to Renato Pucci on the Occasion of his 70th Birthday* (Springer, Berlin, 2017). https://doi.org/10.1007/978-3-319-53664-4, ISBN 9783319536637

Chapter 21
Orbital-Free Density Functional Theory: Pauli Potential and Density Scaling

Á. Nagy

Abstract In orbital-free density functional theory only a single equation, the so-called Euler equation, has to be solved for any system instead of the Kohn–Sham equations. The Euler equation is a Schrödinger-like equation for the square root of the density. This equation contains an extra potential, the so-called Pauli potential, in addition to the usual Kohn–Sham potential. Equations for the Pauli potential, the relationship of the Pauli potential and Pauli energy are reviewed. A derivation of the Euler equation via density scaling is presented.

21.1 Introduction

Nowadays, electron structure calculations are usually done with density functional theory. The history of density functional theory started with the fundamental works of Thomas [1], Fermi [2], Dirac [3] and Gombás [4]. The theory was rigorously established by Hohenberg and Kohn [5]. They derived the Euler equation

$$\frac{\delta E}{\delta n} = \mu, \tag{21.1}$$

where $E[n]$ is the total energy functional and the Lagrange multiplier μ is the chemical potential. The solution of the variational problem gives the electron density. However, the energy functional $E[n]$ is unknown and even accurate approximations are unavailable. $E[n]$ can be regarded as a sum of several terms. The most troublesome is the kinetic energy term. Both the kinetic energy functional and its functional derivative appearing in the Euler equation are difficult to approximate. Kohn and Sham (KS) [6] gave a genuine solution to this problem with the invention of the non-interacting system. In this fictitious system the electrons move independently in a common, local potential. The density

Á. Nagy (✉)
Department of Theoretical Physics, University of Debrecen, Debrecen 4002, Hungary
e-mail: anagy@phys.unideb.hu

G. G. N. Angilella and C. Amovilli (eds.), *Many-body Approaches at Different Scales*, https://doi.org/10.1007/978-3-319-72374-7_21

$$n(\mathbf{r}) = \sum_i^N |\phi_i(\mathbf{r})|^2 \tag{21.2}$$

is the same as the true interacting electron density. The orbitals ϕ_i satisfy the Kohn–Sham equations

$$\left[-\frac{1}{2}\nabla^2 + v_{\mathrm{KS}}(\mathbf{r})\right]\phi_i(\mathbf{r}) = \varepsilon_i\phi_i(\mathbf{r}), \tag{21.3}$$

where N, ε_i, ϕ_i, and v_{KS} are the number of electrons, the one-electron energies, orbitals and the Kohn–Sham potential, respectively.

Nowadays, in the great majority of density functional calculations, the Kohn–Sham equations are solved. The original Hohenberg-Kohn theory would have the great advantage that only one equation, the Euler equation, Eq. (21.1), should be solved instead of several Kohn–Sham equations. It is very important in case the system considered has a lot of electrons. Therefore, there is a growing interest in this so-called orbital-free density functional theory.

The non-interacting kinetic energy

$$T_s = -\frac{1}{2}\sum_i^N \int \phi_i^* \nabla^2 \phi_i d\mathbf{r} \tag{21.4}$$

can be partitioned as $T_s = T_w + T_p$. T_w is the Weizsäcker kinetic energy [7]

$$T_w = -\frac{1}{2}\int n^{1/2}\nabla^2 n^{1/2} d\mathbf{r}. \tag{21.5}$$

The Pauli energy is defined as $T_p = T_s - T_w$ [8–24]. The Euler equation, Eq. (21.1), can be rewritten as

$$\left[-\frac{1}{2}\nabla^2 + v_P + v_{\mathrm{KS}}\right] n^{1/2} = \mu n^{1/2}, \tag{21.6}$$

where

$$v_p = \frac{\delta T_p}{\delta n} \tag{21.7}$$

is the Pauli potential, the functional derivative of the Pauli energy T_p. It was Norman March who first wrote the Euler equation in the form of Eq. (21.6) [10]. The Schrödinger like equation for the square root of the density n appeared a bit earlier in the literature [8, 9], but the partition of the effective potential as sum of the Kohn–Sham and Pauli potentials was first presented in [10]. Norman March used first the notation Pauli potential, because this term emergies owing to the Pauli principle.

As he wrote this term "... distinguishes the fermionic system, with its associated Exclusion Principle, from the Boson problem..." [10].

The Pauli potential has a very important role in the orbital-free density functional theory. Professor March addressed this issue in several important papers [9–11, 14, 15, 17–21, 23].

21.2 Differential Virial and Force-Balance Equations

The differential virial theorem goes back to March and Young [25] in one dimension and is generalized first to spherically symmetric systems by Nagy and March [26], then to three dimensions by Holas and March [27]. In the non-interacting system it reads

$$-\frac{\partial v_{\mathrm{KS}}}{\partial r} = -\frac{1}{4n(r)}\frac{\partial}{\partial r}\nabla^2 n(r) + \frac{\hat{\mathbf{r}} \cdot \mathbf{z}^{(s)}(\mathbf{r})}{n(r)}. \tag{21.8}$$

Here, $\hat{\mathbf{r}}$ denotes the radial unit vector, while the vector field $\mathbf{z}^{(s)}(\mathbf{r})$ is defined via the non-interacting kinetic energy density tensor $t_{\alpha\beta}^{(s)}(\mathbf{r})$ [27]

$$t_{\alpha\beta}^{(s)}(\mathbf{r}) = \frac{1}{4}\left[\frac{\partial^2}{\partial r'_\alpha \partial r''_\beta}\gamma^{(s)}(\mathbf{r}', \mathbf{r}'') + \frac{\partial^2}{\partial r'_\beta \partial r''_\alpha}\gamma^{(s)}(\mathbf{r}', \mathbf{r}'')\right]_{\mathbf{r}''=\mathbf{r}'=\mathbf{r}} \tag{21.9}$$

as

$$z_\alpha^{(s)}(\mathbf{r}) = 2\sum_\beta \frac{\partial t_{\alpha\beta}^{(s)}(\mathbf{r})}{\partial r_\beta}. \tag{21.10}$$

Here, $z_\alpha^{(s)}$ is the α component of the vector $\mathbf{z}(\mathbf{r})$. The non-interacting kinetic energy density tensor $t_{\alpha\beta}^{(s)}(\mathbf{r})$ is defined in terms of the non-interacting one-particle density matrix $\gamma^{(s)}(\mathbf{r}', \mathbf{r}'')$ in Eq. (21.9). Using the differential virial theorem, Eq. (21.8), the force $\langle F \rangle$ can be calculated

$$\langle F \rangle = -\int n(r)\frac{\partial v_{\mathrm{KS}}(r)}{\partial r}d\mathbf{r} = -\frac{1}{4}\int \frac{\partial}{\partial r}\nabla^2 n(r)d\mathbf{r} + \int \hat{\mathbf{r}} \cdot \mathbf{z}^{(s)}(\mathbf{r})d\mathbf{r}. \tag{21.11}$$

One can find useful equations for the external and exchange-correlation forces in [28].

Consider now spherically symmetric systems. Using the Laplacial form $t_L(r)$ of the kinetic energy density for the appropriate general level occupancy, we multiply Eq. (21.3) by ϕ_i^* and sum over occupied levels to find

$$t_L(r) + n v_{\mathrm{KS}}(r) = \sum_{\text{occupied } i} \varepsilon_i |\phi_i|^2 \equiv g(r). \tag{21.12}$$

Forming the gradient of Eq. (21.12) we find

$$\nabla t_L(r) + n(r)\nabla v_{\mathrm{KS}}(r) + v_{\mathrm{KS}}(r)\nabla n(r) = \nabla g(r). \tag{21.13}$$

Equation (21.8) can be rewritten as

$$z_s(r) = \frac{1}{4}\frac{\partial}{\partial r}\nabla^2 n(r) + \frac{\partial t_L}{\partial r} + v_{\mathrm{KS}}(r)\frac{\partial n(r)}{\partial r} - \frac{\partial g(r)}{\partial r}. \tag{21.14}$$

Replacing $t_L(r)$ by the positive definite gradient form $t_G(r)$ of kinetic energy density, we obtain

$$z_s(r) = \frac{\partial t_G(r)}{\partial r} + v_{\mathrm{KS}}(r)\frac{\partial n(r)}{\partial r} - \frac{\partial g(r)}{\partial r}. \tag{21.15}$$

Equation (21.15) can also be written as

$$z_s(r) = 2\frac{\partial t_W}{\partial r} + 4\frac{t_W}{r} + \frac{\partial t_P}{\partial r} + \mu n' - \frac{\partial g(r)}{\partial r} - v_P(r)n'. \tag{21.16}$$

The Pauli potential can also be expressed as

$$v_P(r) = \frac{t_P(r)}{n(r)} + \mu - \frac{g(r)}{n(r)}, \tag{21.17}$$

therefore Eq. (21.16) takes the form

$$z_s(r) = 4\frac{t_W}{r} + 2t'_W(r) + n(r)v'_P(r). \tag{21.18}$$

The final expression, Eq. (21.18), connects z_s to the Weizsäcker kinetic energy density and the Pauli potential. Equation (21.17) establishes a relation between the Pauli potential and the Pauli energy density.

21.3 Pauli Potential via Density Scaling

It has recently been shown that Pauli potential can be constructed via density scaling [29]. In density scaling we immagine another non-interacting system having a scaled density $n_\zeta(\mathbf{r}) = n(\mathbf{r})/\zeta$. Here, $\zeta = N/N_\zeta$ is a positive number. We recover the original non-interacting (Kohn–Sham) system if $\zeta = 1$. Suppose that the original real system has N electrons. Then the Kohn–Sham system with the scaled density n_ζ has N_ζ electrons:

$$\int n_\zeta(\mathbf{r})d\mathbf{r} = N_\zeta. \tag{21.19}$$

While N is always an integer, N_ζ is generally non-integer. Therefore the grand canonical ensemble [30–34] is constructed. A zero temperature grand canonical density matrix Γ in the Fock space takes the form

$$\Gamma = \sum_N \sum_i f_{Ni} |\Psi_{Ni}\rangle\langle\Psi_{Ni}|, \tag{21.20}$$

where Ψ_{Ni} is the ith N-particle eigenfunction of the Hamiltonian. The occupation numbers f_{Ni} should satisfy the conditions $0 \leq f_{Ni} \leq 1$ and $\sum_N \sum_i f_{Ni} = 1$. Apply now the constrained search for the kinetic energy over the density matrices Γ as

$$T_\zeta[n] = \zeta \min_{\Gamma \to n_\zeta} Tr[\hat{\Gamma}\hat{T}], \tag{21.21}$$

where the scaled density is given by

$$n_\zeta = Tr[\hat{\Gamma}\hat{n}]. \tag{21.22}$$

$T_\zeta[n]$ is a convex functional and the functional derivative exists [30, 35, 36].

Consider the case $N_\zeta = 2$ and denote this value of ζ as $\zeta_d = N/2$. It corresponds to a non-interacting system with two electrons. The constrained search [37, 38] minimizes the scaled kinetic energy

$$-2\frac{1}{2}\int \phi^*(\mathbf{r})\nabla^2\phi(\mathbf{r})d\mathbf{r} \tag{21.23}$$

with a fixed scaled density

$n_{\zeta_d} = 2|\phi|^2$:

$$\min\left(-\int \phi^*(\mathbf{r})\nabla^2\phi(\mathbf{r})d\mathbf{r} + \int n_{\zeta_d}(\mathbf{r})v_{\zeta_d}(\mathbf{r})d\mathbf{r} + \mu\int n_{\zeta_d}(\mathbf{r})d\mathbf{r}\right). \tag{21.24}$$

The minimization is done with fixing the density n_{ζ_d} and its norm, Eq. (21.19), using the Lagrange multipliers $v_{\zeta_d}(\mathbf{r})$ and μ, respectively. The minimization leads to the equation

$$-\frac{1}{2}\nabla^2\phi + v_{\zeta_d}\phi = \mu\phi \tag{21.25}$$

that can also be written as

$$\left(-\frac{1}{2}\nabla^2 + v_{\zeta_d}\right)n_{\zeta_d}^{1/2} = \mu n_{\zeta_d}^{1/2} \tag{21.26}$$

or

$$\left(-\frac{1}{2}\nabla^2 + v_{\zeta_d}\right) n^{1/2} = \mu n^{1/2}. \tag{21.27}$$

Comparing Eq. (21.27) with the Euler equation, Eq. (21.6), we obtain the Pauli potential as

$$v_P = v_{\zeta_d} - v_{\mathrm{KS}}. \tag{21.28}$$

According to this remarkable expression the Pauli potential is the difference of the scaled and the original Kohn–Sham potential. To derive another important expression for the Pauli potential consider now the relation between the original non-interacting kinetic (T_s) and exchange-correlation (E_{xc}) and the scaled non-interacting kinetic (T_ζ) and exchange-correlation ($E_{\zeta xc}$) energies [30–32]:

$$T_s + E_{xc} = T_\zeta + E_{\zeta xc}. \tag{21.29}$$

The functional derivation provides an expression between the original and scaled exchange-correlation potentials:

$$\frac{\delta T_s}{\delta n} + v_{xc} = \frac{\delta T_\zeta}{\delta n} + v_{\zeta xc}. \tag{21.30}$$

In our case $\zeta = \zeta_d$, and using the partition $T_s = T_w + T_p$ we obtain

$$T_w + T_p + E_{xc} = T_w + E_{\zeta_d xc}, \tag{21.31}$$

that is, the Pauli energy is the difference of the scaled and the original exchange-correlation energies:

$$T_p = E_{\zeta_d xc} - E_{xc}. \tag{21.32}$$

The functional derivative of Eq. (21.32) gives the Pauli potential

$$v_p = v_{\zeta_d xc} - v_{xc} \tag{21.33}$$

as the difference of the scaled and original exchange-correlation potentials.

21.4 Discussion

Nowadays, there is a growing interest in orbital-free density functional theory. Avoiding the solution of the Kohn–Sham equations – solving the Euler equation instead – would result an enormous simplification and would make it possible to treat systems with large number of electrons.

Pauli potential is a key quantity in orbital-free density functional theory. As it is responsible for the fulfillment of the Pauli principle, its presence in the Euler equation is essential. Unfortunately, its exact form is unknown, therefore, we have to approximate it in calculations. Exact relations might be very useful in constructing adequate approximate potentials. The exchange-correlation part of the Kohn–Sham potential is also unknown. There are, however, several important exact relations for the exchange, correlation and exchange-correlation potentials and energies. Some of them turned to be extremely useful in constructing and improving approximate functionals. It is expected that exact relations for the Pauli potential and energy might be similarly valuable in designing approximate forms. Equations (21.18) and (21.17) seem to be beneficial in modeling the Pauli potential and the Pauli energy density or checking approximations.

Density scaling provides a constructive way of obtaining approximations for the Pauli potential. The Pauli potential (energy) of the density functional theory is expressed as the difference of the scaled and original exchange-correlation potentials (energies). Further, Eqs. (21.32) and (21.33) make it possible to seek alternative approximations for the Pauli energy and potential.

Finding adequate approximation for the Pauli potential is a very hard problem. Density scaling induces a hope of constructing good approximate Pauli potentials. It should be the subject of further research.

Acknowledgements This research was supported by the EU-funded Hungarian grant EFOP-3.6.2-16-2017-00005 and the National Research, Development and Innovation Fund of Hungary, financed under 123988 funding scheme.

References

1. L.H. Thomas, Math. Proc. Camb. Philos. Soc. **23**, 542 (1926). https://doi.org/10.1017/S0305004100011683
2. E. Fermi, Z. Phys. **48**, 73 (1928). https://doi.org/10.1007/BF01351576
3. P.A.M. Dirac, Proc. Camb. Philos. Soc. **26**, 376 (1930). https://doi.org/10.1017/S0305004100016108
4. P. Gombás, *Die statistische Theorie des Atoms und ihre Anwendungen* (Springer, Vienna, 1949)
5. P.C. Hohenberg, W. Kohn, Phys. Rev. **136**, B864 (1964). https://doi.org/10.1103/PhysRev.136.B864
6. W. Kohn, L.J. Sham, Phys. Rev. **140**, A1133 (1965). https://doi.org/10.1103/PhysRev.140.A1133
7. C.F. von Weizsäcker, Z. Phys. **96**, 431 (1935). https://doi.org/10.1007/BF01337700
8. M. Levy, J.P. Perdew, V. Sahni, Phys. Rev. A **30**, 2745 (1984). https://doi.org/10.1103/PhysRevA.30.2745
9. N.H. March, Phys. Lett. A **113**(2), 66 (1985); Reprinted in Ref. [39]. https://doi.org/10.1016/0375-9601(85)90654-1
10. N.H. March, Phys. Lett. A **113**(9), 476 (1986), Reprinted in Ref. [39]. 10.1016/0375-9601(86)90123-4
11. N.H. March, J. Comput. Chem. **8**(4), 375 (1987). https://doi.org/10.1002/jcc.540080414
12. M. Levy, H. Ou-Yang, Phys. Rev. A **38**, 625 (1988). https://doi.org/10.1103/PhysRevA.38.625
13. C. Herring, M. Chopra, Phys. Rev. A **37**, 31 (1988). https://doi.org/10.1103/PhysRevA.37.31

14. A. Holas, N.H. March, Phys. Rev. A **44**, 5521 (1991). https://doi.org/10.1103/PhysRevA.44.5521
15. A. Nagy, N.H. March, Phys. Chem. Liq. **22**(1–2), 129 (1990). https://doi.org/10.1080/00319109008036419
16. Á. Nagy, Acta Phys. Hung. **70**(4), 321 (1991). https://doi.org/10.1007/BF03054145
17. A. Nagy, N.H. March, Int. J. Quantum Chem. **39**(4), 615 (1991). https://doi.org/10.1002/qua.560390408
18. A. Nagy, N.H. March, Phys. Chem. Liq. **25**(1), 37 (1992). https://doi.org/10.1080/00319109208027285
19. A. Nagy, N.H. March, Phys. Chem. Liq. **38**(6), 759 (2000). https://doi.org/10.1080/00319100008030321
20. N.H. March, A. Nagy, Phys. Rev. A **78**, 044501 (2008). https://doi.org/10.1103/PhysRevA.78.044501
21. N.H. March, A. Nagy, Phys. Rev. A **81**, 014502 (2010). https://doi.org/10.1103/PhysRevA.81.014502
22. N.H. March, A. Nagy, F. Bogár, F. Bartha, Phys. Chem. Liq. **50**(3), 412 (2012). https://doi.org/10.1080/00319104.2012.673721
23. N.H. March, J. Mol. Struct.: THEOCHEM **943**(1), 77 (2010) (Conceptual Aspects of Electron Densities and Density Functionals). https://doi.org/10.1016/j.theochem.2009.10.030
24. V.G. Tsirelson, A.I. Stash, V.V. Karasiev, S. Liu, Comput. Theor. Chem. **1006**, 92 (2013). https://doi.org/10.1016/j.comptc.2012.11.015
25. N.H. March, W.H. Young, Nucl. Phys. **12**(3), 237 (1959). https://doi.org/10.1016/0029-5582(59)90169-5
26. A. Nagy, N.H. March, Phys. Rev. A **40**, 554 (1989). https://doi.org/10.1103/PhysRevA.40.554
27. A. Holas, N.H. March, Phys. Rev. A **51**, 2040 (1995); Reprinted in Ref. [39]. https://doi.org/10.1103/PhysRevA.51.2040
28. N.H. March, A. Nagy, J. Chem. Phys. **129**(19), 194114 (2008). https://doi.org/10.1063/1.3013808
29. A. Nagy, Phys. Rev. A **84**, 032506 (2011). https://doi.org/10.1103/PhysRevA.84.032506
30. G.K.L. Chan, N.C. Handy, Phys. Rev. A **59**, 2670 (1999). https://doi.org/10.1103/PhysRevA.59.2670
31. A. Nagy, Chem. Phys. Lett. **411**(4), 492 (2005). https://doi.org/10.1016/j.cplett.2005.06.078
32. A. Nagy, J. Chem. Phys. **123**(4), 044105 (2005). https://doi.org/10.1063/1.1979473
33. J.P. Perdew, R.G. Parr, M. Levy, J.L. Balduz, Phys. Rev. Lett. **49**, 1691 (1982). https://doi.org/10.1103/PhysRevLett.49.1691
34. R.G. Parr, W. Yang, *Density Functional Theory of Atoms and Molecules* (Oxford University Press, Oxford, 1989). ISBN 9780195092769
35. I. Ekeland, R. Teman, *Convex Analysis and Variational Problems* (North-Holland, Amsterdam, 1976)
36. R. van Leeuwen, Adv. Quantum Chem. **43**, 25 (2003). https://doi.org/10.1016/S0065-3276(03)43002-5
37. M. Levy, Proc. Natl. Acad. Sci. **76**(12), 6062 (1979)
38. E.H. Lieb, Int. J. Quantum Chem. **24**(3), 243 (1983). https://doi.org/10.1002/qua.560240302
39. N.H. March, G.G.N. Angilella (eds.), *Many-Body Theory of Molecules, Clusters, and Condensed Phases* (World Scientific, Singapore, 2009)

Chapter 22
The Role of the N-Representability in One-Particle Functional Theories

M. Piris

22.1 Introduction

The purpose of this chapter is to analyze the role of the N-representability in one-particle functional theories, that is, in theories where the ground-state energy is represented in terms of the first-order reduced density matrix (1RDM) Γ or simply its diagonal part: the density ρ. I have chosen to write on this topic to honor Norman H. March since he has always been interested on the subject. Throughout these years during his visits to the Donostia International Physics Center, Professor March has encouraged me to emphasize the importance of the *functional* N-representability, an issue that has not received enough attention in the literature. This has led us to recently write several articles together [1–5] using what is so far the only known natural orbital functional, namely PNOF5 [6, 7], which even including the electronic correlation, maintains a one-to-one correspondence with the energy obtained from an N-particle wavefunction [8, 9].

The term N-representability was coined by John Coleman in 1963 [10]. Already in the 1940s [11] it was known that for an N-particle quantum system with a Hamiltonian involving not more than two-body interactions, the energy is an exact functional $E[D]$ of the second-order reduced density matrix (2RDM) D. Therefore, it was frequently pointed out that the N-particle wavefunction tells us more than we need to know and its role can be assumed by the 2RDM in the discussion of physical systems. Coleman attempted this in 1951 [12] and realized that it is necessary to impose some limitations on the allowed two-matrices, in addition to general properties, to

M. Piris (✉)
Donostia International Physics Center (DIPC), 20018 Donostia, Euskadi, Spain
e-mail: mario.piris@ehu.eus

M. Piris
Euskal Herriko Unibertsitatea (UPV/EHU), 20018 Donostia, Euskadi, Spain

M. Piris
Basque Foundation for Science (IKERBASQUE), 48011 Bilbao, Euskadi, Spain

G. G. N. Angilella and C. Amovilli (eds.), *Many-body Approaches at Different Scales*, https://doi.org/10.1007/978-3-319-72374-7_22

ensure a physical value of ground-state energy. The needed conditions [10] were that the two-matrix be derived from an N-particle wavefunction that is symmetric or antisymmetric with respect to the interchange of similar bosons or fermions, respectively. From that moment on, the search for the necessary and sufficient conditions for ensuring that D corresponds to an N-particle wavefunction became known as the N-representability problem of the 2RDM. In what follows, I will limit our attention to the case of electrons.

Many necessary conditions on the N-representability of the 2RDM were obtained in the last half-century [13], and in principle the problem was formally solved [14]. Recently, the so called $(2, q)$-positivity conditions have been proposed by Mazziotti [15], where the number q corresponds to the higher qRDM that serves as the starting point for the derivation of the condition. The (2,2)- and (2,3)-positivity conditions correspond to the already known D, Q, G, $T1$, and $T2$ conditions [10, 16, 17], whereas when $q = r$, being r the rank of the one-electron basis set, the positivity conditions are complete. Unfortunately, a complete set of N-representability conditions that do not depend on higher-order RDMs remains unknown, so a tractable solution to the N-representability problem of the 2RDM has been not found.

On the contrary, the necessary and sufficient conditions that guarantee the ensemble N-representability of Γ and ρ are well established and are very easy to implement [10, 18, 19], hence an alternative would be to develop a functional theory based on them. Like D, Γ and ρ are simpler objects than the N-particle wavefunction and further reduce the number of coordinates on which the fundamental variable depends, namely, only six or three coordinates, respectively. Hence one-particle theories are very attractive, but can we achieve exact functionals of Γ and ρ?

Starting with the Thomas–Fermi theory [20, 21], extended later by Dirac [22], the beginnings of one-particle theories go back to the time of the appearance of quantum mechanics. Important developments were made in the Thomas–Fermi theory, including those of Professor March [23], until it reached the status of density functional theory (DFT) in 1964 [24]. That year, Hohenberg and Kohn (HK) showed that the ground-state electron density for some external potential determines every property of an electronic system. An extension of the original HK theorem that eliminates the v-representability requirement on ρ was later given by Levy [25], and extensively mathematically treated by Lieb [26].

A decade after the appearance of the HK theorem, Gilbert [18] proved its analog for Γ. This work together with those of Donnelly and Parr [27], Levy [25] and Valone [28] laid the groundwork for the 1RDM functional theory (1RDMFT). However, the computational schemes based on these formulations of DFT and 1RDMFT are several times more expensive than solving directly the Schrödinger equation. Accordingly, the answer to the previous question about the existence of exact functionals is affirmative, but not in a practical sense because these exact functionals do not have an appropriate form for computation. Practical applications of one-particle theories require another approach in the construction of functionals $E[\Gamma]$ or $E[\rho]$ for the ground-state energy.

In 1967 [29], Rosina had already demonstrated that there is a one-to-one mapping from the 2RDM to the N-particle wavefunction in the case of the ground state of a

Hamiltonian with at most two-body interactions. Taking advantage of the Rosina's theorem, Mazziotti defined [30] the universal functionals of Levy [25] and Valone [28] restricting the 2RDM to be pure or ensemble N-representable, respectively. Therefore, the existence theorems of one-particle functionals implicitly establish a one-to-one correspondence between the ground-state D and ground-state Γ and ρ, respectively. Consequently, functionals $E[\Gamma]$ and $E[\rho]$ must match the above-mentioned exact functional $E[D]$ as expected.

It is important to note that $E[D]$ reconstructions in DFT require greater effort than in 1RDMFT, since the non-interacting part of the Hamiltonian is actually a single-particle operator, so it has an explicit dependence on Γ. The unknown functional in a Γ-based theory only needs to reconstruct the electron-electron potential energy. This reflects an undeniable advantage of 1RDMFT with respect to DFT in reconstructing the exact functional $E[D]$, and it is not surprising that the main source of problems in DFT is related to the construction of kinetic energy functional.

It is evident that if we have the exact reconstruction of $E[D]$, either in terms of Γ or ρ, ensuring the N-representability of the fundamental variable will guarantee the N-representability also of the functional. However, this exact reconstruction has been an unattainable goal until now, and we have to settle for making approximations. One possibility may be to employ the exact energy expression $E[D]$ but using solely a reconstruction functional $D[\Gamma]$ or $D[\rho]$ as required. This implies that the exact ground-state energy will not, in general, be entirely rebuilt.

Approximating the energy functional has important consequences. First, the theorems obtained for the exact functionals $E[\Gamma]$ and $E[\rho]$ are no longer valid. The point is that an approximate functional still depends on the 2RDM [31]. This situation is completely analogous to that arising when approximate wavefunctions are used instead of the exact wavefunction. An undesired implication of the 2RDM dependence is that the functional N-representability problem arises, that is, we have to comply the requirement that D reconstructed in terms of Γ or ρ must satisfy the same N-representability conditions as those imposed on unreconstructed 2RDMs to ensure a physical value of the approximate ground-state energy. Otherwise, the functional approximation will not be correct since there will not be an N-electron system with an energy value $E[D]$. In summary, we are no longer really dealing with the 1RDMFT or DFT, but with approximate one-particle theories, where the 2RDM continues to play the dominant role.

Unfortunately, most of the approximate functionals currently in use are not N-representable, and that is why energy is often obtained far below true energy. It has been generally assumed that there is no N-representability problem of the functional, as it is believed that only N-representable conditions on the 1RDM or density are sufficient. The ensemble N-representability constraints for acceptable Γ or ρ are easy to implement, but are insufficient to guarantee that the reconstructed 2RDM is N-representable, and thereby the approximate functional either. To date, only a few papers have drawn attention to this problem. Among these exceptions are the work of Ayers and Liu [32] on N-representability in DFT, and the more recent work by Ludeña, Torres, and Costa [33] who also deals with N-representability in 1RDMFT.

In case of the density, the construction of approximate functionals through the reconstruction of the 2RDM has not been the norm. At the moment we only know some attempts like the one of Colle and Salvetti [34], complemented with the reconstruction of Lee, Yang and Parr [35], but it is not N-representable [36]. A similar situation is found in case of the 1RDM, where the approximate functionals have been proposed using heuristic or reasonable physical arguments [37]. Only the PNOFi ($i =$ 1–7) [38–41] family of functionals relies on the reconstruction of the 2RDM subject to necessary N-representability conditions. Remarkable is the case of PNOF5 [6, 7] which turned out to be strictly pure N-representable [8, 9].

Apart from the special case of the Hartree–Fock (HF) approximation that may be viewed as the simplest approximate $\Gamma-$functional, none of the known approximate functionals are explicitly given in terms of the 1RDM, including the familiar functional that accurately describes two-electron closed-shell systems [42, 43]. There are energy expressions, including those proposed by Muller [44], Csanyi and Arias [45], Sharma, Dewhurst, Lathiotakis, and Gros [46], that avoid the well-known phase dilemma [47] of the 1RDMFT, so they seem to depend properly on the 1RDM. However, these functionals violate the antisymmetric requirement for the 2RDM, consequently none of these functionals affords an N-representable 2RDM [48], nor can they reproduce the simplest two-electron case. Extensive N-representability violations have been recently reported [49] for these functionals. One can obtain quite reasonable results for some systems using them, but this does not guarantee that the calculations made are accurate since there is no N-particle density matrix that support their existence.

The functionals currently in use are constructed in the basis where the 1RDM is diagonal, which is the definition of a natural orbital functional (NOF). Accordingly, it is more appropriate to speak of a NOF rather than a functional of the 1RDM due to the existing dependence on the 2RDM. In this vein, in the NOF theory (NOFT) [50, 51], the natural orbitals (NOs) are the orbitals that diagonalize the 1RDM corresponding to an approximate energy expression, such as those obtained from an approximate wavefunction. This energy is not invariant with respect to a unitary transformation of the orbitals, and the resulting functional is only implicitly dependent on Γ or ρ through the contraction relations that determine them from D.

So far only the NOFT has proven to be able to take into account the functional N-representability in one-particle theories, thereby from now on we focus on it. This chapter continues with a presentation of the basic concepts and notations relevant to NOFT (Sect. 22.2). The following Sect. 22.3 is devoted to present our theory. Here, I discuss in details the reconstruction of the 2RDM that leads to PNOF approximations. The independent pair model PNOF5, as well as two results obtained with Professor March using PNOF5, namely, the behavior of the von Weizsäcker kinetic energy with the increase of N, and the calculation of chemical potential in neutral atoms, are analyzed in Sects. 22.4, 22.5 and 22.6, respectively. The chapter is ended with a discussion on the pure-state N-representability, which so far has been only accomplished for PNOF5 (Sect. 22.7).

22.2 Natural Orbital Functional Theory (NOFT)

The density matrix is the suitable mathematical object for describing an N-particle quantum mechanical system, since it is equally applicable to pure states and statistical ensembles [52]. Thus, consider an N-electron system described by the density matrix

$$\mathfrak{D} = \sum_i \omega_i \Psi_i \left(\mathbf{x}_1', \ldots, \mathbf{x}_\mathbf{N}'\right) \Psi_i^* \left(\mathbf{x}_1, \ldots, \mathbf{x}_\mathbf{N}\right) \tag{22.1}$$

In Eq. (22.1), ω_i are positive real numbers with sum one, so that $\mathfrak{D}$ corresponds to a sum of pure states with weight ω_i. Here and in the following $\mathbf{x} \equiv (\mathbf{r}, \mathbf{s})$ stands for the combined spatial and spin coordinates, $\mathbf{r}$ and $\mathbf{s}$, respectively.

The electronic energy E, in atomic units, for such N-electron system subject to an external potential $v(\mathbf{r})$ is an exactly and explicitly known functional of Γ and D, namely,

$$E = -\frac{1}{2}\int \nabla_1^2 \Gamma\left(\mathbf{r}_1', \mathbf{r}_1\right)|_{\mathbf{r}_1'=\mathbf{r}_1} d\mathbf{r}_1 + \int \Gamma\left(\mathbf{r}_1, \mathbf{r}_1\right) v\left(\mathbf{r}_1\right) d\mathbf{r}_1 + \int \frac{D\left(\mathbf{r}_1, \mathbf{r}_2; \mathbf{r}_1, \mathbf{r}_2\right)}{\left|\mathbf{r}_1 - \mathbf{r}_2\right|} d\mathbf{r}_1 d\mathbf{r}_2 \tag{22.2}$$

where Γ and D are obtained by contraction of $\mathfrak{D}$,

$$\Gamma\left(\mathbf{r}_1', \mathbf{r}_1\right) = N \sum_{\sigma_1} \int \mathfrak{D}\left(\mathbf{r}_1'\sigma_1 \ldots, \mathbf{r}_1\sigma_1 \ldots\right) d\mathbf{x}_2 \ldots d\mathbf{x}_N \tag{22.3a}$$

$$D\left(\mathbf{r}_1, \mathbf{r}_2; \mathbf{r}_1, \mathbf{r}_2\right) = \frac{N(N-1)}{2} \sum_{\sigma_1,\sigma_2} \int \mathfrak{D}\left(\mathbf{x}_1, \mathbf{x}_2, \ldots, \mathbf{x}_\mathbf{N}; \mathbf{x}_1, \mathbf{x}_2, \ldots, \mathbf{x}_\mathbf{N}\right) d\mathbf{x}_3 \ldots d\mathbf{x}_N. \tag{22.3b}$$

In Eq. (22.3), we employ Löwdin's normalization convention in which the trace of the 1RDM equals the number of electrons, and the trace of the 2RDM gives the number of electron pairs in the system.

The N-electron Hamiltonian does not contain any spin coordinates, hence both operators $\widehat{S}_z$ and $\widehat{S}^2$ commute with it. Consequently, the eigenfunctions of the Hamiltonian are also eigenfunctions of these two spin operators. For $\widehat{S}_z$ eigenstates, only density matrix blocks that conserve the number of each spin type are non-vanishing. Specifically, the 1RDM has two nonzero blocks $\Gamma^{\alpha\alpha}$ and $\Gamma^{\beta\beta}$, whereas the 2RDM has three independent nonzero blocks, $D^{\alpha\alpha}$, $D^{\alpha\beta}$, and $D^{\beta\beta}$. The parallel-spin components of the two-matrix must be antisymmetric, but $D^{\alpha\beta}$ possess no special symmetry [50].

In NOFT, the spectral decomposition of the 1RDM

$$\Gamma\left(\mathbf{x}', \mathbf{x}\right) = \sum_i n_i \phi_i \left(\mathbf{x}'\right) \phi_i^* \left(\mathbf{x}\right) \tag{22.4}$$

is used to approximate the electronic energy in terms of the NOs and their occupation numbers (ONs), namely,

$$E = \sum_{i} n_i \mathscr{H}_{ii} + \sum_{ijkl} D[n_i, n_j, n_k, n_l] \langle kl|ij \rangle \tag{22.5}$$

Here, $\mathscr{H}_{ii}$ denotes the diagonal elements of the core-Hamiltonian, $\langle kl|ij \rangle$ are the matrix elements of the two-particle interaction, and $D[n_i, n_j, n_k, n_l]$ represents the reconstructed 2RDM from the ONs.

Restriction of the ONs to the range $0 \leq n_i \leq 1$ represents a necessary and sufficient condition for ensemble N-representability of the 1RDM [10] under the normalization condition $\sum_i n_i = N$. The NOs $\{\phi_i(\mathbf{x})\}$ constitute a complete orthonormal set of single-particle functions,

$$\langle \phi_k | \phi_i \rangle = \int d\mathbf{x} \phi_k^* (\mathbf{x}) \, \phi_i (\mathbf{x}) = \delta_{ki} \tag{22.6}$$

with an obvious meaning of the Kronecker delta δ_{ki}.

For simplicity, we will address only singlet states in this chapter. The spin-orbital set $\{\phi_i(\mathbf{x})\}$ may be split into two subsets: $\{\varphi_p^\alpha(\mathbf{r})\, \alpha(\mathbf{s})\}$ and $\left\{\varphi_p^\beta(\mathbf{r})\, \beta(\mathbf{s})\right\}$. In order to avoid spin contamination effects, the spin restricted theory will be employed, in which a single set of orbitals is used for α and β spins: $\varphi_p^\alpha(\mathbf{r}) = \varphi_p^\beta(\mathbf{r}) = \varphi_p(\mathbf{r})$, and the parallel spin blocks of the RDMs are equal as well.

It should be noted that the first term of Eq. (22.5) has an explicit dependence on Γ, in contrast to ρ, so we do not need to reconstruct the non-interacting part of the electronic energy. In addition, we neglect any explicit dependence of D on the NOs themselves because the energy functional has already a strong dependence on the NOs via the one- and two-electron integrals. Consequently, the resulting approximate functional $E[N, \{n_i, \phi_i\}]$ can solely be implicitly dependent on Γ since the theorems on the existence of the functional $E[\Gamma]$ are valid only for the exact ground-state energy. In this vein, NOs are the orbitals that diagonalize the 1RDM corresponding to an approximate energy that still depends on the 2RDM *ergo* the energy is not invariant with respect to a unitary transformation of the orbitals. If we remember that D determines Γ, then it becomes clear that the construction of an N-representable functional given by Eq. (22.5) is related to the N-representability problem of $D[n_i, n_j, n_k, n_l]$.

22.3 PNOF Theory

A systematic application of the ensemble N-representability conditions in the reconstruction of $D[n_i, n_j, n_k, n_l]$ by means of an explicit approximation [38] of the two-particle cumulant has led to the series of Piris NOFs. Consider the cumulant expansion [53] of the 2RDM in the NO representation,

$$D_{kl,ij} = \frac{n_i n_j}{2}\left(\delta_{ki}\delta_{lj} - \delta_{li}\delta_{kj}\right) + \lambda_{kl,ij}. \tag{22.7}$$

Here, the 2RDM has been partitioned into an antisymmetric product of the 1RDMs, which is simply the HF approximation, and a correction λ to it. The latter is called the cumulant or correlation matrix. It is worth noting that matrix elements of $\boldsymbol{\lambda}$ are non-vanishing only if all its labels refer to partially occupied NOs with ONs different from 0 or 1.

The cumulant matrix must fulfill as many necessary N-representability conditions as possible to ensure the N-representability of D since requiring all conditions is not practicable due to their dependence on higher-order RDMs. The use of the (2,2)-positivity N-representability conditions [15] for reconstructing $\boldsymbol{\lambda}$ was proposed in reference [38]. This particular reconstruction is based on the introduction of two auxiliary matrices $\boldsymbol{\Delta}$ and Π expressed in terms of the ONs. In a spin restricted formulation, the structure of the two-particle cumulant is

$$\lambda^{\sigma\sigma}_{pq,rt} = -\frac{1}{2}\Delta_{pq}\left(\delta_{pr}\delta_{qt} - \delta_{pt}\delta_{qr}\right) \tag{22.8a}$$

$$\lambda^{\alpha\beta}_{pq,rt} = -\frac{1}{2}\Delta_{pq}\delta_{pr}\delta_{qt} + \frac{1}{2}\Pi_{pr}\delta_{pq}\delta_{rt}. \tag{22.8b}$$

$\boldsymbol{\Delta}$ is a real symmetric matrix, whereas Π is a spin-independent Hermitian matrix. The N-representability D and Q conditions of the 2RDM impose the following inequalities on the off-diagonal elements of $\boldsymbol{\Delta}$ [38],

$$\Delta_{qp} \leq n_q n_p, \qquad \Delta_{qp} \leq h_q h_p, \tag{22.9}$$

while to fulfill the G condition, the elements of the Π-matrix must satisfy the constraint [54]

$$\Pi^2_{qp} \leq \left(n_q h_p + \Delta_{qp}\right)\left(h_q n_p + \Delta_{qp}\right), \tag{22.10}$$

where h_p denotes the hole $1 - n_p$ in the spatial orbital p. Furthermore, the sum rules that must fulfill the blocks of the cumulant yield a sum rule for $\boldsymbol{\Delta}$,

$$\sum_{p,q;\,p\neq q} \Delta_{qp} = n_p h_p. \tag{22.11}$$

Within this reconstruction, the energy for singlet states reads as

$$E = \sum_p n_p\left(2\mathcal{H}_{pp} + \mathcal{J}_{pp}\right) + \sum_{p,q;\,p\neq q} \Pi_{qp}\mathcal{L}_{pq} + \sum_{p,q;\,p\neq q}\left(n_q n_p - \Delta_{qp}\right)\left(2\mathcal{J}_{pq} - \mathcal{K}_{pq}\right), \tag{22.12}$$

where $\mathcal{J}_{pq} = \langle pq|pq\rangle$ and $\mathcal{K}_{pq} = \langle pq|qp\rangle$ are the usual direct and exchange integrals, respectively. $\mathcal{J}_{pp}$ is the Coulomb interaction between two electrons with opposite spins at the spatial orbital p, whereas $\mathscr{L}_{pq} = \langle pp|qq\rangle$ is the exchange and time-inversion integral [55], so the functional (22.12) belongs to the $\mathscr{J}\mathscr{K}\mathscr{L}$-only family of NOFs.

The conservation of $\widehat{S}^2$ allows to derive the diagonal elements $\Delta_{pp} = n_p^2$ and $\Pi_{pp} = n_p$ [56]. Appropriate forms of matrices $\boldsymbol{\Delta}$ and $\boldsymbol{\Pi}$ lead to different implementations of the NOF known in the literature as PNOFi (i = 1–7) [38–41]. The performance of these functionals is comparable to those of best quantum chemistry methods in many cases, being particularly interesting the case of PNOF3 [57] in relation with the functional N-representability.

PNOF3 showed [57] exceptional performance for atoms and molecules, both for spin-compensated and spin-non-compensated. This NOF can describe the correct topology of potential energy surfaces highly sensitive to electron correlation, giving reaction barriers and isomerization energies with an accuracy of less than 1 kcal/mol [58]. However, a closer analysis of the dissociation curves for various diatomics [54], as well as the description of diradicals and diradicaloids [59], revealed that PNOF3 overestimates the amount of electron correlation, when orbital near-degeneracy effects become important.

This is a paradigmatic case of how highly accurate results achieved with a functional can be misleading. The PNOF3 satisfies the D and Q N-representability conditions given by Eq. (22.9), but violates the G condition (22.10). It was demonstrated [54] that the ill behavior of PNOF3 is related to the violation of the latter conditions. Only the progressive inclusion of N-representability conditions can lead to reconstructions with physical meaning since it is the only way to ensure that the energy corresponds to a density matrix of N electrons. This approach has been called the bottom-up method [33].

22.4 Independent Pair Approximation

Let's divide the orbital space Ω into $N/2$ mutually disjoint subspaces Ω_g, so each orbital belongs only to one subspace. Consider each subspace contains one orbital g below the Fermi level ($N/2$), and N_g orbitals above it, which is reflected in additional sum rules for the ONs:

$$\sum_{p \in \Omega_g} n_p = 1; \quad g = 1, 2, \ldots, N/2. \tag{22.13}$$

Taking into account the spin, each subspace contains solely an electron pair, and the normalization condition for Γ ($2\sum_p n_p = N$) is automatically fulfilled. Coupling each orbital g below the Fermi level with only one orbital above it ($N_g = 1$) also leads to an orbital perfect pairing. It is important to note that orbitals satisfying the pairing conditions (22.13) are not required to remain fixed throughout the orbital optimization process [60].

The simplest way to comply with all required constraints leads to an independent pair model (PNOF5) [6, 7]:

$$\Delta_{qp} = n_p^2 \delta_{qp} + n_q n_p \left(1 - \delta_{qp}\right) \delta_{q\Omega_g} \delta_{p\Omega_g} \tag{22.14a}$$

$$\Pi_{qp} = n_p \delta_{qp} + \Pi_{qp}^g \left(1 - \delta_{qp}\right) \delta_{q\Omega_g} \delta_{p\Omega_g} \tag{22.14b}$$

$$\Pi_{qp}^g = \begin{cases} -\sqrt{n_q n_p} & p = g \text{ or } q = g \\ +\sqrt{n_q n_p} & p, q > N/2 \end{cases} \tag{22.14c}$$

$$\delta_{q\Omega_g} = \begin{cases} 1, & q \in \Omega_g \\ 0, & q \notin \Omega_g \end{cases} \quad g = 1, 2, \ldots, N/2. \tag{22.14d}$$

Given this functional form of the auxiliary matrices $\boldsymbol{\Delta}$ and Π, the energy (22.12) of the PNOF5 can be conveniently written as

$$E = \sum_{g=1}^{N/2} E_g + \sum_{f \neq g}^{N/2} E_{fg} \tag{22.15a}$$

$$E_g = \sum_{p \in \Omega_g} n_p \left(2\mathscr{H}_{pp} + \mathscr{J}_{pp}\right) + \sum_{p,q \in \Omega_g, p \neq q} \Pi_{qp}^g \mathscr{L}_{pq} \tag{22.15b}$$

$$E_{fg} = \sum_{q \in \Omega_f} \sum_{p \in \Omega_g} n_q n_p \left(2\mathscr{J}_{pq} - \mathscr{K}_{pq}\right). \tag{22.15c}$$

The first term of the energy draws the system as $N/2$ independent electron pairs, whereas the second term contains the contribution to the HF mean-field of the electrons belonging to different pairs.

Several performance tests have shown that PNOF5 yields remarkably accurate descriptions of systems with near-degenerate one-particle states and dissociation processes [4, 61–63]. In this sense, the results obtained with PNOF5 for the electronic structure of transition metal complexes are probably the most relevant [64]. This functional correctly takes into account the multiconfigurational nature of the ground state of the chromium dimer, known as a benchmark molecule for quantum chemical methods due to the extremely challenging electronic structure of the ground state and potential energy curve.

PNOF5 has also been successfully used to predict vertical ionization potentials and electron affinities of a selected set of organic and inorganic spin-compensated molecules, by means of the extended Koopmans' theorem [65]. The improvement due to the inclusion of more orbitals in the description of each pair was also observed by visualizing the electron densities by means of the Bader's theory of atoms in molecules in the case of a set of light atomic clusters: Li_2, Li_3^+, Li_4^{2+} and H_3^+ [7]. The size-consistency property, and the fact that the functional tends to localize spatially the NOs, make PNOF5 an exceptional candidate for fragment calculations. The latter showed a fast convergence, which allowed the treatment of extended system at a fractional cost of the whole calculation [66].

As mentioned above, NOFs still depend explicitly on the 2RDM, so the energy is not invariant with respect to a unitary transformation of the orbitals. Because of this, the NOFT provides two complementary representations of the one-electron picture, namely, the NO representation and the canonical orbital representation [67]. The former arises directly from the optimization process solving the corresponding Euler equations, whereas the latter is attained from the diagonalization of the matrix of Lagrange multipliers obtained in the NO representation. Both set of orbitals represent unique correlated one-electron pictures of the same energy minimization problem, *ergo,* they complement each other in the analysis of the molecular electronic structure. The orbitals obtained in both representations, using the electron pairing approaches in NOFT, have shown [63, 68–70] that the electron pairs with opposite spins continue to be a suitable language for the chemical bond theory.

In this tribute to Professor March, two results obtained with PNOF5 deserve special attention: the behavior of the von Weizsäcker kinetic energy with the increase of N [1], and the calculation of chemical potential in neutral atoms [2].

22.5 von Weizsäcker Kinetic Energy Term for Diatomic Molecules

In 1983, Mucci and March reported [71] on the importance of density gradients for molecular binding in diatomics. The simplest density gradient kinetic energy term was introduced by von Weizsäcker [72] as follows

$$T_w = \frac{\hbar^2}{8m} \int \frac{|\nabla \rho(\mathbf{r})|^2}{\rho(\mathbf{r})} d\mathbf{r} \tag{22.16}$$

Recently [1], Eq. (22.16) was used to calculate the von Weizsäcker energy for some 30 homonuclear diatomic molecules at their experimental equilibrium geometry employing densities obtained from PNOF5. We considered only spin singlets for these molecules, although in some cases their ground state are triplets. In addition, all calculations were made at the non-relativistic level of theory.

Our results, from H_2 to Ge_2, reveal [1] a slowly varying character of T_w/N^2, where N is the total number of electrons in the molecule under consideration. For example, from N_2 to Ge_2, the variation was from 0.44 to 0.49 atomic units. This led us to conclude that the Schrödinger's non-relativistic equation predicts, at the equilibrium, the following relationship

$$T_w \rightarrow \frac{1}{2} N^2 \tag{22.17}$$

as N becomes large, say at Cr_2 and higher N cases.

22.6 The Chemical Potential

The solution in PNOF5 theory is established by optimizing the energy functional (22.15) with respect to the ONs and NOs separately, for which the iterative diagonalization procedure proposed by Piris and Ugalde [60] is employed. The orbitals must satisfy the orthonormality requirement, Eq. (22.6), whereas the occupancies have to conform to the electron pairing constraints (22.13) and ensemble N-representability conditions for the 1RDM ($0 \leq n_p \leq 1$). The latter bounds can be easily enforced by setting $n_p = \cos^2 \gamma_p$, and varying $\{\gamma_p\}$ without these constraints. The other conditions may be easily taken into account by the method of Lagrange multipliers.

Let us focus on the minimization of $\{\gamma_p\}$ for a fixed set of NOs. We associate the Lagrange multipliers $\{\mu_g\}$ with conditions (22.13), and define the auxiliary functional Ω by the formula

$$\Omega\left[N, \{\gamma_p\}\right] = E - 2\sum_{g=1}^{N/2} \mu_g \left(\sum_{p \in \Omega_g} \cos^2 \gamma_p - 1 \right) \tag{22.18}$$

The functional (22.18) has to be stationary with respect to variations in $\{\gamma_p\}$,

$$\delta\Omega = \sum_{g=1}^{N/2} \sum_{p \in \Omega_g} \sin\left(2\gamma_p\right) \left[2\mu_g - \frac{\partial E}{\partial n_p} \right] d\gamma_p = 0 \tag{22.19}$$

The partial derivative $\left(\partial E / \partial n_p\right)$ is taken holding the orbitals fixed. It satisfies the relation

$$\frac{\partial E}{\partial n_p} = 2\mathcal{H}_{pp} + \frac{\partial V_{ee}}{\partial n_p} = 2\mu_g, \quad \forall p \in \Omega_g, \tag{22.20}$$

where the partial derivative of electron-electron repulsion energy V_{ee} is given by the expression

$$\frac{\partial V_{ee}}{\partial n_p} = \mathcal{J}_{pp} + 2 \sum_{q \in \Omega_g, q \neq p} \frac{\partial \Pi_{qp}^g}{\partial n_p} \mathcal{L}_{qp} + 2 \sum_{f \neq g}^{N/2} \sum_{q \in \Omega_f} n_q \left(2\mathcal{J}_{qp} - \mathcal{K}_{qp}\right), \ \forall p \in \Omega_g. \tag{22.21}$$

Accordingly, the Lagrange multipliers $\{\mu_g\}$ can be written as

$$\mu_g = \mathcal{H}_{pp} + \frac{1}{2} \frac{\partial V_{ee}}{\partial n_p}, \quad \forall p \in \Omega_g \tag{22.22}$$

It follows from Eq. (22.22) that we have $N/2$ Lagrange multipliers which can be identified as chemical potentials in virtue of the result obtained in [73]. An infinitesimal change in the number of electrons is energetically more advantageous when added to an orbital associated with the smallest μ_g. Therefore, we identify [2] the latter with the chemical potential μ of an N-electron open system in a singlet ground state.

In Ref. [2], calculations of the chemical potential in 36 neutral atoms of the first three rows of periodic table (H-Kr) were presented and compared and contrasted with available experimental values. The chemical potential of a single atom in its multiplet ground state was obtained from the calculation made in the corresponding dimer in a singlet state considering the homonuclear diatomics at infinite separation (50 Å) between atoms. This is possible because the functional PNOF5 is size-extensive and size-consistent [7]. In general, our results showed that the chemical potentials overestimate the experimental values of the ionization (I) although the expected oscillatory behavior was maintained. There were exceptions where the value of μ lies quite close to the experimental marks, namely, H, Li, B, C, N, Na, Al, Si, P, K, Ga, Ge, and As.

In case of noble gas atoms, the chemical potential doubled approximately the value of $-I$. It is known [73] that μ presents a discontinuity at integer particle number equals to the fundamental gap, i.e., the difference between the electron affinity and the ionization potential for a given number of electrons. In closed-shell systems, like noble gas atoms, no energy is given off when a neutral atom of these elements picks up an electron, hence, there is no electron affinity, or these atoms have electron affinities with opposite sign. Our estimation of the chemical potentials reflected this situation by increasing the expected value referred to the ionization potential.

22.7 Pure N-Representability

So far in this work, we have focused on the N-representability problem for statistical ensembles. The fact that Coleman obtained [10] the restrictions on the ONs ($0 \leq n_i \leq 1$, $\sum_i n_i = N$), also known as Pauli constraints, so that the 1RDM is representable by at least one N-electron density matrix, prompted the search for similar conditions on the pure-state N-representability. In 1972, Borland and Dennis showed [74] computationally that additional constraints on Γ are necessary for it to be representable by at least one pure N-electron density matrix. It was not until very recently that these conditions named generalized Pauli constraints (GPCs) have been obtained [75, 76], taking the form

$$\kappa_{0j} + \sum_{i=1}^{r} \kappa_{ij} n_i \geq 0, \qquad j = 1, 2, \ldots < \infty, \tag{22.23}$$

where $\{\kappa_{ij}\}$ are integer constants, r is the rank of the orbital space Ω, and the ONs are decreasingly ordered. The Pauli conditions define the set $E(N, r)$ of ensemble N-representable 1RDMs, whereas the inequalities (22.23) in combination with the formers define the set $P(N, r)$ of pure N-representable 1RDMs. Spectra $\{n_i\}$ lying outside of the pure set can only correspond to a mixed state.

A relevant result on the necessity of using the pure-state N-representability conditions in the minimization of the exact functional $E[\Gamma]$, was obtained by Nguyen-Dang, Ludeña and Tal [77]. They demonstrated that if we define an universal energy functional in the domain $E(N,r)$ then it will be exactly equal to the universal energy functional defined in the domain $P(N,r)$ over this last set, that is, over the set of pure-state N-representable 1RDMs. This equality justified the use of the ensemble N-representability necessary and sufficient conditions on Γ, but it required an adequate construction of the functional which would guarantee that $\Gamma \in P(N,r)$, as they showed. In fact, they corroborated Valone's thesis [28] that in order to guarantee the pure-state N-representability conditions in the minimization of the energy only the ensemble constraints are necessary if the functional is an appropriate one. But what will happen if the functional is approximate?

If the approximate NOF is strictly pure-state N-representable, i.e., it is obtained from the reconstruction of a strictly pure-state N-representable 2RDM, then the 1RDM that D determines by contraction will also be automatically pure-state N-representable. Therefore, for approximate functionals it is also valid to take into account solely the ensemble constraints in the minimization with respect to Γ if the functional is an appropriate one. A palpable example is PNOF5.

Two years after PNOF5 was proposed using the bottom-up method (vide supra Sect. 22.4), the natural geminals of PNOF5 were analyzed [78] and it was realized by Pernal [8] that this NOF corresponds to the energy obtained from a wavefunction of an antisymmetrized product of strongly orthogonal geminals if the expansion of the $N/2$ geminals is limited to two-dimensional subspaces with fixed signs for the expansion coefficients of the corresponding geminals. Shortly after this ansatz was extended to include more orbitals in the description of the electron pairs [7]. This finding demonstrated that PNOF5 is strictly a pure-state N-representable functional.

Looking more closely at the Eq. (22.13), we realize that the pairing conditions meet the requirements to be GPCs (22.23). Indeed, the equality conditions obtained [79] for 3 electrons in a space of 6 natural spin-orbitals are exactly the perfect-pairing conditions proposed in Ref. [6] to satisfy the sum rules (22.11) for Δ-matrix, i.e., for the two-electron cumulant in PNOF5. It is important to note that the success of PNOF5 relies on the use of GPCs in the formulation of the reconstruction $D[n_i, n_j, n_k, n_l]$, and not in its later use to limit the domain of the trial one-matrices.

GPCs have recently [80] received increased attention in NOFT following the discovery of a systematic way to derive them for any number of electrons and any finite dimensionality of the Hilbert space. Unfortunately, it has been found that the number of conditions increases dramatically with the number of NOs, so it can be quite difficult to handle them in practical implementations. On the other hand, the verification of the fulfillment of these conditions for some current non-N-representable approximations in 3-electron systems showed [80] that, in all cases, some of the pure-state conditions were violated. The enforcement of additional GPCs in the minimization led to an improvement of the total energy and of the optimal 1RDM with respect to those obtained by imposing only the Pauli constraints.

There is no doubt that the application of GPCs restricts the 1RDM variational space leading to improvements in energy and Γ, but it does not improve the reconstruction of the approximate functional per se. A 1RDM that represents a pure state does not guarantee that the reconstructed electron-electron potential energy will be pure-state N-representable, except that the reconstruction is the exact one, something that until now has not been possible to reach. The functional N-representability problem continues to exist for pure states when we make approximations for the functional, and it is still related to the N-representability, in this case for pure states, of the 2RDM.

22.8 Closing Remarks

In this chapter, the N-representability problem in one-particle functional theories has been analyzed. It turns out that here this problem is twofold. First, we have the N-representability problem of the fundamental variable, Γ or ρ, as the case may be. Fortunately, the necessary and sufficient conditions that guarantee the ensemble N-representability of Γ and ρ are well established and are very easy to implement, which makes one-particle theories extremely attractive. The other side however is trickier.

On the one hand we have that for the exact functional of the density or 1RDM, it suffices to require the N-representability of the fundamental variable to guarantee that the ground-state energy corresponds to an N-electron density matrix. However, exact formulations of DFT and 1RDMFT do not have an appropriate form for computation and we have to settle for making approximations, which imply that the theorems obtained for the exact functionals $E[\Gamma]$ and $E[\rho]$ are no longer valid.

The main approach route is to use the well-known exact energy expression $E[D]$ with a reconstruction $D[\Gamma]$ or $D[\rho]$ as required. Such an approximate functional still depends on the 2RDM. An eventual outcome of this D-dependence is the functional N-representability problem, that is, for a 2RDM reconstructed in terms of Γ or ρ we have to impose the same N-representability conditions that we enforce on unreconstructed 2RDMs to ensure a physical value of the approximate ground-state energy. This is the only way of assurance that there is an N-electron system with an energy value $E[D]$. We are no longer really dealing with the 1RDMFT or DFT, but with approximate one-particle theories, where the 2RDM continues to play the dominant role.

Most of the approximate functionals currently in use are not N-representable hence this second side of the N-representability in one-particle functional theories has not received sufficient attention. Breaking the particle number symmetry can sometimes give qualitative descriptions of the phenomena at cost of the good quantum number N. These types of methods are well known in physics but a projection is always imposed to the correct N-conserving quantum state.

So far only the NOF theory, where functionals are constructed in the basis where the 1RDM is diagonal, has proven to be able to take into account the functional N-representability in one-particle theories. In particular, the PNOF family of functionals relies on the reconstruction of the 2RDM subject to necessary

N-representability conditions. Remarkable is the case of PNOF5 which turned out to be strictly pure N-representable since it maintains a one-to-one correspondence with the energy obtained from an N-particle wavefunction.

The most direct method to generate a pure-state N-representable NOF is by reducing the energy expression obtained from an approximate N-particle wavefunction to a functional of the occupation numbers and natural orbitals. By doing this, we automatically avoid the N-representability problem of the 2RDM, or what is the same, of the functional. However, this is a formidable task that is far from being something attainable in most cases. Only PNOF5 has achieved this goal including the electron correlation.

The success of PNOF5 relies on the use of additional pairing conditions in the reconstruction of the 2RDM, which can be seen as generalized Pauli constraints. The use of these conditions when reconstructing the 2RDM will give rise to N-representable functionals that will go beyond the independent pair model.

Finally, if the approximate NOF is strictly pure-state N-representable, i.e., it is obtained from the reconstruction of a strictly pure-state N-representable 2RDM, then the 1RDM will also be automatically pure-state N-representable. Therefore, for approximate functionals it is valid to take into account solely the ensemble constraints in the minimization with respect to the 1RDM if a pure-state N-representable functional is employed.

Acknowledgements Financial support comes from Eusko Jaurlaritza (Ref. IT588-13) and Ministerio de Economía y Competitividad (Ref. CTQ2015-67608-P). The SGI/IZO–SGIker UPV/EHU is gratefully acknowledged for generous allocation of computational resources.

References

1. M. Piris, N.H. March, Phys. Chem. Liq. **52**(6), 804 (2014). https://doi.org/10.1080/00319104.2014.937865
2. M. Piris, N.H. March, Phys. Chem. Liq. **53**(6), 696 (2015). https://doi.org/10.1080/00319104.2015.1029478
3. M. Piris, N.H. March, J. Phys. Chem. A **119**(40), 10190 (2015). https://doi.org/10.1021/acs.jpca.5b02788
4. M. Piris, N.H. March, Phys. Chem. Liq. **54**(6), 797 (2016). https://doi.org/10.1080/00319104.2016.1166364
5. M. Piris, N.H. March, Int. J. Quantum Chem. **116**(11), 805 (2016). https://doi.org/10.1002/qua.25039
6. M. Piris, X. Lopez, F. Ruipérez, J.M. Matxain, J.M. Ugalde, J. Chem. Phys. **134**(16), 164102 (2011). https://doi.org/10.1063/1.3582792
7. M. Piris, J.M. Matxain, X. Lopez, J. Chem. Phys. **139**(23), 234109 (2013). https://doi.org/10.1063/1.4844075
8. K. Pernal, Comput. Theor. Chem. **1003**, 127 (2013). https://doi.org/10.1016/j.comptc.2012.08.022
9. M. Piris, J. Chem. Phys. **139**(6), 064111 (2013). https://doi.org/10.1063/1.4817946
10. A.J. Coleman, Rev. Mod. Phys. **35**(3), 668 (1963). https://doi.org/10.1103/RevModPhys.35.668

11. K. Husimi, Proc. Phys. Math. Soc. Jpn. **22**(4), 264 (1940). https://doi.org/10.11429/ppmsj1919.22.4_264
12. A.J. Coleman, in Mazziotti [81], Chap. 1, pp. 3–9
13. M. Rosina, in Mazziotti [81], pp. 11–18
14. A.J. Coleman, V.I. Yukalov, *Reduced Density Matrices: Coulson's Challenge*. Lecture Notes in Chemistry, vol. 72 (Springer, New York, 2000). ISBN 9783540671480
15. D.A. Mazziotti, Phys. Rev. Lett. **108**, 263002 (2012). https://doi.org/10.1103/PhysRevLett.108.263002
16. C. Garrod, J.K. Percus, J. Math. Phys. **5**(12), 1756 (1964). https://doi.org/10.1063/1.1704098
17. R.M. Erdahl, Int. J. Quantum Chem. **13**(6), 697 (1978). https://doi.org/10.1002/qua.560130603
18. T.L. Gilbert, Phys. Rev. B **12**, 2111 (1975). https://doi.org/10.1103/PhysRevB.12.2111
19. J.E. Harriman, Phys. Rev. A **24**, 680 (1981). https://doi.org/10.1103/PhysRevA.24.680
20. L.H. Thomas, Math. Proc. Camb. Philos. Soc. **23**, 542 (1926). https://doi.org/10.1017/S0305004100011683
21. E. Fermi, Rendiconti dell'Accademia Nazionale dei Lincei **6**, 602 (1927)
22. P.A.M. Dirac, Proc. Camb. Philos. Soc. **26**, 376 (1930). https://doi.org/10.1017/S0305004100016108
23. N.H. March, Adv. Phys. **6**(21), 1 (1957). https://doi.org/10.1080/00018735700101156
24. P.C. Hohenberg, W. Kohn, Phys. Rev. **136**, B864 (1964). https://doi.org/10.1103/PhysRev.136.B864
25. M. Levy, Proc. Natl. Acad. Sci. **76**(12), 6062 (1979)
26. E.H. Lieb, Int. J. Quantum Chem. **24**(3), 243 (1983). https://doi.org/10.1002/qua.560240302
27. R.A. Donnelly, R.G. Parr, J. Chem. Phys. **69**(10), 4431 (1978). https://doi.org/10.1063/1.436433
28. S.M. Valone, J. Chem. Phys. **73**(3), 1344 (1980). https://doi.org/10.1063/1.440249
29. M. Rosina, in *Reduced Density Operators With Application to Physical and Chemical Systems*, ed. by A.J. Coleman, R.M. Erdhal. Queens Papers in Pure and Applied Mathematics, vol. 11 (Queens University, Kingston, 1967), p. 369
30. D.A. Mazziotti, Chem. Phys. Lett. **338**(4), 323 (2001). https://doi.org/10.1016/S0009-2614(01)00251-2
31. R.A. Donnelly, J. Chem. Phys. **71**(7), 2874 (1979). https://doi.org/10.1063/1.438678
32. P.W. Ayers, S. Liu, Phys. Rev. A **75**, 022514 (2007). https://doi.org/10.1103/PhysRevA.75.022514
33. E.V. Ludeña, F.J. Torres, C. Costa, J. Mod. Phys. **4**(3A), 391 (2013). https://doi.org/10.4236/jmp.2013.43A055
34. R. Colle, O. Salvetti, Theor. Chim. Acta **37**, 329 (1975). https://doi.org/10.1007/BF01028401
35. C. Lee, W. Yang, R.G. Parr, Phys. Rev. B **37**, 785 (1988). https://doi.org/10.1103/PhysRevB.37.785
36. R.C. Morrison, Int. J. Quantum Chem. **46**(4), 583 (1993). https://doi.org/10.1002/qua.560460406
37. K. Pernal, K.J.H. Giesbertz, in *Density-Functional Methods for Excited States*, ed. by N. Ferré, M. Filatov, M. Huix-Rotllant (Springer International Publishing, Cham, 2016), pp. 125–183. https://doi.org/10.1007/128_2015_624. [Top. Curr. Chem. **368**], ISBN 978-3-319-22081-9
38. M. Piris, Int. J. Quantum Chem. **106**(5), 1093 (2006). https://doi.org/10.1002/qua.20858
39. M. Piris, Int. J. Quantum Chem. **113**(5), 620 (2013). https://doi.org/10.1002/qua.24020
40. M. Piris, J. Chem. Phys. **141**(4), 044107 (2014). https://doi.org/10.1063/1.4890653
41. M. Piris, Phys. Rev. Lett. (2017), to be published
42. P.O. Löwdin, H. Shull, Phys. Rev. **101**, 1730 (1956). https://doi.org/10.1103/PhysRev.101.1730
43. S. Goedecker, C.J. Umrigar, in *Many-electron Densities and Reduced Density Matrices*, ed. by J. Cioslowski (Kluwer Academic/Plenum Press, New York, 2000), pp. 165–181
44. A.M.K. Müller, Phys. Lett. A **105**(9), 446 (1984). https://doi.org/10.1016/0375-9601(84)91034-X

45. G. Csányi, T.A. Arias, Phys. Rev. B **61**, 7348 (2000). https://doi.org/10.1103/PhysRevB.61.7348
46. S. Sharma, J.K. Dewhurst, N.N. Lathiotakis, E.K.U. Gross, Phys. Rev. B **78**, 201103(R) (2008). https://doi.org/10.1103/PhysRevB.78.201103
47. J. Cioslowski, K. Pernal, J. Chem. Phys. **120**(22), 10364 (2004). https://doi.org/10.1063/1.1738411
48. J.M. Herbert, J.E. Harriman, J. Chem. Phys. **118**(24), 10835 (2003). https://doi.org/10.1063/1.1574787
49. M. Rodriguez-Mayorga, E. Ramos-Cordoba, M. Via-Nadal, M. Piris, E. Matito, Phys. Chem. Chem. Phys. (2017)
50. M. Piris, in Mazziotti [81], Chap. 14, pp. 387–427
51. M. Piris, J.M. Ugalde, Int. J. Quantum Chem. **114**(18), 1169 (2014). https://doi.org/10.1002/qua.24663
52. J. von Neumann, *Mathematische Grundlagen der Quantenmechanik* (Springer, Berlin, 1932)
53. D.A. Mazziotti, Chem. Phys. Lett. **289**(5), 419 (1998). https://doi.org/10.1016/S0009-2614(98)00470-9
54. M. Piris, J.M. Matxain, X. Lopez, J.M. Ugalde, J. Chem. Phys. **133**(11), 111101 (2010). https://doi.org/10.1063/1.3481578
55. M. Piris, J. Math. Chem. **25**(1), 47 (1999). https://doi.org/10.1023/A:1019111828412
56. M. Piris, J.M. Matxain, X. Lopez, J.M. Ugalde, J. Chem. Phys. **131**(2), 021102 (2009). https://doi.org/10.1063/1.3180958
57. M. Piris, J.M. Matxain, X. Lopez, J.M. Ugalde, J. Chem. Phys. **132**(3), 031103 (2010). https://doi.org/10.1063/1.3298694
58. X. Lopez, M. Piris, J.M. Matxain, J.M. Ugalde, Phys. Chem. Chem. Phys. **12**, 12931 (2010). https://doi.org/10.1039/C003379K
59. X. Lopez, F. Ruipérez, M. Piris, J.M. Matxain, J.M. Ugalde, Chem. Phys. Chem. **12**(6), 1061 (2011). https://doi.org/10.1002/cphc.201100136
60. M. Piris, J.M. Ugalde, J. Comput. Chem. **30**(13), 2078 (2009). https://doi.org/10.1002/jcc.21225
61. J.M. Matxain, M. Piris, F. Ruiperez, X. Lopez, J.M. Ugalde, Phys. Chem. Chem. Phys. **13**, 20129 (2011). https://doi.org/10.1039/C1CP21696A
62. X. Lopez, F. Ruipérez, M. Piris, J.M. Matxain, E. Matito, J.M. Ugalde, J. Chem. Theory Comput. **8**(8), 2646 (2012). https://doi.org/10.1021/ct300414t
63. J.M. Matxain, F. Ruipérez, I. Infante, X. Lopez, J.M. Ugalde, G. Merino, M. Piris, J. Chem. Phys. **138**(15), 151102 (2013). https://doi.org/10.1063/1.4802585
64. F. Ruipérez, M. Piris, J.M. Ugalde, J.M. Matxain, Phys. Chem. Chem. Phys. **15**, 2055 (2013). https://doi.org/10.1039/C2CP43559D
65. M. Piris, J.M. Matxain, X. Lopez, J.M. Ugalde, J. Chem. Phys. **136**(17), 174116 (2012). https://doi.org/10.1063/1.4709769
66. X. Lopez, M. Piris, Theor. Chem. Acc. **134**(12), 151 (2015). https://doi.org/10.1007/s00214-015-1756-x
67. M. Piris, J.M. Matxain, X. Lopez, J.M. Ugalde, Theor. Chem. Acc. **132**(2), 1298 (2013). https://doi.org/10.1007/s00214-012-1298-4
68. J.M. Matxain, M. Piris, J.M. Mercero, X. Lopez, J.M. Ugalde, Chem. Phys. Lett. **531**, 272 (2012). https://doi.org/10.1016/j.cplett.2012.02.041
69. J.M. Matxain, M. Piris, J. Uranga, X. Lopez, G. Merino, J.M. Ugalde, Chem. Phys. Chem. **13**(9), 2297 (2012). https://doi.org/10.1002/cphc.201200205
70. M. Piris, X. Lopez, J.M. Ugalde, Chem. Eur. J. **22**(12), 4109 (2016). https://doi.org/10.1002/chem.201504491
71. J.F. Mucci, N.H. March, J. Chem. Phys. **78**(10), 6187 (1983). https://doi.org/10.1063/1.444582
72. C.F. von Weizsäcker, Z. Phys. **96**, 431 (1935). https://doi.org/10.1007/BF01337700
73. N. Helbig, N.N. Lathiotakis, M. Albrecht, E.K.U. Gross, Europhys. Lett. **77**(6), 67003 (2007)
74. R.E. Borland, K. Dennis, J. Phys. B At. Mol. Phys. **5**(1), 7 (1972). https://doi.org/10.1088/0022-3700/5/1/009

75. A.A. Klyachko, J. Phys. Conf. Ser. **36**(1), 72 (2006). https://doi.org/10.1088/1742-6596/36/1/014
76. M. Altunbulak, A. Klyachko, Commun. Math. Phys. **282**(2), 287 (2008). https://doi.org/10.1007/s00220-008-0552-z
77. T.T. Nguyen-Dang, E.V. Ludeña, Y. Tal, J. Mol. Struct. THEOCHEM **120**, 247 (1985). https://doi.org/10.1016/0166-1280(85)85114-9
78. M. Piris, Comput. Theor. Chem. **1003**, 123 (2013). https://doi.org/10.1016/j.comptc.2012.07.016
79. M.B. Ruskai, J. Phys. A Math. Theor. **40**(45), F961 (2007). https://doi.org/10.1088/1751-8113/40/45/F01
80. I. Theophilou, N.N. Lathiotakis, M.A.L. Marques, N. Helbig, J. Chem. Phys. **142**(15), 154108 (2015). https://doi.org/10.1063/1.4918346
81. D.A. Mazziotti (ed.), *Reduced-Density-Matrix Mechanics: With Applications to Many-electron Atoms and Molecules* (Wiley, Hoboken, 2007)

Part III
Theoretical Physics

Chapter 23
Energy Density Functional Theory in Atomic and Nuclear Physics

M. Baldo

Abstract Ab initio calculations have been developed in atomic and molecular physics, as well as in nuclear physics. They are based on theoretical and numerical methods that are common to all fields. One may mention the Monte Carlo method, in different formulations, and the Configuration Interaction scheme. Despite the enormous progress achieved along the years, these methods are limited to systems of not too large number of particles, because the numerical effort at increasing number of particles becomes rapidly too demanding even for the most advanced computers. At the same time, several accurate approximations to the many-body problem have been perfected and applied to different systems. One may mention the cluster expansion method and the variational method. The numerical complexity of these schemes becomes more and more demanding as the number of particle increases. At a more phenomenological level the Energy Density Functional (EDF) approach offers a simpler scheme that can be applied to systems of virtually arbitrary number of particles. For instance, in nuclear physics it is commonly applied throughout the nuclear mass table. In this paper, we will present a general discussion on the EDF method, the analogy and the differences between the atomic and molecular EDF on one hand and the nuclear EDF on the other. Special emphasis is devoted to the Kohn–Sham method and to nuclear structure studies.

23.1 Introduction

One of the most used method to handle complex many-body systems is the Energy Density Functional (EDF) scheme. Its aim is to keep the numerical and analytical complications to a relatively simple level, but at the same time to reach an accuracy that can be satisfactory for many purposes. The method is mainly based on the Hohenberg and Kohn (HK) theorem [1], which proves the existence of a functional only of the density profile that acquires its minimal value for the ground state density

---

M. Baldo (✉)
Istituto Nazionale di Fisica Nucleare, Sezione di Catania, Via S. Sofia 64, 95123 Catania, Italy
e-mail: marcello.baldo@ct.infn.it

G. G. N. Angilella and C. Amovilli (eds.), *Many-body Approaches at Different Scales*, https://doi.org/10.1007/978-3-319-72374-7_23

profile. The value of the functional at the minimum is just the ground state energy. Originally it was proven for systems that are bound by an external potential, like electrons in atoms, molecules and crystals. Later it was extended to self-bound systems [2–5] like nuclei and droplets of e.g. Helium atoms. This functional gives the exact ground state density, and it is therefore essentially unknown. It is expected to be non-local and of intricate structure. In practical applications it is therefore assumed that it can be approximated by a simpler functional, which will provide an approximate density profile and approximate energy of the ground state. The particular form of the approximate functional can be dictated by physical considerations, but it can contain phenomenological parameters that have to be optimized by fitting a certain set of physical systems. This is justified by the fact that the exact EDF is universal, so that the approximate one should be also universal to a certain extent. Usually the binding energy of a selected set of atoms or molecules is used to fix the parameters.

A general practical method to device the EDF is the Kohn–Sham scheme. In this case the functional is written as the sum of the free kinetic energy, the Hartree potential and an additional term, the so called exchange-correlation energy, which incorporates the Fock term and the correlations beyond the Hartree–Fock approximation. In the Kohn–Sham method the exchange-correlation term is assumed to be a function only of the local density and equal to the correlation energy of the homogeneous system at the given density. The latter can be obtained by Monte Carlo calculations on the homogeneous electron gas, and can be considered very accurate. This assumption correspond to an extreme local density approximation (LDA). Corrections and refinements to LDA can be devised, in a phenomenological way or from first principles.

In nuclear physics the introduction of EDF for the description of the ground state of nuclei along the nuclear mass table followed historically a different scheme. In fact the nucleon-nucleon interaction is quite strong and characterized by a repulsive hard core that prevents the use of the Hartree or Hartree–Fock approximation as a starting point. It is therefore assumed that the effect of the strong nuclear correlations could be incorporated in an effective force, whose parameters are to be fitted to the binding of a selected set of nuclei or a large set throughout the mass table. In the latter case, the so fitted effective force is then used to calculate other quantities like the nuclear radii or the excited sates energy. Since the nuclear interaction is short range, also the effective forces are considered short range or even zero range. In this approach the ground state energy is expressed as the Hartre or Hartree–Fock (HF) approximation for the effective force, which then is assumed to include in an effective way the correlations beyond HF. In this way the energy functional coincides with the HF energy calculated with the effective force. In general the effective force is density dependent and the minimization of the functional introduces extra terms coming from the density derivative of the effective force. The zero range forces that have been devised along the years are generally referred to as Skyrme forces.

An introduction to Skyrme forces and their properties together with an extended compilation can be found in [6]. Also non-local or partly non-local forces have been devised, in particular the set of so-called Gogny forces [7].

Only recently a functional of the Kohn–Sham type has been constructed, and one of the aims of this work is to discuss the similarity and differences between the atomic and molecular electronic systems and the nuclear systems when they are treated following the Kohn–Sham method.

The plan of the paper is as follows. In Sect. 23.2 we describe briefly the Kohn–Sham functional for the electron systems and introduce the corresponding functional that can be devised for nuclei, and the particular features of the latter will be discussed. In Sect. 23.3 we describe in some detail the nuclear Kohn–Sham functional that has been developed in the last few years and illustrate the results that have been obtained. We also discuss briefly the refinements that are needed beyond the EDF scheme in general. Section 23.4 is devoted to the conclusions and prospects.

23.2 The Kohn–Sham Functional

We illustrate the Kohn–Sham (KS) method in general term for a generic many-body system. We take for the particle-particle bare interaction a simple scalar potential v, the Coulomb interaction in the atomic case. The method is based on approximating the Hohenberg-Kohn (HK) functional by a simplified form

$$E[\rho] = T_s[\rho] + \int d\mathbf{r}\; v_{\text{ext}}(\mathbf{r})\rho(\mathbf{r}) + E_H[\rho] + E_{\text{xc}}[\rho] \tag{23.1}$$

where E_H is the Hartree mean field term,

$$E_H = \frac{1}{2}\int d\mathbf{r} \int d\mathbf{r}' v(\mathbf{r}, \mathbf{r}')\rho(\mathbf{r})\rho(\mathbf{r}'), \tag{23.2}$$

and v_{ext} is a possible external potential (the central Coulomb potential in the atomic case). The term $E_{\text{xc}}[\rho]$ is the so-called exchange and correlation term, which contains all the correlations beyond the Hartree approximation. According to HK theorem this term must exist in principle, but it is of course unknown. One then makes the following assumptions.

1. One can approximate the exact HK functional, which in general is non-local, by a local one, i.e. one assumes that the exchange and correlation term is a function that depends only on the local density.
2. The function that defines the exchange and correlation term can be taken from accurate calculations in homogeneous matter at the local density $\rho(\mathbf{r})$. In the atomic case one can use the Monte Carlo calculations for an electron gas, which can be considered numerically exact.
3. One assumes that the density can be written as a sum over the contributions from a set of A orbitals ϕ_i, A being the total number of particles,

$$\rho(r) = \sum_i^A |\phi_i(r)|^2. \tag{23.3}$$

Assumption 2 connects in general the functional with the bare particle-particle interaction, since the ground state energy of the homogeneous system can be calculated from accurate ab initio many-body calculations. Minimization of the functional with respect to the orbitals ϕ_i produces Hartree-like equations

$$\left(-\frac{\hbar^2}{2m} \nabla^2 + v_{\text{ext}}(\mathbf{r}) \right) \phi_i(\mathbf{r}) = \varepsilon_i \phi(\mathbf{r}), \tag{23.4}$$

where

$$v_{\text{eff}}(\mathbf{r}) = v_{\text{ext}}(\mathbf{r}) + \int v(\mathbf{r}, \mathbf{r}') \rho(\mathbf{r}') d\mathbf{r}' + \frac{\delta E_{\text{xc}}[\rho]}{\delta \rho(\mathbf{r})} \tag{23.5}$$

is an effective local potential.

The single particle wave functions $\phi_i(\mathbf{r})$ can be taken orthonormal, but it is important to notice that they cannot be interpreted as the Hartree or Hartree–Fock orbitals. In particular, the single particle density matrix cannot be written as

$$\rho(r, r') = \sum_i^A \phi_i(r)^* \phi_i(r'), \tag{23.6}$$

since otherwise this would imply that the ground state wave function is just a Slater determinant of the orbitals $\phi_i(r)$, while the functional contains in an effective way the correlations beyond the Hartree or Hartree–Fock mean field and cannot be a simple Slater determinant. Otherwise stated, if one keeps the density matrix of Eq. (23.6) the corresponding Slater determinant can be interpreted as a component of the exact ground state, and therefore the orbitals should be renormalized if they must describe the exact density matrix.

To improve the functional with respect to the local density approximation one can introduce gradient terms, like $\nabla\rho$, $\nabla^2\rho$, $(\nabla\rho)^2$ ..., which can be phenomenological, but in principle they could be extracted from microscopic calculations on slightly inhomogeneous infinite system. This can be justified if the density profile is smooth enough. This is hardly the case in atomic physics and extensive studies of the correction to LDA can be found in the literature, see e.g. [8, 9]. In general this correction is formulated as an ordinary function of the dimensionless gradient term

$$t = |\nabla n| / k_F n, \tag{23.7}$$

where $n(\mathbf{r})$ and $k_F(\mathbf{r})$ are the local density and Fermi momentum, respectively. The physical meaning of such a quantity is the relative variation of density within a distance of the order of the average inter-particle distance. In principle it should be much less than 1 if the density profile has to be considered smooth. However it can be useful to construct an interpolation between the slow variation limit, where the first quadratic term in t is the dominant one, and the limit of extreme sharp density profile, where the correction is expected to vanish [9]. There are innumerable applications of the KS method and its refinements in condensed matter physics. Here we want to discuss how this method can be adapted to the nuclear structure case and which modifications are needed. For the nuclear systems the main peculiarities can be summarized as follows.

1. The nucleon-nucleon (NN) interaction is short range, at variance with the Coulomb interaction.
2. It has a quite complex operatorial structure, being the central potential term only one of the components.
3. The NN interaction is characterized by a strong hard core at short distance, so the matrix element of the potential is diverging or very large. This has the consequence that the Hartree term must be necessarily absorbed in the exchange and correlation term, i.e. the whole correlation energy must be added to the kinetic energy term.
4. The nuclei have not a smooth surface, but also not so sharp to neglect the surface energy. If one wants to keep gradients terms in addition to the bulk energy, they must be completely phenomenological.

The total correlation energy can be extracted form the nuclear Equation of State (EOS) calculated in nuclear matter within an accurate many-body theory and treated at LDA level, according to the KS method. One can then express the nuclear KS functional E_N as

$$E_N = T_0 + E^{s.o.} + E_{int}^{\infty} + E_{int}^{FR} + E_C, \tag{23.8}$$

where the kinetic energy term T_0 is treated at the quantal level and it can be written in terms of the orbitals of Eq. (23.3),

$$T_0 = \frac{\hbar^2}{2m} \sum_i \int d\mathbf{r} |\nabla \phi_i(\mathbf{r})|^2, \tag{23.9}$$

E_{int}^{∞} is the correlation energy of the nuclear EOS at a given density and E_{int}^{FR} the surface correction, which eventually can be written as a function of gradient terms. The term $E^{s.o.}$ is the single particle spin-orbit interaction, which is absent in homogeneous nuclear matter and it is crucial in nuclear structure, in particular for the correct shell sequence in nuclei. As already mentioned the surface term must be taken on a phenomenological basis, while the spin-orbit one has a standard expression and strength. Finally, E_C is the Coulomb energy due to protons, and it contains both a Hartree and an exchange term.

Another relevant difference with respect to the condensed matter EOS is the fact that the nuclear EOS is not known with the accuracy that has been reached in the electron gas case. On the other hand one of the peculiarity of the nuclear systems is the presence of a saturation point, i.e. the density and energy at which uncharged nuclear matter is self-bound. The saturation point is known phenomenologically to a certain accuracy and this constraints the nuclear EOS that can be adopted.

23.3 Performance of the Nuclear Kohn–Sham Functional

The nuclear KS method was partly followed in [10, 11], where the physical quantities that have been fitted to fix the force parameters included also a microscopic nuclear matter EOS, which therefore determines mainly the local part of the functional. A strict KS procedure was followed in [12–18], where the local part of the functional was identified, once for all, with the correlation energy of the microscopic EOS calculated within the Brueckner–Hartree–Fock (BHF) many-body theory, including three-body forces. The EOS was not entering the fitting procedure, but, following the KS scheme, only the parameters of the surface term, i.e. the non-local component of the functional, were fitted to reproduce the binding energy of a large set of nuclei throughout the mass table. As in [10, 11], the single particle effective mass was taken at the bare value, in agreement with the scheme of the original KS functional. The surface term E_{int}^{FR} of the functional Eq. (23.8) is taken of the form

$$E_{int}^{FR}[\rho_n, \rho_p] = \frac{1}{2}\sum_{t,t'}\int\int d^3r d^3r' \rho_t(\mathbf{r}) v_{t,t'}(\mathbf{r}-\mathbf{r}')\rho_{t'}(\mathbf{r}') - \frac{1}{2}\sum_{t,t'}\gamma_{t,t'}\int d^3r \rho_t(\mathbf{r})\rho_{t'}((\mathbf{r}),$$

where the index t is the label for neutron and proton, i.e. $t = n, p$, and $\gamma_{t,t'}$ the volume integral of $v_{t,t'}(r)$. The subtraction term is introduced not to contaminate the bulk part, determined from the microscopic infinite matter calculation. For the finite range form factor $v_{t,t'}(r)$ a simple Gaussian ansatz is adopted,

$$v_{t,t'}(r) = V_{t,t'} e^{-r^2/r_0^2}. \tag{23.10}$$

The strength parameters are chosen to distinguish only between like and unlike particles, i.e.

$$V_{p,p} = V_{n,n} = V_L; \quad V_{n,p} = V_{p,n} = V_U, \tag{23.11}$$

so that the finite range term contained three adjustable parameters: V_L, V_U, and r_0. The number of these parameters can be reduced if the interaction energy of the nuclear EOS is expressed as a polynomial of the density. In this case one can fix two of the parameters in order to reproduce the first two powers in the density and no subtraction is needed. In that way only one parameter of the surface term remains free to be fitted to the data. In the applications it was found convenient to choose the range r_0 as free parameter. It turns out that r_0 is determined with great accuracy by the fit. To get the interaction energy of the EOS in polynomial form one can consider an accurate microscopic EOS, compatible with the phenomenological constraints, and fit it with a polynomial of large enough degree. Such a fit is shown in Fig. 23.1, redrawn from [17], where further details can be found. Both symmetric matter and pure neutron matter are fitted with polynomials of fifth degree. It has to be stressed that the fit is used only to the purpose of having the EOS in analytical

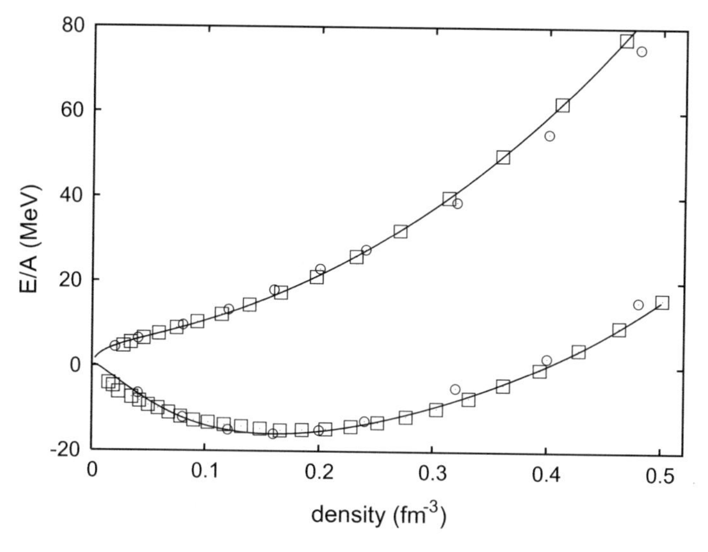

Fig. 23.1 EOS of symmetric and neutron matter obtained by the microscopic calculation (squares) and the corresponding polynomial fits (solid lines). For comparison, the microscopic EOS of Ref. [19] are also displayed by open circles

form, since the microscopic calculations can give the EOS only in numerical form as a series of discrete points. The Gaussian non-local surface term corresponds to the gradient expansion of the atomic KS functional, and together with the bulk EOS forms the nuclear KS functional, in close agreement with the original KS functional of condensed matter physics.

An overall view of the performance of the functional is illustrated in Fig. 23.2, where the differences between the experimental total energy end the fitted one is reported for each one of the 597 nuclei considered. In the lower panel the same comparison is reported for the extrapolated energies of nuclei not available in laboratory.

The average error in the binding is 1.58 MeV, over total bindings that can be of several hundreds of MeV in the region of heavy nuclei. It turns out that with the same parameters and no additional fitting the radii of nuclei are reproduced within 0.027 fm. Notice that the only free parameter can be considered the range r_0 of the surface term, while the spin-orbit and pairing strengths were taken at standard values, with no adjustment.

In conclusion, the nuclear KS functional looks performing at the same level as the corresponding functional in condensed matter physics. However there are physical quantities that need additional refinement. The energies of the monopole and quadrupole Giant Resonances turn out to be well reproduced only if the effective mass is introduced [18], again using values from the microscopic calculations. The quantities that are not yet reproduced at a satisfactory level are the energies of the

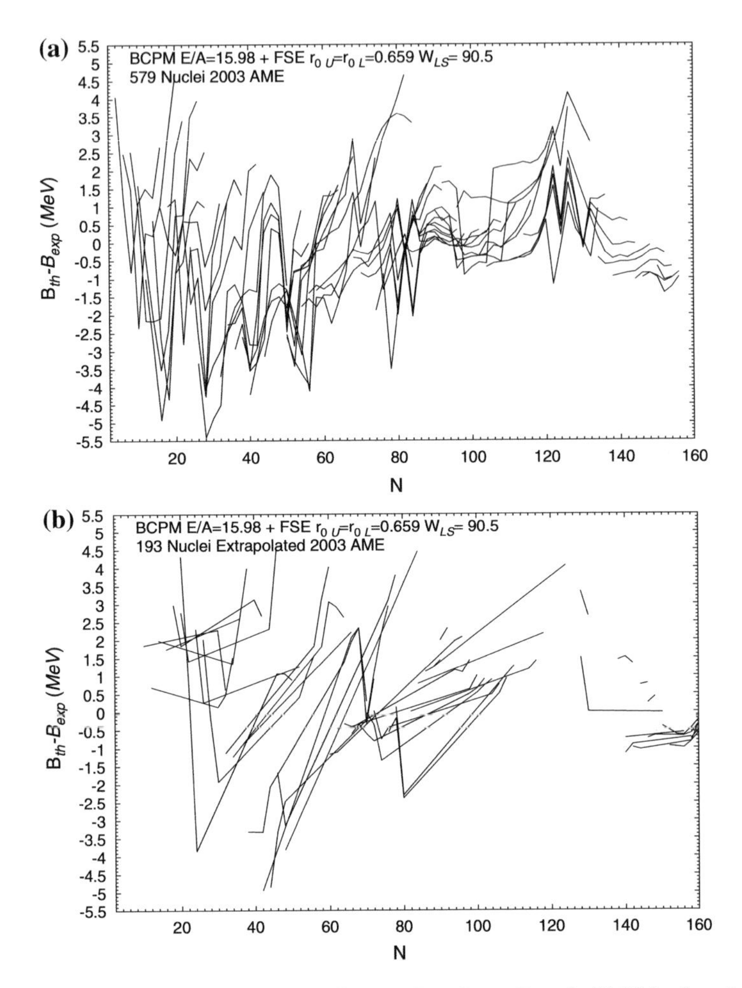

Fig. 23.2 In panel **a**, the binding energy difference $B = B_{th} - B_{exp}$ (in MeV) is plotted with our optimal set of parameters (BCPM) for the 579 nuclei of Audi's AME 2003 [20] as a function of neutron number N. Points corresponding to the same isotope are connected by straight lines. In panel **b**, the same quantity is plotted but this time for the extra set of 193 nuclei in Audi's compilation with 'experimental' values obtained from extrapolation and/or systematics

single particle states. Indeed the energies of the KS orbitals in Eq. (23.3) cannot be considered the energies of the physical single particle states. Hints for solving this problem in general can be found along the lines of Ref. [21] and references therein.

23.4 Conclusions

We have shown that the Kohn–Sham approach to the Energy Density functional method in Nuclear Structure is quite performing and can be a quite accurate basis towards a semi-microscopic many-body theory. The number of parameters was reduced to a minimum, with only one as really free for the surface term, while the bulk part was adopted from microscopic calculations in nuclear matter. This indicates that the functional includes the right balance between bulk and surface energy, an essential feature of the nuclear binding energy. Further refinements are needed for a theory that can cover all the different aspects of nuclear structure.

References

1. P.C. Hohenberg, W. Kohn, Phys. Rev. **136**, B864 (1964). https://doi.org/10.1103/PhysRev.136.B864
2. J. Messud, M. Bender, E. Suraud, Phys. Rev. C **80**, 054314 (2009). https://doi.org/10.1103/PhysRevC.80.054314
3. J. Engel, Phys. Rev. C **75**, 014306 (2007). https://doi.org/10.1103/PhysRevC.75.014306
4. N. Barnea, Phys. Rev. C **76**, 067302 (2007). https://doi.org/10.1103/PhysRevC.76.067302
5. T. Lesinski, Phys. Rev. C **89**, 044305 (2014). https://doi.org/10.1103/PhysRevC.89.044305
6. M. Dutra, O. Lourenço, J.S. Sá Martins, A. Delfino, J.R. Stone, P.D. Stevenson, Phys. Rev. C **85**, 035201 (2012). https://doi.org/10.1103/PhysRevC.85.035201
7. J. Dechargé, D. Gogny, Phys. Rev. C **21**, 1568 (1980). https://doi.org/10.1103/PhysRevC.21.1568
8. J.P. Perdew, K. Burke, Y. Wang, Phys. Rev. B **54**, 16533 (1996). https://doi.org/10.1103/PhysRevB.54.16533
9. J.P. Perdew, K. Burke, M. Ernzerhof, Phys. Rev. Lett. **77**, 3865 (1996); [Erratum Phys. Rev. Lett. **78**, 1396 (1997)]. https://doi.org/10.1103/PhysRevLett.77.3865
10. S.A. Fayans, S.V. Tolokonnikov, E.L. Trykov, D. Zawischa, Nucl. Phys. A **676**(1), 49 (2000). https://doi.org/10.1016/S0375-9474(00)00192-5
11. S.A. Fayans, D. Zawischa, Int. J. Mod. Phys. B **15**(10–11), 1684 (2001). https://doi.org/10.1142/S0217979201006203
12. M. Baldo, P. Schuck, X.V. nas, Phys. Lett. B **663**(5), 390 (2008). https://doi.org/10.1016/j.physletb.2008.04.013
13. L.M. Robledo, M. Baldo, P. Schuck, X. Viñas, Phys. Rev. C **77**, 051301 (2008). https://doi.org/10.1103/PhysRevC.77.051301
14. X. Vinãs, L.M. Robledo, M. Baldo, P. Schuck, Int. J. Mod. Phys. E **18**(04), 935 (2009). https://doi.org/10.1142/S0218301309013075
15. L.M. Robledo, M. Baldo, P. Schuck, X. Viñas, Phys. Rev. C **81**, 034315 (2010). https://doi.org/10.1103/PhysRevC.81.034315
16. M. Baldo, L. Robledo, P. Schuck, X.V. nas, J. Phys. G **37**(6), 064015 (2010). https://doi.org/10.1088/0954-3899/37/6/064015
17. M. Baldo, L.M. Robledo, P. Schuck, X. Viñas, Phys. Rev. C **87**, 064305 (2013). https://doi.org/10.1103/PhysRevC.87.064305
18. M. Baldo, L.M. Robledo, P. Schuck, X. Viñas, Phys. Rev. C **95**, 014318 (2017). https://doi.org/10.1103/PhysRevC.95.014318
19. A. Akmal, V.R. Pandharipande, D.G. Ravenhall, Phys. Rev. C **58**, 1804 (1998). https://doi.org/10.1103/PhysRevC.58.1804

20. G. Audi, A.H. Wapstra, C. Thibault, Nucl. Phys. A **729**(1), 337 (2003); The 2003 NUBASE and Atomic Mass Evaluations. https://doi.org/10.1016/j.nuclphysa.2003.11.003
21. M. Baldo, P.F. Bortignon, G. Colò, D. Rizzo, L. Sciacchitano, J. Phys. G **42**(8), 085109 (2015). https://doi.org/10.1088/0954-3899/42/8/085109

Chapter 24
Second Order Exchange Energy of a d-Dimensional Electron Fluid

M. L. Glasser

Abstract A method is presented for reducing a $3d$-fold integral occurring in higher order many-body integrals for a d-dimensional electron gas to a double integral. The result is applied to the second order exchange energy for a d-dimensional uniform electron fluid. The cases $d = 2, 3$ are examined in detail.

24.1 Introduction

In their classic work on the ground-state energy of an interacting electron gas [1] Gell-Mann and Brueckner encountered the second order exchange term

$$E_{2b} = \frac{3p_F^3 e^4}{16\pi^5} \int d\mathbf{p}\, d\mathbf{k}\, d\mathbf{q}\, \frac{f_{\mathbf{p}} f_{\mathbf{k}}(1 - f_{\mathbf{p}+\mathbf{q}})(1 - f_{\mathbf{k}+\mathbf{q}})}{q^2(\mathbf{p}+\mathbf{k}+\mathbf{q})^2 \mathbf{q} \cdot (\mathbf{p}+\mathbf{k}+\mathbf{q})}, \tag{24.1}$$

where f_p is the Fermi distribution function and p_F denotes the Fermi momentum. Gell-Mann's assistant, H. Kahn, estimated by Monte-Carlo integration the value as -0.044 and in a 1965 lecture in Istanbul [2] L. Onsager claimed that the exact value is $(\ln 2)/3 - 3\zeta(3)/2\pi^2$, which remained unproven till eight years later when Onsager, Mittag and Stephen published a lengthy derivation [2]. In 1980, Ishihara and Ioriatti [3] evaluated the two-dimensional analogue of Eq. (24.1) and in 1984 the author published a note [4] indicating how such integrals might be handled in d-dimensions. But, due to a number of misprints [4] is difficult to follow and it seems appropriate to present a simplified and corrected version, particularly since

M. L. Glasser (✉)
Department of Physics, Clarkson University, Potsdam, NY 13699-5820, USA
e-mail: lglasser@clarkson.edu

M. L. Glasser
Department of Theoretical Physics, University of Valladolid,
47003 Valladolid, Spain

M. L. Glasser
Donostia International Physics Center, P. Manuel de Lardizabal 4,
20018 San Sebastián, Spain

G. G. N. Angilella and C. Amovilli (eds.), *Many-body Approaches at Different Scales*, https://doi.org/10.1007/978-3-319-72374-7_24

the method has been found useful in other contexts [5] and, due to an oversight, it erroneously stated that the value given in [3] was confirmed. The dimension d will be treated as continuous by means of the expedient integration rule for an azimuthally symmetric integrand

$$\int d^d\mathbf{k} = \frac{2\pi^{(d-1)/2}}{\Gamma\left[\frac{1}{2}(d-1)\right]}\int_0^\infty dk\, k^{d-1}\int_0^\pi d\theta\,\sin^{d-2}\theta. \tag{24.2}$$

The following section covers the reduction of a basic $9d$-dimensional integral to more manageable $3d+2$-dimensional form which, in Sect. 24.3, is applied to the second order exchange energy. The last section gives the results for $d=2$ and $d=3$.

24.2 Basic Integral Identity

The units $\hbar = 2m = 1$, will be used along with the notation

$$f_p = [1+\exp[\beta(p^2-p_F^2)]]^{-1}, \tag{24.3a}$$

$$Q(p) = f_p(1-f_{\mathbf{p}+\mathbf{q}}), \tag{24.3b}$$

$$Q'(p) = f_{\mathbf{p}+\mathbf{q}}(1-f_p), \tag{24.3c}$$

$$\Delta(p) = f_{\mathbf{p}+\mathbf{q}} - f_{\mathbf{p}}, \tag{24.3d}$$

$$\delta(p) = (\mathbf{p}+\mathbf{q})^2 - p^2. \tag{24.3e}$$

All vectors are d-dimensional, and vector integrals are over all space.

Lemma 24.1 *In the zero temperature limit*

$$\frac{Q(p)Q(k)-Q'(p)Q'(k)}{\mathbf{q}\cdot(\mathbf{p}+\mathbf{k}+\mathbf{q})} = -\frac{1}{\pi}\int_{-\infty}^{\infty} dz\,\frac{\Delta(p)}{z-i\delta(p)}\frac{\Delta(k)}{z+i\delta(k)}. \tag{24.4}$$

Proof The proof follows closely the derivation of a similar result in Appendix A of [3].

Theorem 24.1 *For real* $\mathbf{r}$ *and* $t \geq 0$

$$\int d\mathbf{p}\, e^{i[\mathbf{r}\cdot\mathbf{p}+\delta(p)t]}\Delta(p) = -2i\left(\frac{2\pi p_F}{\xi}\right)^{d/2} e^{-\frac{1}{2}i\mathbf{r}\cdot\mathbf{q}}\sin\left(\frac{1}{2}\mathbf{q}\cdot\boldsymbol{\xi}\right)J_{d/2}(p_F\xi), \tag{24.5}$$

where $\boldsymbol{\xi} = \mathbf{r}+2t\mathbf{q}$.

Proof First of all note that $\Delta(p)$ is simply a rectangular pulse with height 1 and width q, so has the inverse Laplace transform representation

$$\Delta(p) = \int_{c-i\infty}^{c+i\infty} \frac{ds}{2\pi i s} e^{sp_F^2}[e^{-s(\mathbf{p}+\mathbf{q})^2} - e^{-sp^2}], \quad c > 0. \tag{24.6}$$

By substituting Eq. (24.6) into (24.5) one obtains the difference of two integrals. In the first make the change of variable $\mathbf{p} \to -\mathbf{p} - \mathbf{q}$. This gives

$$\int_{c-i\infty}^{c+i\infty} \frac{ds}{2\pi i s} e^{sp_F^2} \{e^{-i\mathbf{r}\cdot\mathbf{q}} e^{-itq^2} C(-\mathbf{r} - 2t\mathbf{q}) - e^{itq^2} C(\mathbf{r} + 2t\mathbf{q})\} \tag{24.7}$$

$$C(\boldsymbol{\xi}) = \int d\mathbf{p}\, e^{i\mathbf{p}\cdot\boldsymbol{\xi}} e^{-sp^2} = \left(\frac{\pi}{s}\right)^{d/2} e^{-\xi^2/4s}. \tag{24.8}$$

Next, one has

$$\int_{c-i\infty}^{c+i\infty} \frac{ds}{2\pi i} e^{p_F^2 s} s^{-1-d/2} e^{-\xi^2/4s} = \left(\frac{2p_F}{\xi}\right)^{d/2} J_{d/2}(p_F \xi), \tag{24.9}$$

which gives for Eq. (24.7)

$$\left(\frac{2p_F\pi}{\xi}\right) J_{d/2}(p_F\xi) e^{itq^2} [e^{-i\mathbf{q}\cdot\boldsymbol{\xi}} - 1] = -2i \left(\frac{2p_F\pi}{\xi}\right) J_{d/2}(p_F\xi)\, e^{-\frac{1}{2}i\mathbf{r}\cdot\mathbf{q}} \sin\left(\frac{1}{2}\mathbf{q}\cdot\boldsymbol{\xi}\right). \tag{24.10}$$

Now, we choose, from among other possibilities,

$$\alpha(\mathbf{q}) = \int \frac{e^{i\mathbf{r}\cdot\mathbf{q}}}{r} d\mathbf{r} \tag{24.11}$$

and define

$$A(\mathbf{q}) = \int d\mathbf{p}\, d\mathbf{k}\, \alpha(\mathbf{p}+\mathbf{k}+\mathbf{q}) \frac{Q(p)Q(k)}{\mathbf{q}\cdot(\mathbf{p}+\mathbf{k}+\mathbf{q})}. \tag{24.12}$$

By making the substitution $\mathbf{p} \to -\mathbf{p} - \mathbf{q}$, $\mathbf{k} \to -\mathbf{k} - \mathbf{q}$, and adding the result back to Eq. (24.12), we find, using the identity in Lemma 24.1,

$$\begin{aligned} A(q) &= -\frac{1}{2\pi} \int_{-\infty}^{\infty} dz \int d\mathbf{p}\, d\mathbf{k}\, \alpha(\mathbf{p}+\mathbf{k}+\mathbf{q}) \frac{\Delta(p)\Delta(k)}{(z + i\delta(p))(z - i\delta(k))} \\ &= -\frac{1}{2\pi} \int \frac{d\mathbf{r}}{r} \int_{-\infty}^{\infty} dz\, B(\mathbf{r}, z) B(-\mathbf{r}, z) \end{aligned} \tag{24.13}$$

with

$$B(\mathbf{r}, z) = \int d\mathbf{p}\, e^{i\mathbf{r}\cdot\mathbf{p}} \frac{\Delta(p)}{z + i\delta(p)} = \int_0^{\infty} dt\, e^{-zt} \int d\mathbf{p}\, e^{i[\mathbf{r}\cdot\mathbf{p}+t\delta(p)]} \Delta(p). \tag{24.14}$$

By applying Theorem 24.1 and performing the elementary z integration, we have, after scaling q out of t_j,

Theorem 24.2

$$A(q) = \frac{2}{\pi q}(2\pi p_F)^d \int \frac{d\mathbf{r}}{r} \int_0^\infty \int_0^\infty \frac{dt_1 dt_2}{t_1 + t_2} \frac{\sin\left(\frac{1}{2}q\xi_1\right)\sin\left(\frac{1}{2}q\xi_2\right)}{(\xi_1\xi_2)^{d/2}} J_{d/2}(p_F\xi_1)J_{d/2}(p_F\xi_2), \tag{24.15}$$

where $\boldsymbol{\xi}_1 = \mathbf{r} + 2t_1\hat{q}$, $\boldsymbol{\xi}_2 = \mathbf{r} - 2t_2\hat{q}$.

24.3 Application to Second Order Exchange

For our choice of Coulomb interaction

$$\alpha(q) = e^2 \frac{(4\pi)^{(d-1)/2}}{q^{d-1}} \Gamma\left(\frac{d-1}{2}\right) \tag{24.16}$$

which requires $d > 1$, the second order exchange contribution to the ground-state energy per unit volume of a d-dimensional electron fluid is

$$E_{2x} = \frac{1}{(2\pi)^{2d}} \int \alpha(q)A(q)d\mathbf{q}. \tag{24.17}$$

For $d > 2$ we take the polar axis as the $\hat{q}$-direction and apply Theorem 24.2. The q integration is elementary and we have

$$E_{2x} = K_d \int d\Omega_q \int \frac{d\mathbf{r}}{r} \int_0^\infty \int_0^\infty \frac{dt_1 dt_2}{t_1 + t_2} \ln\left|\frac{\xi_1 + \xi_2}{\xi 1 - \xi_2}\right| \frac{J_{d/2}(p_F\xi_1)J_{d/2}(p_F\xi_2)}{(\xi_1\xi_2)^{d/2}}, \tag{24.18}$$

where K_d collects all the numerical prefactors and powers of p_F (for $d = 2 \int d\Omega_q = 2\pi$) and will be made explicit in the final result. Now set $t_2 = ut_1$ and $\mathbf{r} \to t_1\mathbf{r}$, so

$$E_{2x} = K_d \int d\Omega_q \int_0^\infty \frac{du}{u+1} \int \frac{d\mathbf{r}}{r(\eta_1\eta_2)^{d/2}} \ln\left|\frac{\eta_1 + \eta_2}{\eta_1 - \eta_2}\right| \int_0^\infty \frac{dt}{t} J_{d/2}(p_F t\eta_1)J_{d/2}(p_F t\eta_2), \tag{24.19}$$

where $\eta_1 = |\mathbf{r} + 2\hat{q}|$, $\eta_2 = |\mathbf{r} - 2u\hat{q}|$. The $\theta, t-$ integrals can be done next, yielding

$$E_{2x} = K_d \int_0^\infty \frac{du}{u+1} \int \frac{d\mathbf{r}}{r\eta_>^d} \ln\left|\frac{\eta_1 + \eta_2}{\eta_1 - \eta_2}\right|, \quad (d > 2). \tag{24.20}$$

For $d > 2$ we can switch to $d-$dimensional cylindrical coordinates with axis along $\hat{q}$. Since the integrand is independent of the azimuthal angle

$$\int d\mathbf{r} = \frac{2\pi^{(d-1)/2}}{\Gamma[\frac{1}{2}(d+1)]} \int_{-\infty}^{\infty} dz \int_0^{\infty} \rho^{d-2} d\rho. \tag{24.21}$$

Next, after the successive transformations $t = (z-1)/(z+1)$ and $\rho = 2s/(1-t)$ we have

$$E_{2x} = K_d \int_{-1}^{1} \frac{dt}{1-t} \ln\left(\frac{1+t}{1-t}\right) F(t),$$
$$F(t) = \int_0^{\infty} \frac{s^{d-2} ds}{(s^2+1)^{d/2}} \int_t^1 \frac{dy}{\sqrt{y^2+s^2}}. \tag{24.22a}$$

Carrying out the s integration, we come to

$$F(t) = \frac{1}{d-1} \int_t^1 \frac{dy}{|y|} \, {}_2F_1\left(\frac{1}{2}, \frac{1}{2}(d-1); \frac{1}{2}(d+1); 1-y^{-2}\right). \tag{24.23}$$

By integrating by parts and noting that

$${}_2F_1\left(\frac{1}{2}, a; a+1; 1-y^{-2}\right) = \frac{2|y|}{1+|y|} \, {}_2F_1\left(1, 1-a; a+1; \frac{1-|y|}{1+|y|}\right),$$

we arrive at the principal result

Theorem 24.3 *The second order exchange contribution to the ground-state energy of a $d > 2$–dimensional electron fluid is*

$$E_{2x} = K_d G(d), \tag{24.24a}$$
$$G(d) = \int_0^1 \frac{dy}{y+1} \left[\frac{\pi^2}{3} - \ln^2 y\right] {}_2F_1\left[1, \frac{1}{2}(3-d); \frac{1}{2}(1+d); y\right]. \tag{24.24b}$$

24.4 Discussion

Equation (24.24a) is as far as one can proceed without specifying the dimensionality. For $d = 3$, we find, since the hypergeometric function reduces to unity,

$$E_{2x} = K_3 G(3) = \frac{e^4 p_F^3}{4\pi^2} \int_0^1 \frac{dy}{1+y} \left(\frac{\pi^2}{3} - \ln^2 y\right) = \frac{1}{6}[\pi^2 \ln 4 - 9\zeta(3)]. \tag{24.25}$$

which is exactly the Onsager–Stephen–Mittag value, since they have $e^2 = 2$ and $p_F = 1$.

For the case $d = 2$ we take the limit of Eq. (24.24a) which gives

$$G(2) = 2\int_0^1 \left(\frac{\pi^2}{3} - 4\ln^2 y\right)\frac{\tan^{-1} y}{y^2+1}dy, \tag{24.26}$$

which, unlike the corresponding integral in [3] does not seem to be analytically evaluable. This gives

$$E_{2x} = K_2 G(2) = 18.0586 \cdot \frac{p_F^2 e^4}{32\pi^4}, \tag{24.27}$$

about 30% less than the value $28.3664 \cdot (p_F^2 e^4/32\pi^4)$ in [3]. A possible reason is that in Eq. (14) of Ref. [3] the argument of the second Bessel function is $|\mathbf{r} - 2\hat{u}t|$ and after making the substitution $\mathbf{r} \to (t+x)\mathbf{r} + \hat{u}(x-t)$, in Eq. (16) of Ref. [3] the authors present it as $(x+t)|\mathbf{r} - \hat{u}|$, which is incorrect. It is this error which renders the remainder of the evaluation analytically tractable. An attempt to continue the calculation after correcting this was stymied by a further difficulty in Eq. (14) of Ref. [3]; the factor of 2 in the numerator of the argument of the logarithm means that, as $x \to \infty$ this argument tends to 2, rather than unity as required for convergence at the upper limit of the x integration.

Acknowledgements Financial support of MINECO (Project MTM2014-57129-C2-1-P) and Junta de Castilla y León (VA057U16) is acknowledged.

References

1. M. Gell-Mann, K.A. Brueckner, Phys. Rev. **106**, 364 (1957). https://doi.org/10.1103/PhysRev.106.364
2. L. Onsager, L. Mittag, M.J. Stephen, Ann. Phys. **473**(1–2), 71 (1966). https://doi.org/10.1002/andp.19664730108
3. A. Isihara, L. Ioriatti, Phys. Rev. B **22**, 214 (1980). https://doi.org/10.1103/PhysRevB.22.214
4. M.L. Glasser, J. Comp. Appl. Math. **10**(3), 293 (1984). https://doi.org/10.1016/0377-0427(84)90041-4
5. M.L. Glasser, G. Lamb, J. Phys. A: Math. Theor. **40**(6), 1215 (2007). https://doi.org/10.1088/1751-8113/40/6/002

Chapter 25
Nonlocal Quantum Kinetic Theory and the Formation of Correlations

K. Morawetz

Abstract The quantum version of the Boltzmann equation remains still the basis of modern transport theories. Extensions become necessary for transient-time effects like the femtosecond response and for strongly correlated systems. At short time scales higher correlations have no time to develop yet and femto-second laser excitation of collective modes in semiconductors as well as quenches of cold atoms in optical lattices can be described even analytically by fluctuations of the meanfield. For plasma systems exposed to a sudden switching, analytical results are available from the time-dependent Fermi's Golden Rule in good agreement with the results of two-time Green's functions solving the Kadanoff and Baym equation. At later times when correlations develop, a kinetic equation of nonlocal and non-instantaneous character unifies the achievements of the transport in dense quantum gases with the Landau theory of quasiclassical transport in Fermi systems. The numerical solution is not more expensive than solving the Boltzmann equation since large cancellations in the off-shell motion appear which are hidden usually in non-Markovian behaviors. The quasiparticle drift of Landau's equation is connected with a dissipation governed by a nonlocal and non-instant scattering integral in the spirit of Enskog corrections. These corrections are expressed in terms of shifts in space and time that characterize non-locality of the scattering process. In this way quantum transport is possible to recast into a quasi-classical picture. The balance equations for the density, momentum, energy and entropy include besides quasiparticle also the correlated two-particle contributions beyond the Landau theory. The medium effects on binary collisions are shown to mediate the latent heat, i.e., an energy conversion between correlation and thermal energy.

K. Morawetz (✉)
Münster University of Applied Sciences, Stegerwaldstrasse 39,
48565 Steinfurt, Germany
e-mail: morawetz@fh-muenster.de

K. Morawetz
International Institute of Physics (IIP),
Campus Universitário Lagoa Nova, Natal 59078-970, Brazil

K. Morawetz
Max-Planck-Institute for the Physics of Complex Systems,
01187 Dresden, Germany

G. G. N. Angilella and C. Amovilli (eds.), *Many-body Approaches at Different Scales*, https://doi.org/10.1007/978-3-319-72374-7_25

25.1 Introduction

Cold atoms in optical traps [1] and the femtosecond pump and probe experiments [2, 3] allow now to resolve the time-dependent formation of correlations. This has triggered an enormous theoretical activity [4]. Both different physical systems, the long-range Coulomb [5] as well as short-range Hubbard systems can be described by a common theoretical approach leading to a unique formula to describe the formation of correlations at short-time scale [6]. The basic observation here is that correlations need time to be formed such that the meanfield approximation is sufficient to describe the basic features of short-time formation of correlations. Calculating nonequilibrium Green's functions [7, 8] allows one to describe the formation of collective modes [3, 9], screening [7] and even exciton population inversions [10].

This becomes different at later times when essentially strong correlations are formed after a sudden quench. Here, the time-dependent description is covered by various kinetic equation approaches. It started with the foundation of Ludwig Boltzmann's famous equation [11] and has been rapidly developed, from important classical contributions [12–16] to quantum extensions, where the pioneering work along these lines was performed by [17, 18]. In the theory of condensed systems covered by the Landau concept of quasiparticles [19], the quantum Boltzmann-Uhling-Uhlenbeck (BUU) equation, differs from the classical one in the collision term, which takes into account that the final scattering states can be occupied and consequently blocked by the Pauli exclusion principle. Moreover, the quantum mechanical transition rate, rather than the classical one is used. The scattering integral of the Boltzmann equation remains still local in space and time. In other words, the Landau theory does not include a quantum mechanical analogy of virial corrections studied in the theory of gases.

To extend the validity of the Boltzmann equation to moderately dense gases, Clausius and Boltzmann included the space non-locality of binary collisions (cf. Chapter 16 in Ref. [20]). After one century, virial corrections won new interest as they can be incorporated into Monte Carlo simulation methods [21]. The microscopic theory of nonlocal corrections to the collision integral has been pioneered within the theory of gases by many authors [22–38].

In the limit of small scattering rates, the transport equation for the Green's function is converted into the kinetic equation of Boltzmann type by the extended quasiparticle approximation corresponding to the $\rho[f]$ functional. The resulting quantum kinetic theory unifies the achievements of transport in dense gases with the quantum transport of dense Fermi systems [39–43]. The quasiparticle drift of Landau's equation is connected with a dissipation governed by a nonlocal and non-instant scattering integral in the spirit of Enskog corrections. These corrections are expressed in terms of shifts in space and time that characterize the non-locality of the scattering process [44]. In this way quantum transport is possible to recast into a quasiclassical picture suited for simulations. The balance equations for the density, momentum, energy and entropy include quasiparticle contributions and the correlated two-particle contributions beyond the Landau theory as we will demonstrate in Sect. 25.3.2.

First we will discuss the transient time regime where the correlations are formed after a sudden quench with applications to femtosecond pumb and probe as well as cold atom experiments. Then we present the nonlocal kinetic theory in Sect. 25.3, which results from cancellation of off-shell parts by a proper extended quasiparticle picture. Two applications from nuclear and superconducting physics finally illustrate the usefullness of the concept in Sect. 25.4.

25.2 Formation of Correlations

A first guess of the time-dependent formation of correlations can be found from the time-dependent Fermi golden rule

$$P_{nn'} = \frac{1}{\hbar^2} V_{nn'}^2 \left(\frac{\sin \omega_{nn'} \frac{t}{2}}{\omega_{nn'} \frac{t}{2}} \right)^2 = 2V(q)^2 \frac{1 - \cos\left(\Delta E \frac{t}{\hbar}\right)}{(\Delta E t)^2} \tag{25.1}$$

expressing the transition probability between states n and n' which we consider as the state before and after the collision, and $\Delta E = \varepsilon_k + \varepsilon_p - \varepsilon'_{k+q} - \varepsilon'_{p-q}$ denotes the energy difference between initial and final states. Taking the occupation factors into account, the time-dependent formation of kinetic energy is expected to have the form

$$E_{\text{kin}}(t) = \int \frac{dkdpdq}{(2\pi\hbar)^9} V(q)^2 \frac{1 - \cos\left(\Delta E \frac{t}{\hbar}\right)}{\Delta E} \rho_{k+q} \rho_{p-q} (1 - \rho_k)(1 - \rho_p). \tag{25.2}$$

Exactly this expression is obtained if we use the Levinson equation [45] (with $\hbar = 1$)

$$\begin{aligned} \frac{\partial \rho_k(t)}{\partial t} &= 2 \int_0^{t-t_0} d\bar{t} \int \frac{dpdq}{(2\pi)^6} |V(q)|^2 e^{-\frac{\bar{t}}{\tau}} \cos\left[\Delta E \bar{t}\right] \\ &\times \left\{ \rho_{k-q} \rho_{p+q} \left[1 - \rho_k\right] \left[1 - \rho_p\right] - \rho_k \rho_p \left[1 - \rho_{k-q}\right] \left[1 - \rho_{p+q}\right] \right\}_{t-\bar{t}}, \end{aligned} \tag{25.3}$$

and neglect the memory in the distribution functions (finite duration approximation). This memory over-counts correlations [43] resulting into too much off-shell correlations [46].

The solution of the Levinson equation (25.3) in the short-time region $t \ll \tau$ can be written down analytically. It shows how the two-particle and the single-particle concept of the transient behavior is combined in the kinetic equation. The right-hand side of Eq. (25.3) describes how two particles correlate their motion to avoid strong interaction regions. Since the process is very fast, the on-shell contribution to $\delta\rho_k$, proportional to t/τ, can be neglected in the assumed time domain and the $\delta\rho$ has the pure off-shell character as can be seen from the off-shell factor $\sin(t\Delta E)/\Delta E$.

The off-shell character of mutual two-particle correlations is thus reflected in the single-particle Wigner distribution.

Starting with a sudden switching approximation, due to Coulomb interaction the screening is formed during the first transient time period and one finds analytically [47] the quantum result of the time derivative of the formation of correlation for statically as well as dynamically screened potentials

$$\frac{\partial}{\partial t}\frac{E_{\rm corr}^{\rm static}(t)}{n} = -\frac{e^2\kappa T}{2\hbar}\,{\rm Im}\left[(1+2z^2)e^{z^2}(1-{\rm erf}(z)) - \frac{2z}{\sqrt{\pi}}\right] \tag{25.4a}$$

$$\frac{\partial}{\partial t}\frac{E_{\rm corr}^{\rm dynam}(t)}{n} = -\frac{e^2\kappa T}{\hbar}\,{\rm Im}\left[e^{z_1^2}(1-{\rm erf}(z_1))\right], \tag{25.4b}$$

where $z = \omega_p\sqrt{t^2 - it\frac{\hbar}{T}}$ and $z_1 = \omega_p\sqrt{2t^2 - it\frac{\hbar}{T}}$. In Fig. 25.1 these formulas are compared with molecular dynamic simulations [48] in plasmas with $\Gamma = \frac{e^2}{a_e T} = 1$ and the Wigner-Seitz radius a_e.

The characteristic time of formation of correlations in the high temperature limit is the time of the inverse plasma frequency $\tau_c \approx \frac{1}{\omega_p} = \sqrt{2}/v_{\rm th}\kappa$, indicating that the dominant role is played by long range fluctuations. On the other hand, we also see that the correlation time is found to be given by the time a particle needs to travel through the range of the potential with a thermal velocity $v_{\rm th}$ and is not given by the time between successive collisions as one might have thought. For dense Fermi systems, like nuclear matter, one finds the build-up time where the correlation energy reaches its first maximum as the inverse Fermi energy $\tau_c = \hbar/\varepsilon_f$, in agreement with the quasiparticle formation time known as Landau's criterion. Indeed, the quasiparticle formation and the build up of correlations are two alternative views of the same phenomenon.

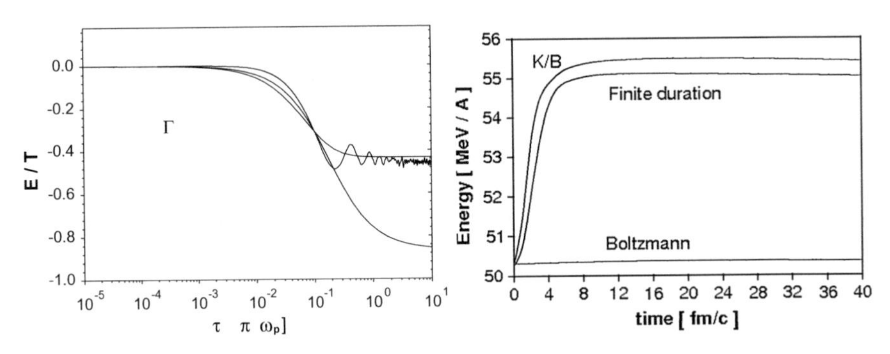

Fig. 25.1 (*Left*) The formation of correlation energy due to molecular dynamic simulations [48] together with the statically screened result of Eq. (25.4a) (curve below) and the dynamically screened result (curve above) of Eq. (25.4b) for a plasma, and (*Right*) a counter-flowing streams of nuclear matter [49] from a solution of the Kadanoff-Baym equation (KB) together with the results from the finite duration approximation and the Boltzmann equation

25.2.1 Quantum Quenches and Sudden Switching

Special preparation of cold atoms in optical lattices allows to study the local relaxation [1, 50] and to explore dissipation mechanisms [51]. We consider the time evolution of the reduced density matrix $\langle p+\frac{1}{2}q|\delta\rho|p-\frac{1}{2}q\rangle=\delta f(p,q,t)$ which is given by linearization $\delta[H,\rho]=[\delta H,\rho_0]+[H_0,\delta\rho]$ of the kinetic equation

$$\dot{\rho}+i[H,\rho]=\frac{\rho^{\text{l.e.}}-\rho}{\tau} \tag{25.5}$$

with respect to an external perturbation δV^{ext}. The effective Hamiltonian consists of the quasiparticle energy, the external and induced mean-field $\langle p+\frac{1}{2}q|\delta H|p-\frac{1}{2}q\rangle=\delta V^{\text{ext}}+V_q\delta n_q$ given by the interaction potential V_q and the density variation δn_q. As possible confining potential we assume a harmonic trap $V^{\text{trap}}=\frac{1}{2}Kx^2$, which leads to $\langle p+\frac{1}{2}q|\delta[V^{\text{trap}},\rho]|p-\frac{1}{2}q\rangle=-K\partial_p\partial_q\delta f(p,q,t)$.

The kinetic equation (25.5) is assumed to relax towards a local equilibrium of Fermi/Bose distribution with an allowed variation of the chemical potential

$$\begin{aligned}\left\langle p+\frac{q}{2}\right|\rho^{\text{l.e.}}-\rho\left|p-\frac{q}{2}\right\rangle&=\langle\rho^{\text{l.e.}}-\rho^0\rangle-\delta f(p,q,t)\\&=-\frac{\Delta f}{\Delta\varepsilon}\delta\mu(q,t)-\delta f(p,q,t).\end{aligned} \tag{25.6}$$

Here, we use the short-hand notation $\Delta f=f_0(p+\frac{q}{2})-f_0(p-\frac{q}{2})$ and $\Delta\varepsilon=\varepsilon_{p+\frac{q}{2}}-\varepsilon_{p-\frac{q}{2}}$. This variation of the chemical potential allows to enforce the density conservation $n=\sum_p f=\sum_p f^{\text{l.e.}}$ [52–54] leading to the Mermin correction, i.e. a relation between density variation $\delta n(q,t)=\tilde{\Pi}(t,\omega=0)\delta\mu(q,t)$ and the polarization in random phase approximation (RPA)

$$\Pi(t,t')=i\sum_p[f_{p+\frac{q}{2}}(t')-f_{p-\frac{q}{2}}(t')]e^{\left(i\varepsilon_{p+\frac{q}{2}}-i\varepsilon_{p-\frac{q}{2}}+\frac{1}{\tau}\right)(t'-t)}, \tag{25.7a}$$

$$\tilde{\Pi}(t,\omega)=\int d(t-t')e^{i\omega(t-t')}\Pi(t,t'). \tag{25.7b}$$

The linearized kinetic equation (25.5) is solved considering the momentum derivatives of the last term as perturbation to obtain [6]

$$\begin{aligned}\delta f(p,q,t)-\delta f(p,q,0)=i\int_{t_0}^{t}dt'e^{[(i\Delta\varepsilon+\frac{1}{\tau})(t'-t)]}\Big\{&\Delta f(t')\left[V_q\delta n(q,t')+V_q^{\text{ext}}(t')\right]\\&+\frac{1}{i\tau\tilde{\Pi}(t',0)}\frac{\Delta f(t')}{\Delta\varepsilon}\delta n(q,t')+K\partial_p\partial_q\delta f(p,q,t')\Big\}.\end{aligned} \tag{25.8}$$

In case of a sudden quench the interaction is switched on suddenly and no external perturbation will be assumed $\delta V^{\rm ext} = 0$. Let us consider the time evolution of an empty place in the lattice if each second place was initially populated. The density $n_t = \frac{n}{2} + \delta n_t$ starts with $n_0 = 0$, which means $\delta n_0 = -n/2$ as initial condition. The solution without a confining trap ($K = 0$) reads

$$\delta n_s = -\frac{n}{2}\frac{(s+\frac{1}{\tau})^2+b^2}{\sqrt{(s+\frac{1}{\tau})^2+4Jb(s^2+\frac{s}{\tau}+nbV_q+b^2)}} \circ\!\!-\!\!\bullet \tag{25.9a}$$

$$\delta n_t = -\frac{n}{2}J_0(\sqrt{4Jbt})e^{-\frac{t}{\tau}} - \frac{n}{4\gamma\tau^2}\int_0^t dx\, J_0(\sqrt{4Jbx})e^{-\frac{t+x}{2\tau}} \times \left(2\gamma\tau\cos\gamma(t-x) + (1-2bnV_q\tau^2)\sin\gamma(t-x)\right), \tag{25.9b}$$

where $\gamma^2 = nbV + b^2 - 1/4\tau^2$, and J_0 the Bessel function. Besides the interaction-free result we obtain an additional contribution due to the interaction and dissipation presented by the relaxation time. Without interaction, $V = 0$, and damping $1/\tau \to 0$, we obtain the exact result of [50].

In Fig. 25.2 we compare Eq. (25.9) with the experimental data [1] where we plot the interaction free evolution together with the interaction one. The main effect of

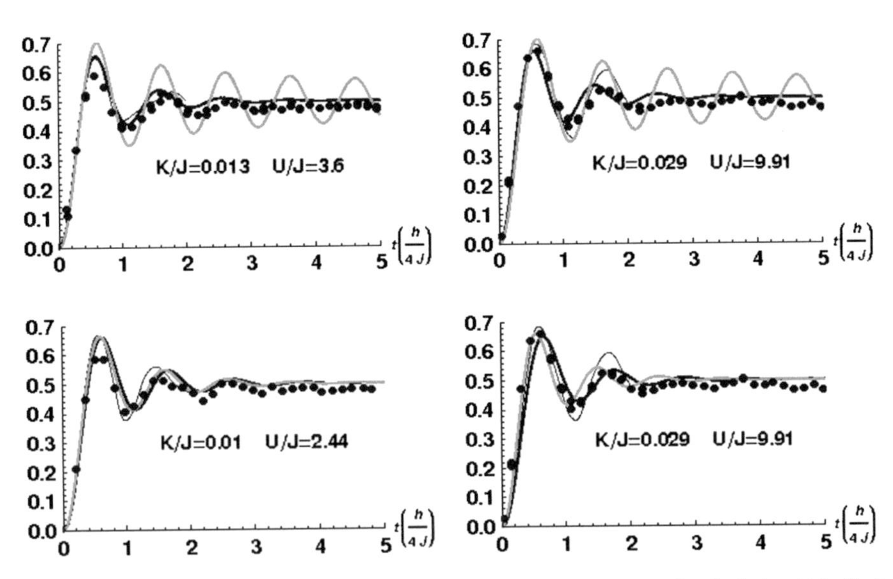

Fig. 25.2 Comparison of the experimental data of [1] (dots) with the RG calculation (thin line) [50] from [6]. Left: Mermin's correction of conserving relaxation time $\tau = 0.6\hbar/J$ approximation, Eq. (25.9), without (gray) and with interaction (black). Right: With (black) and without (gray) the influence of the trapping potential K/J

interaction is the damping which brings the curves nearer to the experimental data. Here, we use the parameter for the lattice constant given by half of the short laser wavelength $a = \lambda/2 = 765$ nm, which provides a wave vector of $q = \pi\hbar/a$, and an initial density $n = 1/2a$ with each second place filled. The relaxation time characterizes dissipative processes which we assume to arise due to polaron scattering. These lattice deformation processes are dominated by hopping transport at high temperatures and band regime transport at low temperature with the transition given by $\hbar/\tau = 2J\exp(-S)$, where S describes the ratio of polaron binding to optical phonon energy. This quantity is generally difficult to calculate [55] but in the order of one. We will use it as fit parameter and find a common value $\tau = 0.6\hbar/J$ for the results in the figures presented here. We see that the analytic result, Eq. (25.9), describes the data slightly better and we can give the time evolution up to more oscillations than it was possible by numerical renormalization group techniques.

25.2.2 *Femtosecond Laser Response*

Now we are interested in the short-time response of the system to an external perturbation V^{ext}. This is different from sudden quench since here we have initially $\delta f(p, q, 0) = 0$ and the system is driven out of equilibrium by V^{ext}. As the result we will obtain the dielectric response which gives microscopic access to optical properties. Integrating Eq. (25.8) over momentum one obtains the time-dependent density response

$$\delta n(q, t) = \int_{t_0}^{t} dt' \chi(t, t') V_q^{\text{ext}}(t') \tag{25.10}$$

describing the response of the system with respect to the external field in contrast to the polarization function, Eq. (25.7), which is the response to the induced field. One obtains the equation for $\chi(t, t')$ from Eq. (25.10) by interchanging integrations in Eq. (25.8)

$$\chi(t, t') = \Pi(t, t') + \int_{t'}^{t} d\bar{t} \left\{ \left[\Pi(t, \bar{t}) V_q + I(t, \bar{t}) \right] \chi(\bar{t}, t') + R(t, \bar{t}) \right\}, \tag{25.11}$$

with the polarization Eq. (25.7) and Mermin's correction

$$I(t, t') = \sum_p \frac{f_{p+\frac{q}{2}}(t') - f_{p-\frac{q}{2}}(t')}{\varepsilon_{p+\frac{q}{2}} - \varepsilon_{p-\frac{q}{2}}} \frac{e^{\left(i\varepsilon_{p+\frac{q}{2}} - i\varepsilon_{p-\frac{q}{2}} + \frac{1}{\tau}\right)(t'-t)}}{\tau \tilde{\Pi}^{\text{RPA}}(t', 0)}. \tag{25.12}$$

For cold atoms on the lattice we have obtained already the solution, Eq. (25.8), which we can use here with $\delta f(0) = 0$ and we have also

$$\Pi(t,t') = ne^{\frac{t'-t}{\tau}} \sin[b(t'-t)], \tag{25.13a}$$

$$I(t,t') = \frac{1}{\tau} e^{\frac{t'-t}{\tau}} \cos[b(t'-t)]. \tag{25.13b}$$

This will lead to the same response formula as a gas of particles with the thermal Fermi/Bose distribution for f_p. For the latter one we work in the limit of long wave lengths $q \to 0$ and the leading terms are $\Pi(t,t') \approx \frac{q^2 n(t')}{m}(t'-t)e^{(t'-t)/\tau}$ and $I(t,t') \approx \frac{1}{\tau}e^{(t'-t)/\tau}$, with the time-dependent density $n(t)$.

We introduce the collective mode of plasma/sound-velocity oscillations for Coulomb gas and for the Hubbard models respectively

$$\omega_p^2 = \begin{cases} \frac{ne^2}{m\varepsilon_0} & \text{for } V_q = \frac{e^2\hbar^2}{\varepsilon_0 q^2},\ \varepsilon_p = \frac{p^2}{2m}, \\ bnaU & \text{for } V_q = Ua,\ \varepsilon_p = 2J(1-\cos\frac{pa}{\hbar}), \end{cases} \tag{25.14}$$

where we have used already $b = 4J\sin^2\frac{aq}{2\hbar}$. For Coulomb interactions one has an optical mode while for atoms on the lattice the mode is acoustic.

For the gas of particles it is convenient to transform Eq. (25.11) into a differential equation

$$\ddot{\chi}(tt') + \frac{1}{\tau}\dot{\chi}(tt') + \omega_p^2 \chi(tt') = 0 \tag{25.15a}$$

$$\chi(t,t) = 0, \quad \dot{\chi}(t,t')|_{t=t'} = -\omega_p^2/V_q, \tag{25.15b}$$

where the influence of the trap can be considered as well [6].

Interestingly, both solutions, the one for the Hubbard lattice and the one for the gas of particles, lead to the same result of the integral equation (25.11) via Eq. (25.15) for the two-time response function

$$V\chi(t,t') = -\frac{\omega_p^2}{\gamma} e^{-\frac{t-t'}{2\tau}} \sin\gamma(t-t'), \tag{25.16}$$

but with a different collective mode $\gamma = \sqrt{\omega_p^2 - \frac{1}{4\tau^2}}$ for the Coulomb gas, and $\gamma = \sqrt{\omega_p^2 + b^2 - \frac{1}{4\tau^2}}$ for cold atoms. In this sense, we consider Eq. (25.16) as universal short-time behavior.

The pump pulse is creating charge carriers in the conduction band and the probe pulse is testing the time evolution of this occupation [3]. The time delay after this probe pulse $T = t - t_0$ is Fourier transformed into frequency. The inverse dielectric function is then given by

$$\frac{1}{\varepsilon(\omega,t)} = 1 + \int_0^{t-t_0} dT e^{i\omega T} V\chi(t,t-T). \tag{25.17}$$

The integral in Eq. (25.17) with Eq. (25.16) describes the experimental time formation of plasma mode quite accurately [6]. The virtue of Eq. (25.17) is also that the long-time limit yields correctly the Drude formula

$$\lim_{t\to\infty} \frac{1}{\varepsilon} = 1 - \frac{\omega_p^2}{\gamma^2 - \omega(\omega + \frac{i}{\tau})}, \tag{25.18}$$

leading to the Drude conductivity which is not easy to achieve within short-time expansions [56], and which had provided the wrong long-time limit $1 - \omega_p^2/[\gamma^2 - (\omega + i/\tau)^2]$ before.

25.3 Nonlocal Kinetic Theory

At later times, when correlations develop, the off-shell motion can be eliminated from the kinetic equation, which requires to introduce an effective distribution (the quasiparticle distribution f) from which the Wigner distribution ρ can be constructed

$$\rho[f] = f + \wp \int \frac{d\omega}{2\pi} \frac{1}{\omega - \varepsilon} \frac{\partial}{\partial \omega} \left((1 - f)\sigma_\omega^< - f\sigma_\omega^>\right). \tag{25.19}$$

Here, $\sigma^>$ and $\sigma^<$ denote the self-energies describing all correlations and ε is the quasiparticle energy. This relation represents the extended quasiparticle picture derived for small scattering rates [41–43, 57]. The limit of small scattering rates has been first introduced by Craig [58]. An inverse relation $f[\rho]$ has been constructed [59]. For equilibrium non-ideal plasmas this approximation has been employed by [60, 61] and has been used under the name of the generalized Beth-Uhlenbeck approach by [62] in nuclear matter for studies of the correlated density. The authors in Ref. [63] have used this approximation with the name 'extended quasiparticle approximation' for the study of the mean removal energy and high-momenta tails of Wigner's distribution. The non-equilibrium form has been derived finally as the modified Kadanoff and Baym ansatz [64].

This extended quasiparticle picture leads to balance equations which include explicit correlation parts analogous to the virial corrections. The firmly established concept of the equilibrium virial expansion has been extended to nonequilibrium systems [42] although a number of attempts have been made to modify the Boltzmann equation so that its equilibrium limit would cover at least the second virial coefficient [24, 65, 66]. The corrections to the Boltzmann equation have the form of gradients or nonlocal contributions to the scattering integral. Please note that the nature of two-particle correlations induces gradients and therefore nonlocal kinetic and exchange energies [67, 68].

25.3.1 Nonlocal Kinetic Equation

The nonlocal quantum kinetic equation following from the extended quasiparticle approximation reads [42]

$$\frac{\partial f_1}{\partial t} + \frac{\partial \varepsilon_1}{\partial \mathbf{k}} \frac{\partial f_1}{\partial \mathbf{r}} - \frac{\partial \varepsilon_1}{\partial \mathbf{r}} \frac{\partial f_1}{\partial \mathbf{k}} = I_1^{\text{in}} - I_1^{\text{out}} \tag{25.20}$$

with the scattering-in

$$\begin{aligned} I_1^{\text{in}} = \sum_b \int \frac{d^3p}{(2\pi)^3} \frac{d^3q}{(2\pi)^3} 2\pi \delta\left(\varepsilon_1 + \bar{\varepsilon}_2 - \bar{\varepsilon}_3 - \bar{\varepsilon}_4 - 2\Delta_E\right) \\ \times \left(1 - \frac{1}{2}\frac{\partial \boldsymbol{\Delta}_2}{\partial \mathbf{r}} - \frac{\partial \bar{\varepsilon}_2}{\partial \mathbf{r}} \frac{\partial \boldsymbol{\Delta}_2}{\partial \omega}\right)_{\omega=\varepsilon_1+\bar{\varepsilon}_2} (1 - f_1 - \bar{f}_2)\bar{f}_3\bar{f}_4 \\ \times \left| t_{\text{SC}}\left(\varepsilon_1 + \bar{\varepsilon}_2 - \Delta_E, \mathbf{k} - \frac{\boldsymbol{\Delta}_K}{2}, \mathbf{p} - \frac{\boldsymbol{\Delta}_K}{2}, \mathbf{q}, \mathbf{r} - \boldsymbol{\Delta}_r, t - \frac{\Delta_t}{2}\right)\right|^2, \end{aligned} \tag{25.21}$$

and the scattering-out by replacing $f \leftrightarrow 1 - f$ and changing the sings of the shifts. All distribution functions and observables have the arguments

$$\varepsilon_1 \equiv \varepsilon_a(\mathbf{k}, \mathbf{r}, t), \tag{25.22a}$$
$$\varepsilon_2 \equiv \varepsilon_b(\mathbf{p}, \mathbf{r} + \boldsymbol{\Delta}_2, t), \tag{25.22b}$$
$$\varepsilon_3 \equiv \varepsilon_a(\mathbf{k} - \mathbf{q} + \boldsymbol{\Delta}_K, \mathbf{r} + \boldsymbol{\Delta}_3, t + \Delta_t), \tag{25.22c}$$
$$\varepsilon_4 \equiv \varepsilon_b(\mathbf{p} + \mathbf{q} + \boldsymbol{\Delta}_K, \mathbf{r} + \boldsymbol{\Delta}_4, t + \Delta_t) \tag{25.22d}$$

and a bar indicates the reversed sign of the Δ's.

In the scattering-out (scattering-in is analogous) one can see the distributions of quasiparticles $f_1 f_2$ describing the probability of a given initial state for the binary collision. The hole distributions giving the probability that the requested final states are empty and the particle distribution of stimulated collisions combine together in the final state occupation factors like $1 - f_3 - f_4 = (1 - f_3)(1 - f_4) + f_3 f_4$. The scattering rate covers the energy-conserving δ-function, and the differential cross section is given by the modulus of the T-matrix, $|t_{\text{SC}}|$, reduced by the wave-function renormalizations $z_1\bar{z}_2\bar{z}_3\bar{z}_4$ [69]. We consider here the linear expansion in small scattering rates, therefore the wave-function renormalization in the collision integral is of higher order.

All Δ's are derivatives of the scattering phase shift ϕ,

$$t_{\text{SC}}^R = |t_{\text{SC}}|\, e^{i\phi}, \tag{25.23}$$

according to the following list

$$\boldsymbol{\Delta}_K = \frac{1}{2}\frac{\partial \phi}{\partial \mathbf{r}}, \tag{25.24a}$$

$$\Delta_E = -\frac{1}{2}\frac{\partial \phi}{\partial t}, \tag{25.24b}$$

$$\Delta_t = \frac{\partial \phi}{\partial \omega}, \tag{25.24c}$$

$$\boldsymbol{\Delta}_2 = \frac{\partial \phi}{\partial \mathbf{p}} - \frac{\partial \phi}{\partial \mathbf{q}} - \frac{\partial \phi}{\partial \mathbf{k}}, \tag{25.24d}$$

$$\boldsymbol{\Delta}_3 = -\frac{\partial \phi}{\partial \mathbf{k}}, \tag{25.24e}$$

$$\boldsymbol{\Delta}_4 = -\frac{\partial \phi}{\partial \mathbf{q}} - \frac{\partial \phi}{\partial \mathbf{k}}, \tag{25.24f}$$

$$\boldsymbol{\Delta}_r = \frac{1}{4}\left(\boldsymbol{\Delta}_2 + \boldsymbol{\Delta}_3 + \boldsymbol{\Delta}_4\right). \tag{25.24g}$$

As special limits, this kinetic theory includes the Landau theory as well as the Beth-Uhlenbeck equation of state [70, 71], which means correlated pairs. The medium effects on binary collisions are shown to mediate the latent heat which is the energy conversion between correlation and thermal energy [42, 72]. In this respect the seemingly contradiction between particle-hole symmetry and time reversal symmetry in the collision integral was solved [73]. Compared to the Boltzmann-equation, the presented form of virial corrections only slightly increases the numerical demands in implementations [74–77] since large cancellations in the off-shell motion appear which are hidden usually in non-Markovian behaviors. Details how to implement the nonlocal kinetic equation into existing Boltzmann codes can be found in [77].

25.3.2 Balance Equations

We multiply the kinetic equation (25.20) with a variable $\xi_1 = 1, \mathbf{k}, \varepsilon_1, -k_B \ln[f_1/(1 - f_1)]$ and integrate over momentum. It results in the equation of continuity, the Navier-Stokes equation, the energy balance and the evolution of the entropy, respectively. All these conservation laws or balance equations for the mean thermodynamic observables have the form

$$\frac{\partial \langle \xi^{\mathrm{qp}} + \xi^{\mathrm{mol}} \rangle}{\partial t} + \frac{\partial (\mathbf{j}_\xi^{\mathrm{qp}} + \mathbf{j}_\xi^{\mathrm{mol}})}{\partial \mathbf{r}} = \mathscr{I}_{\mathrm{gain}}, \tag{25.25}$$

consisting of a quasiparticle part

$$\xi^{\rm qp} = \int \frac{d^3k}{(2\pi)^3} \xi_1 f_1 \tag{25.26}$$

and the correlated or molecular contribution

$$\xi^{\rm mol} = \int \frac{d^3k d^3p d^3q}{(2\pi)^9} |t_{\rm SC}(\varepsilon_1 + \varepsilon_2, k, p, q)|^2 \Delta \frac{\xi_1 + \xi_2}{2} \times 2\pi \delta(\varepsilon_1 + \varepsilon_2 - \varepsilon_3 - \varepsilon_4) f_1 f_2 (1 - f_3 - f_4). \tag{25.27}$$

The latter one has the statistical interpretation of the rate of binary processes $D = |t_{\rm SC}|^2 2\pi\delta(\varepsilon_1 + \varepsilon_2 - \varepsilon_3 - \varepsilon_4)(1 - f_3 - f_4) f_1 f_2$ weighed with the lifetime of the molecule Δ_t, respectively. This has the form of a molecular contribution as if two particles form a molecule.

The quasiparticle currents of the observable reads

$$\mathbf{j}_\xi^{\rm qp} = \int \frac{d^3k}{(2\pi)^3} \xi_1 \frac{\partial \varepsilon_1}{\partial \mathbf{k}} f_1, \tag{25.28}$$

and the molecular currents we have obtained as [78]

$$\mathbf{j}_\xi^{\rm mol} = \frac{1}{2} \int \frac{d^3k d^3p d^3q}{(2\pi)^9} D(\xi_2 \boldsymbol{\Delta}_2 - \xi_3 \boldsymbol{\Delta}_3 - \xi_4 \boldsymbol{\Delta}_4). \tag{25.29}$$

It is the balance of observables carried by the different spatial off-sets.

The additional gain on the right side might be due to an energy or force feed from the outside or the entropy production by collisions.

Due to its intriguing vector character, let us give the Navier-Stokes equation explicitly. The inertial force density is given by the time derivative of the momentum density $\mathcal{Q}$. The deformation force density is given by the divergence of the stress tensor. The stress tensor we derived from the balance between the inertial and the deformation forces

$$\frac{\partial}{\partial t}\left(\mathcal{Q}_j^{\rm qp} + \mathcal{Q}_j^{\rm mol}\right) = -\sum_i \frac{\partial}{\partial r_i}\left(\mathcal{J}_{ij}^{\rm qp} + \mathcal{J}_{ij}^{\rm mol}\right) \tag{25.30}$$

with the momentum density consisting of the quasiparticle

$$\mathcal{Q}_j^{\rm qp} = \int \frac{d^3k}{(2\pi)^3} k_j f_1 \tag{25.31}$$

and molecular part

$$\mathscr{Q}_j^{\rm mol} = \frac{1}{2}\int \frac{d^3kd^3pd^3q}{(2\pi)^9}(k_j + p_j)D\Delta_t, \tag{25.32}$$

which gives the mean momentum carried by a molecule formed with the rate D and lifetime Δ_t.

The total stress tensor formed by the quasiparticles reads

$$\mathscr{I}_{ij}^{\rm qp} = \sum_a \int \frac{d^3k}{(2\pi)^3}\left(k_j\frac{\partial\varepsilon}{\partial k_i} + \delta_{ij}\varepsilon\right) f - \delta_{ij}\mathscr{E}^{\rm qp} \tag{25.33}$$

with quasiparticle energy functional [42]

$$\mathscr{E}^{\rm qp} = \sum_a \int \frac{d^3k}{(2\pi)^3} f_a(k)\frac{k^2}{2m} + \frac{1}{2}\sum_{ab}\int \frac{d^3kd^3p}{(2\pi)^6} f_a(k) f_b(p)\,{\rm Re}\, t_{\rm SC}(\varepsilon_1 + \varepsilon_2, k, p, 0) \tag{25.34}$$

instead of the Landau functional which is valid only in local approximation.

The collision-flux contribution, Eq. (25.29), reads

$$\mathscr{I}_{ij}^{\rm mol} = \frac{1}{2}\sum_{ab}\int \frac{d^3k\, d^3p\, d^3q}{(2\pi)^9} D\left[(k_j - q_j)\Delta_{3i} + (p_j + q_j)\Delta_{4i} - p_j\Delta_{2i}\right]. \tag{25.35}$$

It possesses a statistical interpretation as well. The two-particle state is characterized by the initial momenta **k** and **p** and the transferred momentum **q**. The momentum tensor is the balance of the momenta carried by the corresponding spatial off-sets weighted with the rate to form a molecule D.

For the density $\xi = 1$ we do not have a gain. For momentum gain $\xi = k_j$ we get

$$\mathscr{F}_j^{\rm gain} = \sum_{ab}\int \frac{d^3k\, d^3p\, d^3q}{(2\pi)^9} D\Delta_{Kj}. \tag{25.36}$$

Dividing and multiplying by Δ_t under the integral, we see that the momentum gain is the probability $D\Delta_t$ to form a molecule multiplied by the force $\boldsymbol{\Delta}_K/\Delta_t$ exercised during the delay time Δ_t from the environment by all other particles. This momentum gain, Eq. (25.36), can be exactly recast together with the term of the drift into a spatial derivative [42]

$$\sum_a \int \frac{d^3k}{(2\pi)^3}\varepsilon\frac{\partial f}{\partial \mathbf{r}_j} + \mathscr{F}_j^{\rm gain} = \frac{\partial \mathscr{E}^{\rm qp}}{\partial \mathbf{r}_j} \tag{25.37}$$

of the quasiparticle energy functional, Eq. (25.34). Similarly, the energy gain combines with the drift into the total time derivative of the quasiparticle energy functional, Eq. (25.34)

$$\sum_a \int \frac{d^3k}{(2\pi)^3} \varepsilon \frac{\partial f}{\partial t} - \mathscr{I}^E_{\text{gain}} = \frac{\partial \mathscr{E}^{\text{qp}}}{\partial t}. \tag{25.38}$$

The only remaining explicit gain is the entropy gain

$$\mathscr{I}^S_{\text{gain}} = -\frac{k_B}{2} \sum_{ab} \int \frac{d^3k\, d^3p\, d^3q}{(2\pi)^9} f_1 f_2 (1 - f_3 - f_4) \\ \times 2\pi \delta(\varepsilon_1 + \varepsilon_2 - \varepsilon_3 - \varepsilon_4) |t_{\text{SC}}|^2 \ln \frac{f_3 f_4 (1 - f_1)(1 - f_2)}{(1 - f_3)(1 - f_4) f_1 f_2}, \tag{25.39}$$

while the momentum gain and energy gain are transferring kinetic into correlation parts and do not appear explicitly. In Ref. [79] it is proved that this entropy gain is always positive establishing the H-theorem including single particle and two-particle quantum correlations.

25.4 Applications of Nonlocal Kinetic Theory

25.4.1 Low-Energy Heavy Ion Reactions

During a heavy ion collision we can access a state of matter which gives insight into special aspects of nonequilibrium processes. The dominant features are that we have strong short range interactions of typically 1 fm $= 10^{-15}$ m and the product of the range of interaction with Fermi momentum is in orders of one characteristic for a degenerate quantum system. Since the radius of typical nuclei is $R \approx 1.2A^{1/3}$ fm (where A is the nucleus baryon number) we see that the product of the radius with Fermi momentum is of few $\hbar$ indicating strong spatial inhomogeneity.

Numerical simulations extensively used to interpret experimental data from heavy ion reactions, are based either on the Boltzmann (BUU) equation or on the quantum molecular dynamics (QMD). Due to their quasiclassical character, they offer a transparent picture of the internal dynamics of reactions and allow one to link the spectrum of the detected particles with individual stages of reactions. They fail, however, to describe some energy and angular distributions of neutrons and protons in low and mid energy domain [80–82]. Appreciable values of the collision delay and space displacements show that the nonlocal collisions should be accounted for.

The nonlocal collisions have been implemented in Ref. [74] within the QMD and in Refs. [75, 76] within the BUU equation. Within the local approximation the distribution of high-energy protons is too low to meet the experimental values. The

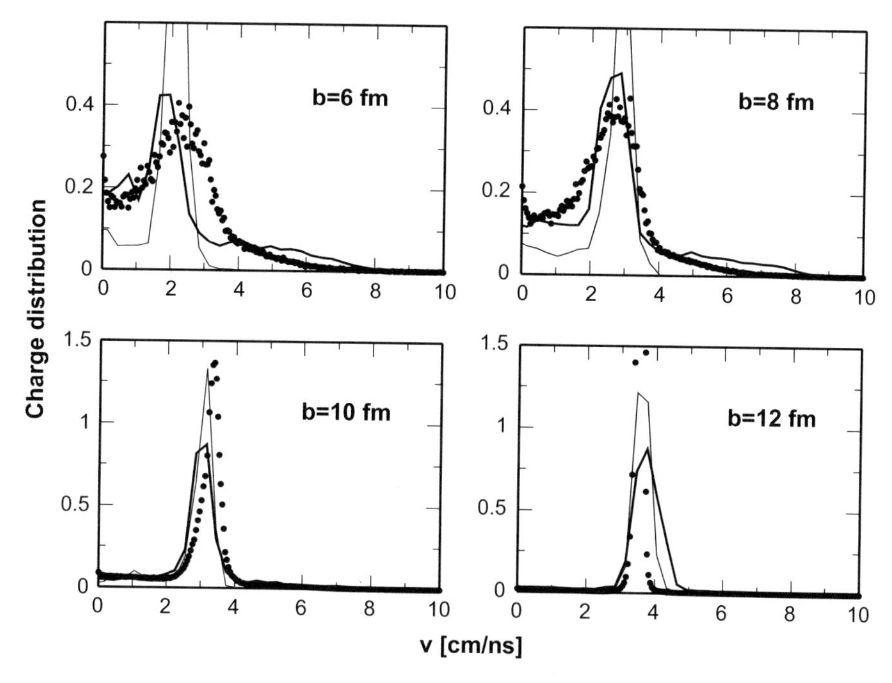

Fig. 25.3 The experimental charge distribution of matter (dotted line) versus velocity in comparison with in the BUU (thin solid line) and the nonlocal model with quasiparticle renormalization (thick line). Redrawn after Ref. [77]

influence of the nonlocal collisions on the reaction of heavy ions has been studied for the $^{181}_{73}$Ta + $^{197}_{79}$Au reaction at 33 MeV [77]. Except for the nonlocal picture, a sufficiently large neck has been achieved by additional inclusion of fluctuations in the Boltzmann (BUU) equation [83, 84], resulting in Boltzmann-Langevin pictures [85–90]. The Boltzmann-Langevin equation has been derived assuming an additional coarse graining of phase-space [91, 92]. Fluctuations to the time-dependent Hartree-Fock (TDHF) equation have been analyzed before in Refs. [93, 94] and tested in Ref. [95].

INDRA observation shows the enhancement of emitted matter in the mid-rapidity region [96, 97]. The simulations can be compared to the experimental data of the Ta + Au collision [77]. In Fig. 25.3, the theoretical and experimental charge density distributions are compared. The experimental charge density distribution has been obtained using the procedure described in Ref. [98]. The data are represented by light gray points, the standard BUU calculation by the thin line and the nonlocal BUU with quasiparticle renormalization calculation by the thick line. A reasonable agreement is found for the nonlocal scenario including quasiparticle renormalization while simple BUU fails to reproduce mid-rapidity matter.

The comparison of the time evolutions of the transverse energy for 8 fm impact parameter can be seen in Fig. 25.4. We recognize that the nonlocal collision scenario

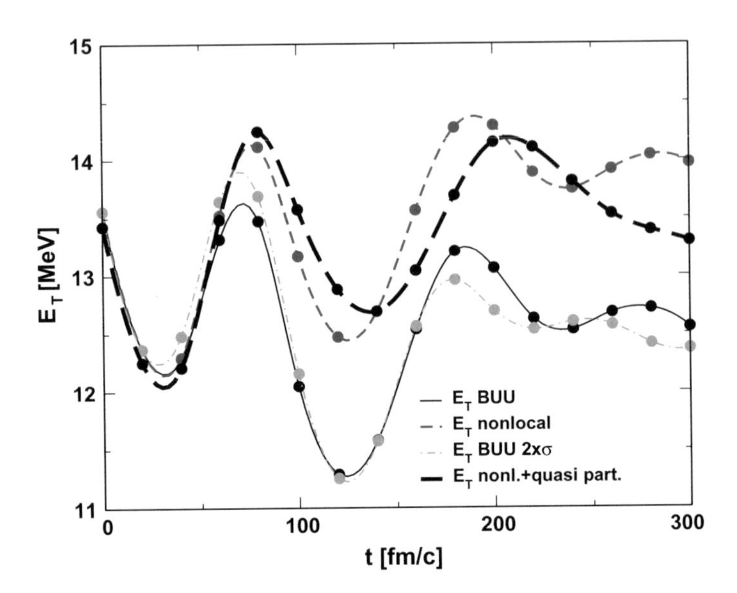

Fig. 25.4 The time evolution of the transverse energy including the Fermi motion for Ta + Au at $E_{\text{lab}}/A = 33$ MeV and 8 fm impact parameter in the local BUU (black line), the nonlocal BUU (dashed line), the local BUU with doubled cross section (dashed dotted line) and the nonlocal scenario with the quasiparticle renormalization (long dashed line)

dissipates much more energy due to the inelastic character than by increasing the collision rate. The transverse energies including quasiparticle renormalization have a different period of oscillation which corresponds to a giant resonance. This period becomes longer for the case of quasiparticle renormalization which means that the compressibility decreases. In other words, the quasiparticle renormalization leads to a softer equation of state.

As documented by the improvement of the high-energy proton production and the midrapidity charge density distribution, the nonlocal treatment of the binary collisions brings a desirable contribution to the dynamics of heavy ion reactions. According to the experience from the theory of gases, one can also expect a vital role of non-localities in the search for the equation of state of the nuclear matter. It is encouraging that the nonlocal corrections are easily incorporated into the BUU and QMD simulation codes and do not increase the computational time.

25.4.2 Relation of Pairing Density to Correlated Density

In superconductors the correlated density which we found as the consequence of the time-nonlocality of the collision process, Eq. (25.27), becomes visible as the difference between the total and normal density $n_{\text{corr}} = n - n_n$. The Wigner distribution function has a two-part structure [99]

$$\rho = \frac{1}{2}\left(1+\frac{\xi}{E}\right) f(E) + \frac{1}{2}\left(1-\frac{\xi}{E}\right) f(-E) = \frac{1}{2} - \frac{\xi}{2E}\tanh\frac{1}{2}\beta E, \quad (25.40)$$

where $\xi = \varepsilon_p - \mu$ is the free-particle energy ε_p minus the chemical potential μ. The quasiparticle energy $E = \sqrt{\xi^2 + \Delta^2}$ describes the influence of the superconducting gap Δ on the excitation spectrum of the superconducting state. The Fermi-Dirac distribution is $f(x) = 1/(e^{\beta x} + 1)$ with the inverse temperature $\beta = 1/k_B T$. The density n is obtained by the momentum integral over Eq. (25.40) and introducing the density of states $h(\xi) = 2\sum_p 2\pi\delta(\xi - \varepsilon_p)$, one has

$$n = \int_{-\bar{\mu}}^{\infty} \frac{d\xi}{2\pi} h(\bar{\mu}+\xi)\left(\frac{1}{2} - \frac{\xi}{2E}\tanh\frac{1}{2}\beta E\right). \quad (25.41)$$

Here, we account for a possible electrostatic potential φ and the velocity v of superconducting electrons by $\bar{\mu} = \mu - e\varphi - mv^2/2$. For a vanishing gap we obtain the corresponding density n_n of normal electrons with the chemical potential $\bar{\mu}$ by $n_n = n(\Delta = 0)$. The difference

$$n_{\mathrm{corr}} = n - n_n \quad (25.42)$$

describes the correlated density. In the ground state the normal density turns into

$$n_n = 2\sum_p \Theta(\bar{\mu} - \varepsilon_p) \approx n_0 - \left(e\varphi + \frac{m}{2}v^2\right)\frac{h(\mu)}{2\pi} \quad (25.43)$$

where we have expanded $\bar{\mu}$ in first order around the Fermi energy, and n_0 describes the number of particles with no motion and no electrostatic potential. The correlated density Eq. (25.42) splits into two parts in the zero-temperature limit of Eq. (25.41)

$$n_{\mathrm{corr}} = \frac{1}{2}\int_0^{\infty} \frac{d\xi}{2\pi} h(\bar{\mu}+\xi)\frac{\sqrt{\xi^2+\Delta^2}-\xi}{\sqrt{\xi^2+\Delta^2}} - \frac{1}{2}\int_{-\bar{\mu}}^{0} \frac{d\xi}{2\pi} h(\bar{\mu}+\xi)\frac{\sqrt{\xi^2+\Delta^2}+\xi}{\sqrt{\xi^2+\Delta^2}}, \quad (25.44)$$

which vanishes for vanishing gap. Since the gap is only nonzero in the vicinity of the Fermi level given by the Debye frequency ω_D we can restrict the integration to the $\pm\omega_D$-range. Expanding the density of states for $\xi < \omega_D$ we obtain finally [100]

$$n_{\text{corr}} = \frac{\partial h}{\partial \mu}\frac{\Delta^2}{4\pi}\left[\ln\left(\frac{\omega_D}{\Delta}+\sqrt{\frac{\omega_D^2}{\Delta^2}+1}\right)-\frac{1}{1+\sqrt{1+\frac{\Delta^2}{\omega_D^2}}}\right] \approx \frac{\partial h}{\partial \mu}\frac{\Delta^2}{4\pi}\ln\left(\frac{2\omega_D}{\sqrt{e}\Delta}\right) \tag{25.45}$$

for $\omega_D \gg \Delta$ in the last step.

Since the total system should stay neutral, we expect $n = n_0$ and the two contributions, $n_n - n_0$ according to Eq. (25.43) and n_{corr} of Eq. (25.45), should cancel. Therefore the required electrostatic potential must read

$$e\varphi = -\frac{m}{2}v^2 + \frac{\partial \ln h}{\partial \mu}\frac{\Delta^2}{2}\ln\left(\frac{2\omega_D}{\sqrt{e}\Delta}\right). \tag{25.46}$$

This resulting electrostatic potential has the form of a Bernoulli potential. Its purpose is to compensate the contribution due to diamagnetic currents and the associated inertial and Lorentz forces. It has a part directly linked to the gap.

The great hope was to measure the Bernoulli potential in order to access directly the gap parameter [101]. The experimental attempts to measure it, however, have yielded no result [102, 103]. Why no signal of thermodynamic corrections is seen remained a puzzle for nearly 30 years. The solution was found by a modification [104] of the Budd-Vannimenus theorem [105] which shows that the surface dipoles cancel the thermodynamical corrections exactly for homogeneous superconductors. The Budd-Vannimenus theorem has been applied also to finite Fermi systems within an exactly solved model to show the BEC-BCS transition in [106].

The electrostatic potential can leak out of a superconductor by three types of charges: (i) The bulk charge which describes the transfer of electrons from the inner to the outer regions of vortices creating a Coulomb force. This force has to balance the centrifugal force by the electrons rotating around the vortex center, the outward push of the magnetic field via the Lorentz force and the outward force coming from the fact that the energy of Cooper pairs is lower than the one of free electrons such that unpaired electrons in the vortex core are attracted towards the condensate around the core [107]. (ii) The surface charge [108] distributed on the scale of the Thomas-Fermi screening length. (iii) The surface dipole which cancels all contributions of pairing forces [109] resulting in an observable surface potential of

$$e\phi_0 = -\frac{f_{\text{el}}}{n}. \tag{25.47}$$

The latter one gives rise to characteristic features predicted for experimental observations. The quadrupole resonance lines in the high-T_c material YBCO have been measured [110] and explained in [111, 112].

25.5 Summary

At short time scales after a sudden quench the correlations need time to be formed. This allows to describe the transient time scales by long-ranged time-dependent meanfield fluctuations. This transient time scale is essentially determined by the off-shell motion. At later times when strong correlations are formed there is a cancellation of off-shell processes in the kinetic equation using a proper extended quasiparticle picture. The remaining modifications of the quantum Boltzmann equation consist in the nonlocal collision scenario where the off-sets are uniquely determined by the phase shift of the T-matrix and the quasiparticle energies modifying the drift. The resulting balance equations show besides the quasiparticle parts of the Landau theory also explicit two-particle contributions of short living molecules. The energy and momentum conservation is ensured due to an internal transfer of energy and momentum analogously to a latent heat. Only for the entropy an explicit gain remains.

The entropy as a measure of complexity, or inversely as the loss of information [113], plays a central role in processes like nuclear or cluster reactions [114], where the kinetic and correlation energy of projectile and target particles transform into heat. In nuclear matter, mainly the single-particle entropy [115–119] is discussed as in ultra-cold atoms [120]. The equilibrium entropy has been given in a form of cluster expansion where the two-particle part is given by the two-particle correlation function [121] which has been calculated numerically for different systems [122, 123]. The majority of approaches calculate the classical entropy in various approximations [124, 125]. Here we have obtained the quantum two-particle entropy explicitly in terms of phase shifts of the scattering T-matrix in nonequilibrium. The second law of thermodynamics holds also in nonlocal kinetic theory. The single-particle entropy can decrease on cost of the molecular part of entropy describing the two-particles in a molecular state. Overall, the H-theorem is maintained [79].

The numerical solution of the nonlocal kinetic equation requires no more time than solving the usual Boltzmann equation. Two distinct examples from nuclear collision and from superconductivity have been shortly discussed as illustrative examples.

References

1. S. Trotzky, Y. Chen, A. Flesch, I.P. McCulloch, U. Schöllwock, J. Eisert, I. Bloch, Nat. Phys. **8**(4), 325 (2012). https://doi.org/10.1038/nphys2232
2. R. Huber, F. Tauser, A. Brodschelm, A. Leitenstorfer, Phys. Status Solidi B **234**(1), 207 (2002). https://doi.org/10.1002/1521-3951(200211)234:1<207::AID-PSSB207>3.0.CO;2-Z
3. R. Huber, C. Kübler, S. Tübel, A. Leitenstorfer, Q.T. Vu, H. Haug, F. Köhler, M.C. Amann, Phys. Rev. Lett. **94**, 027401 (2005). https://doi.org/10.1103/PhysRevLett.94.027401
4. V.M. Axt, T. Kuhn, Rep. Prog. Phys. **67**(4), 433 (2004). https://doi.org/10.1088/0034-4885/67/4/R01
5. K. Morawetz, P. Lipavský, M. Schreiber, Phys. Rev. B **72**, 233203 (2005). https://doi.org/10.1103/PhysRevB.72.233203
6. K. Morawetz, Phys. Rev. B **90**, 075303 (2014). https://doi.org/10.1103/PhysRevB.90.075303

7. L. Bányai, Q.T. Vu, B. Mieck, H. Haug, Phys. Rev. Lett. **81**, 882 (1998). https://doi.org/10.1103/PhysRevLett.81.882
8. P. Gartner, L. Bányai, H. Haug, Phys. Rev. B **60**, 14234 (1999). https://doi.org/10.1103/PhysRevB.60.14234
9. Q.T. Vu, H. Haug, Phys. Rev. B **62**, 7179 (2000). https://doi.org/10.1103/PhysRevB.62.7179
10. M. Kira, S.W. Koch, Phys. Rev. Lett. **93**, 076402 (2004). https://doi.org/10.1103/PhysRevLett.93.076402
11. L. Boltzmann, Wien. Ber. **66**, 275 (1872)
12. S. Chapman, T.C. Cowling, *The mathematical theory of nonuniform gases* (Cambridge University Press, Cambridge, 1939)
13. D. Enskog, *Kinetiske Theorie der Vorgänge in mäßig verdünnten Gasen* (Almqvist & Wiksells, Uppsala, 1917)
14. J.G. Kirkwood, J. Chem. Phys. **14**(3), 180 (1946). https://doi.org/10.1063/1.1724117
15. N.N. Bogoliubov, J. Phys. (USSR) **10**, 256 (1946), Translated in [125]
16. I. Prigogine, *Nonequilibrium Statistical Mechanics* (Wiley, New York, 1962)
17. N.N. Bogolyubov, K.P. Gurov, Zh Eksp, Teor. Fiz. **17**, 614 (1947)
18. H. Mori, S. Ono, Progr. Theor. Phys. **8**(3), 327 (1952). https://doi.org/10.1143/ptp/8.3.327
19. G. Baym, C. Pethick, *Landau Fermi Liquid Theory* (Wiley, New York, 1991)
20. S. Chapman, T.C. Cowling, *The Mathematical Theory of Nonuniform Gases*, 3rd edn. (Cambridge University Press, Cambridge, 1990)
21. F.J. Alexander, A.L. Garcia, B.J. Alder, Phys. Rev. Lett. **74**, 5212 (1995). https://doi.org/10.1103/PhysRevLett.74.5212
22. L. Waldmann, Z. Naturforsch. A **15**, 19 (1960)
23. R.F. Snider, J. Math. Phys. **5**(11), 1580 (1964). https://doi.org/10.1063/1.1931191
24. K. Bärwinkel, Z. Naturforsch. A **24**, 38 (1969)
25. M.W. Thomas, R.F. Snider, J. Stat. Phys. **2**(1), 61 (1970). https://doi.org/10.1007/BF01009711
26. R.F. Snider, B.C. Sanctuary, J. Chem. Phys. **55**(4), 1555 (1971). https://doi.org/10.1063/1.1676279
27. J.C. Rainwater, R.F. Snider, J. Chem. Phys. **65**(11), 4958 (1976). https://doi.org/10.1063/1.432972
28. R. Balescu, *Equilibrium and Nonequilibrium Statistical Mechanis* (Wiley, New York, 1975)
29. J.A. McLennan, *Introduction to nOnequilibrium Statistical Mechanics* (Prentice Hall, Englewood Cliffs, 1989)
30. F. Laloë, J. Phys. (Paris) **50**(14), 1851 (1989). https://doi.org/10.1051/jphys:0198900500140185100
31. P.J. Nacher, G. Tastevin, F. Laloë, J. Phys. (Paris) **50**(14), 1907 (1989). https://doi.org/10.1051/jphys:0198900500140190700
32. D. Loss, J. Stat. Phys. **61**(1), 467 (1990). https://doi.org/10.1007/BF01013976
33. M. De Haan, Physica A **170**(3), 571 (1991). https://doi.org/10.1016/0378-4371(91)90007-Y
34. F. Laloë, W.J. Mullin, J. Stat. Phys. **59**(3), 725 (1990). https://doi.org/10.1007/BF01025848
35. P.J. Nacher, G. Tastevin, F. Laloë, Ann. Phys. (Berlin) **503**(1–3), 149 (1991). https://doi.org/10.1002/andp.19915030114
36. P.J. Nacher, G. Tastevin, F. Laloë, J. Phys. I (Paris) **1**(2), 181 (1991). https://doi.org/10.1051/jp1:1991124
37. R.F. Snider, J. Stat. Phys. **80**(5), 1085 (1995). https://doi.org/10.1007/BF02179865
38. R.F. Snider, W.J. Mullin, F. Laloë, Physica A **218**(1), 155 (1995). https://doi.org/10.1016/0378-4371(95)00124-P
39. V. Špička, P. Lipavský, K. Morawetz, Phys. Rev. B **55**, 5084 (1997). https://doi.org/10.1103/PhysRevB.55.5084
40. V. Špička, P. Lipavský, K. Morawetz, Phys. Rev. B **55**, 5095 (1997). https://doi.org/10.1103/PhysRevB.55.5095
41. V. Špička, P. Lipavský, K. Morawetz, Phys. Lett. A **240**(3), 160 (1998). https://doi.org/10.1016/S0375-9601(98)00061-9

42. P. Lipavský, K. Morawetz, V. Špička, Ann. Phys. Fr. **26**(1), 1 (2001). https://doi.org/10.1051/anphys:200101001
43. K. Morawetz, P. Lipavský, V. Špička, Ann. Phys. **294**(2), 135 (2001). https://doi.org/10.1006/aphy.2001.6197
44. K. Morawetz, P. Lipavský, V. Špička, N.H. Kwong, Phys. Rev. C **59**, 3052 (1999). https://doi.org/10.1103/PhysRevC.59.3052
45. I.B. Levinson, Zh. Eksp. Teor. Fiz. **57**, 660 (1969), [Sov. Phys. JETP **30**, 362 (1970)]
46. H.S. Köhler, K. Morawetz, Phys. Rev. C **64**, 024613 (2001). https://doi.org/10.1103/PhysRevC.64.024613
47. K. Morawetz, V. Špička, P. Lipavský, Phys. Lett. A **246**(3), 311 (1998). https://doi.org/10.1016/S0375-9601(98)00356-9
48. G. Zwicknagel, C. Toeppfer, P.G. Reinhard, in *Physics of Strongly Coupled Plasmas*, ed. by W.D. Kraeft, M. Schlanges, H. Haberland, T. Bornath (World Scientific, Singapore, 1995), p. 45
49. K. Morawetz, H.S. Köhler, Eur. Phys. J. A **4**(3), 291 (1999). https://doi.org/10.1007/s100500050233
50. A. Flesch, M. Cramer, I.P. McCulloch, U. Schollwöck, J. Eisert, Phys. Rev. A **78**, 033608 (2008). https://doi.org/10.1103/PhysRevA.78.033608
51. N. Syassen, D.M. Bauer, M. Lettner, T. Volz, D. Dietze, J.J. García-Ripoll, J.I. Cirac, G. Rempe, S. Dürr, Science **320**(5881), 1329 (2008). https://doi.org/10.1126/science.1155309
52. N.D. Mermin, Phys. Rev. B **1**, 2362 (1970). https://doi.org/10.1103/PhysRevB.1.2362
53. A.K. Das, J. Phys. F **5**(11), 2035 (1975). https://doi.org/10.1088/0305-4608/5/11/015
54. K. Morawetz, Phys. Rev. E **88**, 022148 (2013). https://doi.org/10.1103/PhysRevE.88.022148
55. W. Jones, N.H. March, *Theoretical Solid-State Physics. Non-equilibrium and Disorder*, vol. 2 (Dover, New York, 1986). ISBN 9780486650166
56. K. El Sayed, S. Schuster, H. Haug, F. Herzel, K. Henneberger, Phys. Rev. B **49**, 7337 (1994). https://doi.org/10.1103/PhysRevB.49.7337
57. K. Morawetz (ed.), *Nonequilibrium Physics at Short Time Scales. Formation of Correlations* (Springer, Berlin, 2004)
58. R.A. Craig, Ann. Phys. **40**(3), 416 (1966). https://doi.org/10.1016/0003-4916(66)90143-6
59. B. Bezzerides, D.F. DuBois, Phys. Rev. **168**, 233 (1968). https://doi.org/10.1103/PhysRev.168.233
60. H. Stolz, R. Zimmermann, Phys. Status Solidi B **94**(1), 135 (1979). https://doi.org/10.1002/pssb.2220940114
61. D. Kremp, W.D. Kraeft, A.J.D. Lambert, Physica A **127**(1), 72 (1984). https://doi.org/10.1016/0378-4371(84)90120-1
62. M. Schmidt, G. Röpke, Phys. Status Solidi B **139**(2), 441 (1987). https://doi.org/10.1002/pssb.2221390212
63. H.S. Köhler, R. Malfliet, Phys. Rev. C **48**, 1034 (1993). https://doi.org/10.1103/PhysRevC.48.1034
64. V. Špička, P. Lipavský, Phys. Rev. B **52**, 14615 (1995). https://doi.org/10.1103/PhysRevB.52.14615
65. K. Bärwinkel, in *Proceedings of the 14th International Symposium on Rarefied Gas Dynamics*, ed. by H. Oguchi (University of Tokyo Press, Tokyo, 1984). ISBN 9784130681094
66. R.F. Snider, J. Stat. Phys. **63**(3), 707 (1991). https://doi.org/10.1007/BF01029207
67. N.H. March, R. Santamaria, Int. J. Quantum Chem. **39**(4), 585 (1991). https://doi.org/10.1002/qua.560390405
68. N.H. March, Phys. Rev. A **56**, 1025 (1997). https://doi.org/10.1103/PhysRevA.56.1025
69. H.S. Köhler, Phys. Rev. C **51**, 3232 (1995). https://doi.org/10.1103/PhysRevC.51.3232
70. M. Schmidt, G. Röpke, H. Schulz, Ann. Phys. **202**(1), 57 (1990). https://doi.org/10.1016/0003-4916(90)90340-T
71. K. Morawetz, G. Roepke, Phys. Rev. E **51**, 4246 (1995). https://doi.org/10.1103/PhysRevE.51.4246

72. P. Lipavský, V. Špička, K. Morawetz, Phys. Rev. E **59**, R1291 (1999). https://doi.org/10.1103/PhysRevE.59.R1291
73. V. Špička, K. Morawetz, P. Lipavský, Phys. Rev. E **64**, 046107 (2001). https://doi.org/10.1103/PhysRevE.64.046107
74. K. Morawetz, V. Špička, P. Lipavský, G. Kortemeyer, C. Kuhrts, R. Nebauer, Phys. Rev. Lett. **82**, 3767 (1999). https://doi.org/10.1103/PhysRevLett.82.3767
75. K. Morawetz, Phys. Rev. C **62**, 044606 (2000). https://doi.org/10.1103/PhysRevC.62.044606
76. K. Morawetz, M. Płoszajczak, V.D. Toneev, Phys. Rev. C **62**, 064602 (2000). https://doi.org/10.1103/PhysRevC.62.064602
77. K. Morawetz, P. Lipavský, J. Normand, D. Cussol, J. Colin, B. Tamain, Phys. Rev. C **63**, 034619 (2001). https://doi.org/10.1103/PhysRevC.63.034619
78. K. Morawetz, Phys. Rev. E **96**, 032106 (2017). https://doi.org/10.1103/PhysRevE.96.032106
79. K. Morawetz, *Interacting Systems far From Equilibrium. Quantum Kinetic Theory* (Oxford University Press, Oxford, 2017)
80. J. Tõke, B. Lott, S.P. Baldwin, B.M. Quednau, W.U. Schröder, L.G. Sobotka, J. Barreto, R.J. Charity, D.G. Sarantites, D.W. Stracener, R.T. de Souza, Phys. Rev. Lett. **75**, 2920 (1995). https://doi.org/10.1103/PhysRevLett.75.2920
81. S.P. Baldwin, B. Lott, B.M. Szabo, B.M. Quednau, W.U. Schröder, J. Tõke, L.G. Sobotka, J. Barreto, R.J. Charity, L. Gallamore, D.G. Sarantites, D.W. Stracener, R.T. de Souza, Phys. Rev. Lett. **74**, 1299 (1995). https://doi.org/10.1103/PhysRevLett.74.1299
82. W. Skulski, B. Djerroud, D.K. Agnihotri, S.P. Baldwin, J. Tõke, X. Zhao, W.U. Schröder, L.G. Sobotka, R.J. Charity, J. Dempsey, D.G. Sarantites, B. Lott, W. Loveland, K. Aleklett, Phys. Rev. C **53**, R2594 (1996). https://doi.org/10.1103/PhysRevC.53.R2594
83. S. Chattopadhyay, Phys. Rev. C **52**, R480 (1995). https://doi.org/10.1103/PhysRevC.52.R480
84. S. Chattopadhyay, Phys. Rev. C **53**, R1065 (1996). https://doi.org/10.1103/PhysRevC.53.R1065
85. E. Suraud, S. Ayik, J. Stryjewski, M. Belkacem, Nucl. Phys. A **519**(1), 171 (1990). https://doi.org/10.1016/0375-9474(90)90624-U
86. S. Ayik, C. Grégoire, Nucl. Phys. A **513**(1), 187 (1990). https://doi.org/10.1016/0375-9474(90)90348-P
87. J. Randrup, B. Remaud, Nucl. Phys. A **514**(2), 339 (1990). https://doi.org/10.1016/0375-9474(90)90075-W
88. S. Ayik, E. Suraud, M. Belkacem, D. Boilley, Nucl. Phys. A **545**(1), 35 (1992). https://doi.org/10.1016/0375-9474(92)90444-O
89. M. Colonna, G.F. Burgio, P. Chomaz, M. Di Toro, J. Randrup, Phys. Rev. C **47**, 1395 (1993). https://doi.org/10.1103/PhysRevC.47.1395
90. M. Colonna, P. Chomaz, J. Randrup, Nucl. Phys. A **567**(3), 637 (1994). https://doi.org/10.1016/0375-9474(94)90029-9
91. P. Reinhard, E. Suraud, S. Ayik, Ann. Phys. **213**(1), 204 (1992). https://doi.org/10.1016/0003-4916(92)90289-X
92. P. Reinhard, E. Suraud, Ann. Phys. **216**(1), 98 (1992). https://doi.org/10.1016/0003-4916(52)90043-2
93. R. Balian, M. Vénéroni, Ann. Phys. **164**(2), 334 (1985). https://doi.org/10.1016/0003-4916(85)90020-X
94. H. Flocard, Ann. Phys. **191**(2), 382 (1989). https://doi.org/10.1016/0003-4916(89)90323-0
95. T. Troudet, D. Vautherin, Phys. Rev. C **31**, 278 (1985). https://doi.org/10.1103/PhysRevC.31.278
96. The INDRA Collaboration, F. Bocage, J. Colin, M. Louvel, G. Auger, C. Bacri, N. Bellaize, B. Borderie, R. Bougault, R. Brou, P. Buchet, J.L. Charvet, A. Chbihi, D. Cussol, R. Dayras, N.D. Cesare, A. Demeyer, D. Doré, D. Durand, J.D. Frankland, E. Galichet, E. Genouin-Duhamel, E. Gerlic, D. Guinet, P. Lautesse, J.L. Laville, J.F. Lecolley, R. Legrain, N.L. Neindre, O. Lopez, A.M. Maskay, L. Nalpas, A.D. Nguyen, M. Pârlog, J. Péter, E. Plagnol, M.F. Rivet, E. Rosato, F. Saint-Laurent, S. Salou, J. Steckmeyer, M. Stern, G. Tăbăcaru, B. Tamain, O. Tirel, L. Tassan-Got, E. Vient, M. Vigilante, C. Volant, J.P. Wieleczko, C.L.

Brun, A. Genoux-Lubain, G. Rudolf, L. Stuttgé, Nucl. Phys. A **676**(1), 391 (2000). https://doi.org/10.1016/S0375-9474(00)00193-7
97. The INDRA Collaboration, E. Plagnol, J. Łukasik, G. Auger, C.O. Bacri, N. Bellaize, F. Bocage, B. Borderie, R. Bougault, R. Brou, P. Buchet, J.L. Charvet, A. Chbihi, J. Colin, D. Cussol, R. Dayras, A. Demeyer, D. Doré, D. Durand, J.D. Frankland, E. Galichet, E. Genouin-Duhamel, E. Gerlic, D. Guinet, P. Lautesse, J.L. Laville, J.F. Lecolley, R. Legrain, N. Le Neindre, O. Lopez, M. Louvel, A.M. Maskay, L. Nalpas, A.D. Nguyen, M. Pârlog, J. Péter, M.F. Rivet, E. Rosato, F. Saint-Laurent, S. Salou, J.C. Steckmeyer, M. Stern, G. Tăbăcaru, B. Tamain, L. Tassan-Got, O. Tirel, E. Vient, C. Volant, J.P. Wieleczko, Phys. Rev. C **61**, 014606 (1999). https://doi.org/10.1103/PhysRevC.61.014606
98. The INDRA Collaboration, J. Lecolley, E. Galichet, D.C.R. Guinet, R. Bougault, F. Gulminelli, G. Auger, C. Bacri, F. Bocage, B. Borderie, R. Brou, P. Buchet, J. Charvet, A. Chbihi, J. Colin, D. Cussol, R. Dayras, A. Demeyer, D. Doré, D. Durand, J.D. Frankland, E. Genouin-Duhamel, E. Gerlic, P. Lautesse, J. Laville, T. Lefort, R. Legrain, N. LeNeindre, O. Lopez, M. Louvel, A. Maskay, L. Nalpas, A.D. Nguyen, M. Parlog, J. Péter, E. Plagnol, M. Rivet, E. Rosato, F. Saint-Laurent, J. Steckmeyer, M. Stern, G. Tăbăcaru, B. Tamain, L. Tassan-Got, O. Tirel, E. Vient, C. Volant, J.P. Wieleczko, Nucl. Instr. Meth. Phys. Res. A **441**(3), 517 (2000). https://doi.org/10.1016/S0168-9002(99)00831-1
99. V. Ambegaokar, in *Superconductivity*, vol. 1, ed. by R.D. Parks (Dekker, New York, 1969), chap. 5, p. 259
100. K. Morawetz, P. Lipavský, J. Koláček, E.H. Brandt, M. Schreiber, Int. J. Mod. Phys. B **21**(13n14), 2348 (2007). https://doi.org/10.1142/S0217979207043713
101. G. Rickayzen, J. Phys. C **2**(7), 1334 (1969). https://doi.org/10.1088/0022-3719/2/7/325
102. J. Bok, J. Klein, Phys. Rev. Lett. **20**, 660 (1968). https://doi.org/10.1103/PhysRevLett.20.660
103. T.D. Morris, J.B. Brown, Physica **55**, 760 (1971). https://doi.org/10.1016/0031-8914(71)90330-2
104. P. Lipavský, J. Koláček, J.J. Mareš, K. Morawetz, Phys. Rev. B **65**, 012507 (2001). https://doi.org/10.1103/PhysRevB.65.012507
105. H.F. Budd, J. Vannimenus, Phys. Rev. Lett. **31**, 1218 (1973). https://doi.org/10.1103/PhysRevLett.31.1218
106. K. Morawetz, N.H. March, R.H. Squire, Phys. Lett. A **372**(10), 1707 (2008). https://doi.org/10.1016/j.physleta.2007.10.025
107. P. Lipavský, K. Morawetz, J. Koláček, J.J. Mareš, E.H. Brandt, M. Schreiber, Phys. Rev. B **69**, 024524 (2004). https://doi.org/10.1103/PhysRevB.69.024524
108. P. Lipavský, K. Morawetz, J. Koláček, J.J. Mareš, E.H. Brandt, M. Schreiber, Phys. Rev. B **71**, 024526 (2005). https://doi.org/10.1103/PhysRevB.71.024526
109. P. Lipavský, K. Morawetz, J. Koláček, J.J. Mareš, E.H. Brandt, M. Schreiber, Phys. Rev. B **70**, 104518 (2004). https://doi.org/10.1103/PhysRevB.70.104518
110. K. Kumagai, K. Nozaki, Y. Matsuda, Phys. Rev. B **63**, 144502 (2001). https://doi.org/10.1103/PhysRevB.63.144502
111. P. Lipavský, J. Koláček, K. Morawetz, E.H. Brandt, Phys. Rev. B **66**, 134525 (2002). https://doi.org/10.1103/PhysRevB.66.134525
112. P. Lipavský, J. Koláček, K. Morawetz, E.H. Brandt, T.J. Yang (eds.), *Bernoulli Potential in Superconductors*, Lecture Notes in Physics, vol. 733 (Springer, Berlin, 2008). ISBN 9783540734550. https://doi.org/10.1007/978-3-540-73456-7
113. C. Amovilli, N.H. March, Phys. Rev. A **69**(5), 054302 (2004). https://doi.org/10.1103/PhysRevA.69.054302
114. N.H. March, G.G.N. Angilella, R. Pucci, Int. J. Mod. Phys. B **27**, 1330021 (2013). https://doi.org/10.1142/S0217979213300211
115. Y.B. Ivanov, J. Knoll, D.N. Voskresensky, Phys. Atomic Nuclei **66**(10), 1902 (2003). https://doi.org/10.1134/1.1619502
116. A. Peshier, Phys. Rev. D **70**, 034016 (2004). https://doi.org/10.1103/PhysRevD.70.034016
117. W.M. Alberico, S. Chiacchiera, H. Hansen, A. Molinari, M. Nardi, Eur. Phys. J. A **38**(1), 97 (2008). https://doi.org/10.1140/epja/i2008-10648-8

118. C. Moustakidis, V.P. Psonis, K. Chatzisavvas, C.P. Panos, S.E. Massen, Phys. Rev. E **81**, 011104 (2010). https://doi.org/10.1103/PhysRevE.81.011104
119. E. Suraud, P.G. Reinhard, New J. Phys. **16**(6), 063066 (2014). https://doi.org/10.1088/1367-2630/16/6/063066
120. S.K. Baur, E.J. Mueller, Phys. Rev. A **82**, 023626 (2010). https://doi.org/10.1103/PhysRevA.82.023626
121. J.G. Kirkwood, E. Monroe, Boggs. J. Chem. Phys. **10**(6), 394 (1942). https://doi.org/10.1063/1.1723737
122. B.B. Laird, A.D.J. Haymet, Phys. Rev. A **45**, 5680 (1992). https://doi.org/10.1103/PhysRevA.45.5680
123. D. Nayar, C. Chakravarty, Phys. Chem. Chem. Phys. **15**, 14162 (2013). https://doi.org/10.1039/C3CP51114F
124. M. Puoskari, Physica A **272**(3), 509 (1999). https://doi.org/10.1016/S0378-4371(99)00262-9
125. J.A. Hernando, L. Blum, Phys. Rev. E **62**, 6577 (2000). https://doi.org/10.1103/PhysRevE.62.6577
126. D. de Boer, G.E. Uhlenbeck (eds.), *Studies in Statistical Mechanics*, vol. 1 (North-Holland, Amsterdam, 1962)

Chapter 26
Quantum Lattice Boltzmann Study of Random-Mass Dirac Fermions in One Dimension

Ch. B. Mendl, S. Palpacelli, A. Kamenev and S. Succi

Abstract We study the time evolution of quenched random-mass Dirac fermions in one dimension by quantum lattice Boltzmann simulations. For nonzero noise strength, the diffusion of an initial wave packet stops after a finite time interval, reminiscent of Anderson localization. However, instead of exponential localization we find algebraically decaying tails in the disorder-averaged density distribution. These qualitatively match a $x^{-3/2}$ decay, which has been predicted by analytic calculations based on zero-energy solutions of the Dirac equation.

26.1 Introduction

It is a great pleasure, let alone honor, to present this contribution to a Festschrift volume on the occasion of Prof. Norman H. March 90th birthday. Prof. March made many distinguished contributions across a broad variety of topics in classical and quantum statistical physics; in the following we present a computational investigation

Ch. B. Mendl
Stanford Institute for Materials and Energy Sciences, SLAC National Accelerator Laboratory and Stanford University, Menlo Park, CA 94025, USA
e-mail: mendl@stanford.edu

S. Palpacelli
Hyperlean S.r.l, Via Giuseppe Verdi 4, 60122 Ancona, Italy
e-mail: silvia.palpacelli@hyperlean.eu

A. Kamenev
W. I. Fine Theoretical Physics Institute and School of Physics and Astronomy, University of Minnesota, Minneapolis, MN 55455, USA
e-mail: kamenev@physics.umn.edu

S. Succi (✉)
Istituto Applicazioni Calcolo, CNR, via dei Taurini 19, 00185 Roma, Italy
e-mail: succi@iac.cnr.it

S. Succi
Institute for Applied Computational Science, John Paulson school of Engineering and Applied Sciences, Harvard University, Cambridge, MA 02138, USA

G. G. N. Angilella and C. Amovilli (eds.), *Many-body Approaches at Different Scales*, https://doi.org/10.1007/978-3-319-72374-7_26

along the latter direction, namely the transport properties of random-mass Dirac fermions in $1+1$ dimensions.

Disorder plays an important role in many physical systems, ranging from topological materials [1–4] to transport properties affected by impurities, superconductors [5] and glasses [6]. In condensed matter physics, a prominent effect of disorder is exponential Anderson localization of the electronic wavefunction [7], which has been experimentally observed in Bose-Einstein condensates [8]. Nevertheless, around critical points there can be transitions away from the localized phase [9–11]. In one dimension, similarities between these delocalized phases and classical particle motion in a stationary random potential with a variety of diffusion laws [12–14] have been pointed out, including anomalously slow Sinai diffusion $|x| \propto \log(t)^2$ [4].

In this work, we study the time evolution dynamics governed by a prototypical random-mass Dirac equation in one dimension, and investigate the fate of an initial Gaussian wave packet. The general framework is similar to a recent related work [15], except for the numerical quantum lattice Boltzmann approach pursued here, and different versions of the Dirac equation. Specifically, using the Majorana representation and projecting upon chiral eigenstates (and setting $\hbar = 1$), the Dirac equation considered here reads

$$\left(i\partial_t + ic\sigma^z\partial_x + c^2 m(x)\sigma^y\right)\psi(x,t) = 0, \tag{26.1}$$

where $\psi(x,t)$ is a two-component spinor, σ^α are the Pauli matrices, c the speed of light, and $m(x)$ is the spatially dependent mass. We model quenched disorder by taking $m(x)$ as a Gaussian white noise random variable with mean m_0 and noise strength λ:

$$\langle (m(x) - m_0)(m(x') - m_0)\rangle = 2\lambda\delta(x - x'). \tag{26.2}$$

The spinor $\psi = (u, d)^T$ consists of the chiral right-moving (u) and left-moving (d) states. The stationary version of Eq. (26.1) (without the time derivative) has been identified as an effective theory in a tight-binding model of spinless fermions [9].

The dynamics governed by Eq. (26.1) conserves total density and energy. For example, the local density

$$\rho = |\psi|^2 = |u|^2 + |d|^2 \tag{26.3}$$

obeys the conservation law

$$\partial_t \rho(x,t) + \partial_x J_\rho(x,t) = 0 \tag{26.4}$$

with the density current

$$J_\rho(x,t) = c\left(|u|^2 - |d|^2\right). \tag{26.5}$$

We will see in the numerical simulations that $\psi(x,t)$ converges to a stationary state for $\lambda > 0$; this stationary state can thus be compared to the zero-energy solution studied in [9]: $\psi(x) = \psi_\pm(x)\binom{1}{\mp 1}$, with the scalar function $\psi_\pm(x)$ satisfying

$$\left(\partial_x \pm cm(x)\right)\psi_\pm(x) = 0. \tag{26.6}$$

For "critical" zero average mass ($m_0 = 0$), this results in the log-normally distributed wavefunction

$$\psi_\pm(x) \propto \mathrm{e}^{\pm \int_0^x cm(x')\mathrm{d}x'}, \tag{26.7}$$

which deviates from exponential localization. By a mapping to Liouville field theory, the disorder-averaged spatial correlations of the wavefunction, Eq. (26.7), can be computed analytically [9, 10, 16], resulting in an *algebraic* (instead of exponential) decay with exponent $-3/2$:

$$\left\langle |\psi(x)|^2 |\psi(0)|^2 \right\rangle \propto |x|^{-3/2}. \tag{26.8}$$

Thus, disorder in the random mass distribution does not lead to Anderson localization if the average mass is zero.

26.2 Quantum Lattice Boltzmann Method

Equation (26.1) lends itself to a lattice Boltzmann discretization for the spinor components u and d, as observed in Refs. [17–19]. The propagation step consists of streaming u and d along the x-axis with opposite speeds $\pm c$, while the collision step is performed according to the scattering term $c^2 m(x)\sigma^y\psi$. Integrating Eq. (26.1) along the characteristics of u and d, respectively, and approximating the collision integral by the trapezoidal rule, the following relations are obtained:

$$\begin{aligned} \hat{u} - u &= \tilde{m}(d + \hat{d})/2 \\ \hat{d} - d &= -\tilde{m}(u + \hat{u})/2, \end{aligned} \tag{26.9}$$

where $\hat{u} = u(x + \Delta x, t + \Delta t)$, $\hat{d} = d(x - \Delta x, t + \Delta t)$, $\Delta x = c\Delta t$, and $\tilde{m} = c^2 m \Delta t$. Algebraically solving the linear system, Eqs. (26.9), yields the explicit scheme

$$\begin{pmatrix} \hat{u} \\ \hat{d} \end{pmatrix} = \begin{pmatrix} a & b \\ -b & a \end{pmatrix} \begin{pmatrix} u \\ d \end{pmatrix}, \tag{26.10}$$

with

$$a = (1 - \tilde{m}^2/4)/(1 + \tilde{m}^2/4), \quad b = \tilde{m}/(1 + \tilde{m}^2/4).$$

Note that, since $|a|^2 + |b|^2 = 1$, the collision matrix is unitary, thus the method is unconditionally stable and norm-preserving.

26.3 Numerical Simulation Results

We start from a "wave packet" initial state given by

$$\psi(x,0) \equiv \begin{pmatrix} u \\ d \end{pmatrix} = \left(\sqrt{8\pi}\sigma\right)^{-1/2} \mathrm{e}^{-x^2/4\sigma^2} \begin{pmatrix} 1 \\ 1 \end{pmatrix}, \tag{26.11}$$

with the standard deviation σ measuring the width of the wave packet, and the normalization chosen such that $\int_{-\infty}^{\infty} \rho(x,t)\,\mathrm{d}x = 1$ at $t = 0$. Due to density conservation, this relation holds for all t.

Table 26.1 lists the simulation parameters in detail. The speed of light $c = \Delta x/\Delta t = 1$, and the physical simulation domain is the interval $[-64, 64]$.

Equation (26.2) suggests to draw a random $m(x_i)$ independently at each grid point x_i. However, this would render the simulation sensitive to the grid spacing Δx. Instead, we draw independent Fourier coefficients up to some cut-off Fourier mode n_{cut}, and then transform to real space to obtain a random mass realization. Thus, the grid resolution is much finer than random mass oscillations. The random mass correlations obtained by this procedure decay on a length scale $x - x' = \Delta x L/(2n_{\mathrm{cut}})$. This quantity is chosen small compared to the width of the initial wave packet, in order to approximate the delta function in Eq. (26.2).

Figure 26.1 shows $\langle u(x,t)\rangle$ for various values of λ, for zero average mass ($m_0 = 0$). In the absence of noise ($\lambda = 0$), there is no scattering term in the Dirac equation, and the u and d waves freely propagate to the right and left, respectively. For $\lambda > 0$, the right-moving ray is continuously diminished over time due to scattering. As λ increases, the wave packet remains more and more tied to the origin.

Figure 26.2 visualizes the corresponding density profiles $\langle \rho(z,t)\rangle$ for the same simulations. For any $\lambda > 0$, one observes remnant density centered around the origin. The density profile remains stationary at later times.

To analyze the noise-averaged density quantitatively, Fig. 26.3 shows the density profile on a logarithmic scale at $t = 60$, when it has (almost) reached stationarity between the left- and right-moving sound peaks around $x \simeq \pm 60$. The density decays exponentially with respect to $|x|$ for $0 < \lambda \lesssim 1$, different from the predicted algebraic

Table 26.1 Simulation parameters

L	2048	System size (number of grid points) with periodic boundary conditions
Δx	1/16	Grid spacing
Δt	1/16	Time step
σ	1	Standard deviation of initial spinor
n_{runs}	10^5	Number of random mass realizations (simulation runs) to compute averages $\langle \dots \rangle$
n_{cut}	256	Cut-off Fourier mode of random mass distribution

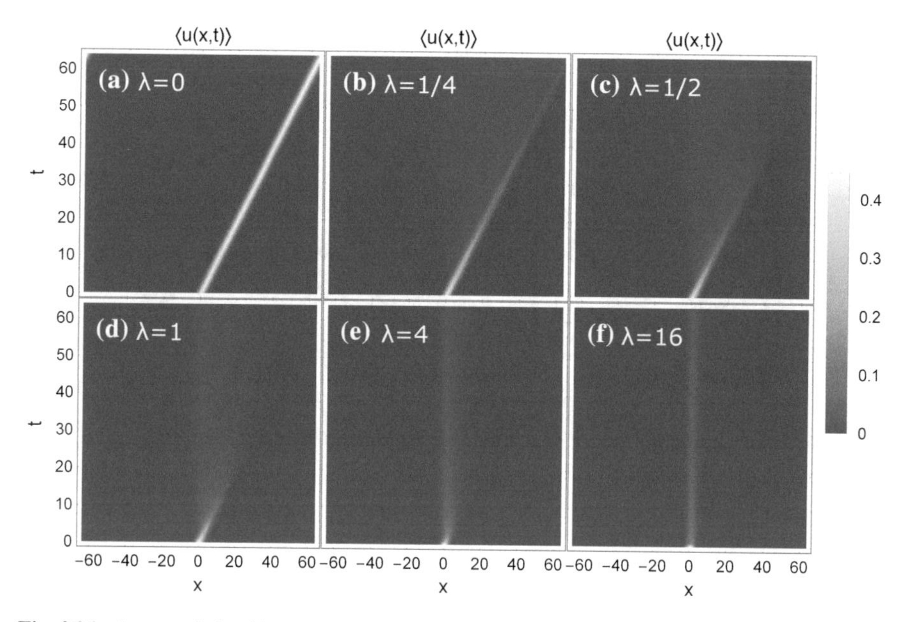

Fig. 26.1 Average $\langle u(x, t)\rangle$ profile for increasing noise strength of the random mass distribution, and $m_0 = 0$

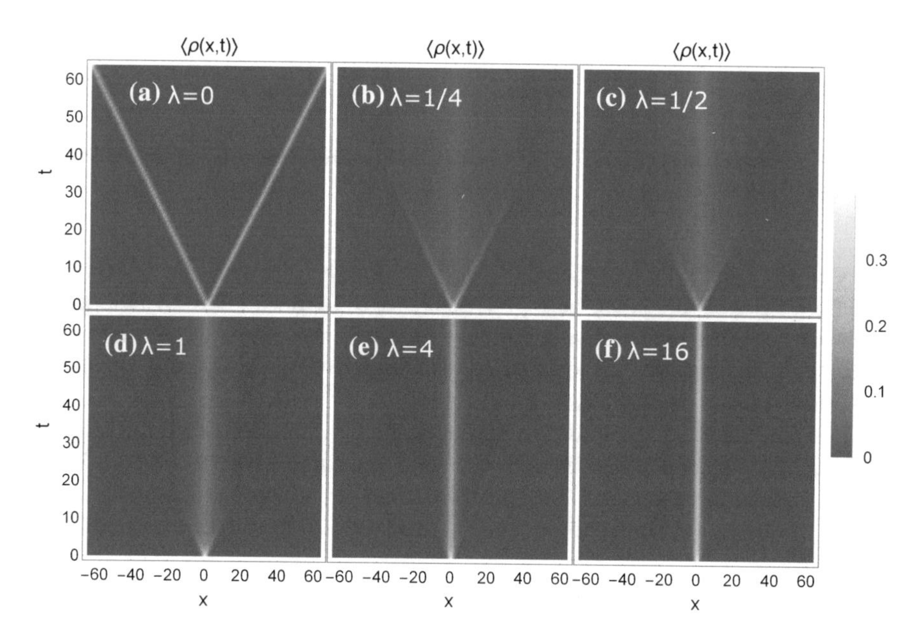

Fig. 26.2 Average density $\langle \rho(x, t)\rangle$ for increasing noise strength of the random mass distribution, and $m_0 = 0$

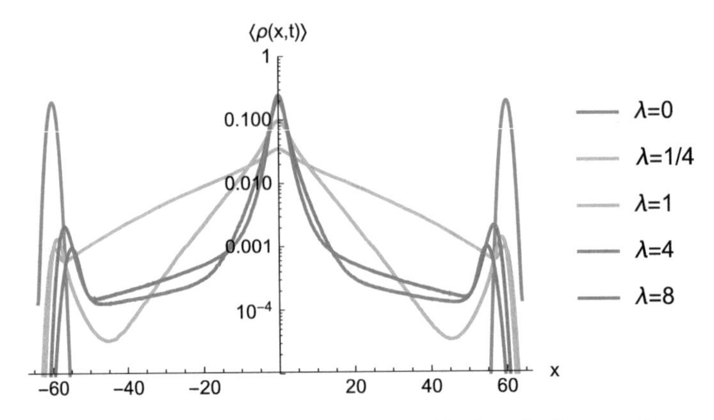

Fig. 26.3 Average density $\langle \rho(x,t)\rangle$ at $t = 60$ on a logarithmic scale, for $m_0 = 0$

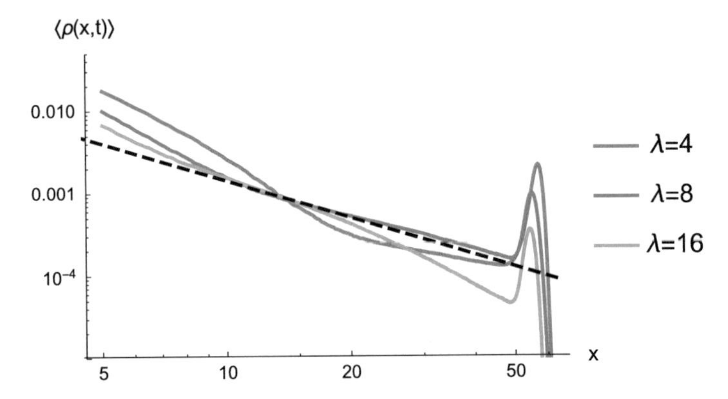

Fig. 26.4 Average density $\langle \rho(x,t)\rangle$ at $t = 60$ on a log-log scale, for $m_0 = 0$. For comparison, the black dashed line is $\propto x^{-3/2}$

decay in Eq. (26.8). One explanation could be that the algebraic decay sets in at larger $|x|$. On the other hand, for $\lambda \gtrsim 4$, one observes a transition from exponential to slower-decaying tails. (Note that for the particular initial condition used in our simulations, we find that the density *correlation* between the the origin and x is proportional to the density profile.)

Figure 26.4 shows these tails on a log-log scale, which indeed ascertains an algebraic decay at larger $|x|$. Between $20 < x < 45$, the curve for noise strength $\lambda = 4$ decays somewhat slower, the $\lambda = 16$ curve somewhat faster, and the $\lambda = 8$ curve almost exactly as the black dashed $\propto x^{-3/2}$ line based on the theoretical prediction, Eq. (26.8).

The logarithmic scale in Fig. 26.3 shows that the outward-moving sound peaks are present also for $\lambda \geq 1$, even though not visible in Fig. 26.2. The effective sound velocity v_{eff} (measured via the peak maximum) monotonically decreases with noise strength, as expected (see Fig. 26.5).

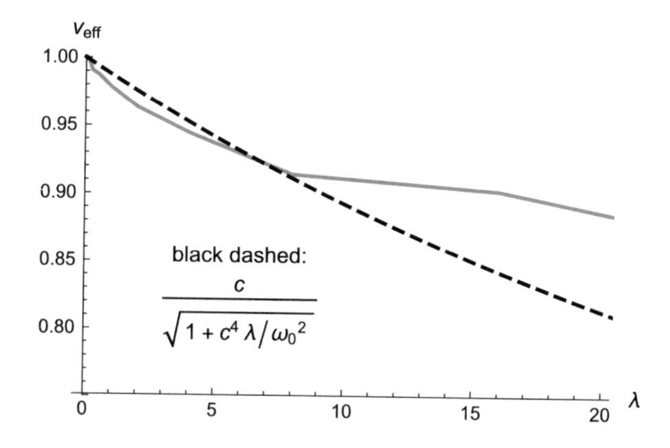

Fig. 26.5 Measured sound velocity in dependence of noise strength λ, for $m_0 = 0$

Solutions of the free Dirac equation also solve the Klein–Gordon equation with dispersion relation $\omega^2 = (ck)^2 + \omega_c^2$, where $\omega_c = c^2 m/\hbar$ is the Compton frequency. The corresponding sound speed is therefore

$$V_{\mathrm{KG}} = \partial_k \omega = \frac{c}{\sqrt{1 + (\omega_c/(ck))^2}}. \tag{26.12}$$

The wave number k should be inversely proportional to the spatial extent of the wave packet; thus we approximate $ck \simeq \omega_0$ with $\omega_0 = 2\pi c/\sigma$. For the Compton

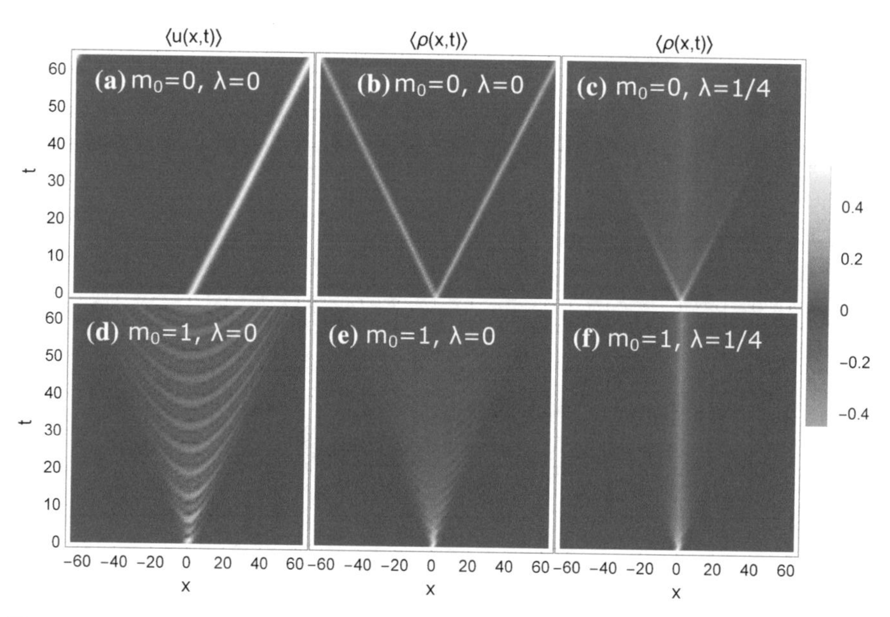

Fig. 26.6 Comparison of the $\langle u(x, t)\rangle$ profile and density $\langle \rho(x, t)\rangle$ for $m_0 = 0$ (top row) with $m_0 = 1$ (bottom row)

frequency, we use $\sqrt{\lambda}$ as proxy for the mass term, and set $\hbar = 1$ as before. This results in the black dashed curve in Fig. 26.5, which indeed qualitatively reproduces the measured sound velocity up to $\lambda \lesssim 8$.

Tuning away from zero average mass should result in "conventional" exponentially localized wavefunctions (see also Eq. (26.7)) at zero-energy. Figure 26.6 directly compares hitherto $m_0 = 0$ simulations with $m_0 = 1$. Without disorder ($\lambda = 0$), the u (and d) component exhibits a parabola-shaped stripe pattern (see Fig. 26.6d), instead of linear propagation. The corresponding density has a more uniform profile. When including disorder ($\lambda = 1/4$), one notices that the average density is more strongly confined for $m_0 = 1$ (Fig. 26.6f) than for $m_0 = 0$ (Fig. 26.6c).

This stronger confinement is confirmed in Fig. 26.7, which compares the densities on a logarithmic scale for $0 \leq \lambda \leq 1$. Besides the oscillatory pattern at $\lambda = 0$, the density for $m_0 = 1$ decays faster than for $m_0 = 0$ at fixed $\lambda > 0$.

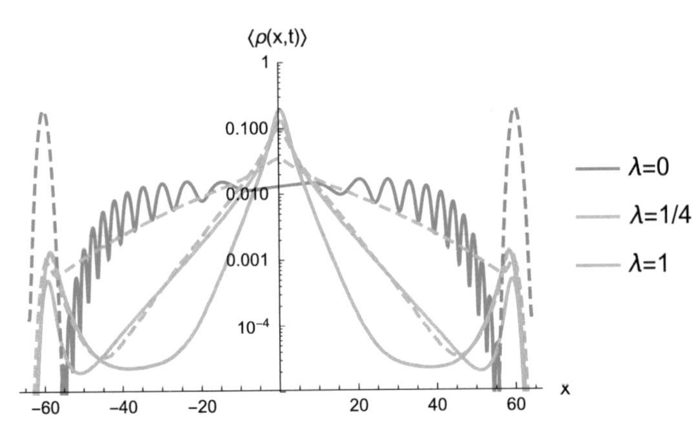

Fig. 26.7 Average density $\langle \rho(x,t) \rangle$ for $m_0 = 1$ (solid lines) compared to $m_0 = 0$ (dashed lines, same data as in Fig. 26.3)

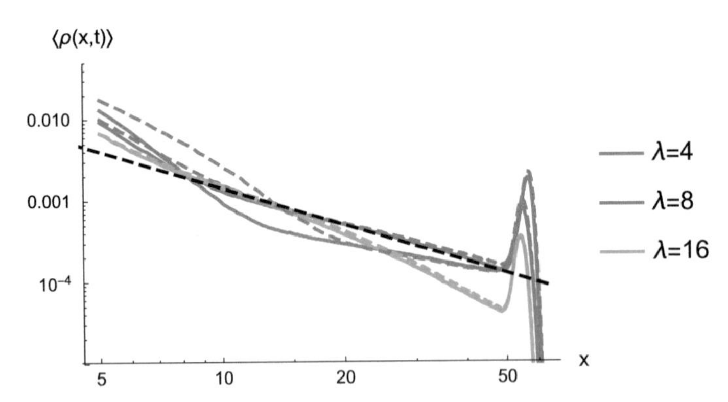

Fig. 26.8 Average density $\langle \rho(x,t) \rangle$ on a log-log scale for $m_0 = 1$ (solid lines) compared to $m_0 = 0$ (dashed lines, same data as in Fig. 26.4)

Figure 26.8 compares the densities on a log-log scale for $\lambda \geq 4$. Somewhat surprisingly, the non-zero average mass $m_0 = 1$ does not affect the algebraic decay, although one would expect exponential decay away from the "critical" $m_0 = 0$. An explanation could be that large values of the noise override small changes in the average mass.

26.4 Conclusions and Outlook

We have shown that quantum lattice Boltzmann methods can efficiently simulate the real-time dynamics of the single-particle Dirac equation, Eq. (26.1), for random-mass fermions in one spatial dimension. Since the quantum lattice Boltzmann scheme is not limited to one-dimensional systems [20], for the future it would be interesting to study the transport properties of random-mass fermions in two and three spatial dimensions. Besides analyzing stationary properties, lattice Boltzmann simulations of the Dirac equation could also be used for investigating the time dynamics of out-of-equilibrium systems, including, e.g., thermalization and quasiparticle lifetime, cf. [21]. Work along the lines is currently underway.

Acknowledgements This work is dedicated to Professor Norman H. March on the occasion of his 90th birthday, with our warmest congratulations on an outstanding career and best wishes for more to come in the future.
C.M. acknowledges support from the Alexander von Humboldt foundation via a Feodor Lynen fellowship, as well as support from the US Department of Energy, Office of Basic Energy Sciences, Division of Materials Sciences and Engineering, under Contract No. DE-AC02-76SF00515. A.K. was supported by NSF grant DMR-1608238. S.S. was supported by the European Research Council under the European Union's Seventh Framework Programme (FP/2007-2013)/ERC Grant Agreement No. 306357 (ERC Starting Grant "NANO-JETS").

References

1. C.W. Groth, M. Wimmer, A.R. Akhmerov, J. Tworzydło, C.W.J. Beenakker, Phys. Rev. Lett. **103**, 196805 (2009). https://doi.org/10.1103/PhysRevLett.103.196805
2. K. Kobayashi, T. Ohtsuki, K.I. Imura, I.F. Herbut, Phys. Rev. Lett. **112**, 016402 (2014). https://doi.org/10.1103/PhysRevLett.112.016402
3. T. Morimoto, A. Furusaki, C. Mudry, Phys. Rev. B **91**, 235111 (2015). https://doi.org/10.1103/PhysRevB.91.235111
4. D. Bagrets, A. Altland, A. Kamenev, Phys. Rev. Lett. **117**, 196801 (2016). https://doi.org/10.1103/PhysRevLett.117.196801
5. S. Seo, X. Lu, J.X. Zhu, R.R. Urbano, N. Curro, E.D. Bauer, V.A. Sidorov, L.D. Pham, T. Park, Z. Fisk, J.D. Thompson, Nat. Phys. **10**, 120 (2014). https://doi.org/10.1038/nphys2820
6. P. Yunker, Z. Zhang, A.G. Yodh, Phys. Rev. Lett. **104**, 015701 (2010). https://doi.org/10.1103/PhysRevLett.104.015701
7. P.W. Anderson, Phys. Rev. **109**, 1492 (1958). https://doi.org/10.1103/PhysRev.109.1492
8. J. Billy, V. Josse, Z. Zuo, A. Bernard, B. Hambrecht, P. Lugan, D. Clément, L. Sanchez-Palencia, P. Bouyer, A. Aspect, Nature **453**, 891 (2008). https://doi.org/10.1038/nature07000

9. L. Balents, M.P.A. Fisher, Phys. Rev. B **56**, 12970 (1997). https://doi.org/10.1103/PhysRevB.56.12970
10. D.G. Shelton, A.M. Tsvelik, Phys. Rev. B **57**, 14242 (1998). https://doi.org/10.1103/PhysRevB.57.14242
11. V.V. Mkhitaryan, M.E. Raikh, Phys. Rev. Lett. **106**, 256803 (2011). https://doi.org/10.1103/PhysRevLett.106.256803
12. Y.G. Sinai, Theory Probab. Appl. **27**, 256 (1982). https://doi.org/10.1137/1127028
13. J.P. Bouchaud, A. Comtet, A. Georges, P. Le Doussal, Ann. Phys. **201**, 285 (1990). https://doi.org/10.1016/0003-4916(90)90043-N
14. A. Comtet, D.S. Dean, J. Phys. A **31**, 8595 (1998). https://doi.org/10.1088/0305-4470/31/43/004
15. A. Yosprakob, S. Suwanna, arXiv:1601.03827 (2016)
16. M. Steiner, M. Fabrizio, A.O. Gogolin, Phys. Rev. B **57**, 8290 (1998). https://doi.org/10.1103/PhysRevB.57.8290
17. S. Succi, R. Benzi, Physica D **69**, 327 (1993). https://doi.org/10.1016/0167-2789(93)90096-J
18. S. Palpacelli, S. Succi, Phys. Rev. E **77**, 066708 (2008). https://doi.org/10.1103/PhysRevE.77.066708
19. F. Fillion-Gourdeau, H.J. Herrmann, M. Mendoza, S. Palpacelli, S. Succi, Phys. Rev. Lett. **111**, 160602 (2013). https://doi.org/10.1103/PhysRevLett.111.160602
20. P.J. Dellar, D. Lapitski, S. Palpacelli, S. Succi, Phys. Rev. E **83**, 046706 (2011). https://doi.org/10.1103/PhysRevE.83.046706
21. S. Succi, EPL **109**, 50001 (2015). https://doi.org/10.1209/0295-5075/109/50001

Chapter 27
Topological Effects and Critical Phenomena in the Three-Dimensional (3D) Ising Model

Zhidong Zhang

Abstract In this work, we analyze in detail the transfer matrices as well as the partition function of the three-dimensional (3D) Ising model, in order to understand the topological effects in three dimensions. We attempt to extract the non-local contributions to the physical properties of the 3D Ising system, by comparing the conjectured exact solution with the results of the conventional high-temperature and low-temperature expansions. This is because the conjectured solution takes into account both the local and non-local behaviors, while these conventional expansions consider only the local environments. Then, we summarize the results of our previous work, most of them being joint work with Professor N. H. March, on the critical exponents of the 3D Ising model and some related topics. This work provides a deep understanding on the topological contributions and the non-local behavior in the 3D Ising model and their effects on the critical exponents in the 3D Ising universality.

27.1 Introduction

The exact solution of 3D Ising model has been a well-known unsolved problem in physics for almost one hundred years [1, 2]. Since it is the simplest model for many-body interacting systems, the lack of an exact solution certainly hinders us to understand in depth many physical systems. Though most of the real systems may be even much more complicated than what the Ising model describes, the Ising model itself serves as a toy model, which can be utilized to study the nature of critical phenomena in various fields, for instance, ferromagnetic-paramagnetic transition in a magnet, the order-disorder transformation in alloys, the freezing and evaporation of liquids, etc. In order to solve the problem, the author proposed two conjectures in 2007 [3], which are repeated here:

Z. Zhang (✉)
Shenyang National Laboratory for Materials Science, Institute of Metal Research, Chinese Academy of Sciences, 72 Wenhua Road, Shenyang 110016, People's Republic of China
e-mail: zdzhang@imr.ac.cn

G. G. N. Angilella and C. Amovilli (eds.), *Many-body Approaches at Different Scales*, https://doi.org/10.1007/978-3-319-72374-7_27

Conjecture 1

The topological problem of a 3D Ising system can be solved by introducing an additional rotation in a four-dimensional (4D) space, since knots in a 3D space can be opened by a rotation in a 4D space. One can find a spin representation in $2^{n \cdot l \cdot o}$-space for this additional rotation in $2n \cdot l \cdot o$-space. Meanwhile, the transfer matrices $\mathbf{V}_1$, $\mathbf{V}_2$, and $\mathbf{V}_3$ have to be represented and rearranged, also in the $2n \cdot l \cdot o$-space.

Conjecture 2

The weight factors w_x, w_y and w_z of the eigenvectors represent the contribution of $e^{2\pi t_x i/n}$, $e^{2\pi t_y i/l}$, and $e^{2\pi t_z i/o}$ in the 4D space to the energy spectrum of the system.

Based on these two conjectures, the partition function, and thermodynamic consequences such as the specific heat, the spontaneous magnetization and the true range κ_x of the correlation, and the correlation functions of the 3D Ising model were obtained. Accordingly, the critical temperature and the critical exponents were derived [3], also based on these two conjectures. The golden (or silver) ratio is the exact solution for the critical temperature of the simple cubic (or square) Ising lattice. The critical exponents for the simple orthorhombic Ising lattices were determined to be $\alpha = 0$, $\beta = 3/8$, $\gamma = 5/4$, $\delta = 13/3$, $\eta = 1/8$, and $\nu = 2/3$, showing the universality behavior and satisfying the scaling laws (see Chap. 9 in [4] for a general introduction to the Ising model, and for a review of some of the results presented here).

Soon after publication of my conjectured exact solution [3], Professor Norman H. March, jointly with Professor Douglas J. Klein, published a paper in Physical Letters A [5], under the title *Critical exponents in d dimensions for the Ising model, subsuming Zhang's proposals for* $D = 3$. In that paper [5], they pointed out that for the future work in this area, the major interest is to establish the precise status of the two conjectures on which the basic statistical mechanics developed in [3] is entirely dependent. This is the first published paper which expresses a positive view on my ideas and solution, which is indeed a great encouragement to me. Although the conjectured exact solution met serious arguments in statistical mechanics community, Professor March invited me to collaborate with him to write a series of papers, investigating the critical exponents in the 3D Ising model and related systems, and also some related topics. 21 papers have been completed [6–26], among them 16 are co-authored by two of us [6–21], and a recent paper is also jointed with Professor Osamu Suzuki [26]. One of the aims of this paper is to review briefly the results obtained in these papers [5–25] (Sect. 27.3). However, some attention will be paid to the topological effects in the 3D Ising model. In Sect. 27.2, I shall first inspect the transfer matrices and the partition function of the 3D Ising model, and discuss the topological effects in this system. I shall extract the contributions of non-local behavior in the 3D Ising model, by comparing the conjectured exact solution and the conventional series expansions. In Sect. 27.4, I shall mention briefly disadvantages of conventional series expansions, renormalization group and Monte Carlo simulations. A summary is given in Sect. 27.5, together with a brief introduction of a recent work

[26], in which we develop a Clifford algebra approach for 3D Ising model, and give a positive answer to Zhang's two conjectures.

27.2 Transfer Matrices and Topological Effects in 3D Ising Model

As in [3, 27, 28], we consider the Ising Hamiltonian on an orthorhombic lattice in 3D Euclidean space, with up-spin or down-spin at each lattice point:

$$H = -J\sum_{\tau=1}^{n}\sum_{\rho=1}^{m}\sum_{\delta=1}^{l} s_{\rho\delta}^{(\tau)} s_{\rho\delta}^{(\tau+1)} - J'\sum_{\tau=1}^{n}\sum_{\rho=1}^{m}\sum_{\delta=1}^{l} s_{\rho\delta}^{(\tau)} s_{\rho+1\delta}^{(\tau)} - J''\sum_{\tau=1}^{n}\sum_{\rho=1}^{m}\sum_{\delta=1}^{l} s_{\rho\delta}^{(\tau)} s_{\rho\delta+1}^{(\tau)}. \tag{27.1}$$

Here, only the nearest neighboring interaction between spins at each lattice point is considered. We follow the notation of Onsager-Kaufman-Zhang for Pauli matrices [2, 3, 26–29]:

$$s'' = \begin{pmatrix} 0 & -1 \\ 1 & 0 \end{pmatrix} \qquad (= i\sigma_2), \tag{27.2a}$$

$$s' = \begin{pmatrix} 1 & 0 \\ 0 & -1 \end{pmatrix} \qquad (= \sigma_3), \tag{27.2b}$$

$$C = \begin{pmatrix} 0 & 1 \\ 1 & 0 \end{pmatrix} \qquad (= \sigma_1), \tag{27.2c}$$

and we have the relation $\tanh K^* = e^{-2K}$. We introduce $K = \beta J$, $K' = \beta J'$, and $K'' = \beta J''$, with $\beta = (k_B T)^{-1}$. The partition function of the 3D Ising model can be given as follows [3, 27, 28]:

$$Z = (2\sinh 2K)^{m\cdot n\cdot l/2}\,\mathrm{Tr}(V_3V_2V_1)^m \equiv (2\sinh 2K)^{m\cdot n\cdot l/2}\sum_{i=1}^{2^{n\cdot l}}\lambda_i^m, \tag{27.3}$$

with

$$V_3 = \prod_{j=1}^{nl}\exp\left(iK''\Gamma_{2j}\left[\prod_{k=j+1}^{j+n-1} i\Gamma_{2k-1}\Gamma_{2k}\right]\Gamma_{2j+2n-1}\right) = \prod_{j=1}^{nl}\exp\left(K''s'_js'_{j+n}\right), \tag{27.4a}$$

$$V_2 = \prod_{j=1}^{nl}\exp\left(iK'\Gamma_{2j}\Gamma_{2j+1}\right) = \prod_{j=1}^{nl}\exp\left(K's'_js'_{j+1}\right), \tag{27.4b}$$

$$V_1 = \prod_{j=1}^{nl}\exp\left(iK^*\Gamma_{2j-1}\Gamma_{2j}\right) = \prod_{j=1}^{nl}\exp\left(K^*C_j\right). \tag{27.4c}$$

Here, we have introduced the following generators of the Clifford algebra for the 3D Ising model:

$$\Gamma_{2j-1} = \underbrace{C \otimes \ldots \otimes C}_{j \text{ times}} \otimes s' \otimes 1 \otimes \ldots \otimes 1, \tag{27.5a}$$

$$\Gamma_{2j} = \underbrace{C \otimes \ldots \otimes C}_{j \text{ times}} \otimes (-is'') \otimes 1 \otimes \ldots \otimes 1. \tag{27.5b}$$

The Clifford algebra is generated by the Pauli matrices. Clearly, we have the following formulas:

$$i\Gamma_{2j-1}\Gamma_{2j} = 1 \otimes \ldots \otimes 1 \otimes C \otimes 1 \otimes \ldots \otimes 1, \tag{27.6a}$$

$$s'_j s'_{j+1} = 1 \otimes \ldots 1 \otimes 1 \otimes s' \otimes s' \otimes 1 \otimes \ldots \otimes 1, \tag{27.6b}$$

$$s'_j s'_{j+n} = 1 \otimes \ldots 1 \otimes 1 \otimes s' \otimes 1 \otimes \ldots \otimes 1 \otimes s' \otimes 1 \otimes \ldots \otimes 1. \tag{27.6c}$$

The problem becomes to how to deal with the trace of the transfer matrices, $V = V_3 V_2 V_1$, and especially of V_3, as it contains non-linear terms of Γ matrices. There are direct products of $(n-1)$ unit matrices in between two Pauli matrices s' in Eq. (27.6c).

It should be noted that the transfer matrix V_1 or V_2 with the product of two Γ matrices possesses a trivial topological structure, which can be represented as the Lie group for a rotation. The transfer matrix V_3 is a source of difficulties, which hinder the application of the Onsager-Kaufman approach to the 3D Ising model, due to the high-order terms of Γ-matrices. The non-trivial topological effect as well as the non-local behavior can be seen also in the language of the spin (σ) variables although all expressions seem local. Although the formulas for the transfer matrices V_2 and V_3, *i.e.* Eqs. (27.4a) and (27.4b), look very similar, they are quite different in nature. The difference in the subscripts of s' in Eqs. (27.4a) and (27.4b) represents different meaning, as shown in Eqs. (27.6b) and (27.6c). $s'_j s'_{j+1}$ represents the interaction between the nearest neighbor spins along the second dimension in a plane, whereas $s'_j s'_{j+n}$ represents the interaction between the nearest neighbor spins along the third dimension. The latter term looks like an interaction between two spins located far from each other, via a chain of n spins in the plane (actually, via nm spins in the plane if one considers the periodic condition already used along the first dimension, as shown in Eq. (27.3)). The reason is that the third transfer matrix V_3 must follow the sequence of the spin (σ) variables arranged and fixed already in the first two transfer matrices V_1 and V_2. According to this fixed order, although the interaction between a spin and one of its neighboring spins along the third dimension is the nearest neighboring one, its effect is correlated with the states of n other spins in the plane. It is equivalent effectively to a long-range and many-body interaction in which $n+1$ spins are involved. The same effect happens for every interaction along the third dimension in the 3D Ising model. Therefore, a non-local correlation and a global effect with non-trivial topological structure indeed exist in the 3D Ising system. It implies that all the spins in the 3D Ising model are entangled. It was understood that the different topological states (*e.g.* knots/links) formed by up or down spins also contribute to the partition function and hence the free energy and other physical properties of the system [3, 27, 28] According to the topology theory [27, 30–33], there are two choices for smoothing a given crossing ($\times$), and thus there are 2^N states for a diagram with N crossings. The bracket polynomial, *i.e.* the state summation, is defined by the formula [27, 30–32]:

$$\langle K \rangle = \sum_{\sigma} \langle K | \sigma \rangle d^{\|\sigma\|}. \tag{27.7}$$

Here, σ runs over all the states of K, and $d = -A^2 - B^2$, with A, B, and d being commuting algebraic variables. The bracket state summation is the analog of a partition function in discrete statistical mechanics, which can be used to express the partition function for the Potts model (including the Ising model) for appropriate choices of commuting algebraic variables [30–33]. One may transform from the basis of $\langle \chi \rangle$ and $\langle \chi^{-1} \rangle$ to the basis of $\langle {}^{\cup}_{\cap} \rangle$ and $\langle \supset\subset \rangle$ by a transformation [27]:

$$\begin{pmatrix} \langle \chi \rangle \\ \langle \chi^{-1} \rangle \end{pmatrix} = \begin{pmatrix} A & B \\ B & A \end{pmatrix} \begin{pmatrix} \langle {}^{\cup}_{\cap} \rangle \\ \langle \supset\subset \rangle \end{pmatrix} \tag{27.8}$$

and vice versa. The bracket with $B = A^{-1}$, $d = -A^2 - A^{-2}$ is invariant under the Reidemeister moves II and III. As long as knots or links exist in a system, a matrix representing the transformation between the trivial and non-trivial topological structures may always exist, no matter how complicated the knots or links are. The transformation also implies that there exist the extra contributions from the non-trivial topological structure to the partition function and the free energy, in addition to the contributions of the trivial topological structures. Therefore, the partition function of the 3D Ising model can be written in the following two parts as [34]:

$$Z = Z_{\text{local}} + Z_{\text{non-local}}, \tag{27.9}$$

where Z_{local} and $Z_{\text{non-local}}$ represent the contributions from the local environment and the non-local behavior to the partition function of the 3D Ising system. Correspondingly, the free energy of the 3D Ising model can be written as the sum of the local and non-local parts:

$$F = F_{\text{local}} + F_{\text{non-local}}. \tag{27.10}$$

Here, F_{local} and $F_{\text{non-local}}$ represent the contributions from the local environment and the non-local behavior to the free energy of the 3D Ising system. Consequently, each of the physical properties, such as specific heat, magnetization, susceptibility, correlation function, etc., can be treated as the sum of local and non-local parts.

For finite temperatures, the partition function of the 3D Ising model obtained based on the two conjectures can be expanded as [3]:

$$\lambda = Z^{1/N} = 2\cosh^3 K\left(1 + \frac{7}{2}\kappa^2 + \frac{87}{8}\kappa^4 + \frac{3613}{48}\kappa^6 + \frac{170209}{384}\kappa^8 + \frac{929761}{256}\kappa^{10}\right.$$
$$\left. + \frac{318741323}{9216}\kappa^{12} + \frac{6705494087}{18432}\kappa^{14} + \frac{411207879769}{98304}\kappa^{16} + \ldots\right) \quad (27.11)$$

which consists of the contributions from the local environment and the non-local behavior. On the other hand, the conventional high-temperature series expansion for the 3D Ising model yields [35, 36]:

$$\lambda = Z^{1/N} = 2\cosh^3 K\left(1 + 3\kappa^4 + 22\kappa^6 + 192\kappa^8\right.$$
$$\left. + 2046\kappa^{10} + 24853\kappa^{12} + 329334\kappa^{14} + 4649601\kappa^{16} + \ldots\right) \quad (27.12)$$

which includes only the local environment of spins in the 3D Ising system. Therefore, one may obtain the non-local contribution to the partition function by comparing these two formulas:

$$\lambda_{\text{non-local}} = 2\cosh^3 K\left(\frac{7}{2}\kappa^2 + \frac{63}{8}\kappa^4 + \frac{2557}{48}\kappa^6 + \frac{96481}{384}\kappa^8\right.$$
$$\left. + \frac{405985}{256}\kappa^{10} + \frac{89696075}{9216}\kappa^{12} + \frac{635209799}{18432}\kappa^{14} - \frac{45866496935}{98304}\kappa^{16} - \ldots\right). \quad (27.13)$$

In the above formula for the non-local contribution, the leading terms, up to the 14th order, have positive signs, while terms with higher order have a negative sign.

At low temperatures, the spontaneous magnetization I obtained on the basis of the two conjectures gives the expansion in power of x as [3]:

$$I = 1 - 6x^8 - 12x^{10} - 18x^{12} - 36x^{14} - 84x^{16}$$
$$- 192x^{18} - 408x^{20} - 864x^{22} - 1970x^{24}\ldots \quad (27.14)$$

On the other hand, the conventional low-temperature series yields [37]:

$$I = 1 - 2x^6 - 12x^{10} + 14x^{12} - 90x^{14}$$
$$+ 192x^{16} - 792x^{18} + 2148x^{20} - 7716x^{22} + 23262x^{24}\ldots \quad (27.15)$$

which takes into account of only the local environment of spins in the 3D Ising model. Thus, the non-local part of the magnetization of the 3D Ising model is obtained by extracting Eq. (27.15) from Eq. (27.14). It is represented as follows:

$$I_{\text{non-local}} = 2x^6 - 6x^8 - 32x^{12} + 54x^{14} - 276x^{16} + 600x^{18} - 2556x^{20} + 6852x^{22} - 25232x^{24} \ldots . \tag{27.16}$$

27.3 Critical Exponents in 3D Ising Model and Related Systems

In this section, we will review briefly some results obtained in our previous work for the critical exponents in the 3D Ising model and related topics.

Klein and March set up a d-dimensional result for the critical exponent $\delta(d)$ which embraces Zhang's value for $d = 3$ as well as known values for $d = 1, 2$, and 4 [5]. Scaling relations yield further critical exponents as a function of d. Finally, a critical exponent defined for random walks was treated. In a further work, Zhang and March refined the semi-empirical formula for liquid-vapour critical exponents [13], also based on the yield of Zhang's values for the 3D Ising model.

It was pointed out in [6] that the experimental data for the critical exponents of ferromagnet $CrBr_3$ are very close to Zhang's predictions for the 3D Ising model. The magnetic equation of state for the 3D Ising model near criticality was proposed for $CrBr_3$ [6, 7]. The experimentally observed magnetic behaviour near criticality for the ferromagnet $CrBr_3$ was found to show fingerprints of the 3D Ising Hamiltonian. The analogy between magnetic behaviour near criticality and the corresponding liquid-vapour behaviour was used to discuss the 2D-3D crossover in the latter case [7]. Moreover, the dimensionality and the critical exponents of the models for polymer growth in solution and the dynamic epidemic model were discussed in [10]. The critical exponents were predicted for the Potts model in three dimensions and the percolation exponents ($q = 1$) for $d = 1, 2, 3, 4, 5$, and $d \geq 6$.

The 3D Ising universality for critical indices in magnets and at fluid-fluid phase transition was reviewed in [15]. Experimental data for critical exponents in magnetic materials were compared with theoretical results on the 3D Ising model, as derived based on two conjectures [3]. It was found that critical exponents in some bulk magnetic materials indeed form a 3D Ising universality. The attention was then focused on the critical indices at the fluid-fluid phase transition. It is commonly accepted that pure fluids at gas-liquid critical points, and binary liquid mixtures at liquid-liquid critical points belong to the same universality class as the Ising model. Furthermore, statistical-mechanical models with separable many-body interactions, especially partition functions and thermodynamic consequences, for a variety of classical and quantum phase transition, were reviewed in [8, 38].

A relationship between the exponents γ and η at the liquid-vapour critical point was proposed via the dimensionality d [12]. Similarities and contrasts between critical point behavior of heavy fluid alkalis and d-dimensional Ising model were discussed in [22]. Crucial combinations of critical exponents were given for liquids-vapour and ferromagnetic second-order phase transitions [23]. A theory of critical exponents was developed in terms of dimensionality d plus universality class n [24]. March also presented a unified theory of critical exponents generated by the Ising Hamiltonian embracing solely the discrete values $d = 2, 3$ and 4, in terms of the critical exponent η [25]. The formula contains the exact values for $d = 2$ and 4, while for $d = 3$ it yields the value obtained earlier by Zhang [3].

Scaling properties near the critical point indicate the existence of a self-similarity behavior for the critical phenomena. Although the 3D Ising system considered is not a truly dynamic one, Zhang and March proposed a specific set of relations between fractal dimensions and critical exponents in the Ising model of statistical mechanics [17]. In particular, we put forward, corresponding to six critical exponents for the Ising model, six fractal dimensions. Assuming the latter proposals, we then derived relationships between such fractal dimensions.

Zhang and March presented a detailed analysis of temperature-time duality in the 3D Ising model, by inspecting the resemblance between the density operator in quantum statistical mechanics and the evolution operator in quantum field theory, with the mapping $\beta = (k_B T)^{-1} \mapsto it$ [14]. We pointed out that in systems like the 3D Ising model, for the nontrivial topological contributions, the time necessary for the time averaging must be infinite, and being comparable with or even much larger than the time of measurement of the physical quantity of interest. The time averaging is equivalent to the temperature averaging. The time is needed to construct the $(3 + 1)$D framework for the quaternionic sequence of Jordan algebras, in order to employ the Jordan-von Neumann-Wigner procedure [39–42].

Based on the quaternionic approach developed in [3] for the three-dimensional (3D) Ising model, we study conformal invariance in three dimensions. The 2D conformal field theory was generalized to be appropriate for three dimensions, within the framework of the quaternionic coordinates with complex weights [21]. The Virasoro algebra still works, but for each complex plane of quaternionic coordinates. The quaternionic geometric phase appears in quaternionic Hilbert space as a result of diagonalization procedure which involves the smoothing of knots/crossings in the 3D many-body interacting spin Ising system. Possibility for application of conformal invariance in three dimensions on studying the behaviour of the world volume of the brane, or the world sheet of the string in 3D or $(3 + 1)$D, is discussed.

Our interests are also paid to other related topics, for instance, the tricritical point in ^{4}He-^{3}He mixtures and some polymeric assemblies [9], the critical exponents for Anderson localization due to disorder [11, 16], critical exponents for marginal Fermi liquids [19, 20], etc. We pointed out that the spin fluctuations at the tricritical point are much weaker than those at the critical point, which leads to the conclusion that the tricritical behaviours can be described well by the mean-field exponents with logarithmic corrections at $d = 3$ [9]. On the other hand, our discussion on the Anderson localization is related with the semiclassic treatment on the effect of disorder on the nature of electron states in the quantum-chemical network model [11]. We also discuss the range of validity of a semiclassical relation between the critical exponents at the Anderson localization transition [16].

27.4 Disadvantages of Several Theoretical Approaches

We will now discuss the disadvantages of several theoretical techniques (for instance, conventional low-temperature expansions, conventional high-temperature expansions, perturbative renormalization group, Monte Carlo simulations, etc.,) as follows (for details, the readers may refer to [3, 27, 28, 34]).

The conventional low-temperature expansions, evaluated by systematically overturning spins from the ground state with all spins "up" [35–37, 43, 44], consider only the local environment of spins and their interactions, neglecting the non-local behavior of the 3D Ising model (*i.e.,* the interaction via the chain of n spins in the plane). Therefore, the conventional low-temperature expansions take into account only the local part of the partition function and the free energy of the system. The conventional high-temperature series expansions account the boundary between the areas of spin up and spin down [35, 43, 45]. A basic difficulty is that in 3D, the boundaries are polyhedrons, not polygons in the 2D lattice. The polygons cannot separate the domains of spin up and spin down for the 3D Ising lattice. Furthermore, the existence of the non-trivial topological objects (such as the Möbius band, the Klein bottle, crosses and knots/links, etc.) in the 3D Ising lattice makes the well-known topological troubles. The conventional high-temperature expansion cannot take into account the non-local behavior of the 3D Ising system.

During the renormalization group procedures for 3D Ising model, various approximations, such as expansions, perturbations, linearizations, normalizations, etc. are performed [46–51]. The real-space renormalization group techniques are concerned with the construction of new models from old by averaging dynamical variables of the old model to form the block variable of the new one [46–51]. The final results depend sensitively on how to divide Kadanoff blocks, define the effective Hamiltonian, determine the details of the block variables, and calculate approximately the partial trace. The final results would approach the exact solution if and only if the size of the Kadanoff blocks were chosen to be infinite and the infinite terms of the expansion were remained. This is impossible to be done, because much more variables would emerge with increasing the size of the Kadanoff blocks and taking into account more terms of the expansion, which make the calculations become extremely difficult. A serious trouble is that the non-local behavior in 3D involves in the interaction of long-range-like type and all the possible states of nm spins in a plane ($n \to \infty$ and $m \to \infty$) even for only one interaction between two spins along the third dimension, which make any procedure with finite blocks cut the plane of nm spins with the non-local effect. The field theoretical or **k**-space renormalization group techniques meets the same difficulty that cannot be overcome, since at the deeper level there is a close connection, because the basic idea of them is the same and because there are connections between ϕ and a set of block variables. It is clear now that the perturbative renormalization group with finite blocks cannot take into account the non-local behavior of the 3D Ising model. Therefore, one has to develop the nonperturbative renormalization group for studying the critical phenomena of the 3D Ising model.

Any computer simulation including Monte Carlo method for the 3D Ising model is limited by the size effects since the number of the configurations of the 3D Ising model increases in a fashion of 2^N, with the number of the spins or atoms, $N \rightarrow \infty$. Usually, the period boundary condition is used as that the simulation is performed on a small system with finite size [50, 52–55]. This technique breaks down the long-range many-body interaction between the two nearest neighboring spins along the third dimension in the 3D Ising model, which is due to the non-local behavior.

There is a new development in the 3D Ising model, using convex optimization of c-parameter within the conformal bootstrap approach to the four-point correlation functions [56]. In particular the estimates of $\eta = 0.036$ and $\nu = 0.629$ were reported in [56]. However, it was pointed out in [57] that these estimates in [56] were obtained based on certain hypothesis (*e.g.* the existence of a sharp kink) and that if these hypothesis are not used, then the conformal bootstrap analysis appears to be consistent with the GFD values $\eta = 1/8$ and $\nu = 2/3$. It is noticed that these GFD values are consistent with the solutions obtained in [3].

It is clear that any approaches based on only local environments, such as conventional low-temperature expansions, conventional high-temperature expansions, perturbative renormalization group, Monte Carlo simulations, etc., cannot be exact for the 3D Ising model, even though they may be exact for the 2D cases [2, 3, 27, 28, 34].

27.5 Summary and a Recent Work

In summary, we have investigated the non-local behavior of the 3D Ising model. The interaction between the nearest neighboring spins along the third dimension is effectively equivalent to the long-range many-body interaction in which $(n + 1)$ spin are involved. This nonlocal property is an intrinsic property of the 3D Ising model, which may imply that any approximation techniques (including the conventional low-temperature expansions, the conventional high-temperature expansions, the perturbative renormalization group and Monte Carlo method) taking into account only the local property cannot be exact for the 3D Ising model, though these approaches work well for other 2D or 1D Ising models. Any approach taking into account only the local environments (such as local alignments of spins in low-temperature expansions, polygons in high-temperature expansions, finite blocks in renormalization group, or finite unit cells plus the periodic boundary condition in Monte Carlo method) is inconsistent with the global effect induced by the chain of n spins. The exact solution of the 3D Ising model must be calculated by evaluating the partition function of the whole system involving all the possible states of all the spins of the whole 3D system. Therefore, approximation techniques are plagued by systematic errors, which are brought about by their disadvantages. Such systematic errors of these approximation techniques are related directly to the physical conceptions/pictures they are based upon, and to the neglect of important non-local factors during the procedures. These systematic errors are intrinsic, which cannot be removed by the efforts of improving technically the precision. Finally, we reviewed briefly some recent advances in the 3D Ising model.

In a more recent work [26], we develop a Clifford algebra approach for 3D Ising model, and give a positive answer to Zhang's two conjectures proposed in [3]. By utilizing some mathematical facts of the direct product of matrices and their trace, we can expand the dimension of the 3D Ising system by adding unit matrices 1 (with compensation of a factor) and adjusting their sequence, which do not change the trace of the transfer matrices V (Trace Invariance Theorem). The transfer matrices V are re-written in terms of the direct product of sub-transfer-matrices $\mathrm{Sub}(V^{(j)}) = [1 \otimes \ldots \otimes 1 \otimes V^{(j)} \otimes 1 \otimes \ldots \otimes 1]$, where each $V^{(j)}$ is surrounded by a large number of unit matrices 1. The sub-transfer-matrices $V^{(j)}$ are thus isolated by the unit matrices, which allow us to perform a linearization process on $V^{(j)}$ (Linarization Theorem). It is found that locally for each j, the internal factor W_j in the transfer matrices $V^{(j)}$ can be treated as a boundary factor, which can be dealt with by a procedure similar to the Onsager-Kaufman approach for the boundary factor U in the 2D Ising model. This linearization process splits sub-transfer matrix into subspaces. Furthermore, a local transformation is employed on each of the sub-transfer matrices (Local Transformation Theorem). The local transformation changes the coordinates of the local system and trivializes the non-trivial topological structure, while it generalizes the topological phases on the eigenvectors. In order to determine the rotation angle for the local transformation, the star-triangle relationship of the 3D Ising model is employed for the Curie temperature, which is the solution of generalized Yang-Baxter equations in the continuous limit. In future work, it will be of interest to inspect the origins of the relation $K''' = K'K''/K$ and the phase factors ϕ_x, ϕ_Y and ϕ_z, from the views of topological and geometrical aspects.

Acknowledgements This work has been supported by the National Natural Science Foundation of China under grant number 51331006 and 51590883.

References

1. E. Ising, Z. Phys. **31**(1), 253 (1925). https://doi.org/10.1007/BF02980577
2. L. Onsager, Phys. Rev. **65**(3-4), 117 (1944). https://doi.org/10.1103/PhysRev.65.117
3. Z.D. Zhang, Philos. Mag. **87**(34), 5309 (2007). https://doi.org/10.1080/14786430701646325
4. N.H. March, G.G.N. Angilella, *Exactly Solvable Models in Many-body Theory* (World Scientific, Singapore, 2016). ISBN 9789813140141
5. D.J. Klein, N.H. March, Phys. Lett. A **372**(30), 5052 (2008). https://doi.org/10.1016/j.physleta.2008.04.073
6. N.H. March, Z.D. Zhang, Phys. Lett. A **373**(23–24), 2075 (2009). https://doi.org/10.1016/j.physleta.2009.04.006
7. N.H. March, Z.D. Zhang, Phys. Chem. Liq. **47**(6), 693 (2009). https://doi.org/10.1080/00319100903079154
8. N.H. March, Z.D. Zhang, J. Math. Chem. **47**(1), 520 (2009). https://doi.org/10.1007/s10910-009-9575-8
9. N.H. March, Z.D. Zhang, Phys. Chem. Liq. **48**(2), 279 (2010). https://doi.org/10.1080/00319100903296121

10. Z.D. Zhang, N.H. March, Phys. Chem. Liq. **48**(3), 403 (2010). https://doi.org/10.1080/00319100903539538
11. Z.D. Zhang, N.H. March, J. Math. Chem. **49**(3), 816 (2011). https://doi.org/10.1007/s10910-010-9778-z
12. Z.D. Zhang, N.H. March, Phys. Chem. Liq. **49**(2), 270 (2011). https://doi.org/10.1080/00319104.2010.499514
13. Z.D. Zhang, N.H. March, Phys. Chem. Liq. **49**(5), 684 (2011). https://doi.org/10.1080/00319104.2010.517206
14. Z.D. Zhang, N.H. March, J. Math. Chem. **49**(7), 1283 (2011). https://doi.org/10.1007/s10910-011-9820-9
15. Z.D. Zhang, N.H. March, Phase Trans. **84**(4), 299 (2011). https://doi.org/10.1080/01411594.2010.535351
16. Z.D. Zhang, N.H. March, J. Phys. Chem. Solids **72**(12), 1529 (2011). https://doi.org/10.1016/j.jpcs.2011.09.012
17. Z.D. Zhang, N.H. March, J. Math. Chem. **50**(4), 920 (2012). https://doi.org/10.1007/s10910-011-9935-z
18. N.H. March, Z.D. Zhang, J. Math. Chem. **51**(7), 1694 (2013). https://doi.org/10.1007/s10910-013-0181-4
19. Z.D. Zhang, N.H. March, Phys. Chem. Liq. **51**(2), 261 (2013). https://doi.org/10.1080/00319104.2012.753545
20. N.H. March, Z.D. Zhang, Phys. Chem. Liq. **51**(6), 742 (2013). https://doi.org/10.1080/00319104.2013.812023
21. Z.D. Zhang, N.H. March, Bull. Soc. Sci. Lettres Łódź Sér. Rech. Déform. **62**(3), 35 (2012), Also available as preprint arXiv:1110.5527 [cond-mat.stat-mech]
22. N.H. March, Phys. Lett. A **378**(3), 254 (2014). https://doi.org/10.1016/j.physleta.2013.10.030
23. N.H. March, Phys. Chem. Liq. **52**(5), 697 (2014). https://doi.org/10.1080/00319104.2014.906600
24. N.H. March, Phys. Lett. A **379**(9), 820 (2015). https://doi.org/10.1016/j.physleta.2014.11.043
25. N.H. March, Phys. Chem. Liq. **54**(1), 127 (2016). https://doi.org/10.1080/00319104.2015.1058943
26. Z.D. Zhang, O. Suzuki, N.H. March, to be published (2018).
27. Z.D. Zhang, Chin. Phys. B **22**(3), 030513 (2013). https://doi.org/10.1088/1674-1056/22/3/030513
28. Z.D. Zhang, Acta Metallurgica Sinica **52**, 1311 (2016)
29. B. Kaufman, Phys. Rev. **76**(8), 1232 (1949). https://doi.org/10.1103/PhysRev.76.1232
30. L.H. Kauffman, *Knots and Physics* (World Scientific, Singapore, 2001)
31. L.H. Kauffman, Rep. Prog. Phys. **68**(12), 2829 (2005). https://doi.org/10.1088/0034-4885/68/12/R04
32. L.H. Kauffman, in *Encyclopedia of Mathematical Physics*, ed. by J.P. Françoise, G.L. Naber, T.S. Tsun (Academic Press, Oxford, 2006), pp. 220–231. ISBN 978-0-12-512666-3. https://doi.org/10.1016/B0-12-512666-2/00240-6
33. F.Y. Wu, Rev. Mod. Phys. **54**, 235 (1982). https://doi.org/10.1103/RevModPhys.54.235
34. Z.D. Zhang, J. Phys.: Conf. Series **827**(1), 012001 (2017). https://doi.org/10.1088/1742-6596/827/1/012001
35. C. Domb, in *Phase Transitions and Critical Phenomena*, vol. 3 : Series Expansions for Lattice Models, ed. by C. Domb, M.S. Green (Academic Press, London, 1974), chap. 1, pp. 1–96. ISBN 0122203038
36. A.J. Guttmann, I.G. Enting, J. Phys. A **26**(4), 807 (1993). https://doi.org/10.1088/0305-4470/26/4/010
37. G. Bhanot, M. Creutz, I. Horvath, J. Lacki, J. Weckel, Phys. Rev. E **49**, 2445 (1994). https://doi.org/10.1103/PhysRevE.49.2445
38. W. Zhang, A.R. Oganov, A.F. Goncharov, Q. Zhu, S.E. Boulfelfel, A.O. Lyakhov, E. Stavrou, M. Somayazulu, V.B. Prakapenka, Z. Konôpková, Science **342**(6165), 1502 (2013). https://doi.org/10.1126/science.1244989

39. J. Ławrynowicz, S. Marchiafava, A. Niemczynowicz, Adv. Appl. Clifford Alg. **20**(3), 733 (2010). https://doi.org/10.1007/s00006-010-0219-7
40. J. Ławrynowicz, O. Suzuki, A. Niemczynowicz, Adv. Appl. Clifford Alg. **22**(3), 757 (2012). https://doi.org/10.1007/s00006-012-0360-6
41. J. Ławrynowicz, M. Nowak-Kępczyk, O. Suzuki, Int. J. Bifurc. Chaos **22**(01), 1230003 (2012). https://doi.org/10.1142/S0218127412300030
42. J. Ławrynowicz, O. Suzuki, A. Niemczynowicz, Int. J. Nonlinear Sci. Numer. Simul. **14**(3–4), 211 (2013). https://doi.org/10.1515/ijnsns-2013-0030
43. G.F. Newell, E.W. Montroll, Rev. Mod. Phys. **25**, 353 (1953). https://doi.org/10.1103/RevModPhys.25.353
44. C. Domb, A.J. Guttmann, J. Phys. C **3**(8), 1652 (1970). https://doi.org/10.1088/0022-3719/3/8/003
45. M.E. Fisher, Rep. Prog. Phys. **30**(2), 615 (1967), [**31**, 418 (1968)]
46. K.G. Wilson, Phys. Rev. B **4**, 3174 (1971). https://doi.org/10.1103/PhysRevB.4.3174
47. K.G. Wilson, J. Kogut, Phys. Rep. **12**(2), 75 (1974). https://doi.org/10.1016/0370-1573(74)90023-4
48. L.P. Kadanoff, W. Götze, D. Hamblen, R. Hecht, E.A.S. Lewis, V.V. Palciauskas, M. Rayl, J. Swift, D. Aspnes, J. Kane, Rev. Mod. Phys. **39**, 395 (1967). https://doi.org/10.1103/RevModPhys.39.395
49. L.P. Kadanoff, Phys. Rev. Lett. **34**, 1005 (1975). https://doi.org/10.1103/PhysRevLett.34.1005
50. J.J. Binney, N.J. Dowrick, A.J. Fisher, M.E.J. Newman, *The Theory of Critical Phenomena* (An introduction to the renormalization group (Clarendon Press, Oxford, 1992)
51. M.E. Fisher, Rev. Mod. Phys. **70**(2), 653 (1998). https://doi.org/10.1103/RevModPhys.70.653
52. K. Binder, E. Luijten, Renormalization group theory in the new millennium. Phys. Rep. **344**(4–6), 179 (2001). https://doi.org/10.1016/S0370-1573(00)00127-7
53. A. Pelissetto, E. Vicari, Phys. Rep. **368**(6), 549 (2002). https://doi.org/10.1016/S0370-1573(02)00219-3
54. H. Müller-Krumbhaar, K. Binder, J. Stat. Phys. **8**(1), 1 (1973). https://doi.org/10.1007/BF01008440
55. E. Stoll, K. Binder, T. Schneider, Phys. Rev. B **8**, 3266 (1973). https://doi.org/10.1103/PhysRevB.8.3266
56. S. El-Showk, M.F. Paulos, D. Poland, S. Rychkov, D. Simmons-Duffin, A. Vichi, J. Stat. Phys. **157**(4), 869 (2014). https://doi.org/10.1007/s10955-014-1042-7
57. J. Kaupužs, R.V.N. Melnik, J. Rimšāns, Int. J. Mod. Phys. C **28**(04), 1750044 (2017), Also available as preprint arXiv:1407.3095 [cond-mat.stat-mech]. https://doi.org/10.1142/S0129183117500449

Chapter 28
Do Two Symmetry-Breaking Transitions in Photosynthetic Light Harvesting Complexes (PLHC) Form One, Two or More Kibble–Zurek (KZ) Topological Defect(s)?

R. H. Squire

Abstract Kibble (J Phys A: Math Gen, 9(8):1387, 1976) [1] and Zurek (Nature, 317(6037):505, 1985) [2] (KZ) proposed that rapid symmetry-breaking transitions in the hot, early universe could result in causally disconnected topological defects such as cosmic strings. This type of first order transition has analogues in certain second order transitions present in condensed matter such as liquid crystals, superfluids and charge density waves in terms of flux tubes or vortices. Recently, we discovered that Rhodopseudomonas acidophila's Photosynthetic Light Harvesting Complex might have different types of coherent ground and excited states, suggesting that there are two different symmetry-breaking transitions. The B850 ground states comprise eight identical rings each containing 18 bacteriochlorophyll components, and each ring has undergone a Bose–Einstein phase transition to a charge density wave that lowers the energy. The excited state coherence results from polariton formation from the non-crossing of bosons, here excitons and photons, an extension of exciton theory (Knox, Theory of excitons. Academic Press, New York, 1963) [3]. The result is short-lived quasiparticles with very low mass that can form an unusual Bose–Einstein condensate (BEC). We suggest that the oriented, circular B850's and enclosed single B875 create new cavity structure with some attributes of toroidal nanopillars (Fan et al., Nano Lett, 10(10):3823, 2010) [4], (Pelton and Bryant, Introduction to metal-nanoparticle plasmonics. Wiley, Hoboken, 2013) [5], (Vahala, Nature, 424(6950):839, 2003) [6]. Since both the ground and excited states should contain solitons, we envisage trARPES (time and angle resolved photoemission spectroscopy) in conjunction with a three-pulse probe with various appropriate time delays should be able to map both the KZ phase transitions and energy transfers as a function of light intensity and time in this complex at room temperature (Mihailovic et al., J Phys: Condens Matt, 25(40):404206, 2013) [7], (Rettig et al., Nat Comms, 7:10459, 2016) [8].

R. H. Squire (✉)
Department of Natural Sciences, West Virginia University,
Institute of Technology, Beckley, WV 25136, USA
e-mail: Richard.Squire@mail.wvu.edu

G. G. N. Angilella and C. Amovilli (eds.), *Many-body Approaches at Different Scales*, https://doi.org/10.1007/978-3-319-72374-7_28

28.1 Introduction

Several years ago, Professor Norman March and I independently studied the photosynthetic process and then we concluded that the efficiency of energy transfer was beyond our understanding. What we are offering here is a model that rationalizes how coherent energy transfer from light absorption to the reaction center (RC) might proceed. Though the Photosynthetic Light-Harvesting Complex (PLHC) ring has been studied for many years, it apparently has not been described as a charge density wave (CDW), despite the 'dimer' feature (confirmed by crystallography) PLHCs are described in Sect. 28.2. Section 28.3.1 discusses symmetry-breaking, Sect. 28.3.2 outlines the Kibble–Zurek (KZ) theory [1, 2], and Sect. 28.3.3 outlines coherence, followed by Sect. 28.3.4, the Geodesic Rule (phase addition) and defect types. Section 28.5 is devoted to polaritons. Certainly, recent advances in optical micro- and nanocavity spectroscopy have permitted a deluge of significant experiments identifying the properties of excited states of CDW's and polaritons, despite their transient nature. Finally, we put the pieces together (Sect. 28.6) in a coherent fashion and discuss the cooperation that may be present to transfer energy efficiently, followed by a summary, conclusions, and possibilities for further work (Sect. 28.7).

28.2 Brief Description of Photosynthetic Light-Harvesting Complexes (PLHC)

In anaerobic photosynthetic prokaryotes (purple bacteria), the process of light capture and transfer is highly efficient [9–11]. Our emphasis is on capture of solar energy and its nearly perfect transfer to the reaction center (RC or P870). A purple bacterium (Rhodopseudomonas acidophila) has a photosynthetic unit comprised of two pigment-protein light-harvesting (LH) complexes called LH1/RC and LH2. These LH's have similar molecular structures comprised of bacteriochlorophyll components ($BChl_a$, designated BChl hereafter) that are comprised of circular conjugated arrays with a coordinated magnesium ion, Mg^{2+} (Fig. 28.1).

Eight LH2s, each with 18 BChl units, surround the RC-LH1. X-ray crystallography has determined the atomic structure and orientations [7, 8], and the associated pigments' groups are separated clearly enough so electronic energy transfer (EET) can be measured. The complexes' objective is to absorb and transfer energy efficiently to the reaction center (RC) where 'charge splitting' takes place to produce a 1.1 eV potential and thus generate energy for the survival and growth of the bacteria. The oxygen-generating bacteria are presumed to be responsible for converting the oceans and atmosphere from the environment to one rich in oxygen (the 'Great Oxygenation Event') a couple of billion years ago [12].

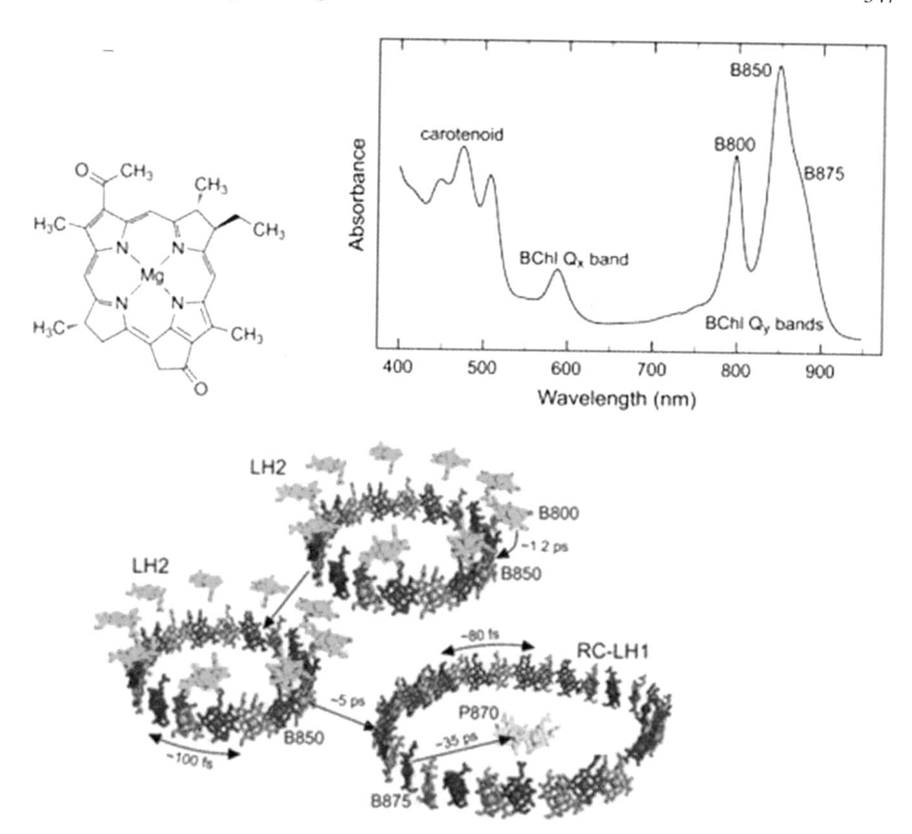

Fig. 28.1 (Top left) A prototypical bacteriochlorophyll *a* $(BChl)_a$; (Top right) Spectrum of BChl complexes B800 (green yellow), B850 and B875 (red and yellow). The Soret bands are located between 300 and 400 nm. (Bottom) Graphical illustration of the crystal structure coordinates for light harvesting (LH) complexes 1, 2 and the reaction center (P870). The orange, red and green structures are comprised of the BChl ring structure surrounding the Mg^{2+} ion that is responsible for the charge density wave that generates the 'dimer' structure of the B850 and B875 rings (see text). The absorption maxima indicate the complexes; energy transfer rates are shown. (Figure courtesy of Professor M. Jones, Bristol)

28.3 Background

28.3.1 Spontaneous Symmetry Breaking (SSB)

SSB is ubiquitous in our daily life; water freezes, and the rotational symmetry is broken by a selection of one out of many rotational possibilities by the newly formed ice crystal. Many crystals in a proper environment can undergo a very slow crystallization to minimize defects. However, in the early universe, the transition was rapid and multiple stable topological defects should have been formed. The stages

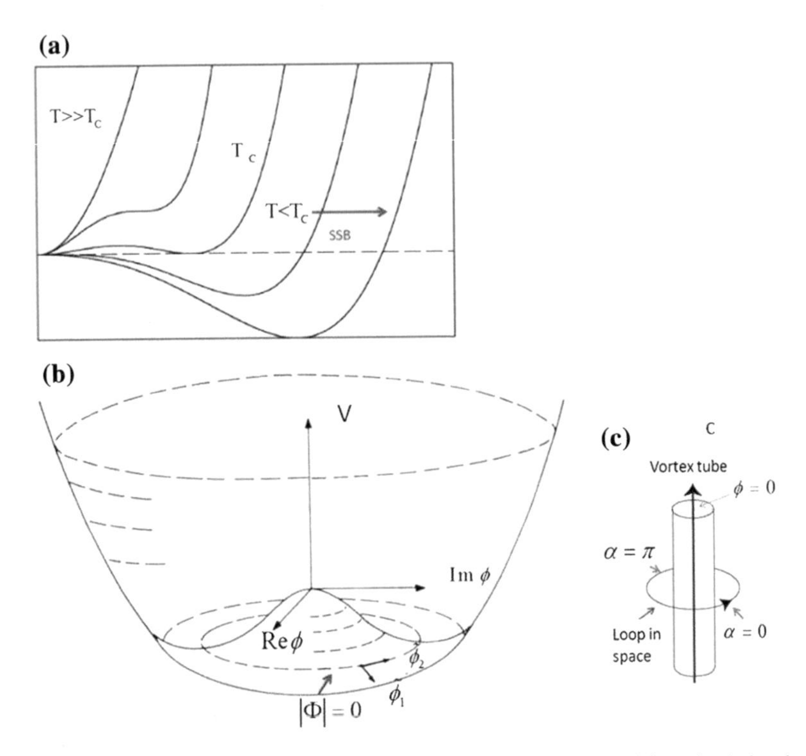

Fig. 28.2 **a** Rapid cooling of the universe changes the parabola potential to the 'wine bottle' structure (**b**). The choice of phase is arbitrary, but once chosen, this state can generate a Bose–Einstein condensation if certain parameters are suitable (temperature, mass, etc.; see Sect. 28.3.2). **c** The simplest order parameter is complex. If we have a system with three separated regions, each with an arbitrary phase, and we trace a large loop in space and add the phase values, if the phase changes by $2n\pi$, there must be at least one vortex present (**c**). The number n is then 'topological'

of the 'wine bottle' SSB potential (Fig. 28.2b) are formed by sufficiently lowering the temperature (Fig. 28.2a).

The potential $V(\psi) = \lambda(|\psi|^2 - \rho^2)^2$ is symmetric about the rotation axis, i.e. $\psi \mapsto \psi e^{i\phi}$, about $V(\psi = 0)$; λ and ρ are constants. Ar $T \gg T_c$, thermal fluctuations are much more energetic than $V(\psi)$, and the average $\psi = 0$. When $T < T_c$, an arbitrary phase is selected and the symmetry is broken. Minima occur on the circle $|\Psi| = 0$ with the same energy because of symmetry, but are different ground states, since their phase angles differ in general. However, a sizeable fraction of, say, cold atoms with the same phase angle can create a Bose condensation if the temperature and/or mass is low enough. Both CDWs and polaritons break the electromagnetic field symmetry, so once ϕ has been selected, SSB prevails: $U(1) = e^{i\phi}$ $(\phi \in [0, 2\pi])$.

28.3.2 Coherence

A Hamiltonian describing bosons such as excitons is

$$H = \sum_i \frac{\hbar^2 k^2}{2m} a_{\mathbf{k}}^{\dagger} a_{\mathbf{k}}, \tag{28.1}$$

where $a_{\mathbf{k}}^{\dagger}$, $a_{\mathbf{k}}$ are the creation and annihilation operators, $\mathbf{k}$ is the wave vector, and m is the mass. The distribution number of *bosonic* particles is different from that of *fermions*

$$N_{\mathbf{k}} = \frac{1}{e^{\beta(\hbar^2 k^2/2m - \mu)} - 1} \tag{28.2}$$

(-1 is changed to $+1$ for fermions), with μ the chemical potential, and $\beta = (k_B T)^{-1}$, where k_B is the Boltzmann constant. In the thermodynamic limit (large number of particles), the density of particles, n, is

$$n = \lim_{V \to \infty} \frac{N}{V} = \frac{1}{V} \sum_{\mathbf{k}} N_{\mathbf{k}} = \frac{1}{(2\pi)^3} \int d^3\mathbf{k} \frac{1}{e^{\beta(\hbar^2 k^2/2m)} - 1} \tag{28.3}$$

so, the integral becomes [13, 14]

$$n = 2.612 \left(\frac{m k_B T}{2\pi \hbar^2} \right)^{3/2}, \quad \text{or} \quad T_c = \frac{2\pi \hbar^2}{k_B m} \left(\frac{n}{2.612} \right)^{3/2}. \tag{28.4}$$

However, the integral has not accounted for all the particles since the zero state, n_0, must be treated separately, so rewriting the density, Eq. (28.3),

$$n = \lim_{V \to \infty} \frac{N}{V} = n_0 + \frac{1}{(2\pi)^3} \int d^3\mathbf{k} \frac{1}{e^{\beta(\hbar^2 k^2/2m)} - 1}. \tag{28.5}$$

When a large number of bosons are cooled below T_c (note the inverse dependence on mass Eq. (28.4)), a large fraction of the bosons occupy the lowest state and become coherent which means that the phase, ϕ, is no longer random ($\phi \in [0, 2\pi]$), but becomes a single value and each boson 'phase-locks' with all of the other bosons as a result of interactions. Identifying bosons as excitons would mean that we could diagonalize the exciton density matrix in a new basis and represent all the participating excitons with a single wave function, $\psi(\mathbf{r}, t) = \sqrt{n_0(\mathbf{r}, t)} e^{iS(\mathbf{r}, t)}$, where n_0 was defined by Eq. (28.5) and $S(\mathbf{r}, t)$ is the common phase. The 'normal' state phases of a wave function are arbitrary from 0 to 2π. In a BEC this phase value is arbitrary, but once chosen, the electromagnetic field symmetry has been 'broken'. Further, the chemical potential has now become zero, so the 'resistance' to adding or removing a particle is now zero (see Fig. 28.3).

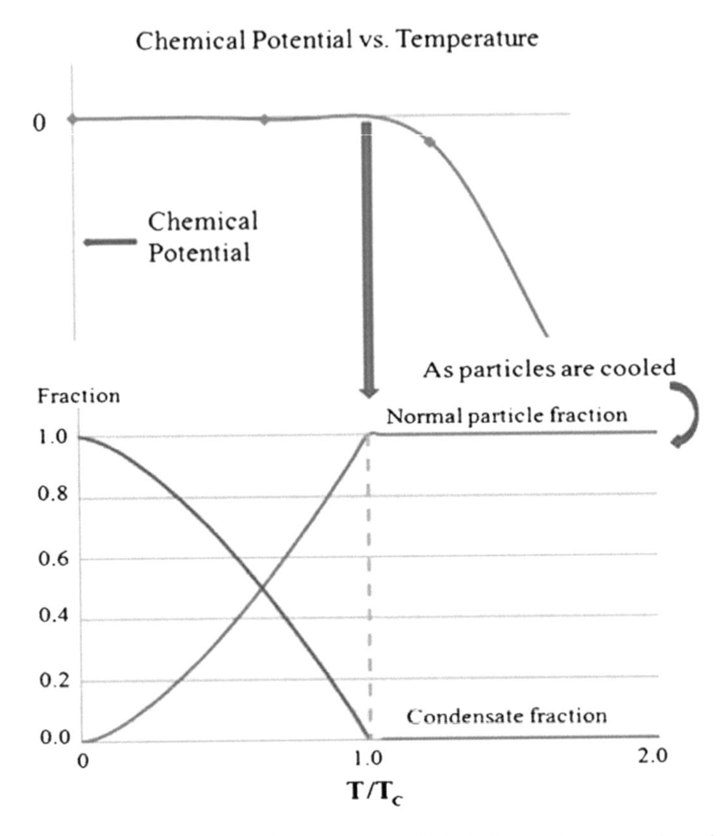

Fig. 28.3 When the chemical potential reaches zero, T_c is defined and Bose condensation begins. Thus, the energy cost to add or remove a particle in a BEC is zero, and the uncertainty principle can be written as $\Delta p \Delta \phi \sim \hbar$, where $\Delta \phi$ is a single value, so Δp has some flexibility

The physical significance of the inverse relationship between T and mass is that the particles are represented by the thermal de Broglie wavelength as the temperature is lowered; the wave packets begin to overlap and quantum effects characterize the system in a matter wave. When the mean de Broglie thermal wavelength of the particles approaches unity, quantum effects characterize the system. The BEC phase transition results from the particle wavelength spanning the interparticle distance, roughly the same point when exchange effects become important. We illustrate pertinent comparisons of BEC systems below [15]:

Systems	Excitons	Polaritons
Effective mass, m^*/m_e	10^{-1}	10^{-5}
Critical temperature, T_c	1 mK – 1 K	1–300 K
Thermalization time/lifetime	10 ps/1 ns	(1–10 ps)/(1–10 ps)
	$\sim 10^{-2}$	$= 0.1–10.$

28.3.3 Kibble–Zurek (KZ) Model

Crucial to the discussion is an estimate as to how many random cosmic strings should there be as the universe rapidly cools in a rapid first order transition [16]. Fortunately, condensed matter offers phase transitions where the relevant speed is not the speed of light, but for example, the second sound in a normal-to-superfluid second-order phase transition in helium. In this transition the order parameter (OP) can be used as a measure of the degree of progress since it ranges continuously from zero in the phase above the critical point to a finite number below it, and the order parameter susceptibility usually diverges. Landau exploited this continuous change property by placing two conditions on the free energy; it needs to be analytic and obey the symmetry of the Hamiltonian. Close to a critical temperature T_c, the free energy can be expanded in a Taylor series in the OP, here, the Ginzburg–Landau potential,

$$V(\psi) = \alpha(t)|\psi(\mathbf{r})|^2 + \frac{1}{2}\beta|\psi(\mathbf{r})|^4, \tag{28.6}$$

with $\psi = \sqrt{\rho}e^{i\phi}$, the Bose–Einstein condensate (BEC) complex order parameter, where ρ is the fluid density and ϕ is the phase. A complex order parameter permits formation of strings (or vortices). Near the critical temperature T_c, the coefficient $\alpha = \alpha'(T - T_c)$, and both α' and β are constants that can be determined by the specific heat data.

The coherence length ξ is the length scale of the order parameter, short for a 'normal' state, and longer for a BEC (coherent) state. The coherence length diverges as the critical temperature is approached. The complex order parameter obeys the Gross–Pitaevskii equation (GPE):

$$i\hbar\frac{\partial\psi}{\partial t} = -\frac{\hbar^2}{2m}\nabla^2\psi + \alpha\psi + \beta|\psi|^2\psi, \tag{28.7}$$

where m is the atomic mass. Applying Eq. (28.7) to the KZ model and rescaling the characteristic equilibrium quantities, for example the correlation length, $\xi_{eq} = \hbar/\sqrt{2m|\alpha|} = \xi_0/|\varepsilon|^{1/2}$, superfluid density, $|\psi_{eq}|^2 = -\alpha/\beta$, and relaxation times, $\tau = \hbar/|\alpha| = \tau_0/\varepsilon$, where we assume ε changes linearly since $\varepsilon = t/\tau_Q$, τ_Q being a quench time. The initial correlation length $\xi(t)$ of ψ has the same value as $\xi_{eq}(t)$ when approaching the phase transition from $t < 0$. The critical slowing forces $\xi_{eq}(t)$ to diverge, which suppresses the velocity of the spreading of the order parameter coherence. Consequently, the regions containing ordered phases begin to form independently in parts of the system. These disconnected domains continue growing and overlapping, eventually coalescing at time $\hat{t} = \sqrt{\tau_0\tau_Q}$ as $\xi(t)$ approaches $\xi_{eq}(t)$. However, in different domains the order parameter phases ϕ are uncorrelated, leaving topological defects (such as vortices) at the *domain boundaries*. An estimated separation between defects can be estimated by the correlation length at $t = \hat{t}$, since

$$\hat{\xi} = \xi_{eq}(\hat{t}) = \xi_0 \left(\frac{\tau_Q}{\tau_0}\right)^{1/4}, \tag{28.8}$$

resulting in an approximate defect density of $n_{def} \approx \hat{\xi}^{-2}$.

28.3.4 The Geodesic Rules and Defect Types

Geodesics are the straightest curve in a manifold. Here, when bubbles of new phases are nucleated with random phases, these new regions follow a geodesic path that interpolates the new phase [17]. A scenario for vortex formation in two dimensional space might be a simultaneous collision between three bubbles with phases of $2\pi/3$, 0, and $4\pi/3$. A geodesic rule would be to move from 0 to $2\pi/3$, then $4\pi/3$, so we have completely covered the vacuum manifold once. Then, the value of the field is zero at the vortex core. Therefore, what happens if only two bubbles collide (a non-geodesic path); the bubbles can decay by nucleating vortices, or if a vortex is created transiently, a stable vortex could be formed. Alternatively, if the third bubble arrives before a vortex has been formed, the future of the vortex depends on the values of the bubble phases. When the sum of the three bubbles' phases generates a complete rotation around the vacuum manifold (to infinity), the vortex is topologically stable; otherwise, it is not.

Defect Types.

In general, the matter symmetry and the type of phase transition determine the type of defect, so focusing on condensed matter, we will limit our discussion to the first two types of the four basic defects defined as:

Planar domain walls. A network of two-dimensional walls formed when a discrete symmetry is broken, resulting in partitions of the 'universe' into cells.

Cosmic strings (vortices). These strings result by the breaking an axial or cylindrical symmetry. A key feature is a set of non-simply connected minima surrounding the high energy state, resulting in a non-trivial winding number. *Our focus.*

Monopoles. Zero-dimensional objects which are diluted by inflation.

Textures. Delocalized topological defects that are unstable.

28.4 Interacting Bose Condensates

28.4.1 Two Bose Condensates

A Josephson Junction (JJ) between a Bose gas trapped in a double-well magnetic trap with a high enough barrier can be solved using the GPE for each well, and the classically forbidden region [18, 19]. A different number of atoms N_1, N_2 will result in two different chemical potentials μ_1, μ_2 localized in their respective traps. Following Dalfovo et al. [19], we take a linear combination of the two solutions to provide a solution of the time-dependent Schrödinger equation:

$$\psi(x,t) = \psi_1(x)\exp\left(-i\frac{\mu_1 t}{\hbar}\right) + \psi_2(x)\exp\left(-i\frac{\mu_2 t}{\hbar}\right), \tag{28.9}$$

and we can show that ψ_1 and ψ_2 overlap significantly in the forbidden region. The current density is calculated and it has the form of a typical Josephson form:

$$I = I_0 \sin\frac{(\mu_1 - \mu_2)t}{\hbar}. \tag{28.10}$$

The Josephson current will result in an oscillation of the number of atoms in the two traps,

$$\frac{dN_1}{dt} = -\frac{dN_2}{dt} = -I_0 \sin\frac{(\mu_1 - \mu_2)t}{\hbar}, \tag{28.11}$$

which has been experimentally verified [20].

28.4.2 Three Homogeneous Interacting Bose Condensates

It is not obvious whether three cyclically and weakly coupled superfluids can result in a supercurrent since the current strength depends on initial relative phases and the strength of the coupling with the two adjacent superfluids [21–23]. An ansatz for the three order parameters is

$$\Psi(\mathbf{r},t) = \sum_i \psi_i(t)\Phi_i(\mathbf{r}), \tag{28.12}$$

where $\Phi_i(\mathbf{r})$ represents each 'domain' with a unique phase. Substituting Eq. (28.12) into the Gross–Pitaevskii equation, Eq. (28.7), we find a cyclical linear relationship between adjacent complexes

$$i\hbar\frac{\partial\psi_i}{\partial t} = (E_i + U_i|\psi_i|^2)\psi_i - K_{i,i-1}\psi_i - K_{i,i+1}\psi_{i+1}. \tag{28.13}$$

While the equations for both cases are similar, only three or more supercurrents can create a topological entity. A key difference here is that the three supercurrents or cyclical Josephson currents (CJC) can enclose a topological entity (vortex) depending on the coupling constants and their initial relative phases. The size of the coupling constant is crucial since it must be small to avoid exciting the plasma state. In the weak coupling limit and assuming only stationary solutions produce CJC,

$$\langle j_c j_c \rangle_n = \frac{1}{(2\pi)^{n-1}} \sum_{k=0}^{n-1} \frac{2}{n} \left| \sin\left(\frac{2\pi k}{n}\right) \right|. \tag{28.14}$$

$\langle j_c j_c \rangle_n$ decreases as $n \to \infty$. The conclusion is that most of the CJCs result from a small value of n at a stationary phase of $\phi_{1,2} = \phi_{2,3} = 2\pi/3$, so the value of $\langle j_{KZ} \rangle = 0.577$ seems exaggerated. However, this value was calculated based on the maximum possible number of vortices. As Nozières has pointed out, a BEC cannot be formed without interactions [24], so, if the separate next Bose interactions were larger, vortex formation is more likely to happen, as in the example.

28.4.3 Merging Multiple Bose Condensates to Form Vortices

There have been several interesting superfluid vortex experiments. Scherer et al. [25] merged three dilute gaseous Bose condensates at both fast and slow rates to form quantized vortices, demonstrating a connection between interference, merging and vortex generation [25]. Ryu et al. [26] created a persistent superfluid quantized rotation in a toroidal trap, initiated by the transfer of one unit of orbital angular momentum [26]. Stable flow required a multiply connected trap and a condensate fraction of at least 20%. When two units of angular momentum were added, the vortices split into two singly charged vortices. While the KZ model regards evolution as completely reversible and only topological winding numbers 'remember' the scaling, the post-transition phase ordering being smoothed might be regarded a quantum analogue [27]. Our final exam focuses on the Meissner state where a planar, circular superconductor contains a hole with circumference, C [28]. The order parameter is a complex field $\psi = \rho e^{i\phi}$, with $|\rho|^2$ the density of Cooper pairs and phase ϕ. Quenching the system from normal to superconducting prevents a uniform phase, so we define a topological winding number density, $n(x) = (2\pi)^{-1} d\phi(x)/dx$, so the normalized magnetic flux through the hole is

$$n = \int_C n(x)dx = \frac{\Delta\phi}{2\pi}. \tag{28.15}$$

Assume $C \gg 2\pi\bar{\xi}$ and Eq. (28.8) follows, then the probability of a single trapped fluxoid should extrapolate to $f_1 \approx \langle n^2 \rangle \approx (r/\xi_0)(\tau_Q/\tau_0)^{-\sigma}$. However, the data indicated that the exponent should be -2σ, suggesting small defects may be difficult to be properly analyzed in simulations.

28.4.4 Three Inhomogeneous Interacting Bose Condensates (CDW)

Recently, Miller et al. [29] have expanded earlier work, particularly by Maki [30] and Bardeen [31] (see [32] for a historical perspective and final results) on 'depinning' a charge density wave (CDW) trapped by impurities some twenty-five years ago. Bardeen showed experimentally that CDWs can tunnel through a potential barrier, like Josephson's discovery of cooperative quantum tunneling in superconductors [33–35]. Coleman broadened the description to include macroscopic quantum tunneling and decay of the 'false vacuum' [36] that characterize instability in a scalar field relative to a lower energy state. We now know that two BECs in proximity can use a Josephson Junction (JJ) to exchange particles.

For the moment, we will assume that this theory allows the possibility of a B850 complex CDW in an excited state (metastable well) to tunnel to adjacent unexcited B850 rings or a ground state B850 to tunnel a lower potential well (B875). The event nucleates a bubble of a 'true vacuum' contained within a soliton description. The quantum charged ($\pm 2e$) solitons, delocalized in the transverse directions, use Josephson tunneling to move to another CDW chain. Following Miller et al. [29], we consider nucleated droplets of kinks and anti-kinks with a charge as quantum fluids with quantum delocalization between CDW chains. The model proposed relates a vacuum angle θ to displacement charge Q between inter-chain contacts $\theta = 2\pi(Q/Q_0)$ where the potential energy of the ith chain is,

$$u_i(\phi_i) = 2u_0\,(1 - \cos\phi_i(x)) + u_E\,(\theta - \phi_i(x))^2\,. \tag{28.16}$$

We ignore the first term, the periodic density-wave pinning energy, to focus on the quadratic electrostatic energy from the net charge displacement; graphs of u versus θ for $\phi \simeq 2\pi n$ when the energy is minimized are shown in Fig. 28.4. Tunneling is coherent into the next well (chain) via tunneling matrix element T as each parabolic branch crosses the next instability point, $\theta = 2\pi(n + 1/2)$.

Since there are eight C850 CDW rings surrounding B875, also a CDW (Sect. 28.1), multiple vortices are very possible since the interaction term K_{CDW} can be considerably larger than the plasma energy level restriction in the superfluid case (Sect. 28.4.2). Based on the eight possible vortices in the B850/B875 complex, it seems likely that at least two vortex tubes are present. In fact, the value of K_{CDW} could be several hundred meV, since the CDW is stable at room temperature, so there could be as many as eight (or nine including B875) coherent rings and vortex tubes (Fig. 28.5).

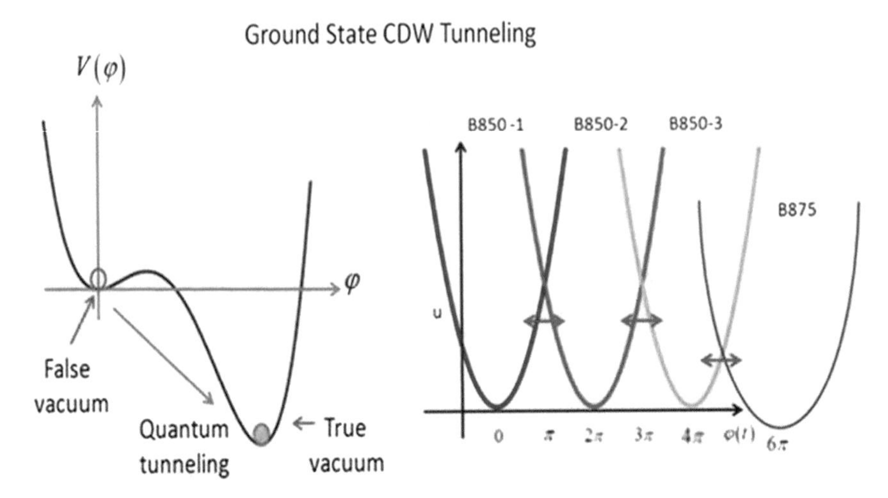

Fig. 28.4 (Left panel) Both tunneling to a lower energy potential well and (right panel) to a series of identical B850 complex rings can take place. Eight B850 complexes surround the B875-RC complex. If the population changes, say, in B850-2, CDW tunneling could rapidly re-equilibrate the population

Fig. 28.5 Eight circular B850 CDWs surround the B875 CDW. The red dots indicate the location of possible vortex tubes

28.4.5 More Than Three CDWs

A model system in this regard could be the quasi-1D tritellurides, $TbTe_3$, where trARPES (time and angle resolved photoemission spectroscopy) experiments after ultrafast optical excitation revealed a coherent oscillation of the occupied CDW band [7, 8, 37, 38], and the Peierls distortion energy gap $\Delta(\mathbf{k})$ was measured by examining the occupied ground state and unoccupied excited state. Here, trARPES (time and angle resolved photoemission spectroscopy) experiments after ultrafast optical excitation revealed a coherent oscillation of the occupied CDW band at about 2.3 THz. The oscillation modulated the CDW peak position, amplitude, and spectral width. The unoccupied band also had a downshift and the two band shifts reduced the gap size. The Peierls distortion energy gap $\Delta(\mathbf{k})$ was measured by examining the occupied ground state and unoccupied excited state. Since the complex order parameter can be expressed as $\Delta = |\Delta|e^{i\phi}$, the variation of the magnitude of the gap (called an 'amplitudon') was measured, while the phase (a 'phason') remained constant [8]. Another experimental scheme measured both the gap and the phases of both bands using trARPES in conjunction with a three-pulse probe with various appropriate time delays that can repopulate the upper, normally unoccupied CDW band over extended times completed the dynamic measurements of the CDW gap in tritellurides.

The experimental setup is a femto-to-picosecond three-pulse excitation scheme. The first pulse excited and created the coherent amplitude mode oscillations; the second mode repopulated the unoccupied states and the third probe pulse, the bands. A Lorentzian line fit of the peaks of the upper and lower bands shows oscillations of the same magnitude in both bands that are clearly anti-correlated. Both single-particle and collective modes can be tracked with femtosecond resolution through various phase transition, hence rapid topological defects can be followed in an ultrafast experiment that opens a new opportunity to explore the collective BEC interactions as well as the Kibble–Zurek mechanism.

28.5 Polaritons

28.5.1 Polariton States and Their Possible Use in PLHC

A polariton is a two-level state created by the superposition of a photon and another boson with a dipole such as an exciton [39, 40] (Fig. 28.6). Polaritons were discovered years ago, but they were difficult to study. Microcavities could overcome this barrier, but it took years to fabricate a microcavity with the dimensions lower than $\lambda/2$. The polariton interaction Hamiltonian is the dot product of the dipole energy (exciton) in the polarization electric field of a photon, $\mathbf{E}$:

$$H_{\text{int}} = -\int d^3\mathbf{r}\mathbf{P}\cdot\mathbf{E}. \tag{28.17}$$

This technology enabled a much larger polariton density that was further increased by the recognition that the large oscillator strength of organic materials leads to a much larger Rabi separation between the levels (1.2 eV). Both advances allow a macroscopic number of polaritons to occupy the same state so a BEC forms even though these quasiparticles are *inherently in non-equilibrium states.* Nonetheless, measured polariton properties include: coherence, off-diagonal long-range order (ODLRO), superfluidity, Bogoliubov excitations, quantized vortices, and solitons (topological 'lumps of energy') that could tunnel from excited B850 complexes or CDWs [15].

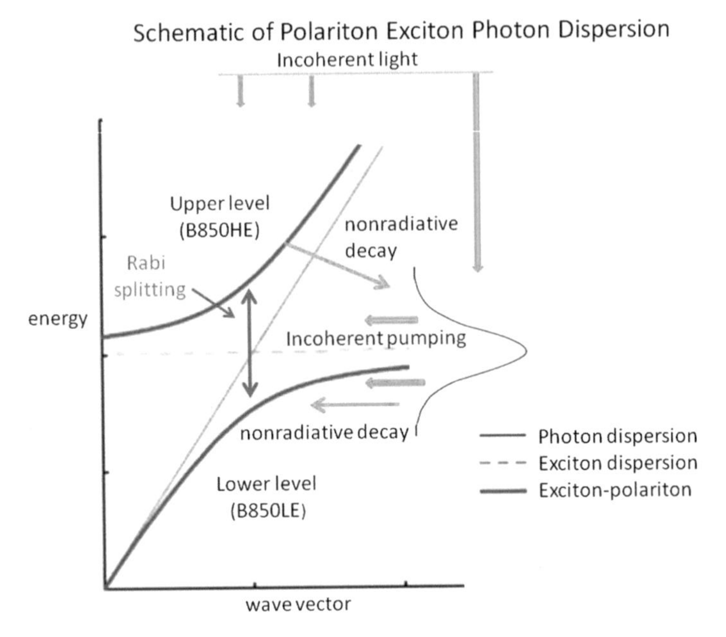

Fig. 28.6 An electric field pulls an exciton comprised of an electron and a hole in different directions affecting the dielectric constant of the medium. The resonance frequency of the exciton is not constant, resulting in variable branches of the exciton-polariton, the strength of which is measured by the Rabi splitting between the two branches. The exciton oscillator strength determines the Rabi splitting. The eigen-modes mix *only* in the region of the crossover

28.5.2 Why Polaritons in the PLHC?

An exciton-polariton should be present simply because it seems to be the only suitable coherent structure at room temperature to absorb incident photons completely (99%) and deliver fast, near perfect, energy transfer over significant distances. The B800-B850-B875 supercomplex (B^3SC) should be able to store energy using the Jaynes–Cummings effect. An examination of the dipole structure of the B^3SC shows attributes of a nanopillar, normally a synthetic device that captures over 99% of incident light [4–6] and coherently transfers the energy to Reaction Centers, either near or far using (coherent) solitons and CDWs in a cooperative energy transport network. Even assuming a nanopillar, the detailed physical description of a PHLC contains several anomalies, which we can rationalize with our model, such as:

1. Dostal et al. [41] using 2D spectroscopy observed a 'rapid loss of excitonic coherence between structural subunits' [41]. After the rapid transfer, the chemical potential becomes zero, hence transfer is in equilibrium.
2. The ability of uphill energy transfer from B875 to B850 [11, 42, 43] (see also Sect. 28.3.2) could result from a highly excited B875 matter wave peak becoming transiently overlapping a non-excited B850 complex.

Energy diagram with CDW's, their DOS modifications and "stacked"
polaritons illustrating the flexibility of the light-harvesting complex.
High light (HL) and Low Light (LL) intensity may choose different routes.

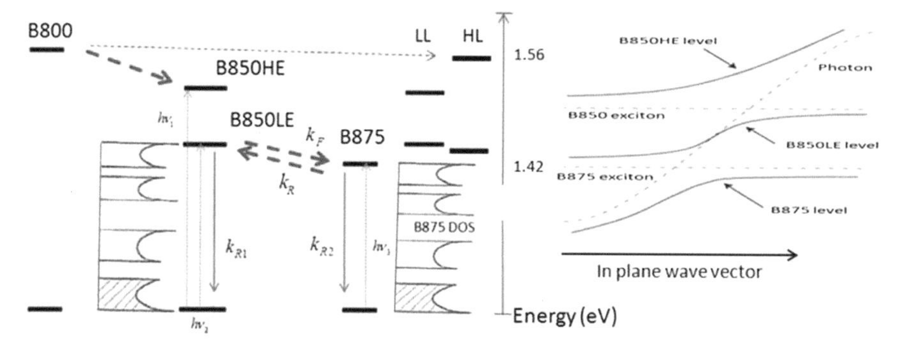

Fig. 28.7 Excited state energy levels and band gaps for B800 molecules (left), then eight B850's surrounding the B875/RC along with possible polariton structures for these states. Most bosons are restricted to the region of $\Delta p = 0$; polaritons can have considerable freedom in this regard, covering distances to 50 μm or more [11]. The breaks in the 1D CDWs are due to a band gap and the Peierls distortion gap and its symmetry partner. The LL and HL illustrate the non-linear splitting, 0.075 eV versus 0.11 eV, respectively of the B850 band (left center top). In LL there would probably be only two polariton levels and B800 would contribute incoherent energy; in HL B800 could transfer coherent energy to B850HL. The splitting of B850HE versus LE is necessary to describe the experimental spectrum properly

3. The non-linear splitting of the B850 band by high and low light intensities (HL and LL) into high (B850HE) and low energy (B850LE) bands to describe the spectra properly [11, 43]. This non-linear splitting could be a Rabi splitting of a polariton (Fig. 28.7).
4. A polariton has the ability to extend its length to the order of the cavity mode [44], and the ability to generate solitons [45] to move energy. Josephson Junctions are also present, so there are several different approaches to obtain coherent energy connections [46].
5. When related bacteria expand the B850 ring, it is two units of B850 that are added, presumably to maintain the Peierls distortion; a single B800 is also added between two could result B850 units.
6. The B850 UL/LL could support energy storage via Jaynes–Cummings effect [47]. This may be essential due to significantly different energy transfer rates.
7. Coles et al. [48] have formed a polariton from strong coupling between a confined optical cavity mode and a photosynthetic bacteria chlorosomes, so polariton formation is possible.
8. Entanglement seems possible in the ground and excited states (Fig. 28.8).

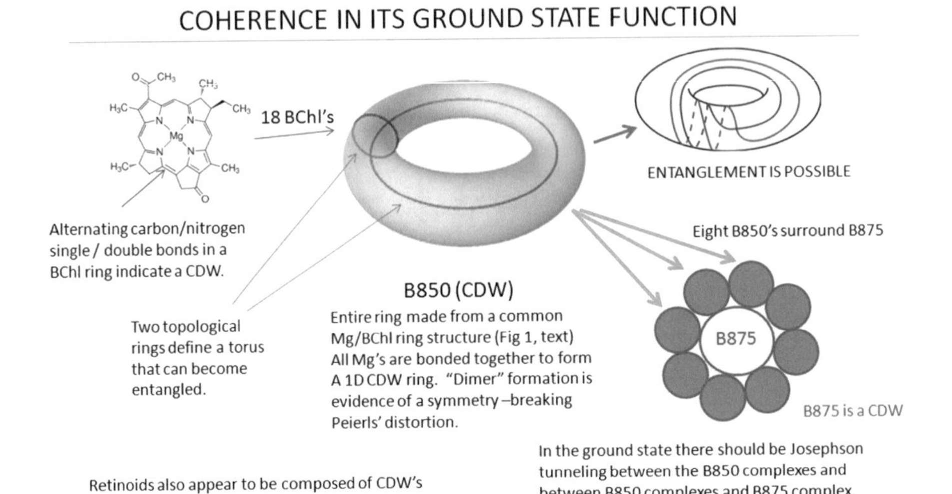

Fig. 28.8 Circular CDWs in a magnetic field have been shown to quantize the magnetic flux in units of $\Phi = nh/2e = n\Phi_0$, where n is the number of flux quanta. The Meissner effect is a rotation of charge in a SC or CDW that opposes the external flux, permitting only the flux in the fundamental units of Φ_0 (or $h/2e$). In principle BChl could generate an n, a B850 complex n', and the eight B850 surrounding B875 n''

28.6 Possible Experiments

28.6.1 *Ground State*

The three-pulse system (D, destruction, P-p, Pump-probe) [7, 8] could focus on a single ground state CDW, two adjacent CDWs, two non-adjacent CDWs to monitor and/or suppress the susceptibility. A second probe might be useful. Is there a real possibility of eight vortex tubes being present? In a magnetic field will there be current rotation of the eight B850 CDWs or will they flow in opposite rotations so the overall effect of the eight B850 complex be 'neutral'. However, if two adjacent B850s have a current, can it rotate in opposite directions and possibly create a flux loop perpendicular to the plane of the rings?

28.6.2 *Excited States*

Many of the ground state experiments can be expanded and repeated in the polariton/CDW complex at low light and high light, these 'normal' light being a fourth energy source. There is considerable expertise in microcavity spectroscopy and nanocavity work is progressing, all of which is specifically designed to avoid or selectively include the continuum.

28.7 Summary and Conclusions

Have we answered the question in the chapter's title? If the array of rings surrounding the B875/RC are CDWs based on their 'dimer' structure, then the answer is yes. Each circular CDW requires a separate symmetry breaking (SB) [1]. Are these SBs correlated? Probably not when the structure was assembled, but almost certainly later. The time scale of PLHC events ranges from pico to microseconds, so a key question is on a high light day, how is the energy distributed so that the various sensitive entities avoid photo bleaching? Based on the discussion in Sect. 28.5.2, we suggest that the coherent entities (chemical potential $\mu = 0$) and using what we believe is the dominate connecting transfer mechanism, solitons, rapidly smooth the chemical potential over the complex [41]. Kim et al. [49] have recently observed a soliton transfer between CDWs following the Bardeen-Miller mechanism [49], and Sich et al. [50] report that giant optical non-linearity enables the formation of mesoscopic polariton solitons numbering in the hundreds [50].

Acknowledgements RHS is grateful to NHM for suggesting the KZ theory approach to analyzing the PLHC and for sharing his ideas for twenty-five years.

References

1. T.W.B. Kibble, J. Phys. A: Math. Gen. **9**(8), 1387 (1976). https://doi.org/10.1088/0305-4470/9/8/029
2. W.H. Zurek, Nature **317**(6037), 505 (1985). https://doi.org/10.1038/317505a0
3. R.S. Knox, *Theory of Excitons* (Academic Press, New York, 1963)
4. Z. Fan, R. Kapadia, P.W. Leu, X. Zhang, Y.L. Chueh, K. Takei, K. Yu, A. Jamshidi, A.A. Rathore, D.J. Ruebusch, M. Wu, A. Javey, Nano Lett. **10**(10), 3823 (2010). https://doi.org/10.1021/nl1010788
5. M. Pelton, G. Bryant, *Introduction to Metal-Nanoparticle Plasmonics* (Wiley, Hoboken, 2013)
6. K.J. Vahala, Nature **424**(6950), 839 (2003). https://doi.org/10.1038/nature01939
7. D. Mihailovic, T. Mertelj, V.V. Kabanov, S. Brazovskii, J. Phys.: Condens. Matt. **25**(40), 404206 (2013). https://doi.org/10.1088/0953-8984/25/40/404206
8. L. Rettig, R. Cortés, J.H. Chu, I.R. Fisher, F. Schmitt, R.G. Moore, Z.X. Shen, P.S. Kirchmann, M. Wolf, U. Bovensiepen, Nat. Comms. **7**, 10459 (2016). https://doi.org/10.1038/ncomms10459
9. G. McDermott, S.M. Prince, A.A. Freer, A.M. Hawthornthwaite-Lawless, M.Z. Papiz, R.J. Cogdell, N.W. Isaacs, Nature **374**(6522), 517 (1995). https://doi.org/10.1038/374517a0
10. J. Koepke, X. Hu, C. Muenke, K. Schulten, H. Michel, Structure **4**(5), 581 (1996). https://doi.org/10.1016/S0969-2126(96)00063-9

11. L. Lüer, V. Moulisová, S. Henry, D. Polli, T.H.P. Brotosudarmo, S. Hoseinkhani, D. Brida, G. Lanzani, G. Cerullo, R.J. Cogdell, Proc. Natl. Acad. Sci. **109**(5), 1473 (2012). https://doi.org/10.1073/pnas.1113080109
12. G. Luo, S. Ono, N.J. Beukes, D.T. Wang, S. Xie, R.E. Summons, Sci. Adv. **2**(5) (2016). https://doi.org/10.1126/sciadv.1600134
13. F. London, Phys. Rev. **54**, 947 (1938). https://doi.org/10.1103/PhysRev.54.947
14. W. Greiner, L. Neise, H. Stöcker, *Thermodynamics and Statistical Mechanics* (Springer, New York, 1995)
15. H. Deng, H. Haug, Y. Yamamoto, Rev. Mod. Phys. **82**, 1489 (2010). https://doi.org/10.1103/RevModPhys.82.1489
16. A. del Campo, T.W.B. Kibble, W.H. Zurek, J. Phys.: Condens. Matt. **25**(40), 404210 (2013). https://doi.org/10.1088/0953-8984/25/40/404210
17. A. Ferrera, Phys. Rev. D **59**, 123503 (1999). https://doi.org/10.1103/PhysRevD.59.123503
18. G. Baym, C.J. Pethick, Phys. Rev. Lett. **76**, 6 (1996). https://doi.org/10.1103/PhysRevLett.76.
19. F. Dalfovo, L. Pitaevskii, S. Stringari, Phys. Rev. A **54**, 4213 (1996). https://doi.org/10.1103/PhysRevA.54.4213
20. M. Saba, T.A. Pasquini, C. Sanner, Y. Shin, W. Ketterle, D.E. Pritchard, Science **307**(5717), 1945 (2005). https://doi.org/10.1126/science.1108801
21. S. Raghavan, A. Smerzi, S. Fantoni, S.R. Shenoy, Phys. Rev. A **59**, 620 (1999). https://doi.org/10.1103/PhysRevA.59.620
22. M. Tsubota, K. Kasamatsu, J. Phys. Soc. Jpn. **69**(7), 1942 (2000). https://doi.org/10.1143/JPSJ.69.1942
23. K. Kasamatsu, M. Tsubota, J. Low Temp. Phys. **126**(1), 315 (2002). https://doi.org/10.1023/A:1013741017477
24. P. Nozières, in *Bose-Einstein Condensation*, ed. by A. Griffin, D.W. Snoke, S. Stringari (Cambridge University Press, Cambridge, 1995), pp. 15–30, chap. 2
25. D.R. Scherer, C.N. Weiler, T.W. Neely, B.P. Anderson, Phys. Rev. Lett. **98**, 110402 (2007). https://doi.org/10.1103/PhysRevLett.98.110402
26. C. Ryu, M.F. Andersen, P. Cladé, V. Natarajan, K. Helmerson, W.D. Phillips, Phys. Rev. Lett. **99**, 260401 (2007). https://doi.org/10.1103/PhysRevLett.99.260401
27. J. Dziarmaga, J. Meisner, W.H. Zurek, Phys. Rev. Lett. **101**, 115701 (2008). https://doi.org/10.1103/PhysRevLett.101.115701
28. R. Monaco, J. Mygind, R.J. Rivers, V.P. Koshelets, Phys. Rev. B **80**, 180501 (2009). https://doi.org/10.1103/PhysRevB.80.180501
29. J.H. Miller, A.I. Wijesinghe, Z. Tang, A.M. Guloy, Phys. Rev. Lett. **108**, 036404 (2012). https://doi.org/10.1103/PhysRevLett.108.036404
30. K. Maki, Phys. Rev. Lett. **39**, 46 (1977). https://doi.org/10.1103/PhysRevLett.39.46
31. J. Bardeen, Phys. Rev. B **39**, 3528 (1989). https://doi.org/10.1103/PhysRevB.39.3528
32. J. Bardeen, Phys. Scr. **1989**(T27), 136 (1989). https://doi.org/10.1088/0031-8949/1989/T27/024
33. B.D. Josephson, Phys. Lett. **1**(7), 251 (1962). https://doi.org/10.1016/0031-9163(62)91369-0
34. B.D. Josephson, Rev. Mod. Phys. **36**, 216 (1964). https://doi.org/10.1103/RevModPhys.36.216
35. B.D. Josephson, Adv. Phys. **14**(56), 419 (1965). https://doi.org/10.1080/00018736500101091
36. S. Coleman, Phys. Rev. D **15**, 2929 (1977). https://doi.org/10.1103/PhysRevD.15.2929
37. N. Ru, C.L. Condron, G.Y. Margulis, K.Y. Shin, J. Laverock, S.B. Dugdale, M.F. Toney, I.R. Fisher, Phys. Rev. B **77**, 035114 (2008). https://doi.org/10.1103/PhysRevB.77.035114
38. F. Schmitt, P.S. Kirchmann, U. Bovensiepen, R.G. Moore, J.H. Chu, D.H. Lu, L. Rettig, M. Wolf, I.R. Fisher, Z.X. Shen, New J. Phys. **13**(6), 063022 (2011). https://doi.org/10.1088/1367-2630/13/6/063022
39. D.W. Snoke, *Solid State Physics Essential Concepts* (Addison-Wesley, New York, 2009)
40. V.M. Agranovich, Y.N. Gartstein, M. Litinskaya, Chem. Rev. **111**(9), 5179 (2011). https://doi.org/10.1021/cr100156x

41. J. Dostál, T. Mančal, R. Augulis, F. Vácha, J. Pšenčík, D. Zigmantas, J. Am. Chem. Soc. **134**(28), 11611 (2012). https://doi.org/10.1021/ja3025627
42. H.W. Trissl, C.J. Law, R.J. Cogdell, Biochim. Biophys. Acta **1412**(2), 149 (1999). https://doi.org/10.1016/S0005-2728(99)00056-0
43. V. Moulisová, L. Lüer, S. Hoseinkhani, T.H. Brotosudarmo, A.M. Collins, G. Lanzani, R.E. Blankenship, R.J. Cogdell, Biophys. J. **97**(11), 3019 (2009). https://doi.org/10.1016/j.bpj.2009.09.023
44. G. Christmann, G. Tosi, N.G. Berloff, P. Tsotsis, P.S. Eldridge, Z. Hatzopoulos, P.G. Savvidis, J.J. Baumberg, Phys. Rev. B **85**, 235303 (2012). https://doi.org/10.1103/PhysRevB.85.235303
45. D. Tanese, H. Flayac, D. Solnyshkov, A. Amo, A. Lemaître, E. Galopin, R. Braive, P. Senellart, I. Sagnes, G. Malpuech, J. Bloch, **4**, 1749 (2013). https://doi.org/10.1038/ncomms2760
46. M. Abbarchi, A. Amo, V.G. Sala, D.D. Solnyshkov, H. Flayac, L. Ferrier, I. Sagnes, E. Galopin, A. Lemaître, G. Malpuech, J. Bloch, Nat. Phys. **9**(5), 275 (2013). https://doi.org/10.1038/nphys2609
47. C.C. Gerry, P.L. Knight, *Introductory Quantum Optics* (Cambridge University Press, Cambridge, 2005)
48. D.M. Coles, Y. Yang, Y. Wang, R.T. Grant, R.A. Taylor, S.K. Saikin, A. Aspuru-Guzik, D.G. Lidzey, J.K.H. Tang, J.M. Smith, **5**, 5561 (2014). https://doi.org/10.1038/ncomms6561
49. T.H. Kim, H.W. Yeom, Phys. Rev. Lett. **109**, 246802 (2012). https://doi.org/10.1103/PhysRevLett.109.246802
50. M. Sich, D.V. Skryabin, D.N. Krizhanovskii, C. R. Acad. Sci. **17**(8), 908 (2016). https://doi.org/10.1016/j.crhy.2016.05.002

Chapter 29
Spacetime as a Quantum Many-Body System

D. Oriti

Abstract Quantum gravity has become a fertile interface between gravitational physics and quantum many-body physics, with its double goal of identifying the microscopic constituents of the universe and their fundamental dynamics, and of understanding their collective properties and how spacetime and geometry themselves emerge from them at macroscopic scales. In this brief contribution, we outline the problem of quantum gravity from this emergent spacetime perspective, and discuss some examples in which ideas and methods from quantum many-body systems have found a central role in quantum gravity research.

29.1 The Problem of Quantum Gravity and Its Evolution

The construction of a complete theory of quantum gravity remains an open issue, despite decades of activities and important progress [1–3]. These have led to a number of approaches [4] and many fascinating suggestions for a radically new basis for understanding the world, which remain, however, still incomplete and not directly tested by observations. To communicate this variety and fascination is the first goal of this contribution. Over the last two decades, the very perspective on the problem has changed, as a result of new insights obtained in different quantum gravity approaches and some surprises found along the way. The new perspective brings the conceptual framework of quantum gravity closer to that of condensed matter theory, with a consequent import of mathematical tools and continuous exchanges between these two areas. To explain this change in perspective is the second goal of this contribution. This evolution is apparent in several corners of current research in quantum gravity, and very concrete: a number of quantum many-body systems have been studied in order to obtain insights for quantum gravity research, quite a few others have been shown to be in very direct correspondence with quantum gravitational phenomena, and quantum gravity formalisms themselves have developed in such a way that the

D. Oriti (✉)
Max Planck Institute for Gravitational Physics (Albert Einstein Institute),
Am Mühlenberg 1, 14476 Potsdam-Golm, EU, Germany
e-mail: daniele.oriti@aei.mpg.de

G. G. N. Angilella and C. Amovilli (eds.), *Many-body Approaches at Different Scales*, https://doi.org/10.1007/978-3-319-72374-7_29

have a closer resemblance to quantum many-body systems. This contribution will therefore present also a quick survey of this multi-facets relation between quantum gravity and quantum many-body systems. The vast amount of literature that has accumulated in recent years on this interface, let alone quantum gravity per se, exceeds greatly our own competence, and certainly the space of this contribution. This will therefore be limited to a summary of interesting developments, without entering into any detail, and it is only meant to be an invitation to dwelling more into this subject and a pointer to the relevant literature.

29.1.1 Quantum Gravity as Quantized GR

The main attitude towards the problem of quantum gravity has been for many years [1–3] (and still is, in many corners) the straightforward one applied to all other fundamental interactions: start from the best classical description we have of the interaction and quantize it, i.e. apply to the corresponding classical theory well-tested techniques for turning it into a quantum theory. The starting point is of course General Relativity (GR), the modern description of gravitational phenomena. The variety of quantization methods (canonical quantization schemes, geometric quantization, path integral methods, etc.) already led to a variety of quantum gravity formalisms. General Relativity is however a very complicated theory at the mathematical level (e.g. it is highly non-linear and it is invariant under spacetime diffeomorphisms, a notoriously involved symmetry group) and this already suggests that progress should be expected to be difficult. Severe challenges come also from the conceptual revolution that GR forces into our understanding of the world. The gravitational field is identified with spacetime geometry, on which all other interactions rely, and this in turn determines the causal structure of all physical events. Spacetime itself becomes therefore a dynamical, physical system, and we have no experience in dealing with physical processes in presence of a dynamical spacetime, at the quantum level. Also diffeomorphism invariance expresses this need to define localization of physical objects in time and space in relational terms, i.e. in terms of one another, without relying on any fixed notion of time and space. This immediately calls into question the notions of time evolution, conservation of probabilities and unitarity of usual quantum field theory, together with the causality restrictions that enter its very foundation.

Some quantum gravity approaches take onboard these lessons from GR and try to apply quantization techniques to the full gravitational field, i.e. to spacetime geometry itself: canonical quantum gravity and its modern incarnation Loop Quantum Gravity [5–8]; path integral formulations of quantum GR and their modern evolution mostly based on lattice structures (e.g. causal dynamical triangulations [9] and spin foam models [10], or group field theories [11–14]) or other forms of discrete substrata like Causal Set Theory [15]; other strategies to quantize spacetime structures, like non-commutative geometry [16, 17].

Researchers coming from a particle physics tradition have instead often relied on different way of tackling the problem of quantizing the gravitational field. The basic intuition comes from splitting the spacetime geometry into a (usually flat) fixed background and a dynamical part, which is then quantized by standard quantum field theory methods. The result is a dynamical quantum theory of 'gravitons', quanta of the gravitational field defined with respect to the fixed background spacetime. Prima facie, this procedure runs contrary to the main lessons of GR, but it allows to take on board most techniques of the usual quantum field theory formulation of other fundamental interactions. However, one could expect that the quantum field theory of gravitons, if taken beyond the perturbative regime to account for all the non-linearities and non-perturbative features of GR, and its full symmetry content, will in the end lead as well to a theory of the whole quantum spacetime giving answers to the same basic issues that GR-based approaches face. If this is achieved, the fixed background geometry that enters the initial formulation would be recognized as an auxiliary tool with no physical content, as in standard background-field methods in QFT. The fact that going beyond the perturbative regime is an absolute necessity is further confirmed by the failure of the perturbative quantum field theory of gravitons to be consistent, having been shown to be non-renormalizable.

The non-perturbative route to a quantum field theory of the gravitational field is for example taken by the asymptotic safety approach [18]. String theory [19, 20] was born as an enrichment of the perturbative approach to quantum gravity in terms of gravitons, by adding to them more degrees of freedom (an infinity of them), naturally resulting from the step from point-like objects to extended, string-like entities propagating on the same fixed background; it has also revealed a vast number of non-perturbative phenomena which involve extended spacetime configurations and quantum aspects of the gravitational field, and a new understanding of quantum field theories themselves (but not yet a clear, if tentative, picture of the fundamental nature of spacetime itself).

All the modern approaches to quantum gravity show that the straightforward perspective on quantum gravity as 'quantized GR', in its many versions, can lead to important results and insights.

At the same time, a slow change in perspective has taken place over the last (15 or so) years, with the result that the very problem of quantum gravity is now understood from a different angle, and many quantum gravity researchers are now inclined to take a more radical standpoint.

This has also led to a closer relation between this research area and condensed matter theory (as well as quantum information).

29.1.2 The Emergent Spacetime Scenario

The first hint that something more radical is in store came already in the seventies (of the last century) with the development of black hole thermodynamics [21]. The discovery, within the semi-classical, very conservative framework of quantum field theory in curved backgrounds, that a finite entropy can be associated to a black hole and that the same can emit thermal radiation, leading to a consistent (if rather mysterious) set of thermodynamical laws, had tremendous implications. Indeed, these laws simply cannot be made sense of within classical physics. The finite entropy, in particular, strongly hints at some underlying discrete microstructure of spacetime, since black holes are in the end particular configurations of it. The task of quantum gravity approaches became (also) to unravel such discrete microstructure and compute this entropy from first principles. In the ensuing decades, semi-classical black hole physics have turned into an unrivalled source of radical suggestions [22]. the end result of black hole evaporation remains a mystery (a Planck-sized exotic object? nothing?) just as a mystery remains the fate of the information stored in the evaporating black holes (forever lost? somehow encoded in the outgoing radiation? stored in the exotic object that is left over?), but any proposed solution to these mysteries calls for relinquishing at least one of the sacred ingredients of modern theoretical physics, be it locality, the equivalence principle, unitarity, causality, quantum monogamy or no-cloning, or the real dimensionality of spacetime. And this is *in addition* to the continuum spacetime structure.

Related to this, a number of developments, starting from Jacobson's derivation of Einstein's equations as an equation of state, have strongly supported the idea that GR itself, and Einstein's equations of motion, can be seen as the macroscopic thermodynamic (or hydrodynamic, depending on the specific framework) description of an underlying set of non-geometric (thus not directly 'gravitational') degrees of freedom [23, 24].

More recently, the main drivers of such change in perspective have been the same approaches to quantum gravity mentioned above. Indeed, the more they were unraveling about the fundamental nature of gravity and spacetime, the more the new insights pushed for a different point of view and for the use of new tools. What happened is that all these different approaches have either identified explicitly or at least strongly suggested a set of fundamental degrees of freedom underlying spacetime which are not simply quantized continuum fields and are nor directly geometric, nor characterized in continuum spatiotemporal manner. Let us mention a few examples.

Loop quantum gravity has proceeded far enough as to identify a candidate Hilbert space of quantum states for the gravitational field; such states admit a characterization in terms of 'generalized continuum fields', under appropriate conditions, but still they are labeled by purely combinatorial and algebraic structures: graphs labeled by data from the representation theory of $SU(2)$ (or the Lorentz group), defining so-called 'spin networks'. Moreover, the continuum interpretation is possible only in a limit; each state can at best, i.e. under additional semi-classical and regularity conditions, be put in correspondence with a piece-wise flat geometry [25], as in lattice gravity, rather singular from the continuum GR perspective.

Group field theories [26] and spin foam models [10] have the same type of quantum states.

In fact, the same type of singular configurations are used as basic entities (even if mostly interpreted just as regularization tools) in lattice approaches to quantum gravity, like (causal) dynamical triangulations [9]. Thus, a generic configuration of fundamental degrees of freedom in these formalisms will not correspond to anything resembling a nice continuum spacetime or geometry, even if a classical limit is taken.

Causal set theory is another formalism in which purely discrete structures (partially ordered sets), even though strongly motivated from continuum geometric considerations, replace entirely the continuum and geometric structures on which GR is based, and in which their generic configurations do not admit any immediate interpretation in terms of spatiotemporal notions like distances or even dimensionality, which have to be instead reconstructed with laborious procedures [15].

Of course, the actual distance between the quantum configurations at the basis of a given approach and the usual notions of spacetime and geometry can vary, depending on the specific version of the formalism adopted and additional conditions that may be imposed on them.

From the other side of the spectrum of quantum gravity approaches, also string theory has produced a large number of challenges to the usual notions of spacetime and geometry, and hints that the fundamental degrees of freedom of the theory (whatever they are) will not be anywhere close to straightforward quantized continuum fields [27]. T-dualities indicate that string theories on a spacetime in which one spatial direction is compactified to a small radius can actually be equivalent to string theories in which the same compact dimension has a very large radius; mirror symmetry shows that some formulations of string theory on a given topology can be physically equivalent to formulations on an entirely different topology; the very notion of spacetime dimension is dynamical, since it can be 'exchanged' with a different field content and internal symmetry, while the background 'geometry' on which strings propagate and interact can be very exotic, going beyond the usual notion of Riemannian geometry of GR, but also beyond the already exotic non-commutative but associative geometries used in other approaches to quantum gravity [28].

Therefore, we have by now many reasons to believe that spacetime, at the fundamental level, 'dissolves' into a new set of quantum entities, call them 'atoms of space', of no direct gravitational or spatiotemporal or geometric interpretation, and from which it has to re-emerge, together with the usual notion of geometry and gravity, in some limit and under suitable approximations [29, 30].

The detailed nature of the fundamental entities, to the extent in which it is identified, varies greatly across different quantum gravity formalisms, just as their 'weirdness' as compared to the usual notions of continuum geometry, spacetime, fields, and so on. At the same time, different formalisms enjoy a number of mutual relationships, both conceptual and mathematical, and their different suggestions need not be seen as incompatible with one another, nor, necessarily, with the traditional view of quantum gravity as 'quantized GR'. However, the fact remains that some more conceptual flexibility is called for, and a new set of tools is required, to tackle the new issue of the emergence of spacetime from this new type of entities. Let us dwell more into the novel aspects of this situation.

The change of fundamental degrees of freedom implies that a classical approximation is not enough to recover GR, which is not only a classical theory but also based on continuum spacetime and geometry, which have to emerge first. Given that such emergent quantities have a continuum character and sustain a field theory framework with its infinite number of degrees of freedom it is reasonable to expect that such continuum approximation (independent, we repeat, from the classical approximation) requires the control over a very large number of the fundamental entities.

The key tool to achieve such control and to define the continuum limit is the renormalization group, which has to help us to map the macroscopic phase diagram of any given quantum gravity model. In particular, we should seek to identify the phase where the continuum *geometric* characterization of the emergent physics is possible, and the phase transition separating it from non-geometric phases, which cannot describe our world. In any geometric phase so identified, we also have to be able to characterize in detail the collective dynamics of the atoms of space and any emergent property coming out of it, translating it in the language of GR and effective field theories, as needed to extract phenomenological predictions.

This list of tasks is common to all approaches to quantum gravity. It is also a more abstract and, admittedly, exotic analogue of the list of tasks that any theory of 'real' atomic, many-body systems has to undertake to go from the identification of the fundamental constituents to the characterization of their collective, macroscopic properties. In this sense, quantum gravity is now forced to take onboard the many lessons and tools of condensed matter theory, solid-state physics and, more generally, many-body quantum theory in order to make further progress.

This change in perspective comes with a new set of questions and fascinating possibilities, actively explored. Among them: is geometry emergent from entanglement (of the fundamental 'atoms of space')? what is the geometric, gravitational counterpart of other quantum properties of the underlying atoms of space? should we drop basic features of effective field theories, like locality or unitarity? what about the very notion of separation of scales, which is based on the existence of a given geometry and it is what suggests that quantum gravity effects should only become relevant at microscopic, Planck scales? if we drop this, and more generally if the very notion of spacetime is emergent, what prevents phenomena that we usually consider macroscopic, large-scale aspects of the world (e.g. the cosmological constant, dark energy and the current accelerated expansion of the universe) to be really of quantum gravity origin? The list could go on for quite a while. Rather than listing possibilities, which we would have no way to discuss properly in this contribution, we move on instead to give a few examples of many-body systems which offer insights for quantum gravity research, that are directly connected to (quantum) gravitational phenomena, or that *are* themselves candidate quantum gravity systems.

29.2 The Interface Between Quantum Gravity and Many-Body Systems: Some Examples

29.2.1 Analogue Gravity in Condensed Matter Systems

The first examples of many-body systems of interest for quantum gravity are actually standard condensed matter systems, of great interest as such, and are instances of so-called 'analogue gravity models'. A very comprehensive account can be found in Ref. [31].

The basic fact about all of them is surprising as it is interesting: propagating excitations over given macroscopic (more or less special) background configurations of the system are described by (quantum) field theories on an emergent curved geometry, whose metric is a function of macroscopic collective variables. The interesting fact is both that these excitations couple to an effective curved metric, rather than the flat one associated to the laboratory and the underlying atomic system, and that this metric is general enough to include configurations mimicking black hole horizons or cosmological spacetimes. Beside the theoretical interest as models for quantum effects on curved spacetimes (but with modified dispersion relations and Lorentz breaking at high energies) and the insights these systems can offer on quantum gravity and the nature of spacetime, they have also raised the hope of actual observation of such effects in a controlled laboratory environment.

In fact, the simplest example of analogue gravity is given by ordinary fluids: the velocity potential for an acoustic perturbation of any barotropic and inviscid fluid with irrotational flow is governed by an equation that can be recast in the form of a D'Alambertian equation for a minimally coupled, massless (real) scalar field propagating on a curved geometry with a metric that is an algebraic function of the fluid density and the flow velocity, and the local speed of sound. The emergent geometry is general enough to reproduce features of black hole horizons and their surface gravity, as well as many of their causal properties. Also, depending on the details of the fluid flow, it may reproduce various geometries of cosmological interest, and it may be used, for example, to simulate cosmological particle production in an expanding universe.

This is the simplest example of a long list of condensed matter and solid-state systems where this type of behaviour occurs. The most studied ones are Bose condensates and fermionic superfluids, which show an even richer emergent gravitational-like physics, thanks to their quantum coherence and low speed of sound, as well as gravity analogues in quantum optics. More recently, analogue gravity models have been developed using graphene, also showing a rich set of emergent phenomena.

The range of topics, at the interface between gravitational physics and quantum field theory, that analogue gravity systems have allowed to investigate from a new angle is impossible to even summarize here (to give just one more example, they have suggested new ways of tackling the cosmological constant problem [32, 33]), so we refer again to the literature [31]. These models have also been quite influential in quantum gravity research even beyond the actual results obtained, since they have provides examples of the emergence of gravitational phenomena and effective spacetimes with non-trivial geometry from systems where gravity played no role and geometry was trivial, and thus realized to some extent the idea of geometry from non-geometry. Therefore, they have provided further support to the whole idea that spacetime and geometry could be emergent notions even at the fundamental level of quantum gravity, beyond simple analogy [34, 35].

Having said this, it is also to be noted that these systems also have strong limitations from the point of view of fundamental quantum gravity: the emergent matter fields propagating on the effective curved geometry do not have the right type of backreaction on it, i.e. they do not 'gravitate' as prescribed by GR; moreover, and perhaps most importantly, it has proven very hard to reproduce in such many-body (quantum) systems the gravitational *dynamics* for the effective metric, i.e. some possibly modified version of the GR equations, characterized by a general covariance that it seems very hard to mimic, even approximately. Many-body quantum physics systems have started playing a prominent role, however, also in actual quantum gravity formalisms and true gravitational theories, to which we now turn.

29.2.2 AdS/CFT Correspondence and Geometry from Entanglement

A most extensive application of many-body systems in quantum gravity has stemmed from the establishment of the Anti-de Sitter/Conformal Field Theory (AdS/CFT) correspondence [36, 37] as one of the most active and fertile research directions in the theory of fundamental interactions, starting from string theory but with an influence extending well beyond it (see Ref. [38] and Chap. 9 in [39] for the relevance of the AdS/CFT in condensed matter theory and statistical mechanics).

In brief, the AdS/CFT correspondence is a map between a conformal field theory (CFT) living on a flat d-dimensional Minkowski spacetime and a gravitational theory living on a $(d+1)$-dimensional curved spacetime with a metric approximating the Anti-deSitter (AdS) one close to its flat boundary, where it reduces to the Minkowski metric on which the CFT depends. The map implies that the partition function of the (quantum) gravitational theory, with the mentioned boundary conditions, its observables and Hilbert space of states have a precise counterpart in the partition function and observables of the boundary CFT: whatever one computes within one theory, one is implicitly performing some correspondent computation within the other. The correspondence has been first proposed to hold between one specific CFT, which is also a specific gauge theory, i.e. $SU(N)$ Yang–Mills theory with $N=4$ supersymmetry in 4 spacetime dimensions, and a rather exotic gravitational theory, i.e. IIB superstring theory living on a 10-dimensional spacetime in which 5 dimensions are compactified to a 5-sphere (and play basically the role of 'internal' dimensions) and the others have AdS boundary conditions. However, it has since been extended to quite a few other CFTs (with varying degree of supersymmetry and with different gauge symmetries) and to other gravitational theories (including other formulations of string theory, but also many other results concerning generic semi-classical gravity and quantum field theories in curved spacetimes), and to different spacetime dimensions, including lower dimensional contexts.

One interesting example is the map of the observable properties of a gas of bosons, fermions and scalars at thermal equilibrium in the CFT to the (semi-classical) observable properties of a black hole living inside an AdS spacetime, with the equilibrium temperature of the gas being the Hawking radiation temperature of the black hole. Another recent and very simple example (possibly the simplest example of AdS/CFT correspondence) is the SYK model (which has been generalized in various ways), a statistical system of Majorana fermions in $0 + 1$ dimensions with random interactions, which is the only known system that is simultaneously solvable at strong coupling, maximally chaotic and with an emergent conformal symmetry [40]; the AdS/CFT correspondence suggests that this model relates to a black hole in a 2-dimensional AdS spacetime; moreover, its relation to quantum gravity is strengthened by the fact that its quantum interaction processes can be mapped to the interaction processes of random tensor models [41–44], themselves models of quantum gravity closely related to group field theories.

The conjecture (still unproven) is that the map defines an exact AdS/CFT *duality*, i.e. the two theories are strictly equivalent despite being so different and living in different spacetimes (with different dimensions!). Some researchers, on the other hand, think of it more as an *approximate* map, involving coarse-graining of degrees of freedom or holding only in appropriate limits of the two theories (most of the evidence supporting it, we remark, concerns only semi-classical gravity), or an exact one but holding only for a subset of observables and configurations.

The hope, supported by some circumstantial evidence, is that this correspondence is only the most evident example of a much more general one between gravity and gauge theory tout court (thus holding for any gauge theory, with any choice of boundary conditions, in any dimension, etc.).

The details of what is known and what is not in this context do not concern us too much. The general idea does. The AdS/CFT correspondence gives us examples of theories in which properties of a standard many-body quantum system, the CFT in a flat spacetime, which a priori does not know about gravity (classical or quantum) and curved geometries, somehow produce full-blown gravitational effects and can be translated in terms of curved *dynamical* geometries. This is still far from answering the most fundamental questions that we may hope a theory of quantum gravity to address. These would concern, in fact, the origin of the very flat spacetime on which the boundary CFT is defined, its emergence from more fundamental non-spatiotemporal entities. Still, the interest for understanding the nature of gravity and geometry remains clear. Of course, the interest is even greater if one believes in an exact duality that extends beyond the AdS boundary conditions, since it would imply that quantum gravity is actually *defined* by some specific quantum many-body system. But even leaving aside this belief, it remains the fact that non-trivial geometric aspects of the world can emerge from quantum mechanical properties of a many-body system living in a flat spacetime. Inspired by and taking advantage of the concrete testbed of the AdS/CFT correspondence, quantum gravity researchers have then explored a variety of possibilities.

One of the most fascinating is that geometry can actually be entirely characterized in terms of the quantum information-theoretic properties of a many-body system, in particular entanglement [45, 46]. To start with, this has prompted the application of powerful tools from quantum information theory and quantum condensed matter systems to gravitational physics, the most notable example being tensor network techniques [47]. The idea of 'geometry from entanglement' has already found support in a number of results. For example, the entanglement entropy between a region A on the flat boundary of an AdS space and the rest of such boundary, computed within a simple CFT, is proportional to the area of the minimal surface inside the bulk AdS space with the same boundary as A; also, the mutual information between two regions of space on the same flat boundary can be shown to be inversely proportional to the geodesic distance between the two regions as measured in the bulk AdS; moreover, the very connectivity between two regions of spacetime, thus the topology of the latter, has been conjectured to be related to the entanglement between the quantum degrees of freedom associated to the two regions [48]. This type of results suggest the possibility that entanglement measures, mutual information and the like can be used to *define* geometric and topological quantities and *reconstruct* spacetime out of the quantum properties of a many-body system that, a priori, is not about spacetime and geometry.

29.2.3 Quantum Gravity Many-Body Systems

The above suggestion may sound radical and rather speculative. However, from the point of view of background independent quantum gravity approaches it is basically a necessity [49–52]. The reason is the point we emphasized from the beginning: modern quantum gravity approaches are based on fundamental entities that are not gravitational or geometric per se, at least not in the standard GR sense of corresponding to continuum metric and matter fields, but are of more abstract, discrete quantum nature. Spacetime with its geometric properties has to emerge from their collective quantum dynamics. For this simple reason, we have anticipated, modern quantum gravity approaches are importing ideas and tools from condensed matter physics and many-body quantum systems. In fact, these formalisms, in their different ways, all propose a description of the microstructure of spacetime as a quantum many-body system. Let us see three examples.

As we mentioned, Loop Quantum Gravity [5–8] (and its covariant version, spin foam models [10]) replace continuum manifolds and fields with purely combinatorial and algebraic data, with quantum states of space associated to (linear combinations of) spin networks: graphs labeled by irreducible representations of the rotation (or Lorentz) group. In turn, the graphs underlying such quantum states are dual to lattices and the algebraic data define for them a quantum geometry that reduces, in a classical limit (and under additional conditions), to a piecewise-flat discrete geometry (i.e. one in which each fundamental cell of the lattice is flat in the interior, with non-trivial curvature associated to their gluings and fully captured by the specification of lengths of the links of the lattice). Looking at such quantum states in more detail, one realizes that they define rather peculiar lattice spin systems, generalized to arbitrary $SU(2)$ representations assigned to the lattice links and without any fixed lattice spacing (since the geometry of the lattice is to be determined by the dynamical algebraic data assigned to it. The quantum dynamics of the theory generically modified the combinatorial structure of the lattice itself and, possibly, its topology. Still, the similarity in mathematical structures with lattice many-body systems is striking, in particular they are very close to string-net condensate models, to spin chains and Ising systems, and of course to lattice gauge theories (which basically share the same type of state space). This implies on the one hand that techniques from quantum gravity have been used to study such statistical systems [53], but also that tools developed for them have been imported in quantum gravity, in particular for the study of coarse graining and continuum limits, i.e. to investigate the macroscopic phases of such quantum gravity models and the emergence of geometry [54, 55]. Tensor network renormalization methods, in particular, seem very promising [56–59].

Another example is provided by random tensor models [41–43], a generalization to higher dimensions of matrix models for 2D quantum gravity and worldsheet string theory. The basic dynamical variables of these models are abstract rank-d tensors over N-dimensional index spaces to which one assigns a non-Gaussian probability distribution depending on a number of coupling constants (they are themselves statistical systems). The connection to quantum gravity arises because their perturbative expansion generates a sum over lattices of topological dimension equal to d, with purely combinatorial amplitudes given as a path integral for gravity discretized on the same lattices interpreted as equilateral. The latter sum actually defines lattice quantum gravity formalisms like (euclidean) dynamical triangulations [9]. Thus random tensor models attempt a purely combinatorial definition of the dynamical microstructure of spacetime, in terms of a statistical system of abstract tensors, in turn equivalent to a theory of random lattices. They have been developed to a great extent in rigorous mathematical terms as statistical systems, with many important results, like a well-defined large-N limit, double-scaling limits, some control over their critical behaviour, etc. They have been quite influential also in giving a solid mathematical basis for other quantum gravity formalisms, like group field theories, but also found some applications in the study of other statistical systems, like the Ising model on a random lattice or dimer models. More recently, as mentioned, they became central in yet another quantum gravity context, i.e. the AdS/CFT correspondence, by capturing the dynamics of the SYK model [44].

Finally, the already cited group field theories [11–14] generalize random tensor models by replacing their finite index set with a Lie group, but maintaining the combinatorial structures defining them. By doing so, the models become bona fide quantum field theories, although the retain their combinatorial peculiarities, like the fact that the Feynman processes of the models are dual to d-dimensional lattices rather than graphs. The other crucial point is that, when the Lie group used is the rotation or the Lorentz group, their Hilbert space of states is a Fock space whose generic states are (linear superpositions of) spin networks, the same quantum states of Loop Quantum Gravity [26], while their Feynman amplitudes are spin foam models (defining the covariant dynamics of spin networks), which can then be shown to be equivalent to discrete gravity path integrals (generalizing the ones of tensor models and dynamical triangulations to the case of dynamical edge lengths). Therefore, they naturally share the connection to quantum many-body and statistical systems that we have highlighted for these related formalisms. The new aspect, however, is that now quantum spacetime is tentatively described in a form that is *literally* a quantum many-body system, although with a more abstract physical interpretation (and of course not defined on any physical spacetime, with the base group manifold playing a different role) with a Fock space of states, a rather conventional field-theoretic definition of the quantum dynamics, etc. This allowed to use many of the conventional tools, suitably adapted, for tackling outstanding quantum gravity problems. Most notably, GFT renormalization [60–64] is by now a thriving area of research, aiming establishing the renormalizability of GFT models for quantum gravity, and their continuum phase diagram, looking for a geometric, spatiotemporal phase. In parallel, the analogy with standard condensed matter systems have suggested a new approach to the extraction of an emergent gravitational dynamics, and of cosmology in particular as the simplest example of it. A very fertile direction of research [65–68] has started from the hypothesis that the universe can be understood as a quantum condensate of the GFT 'atoms of space' (the GFT quanta, forming spin networks and constituting its Hilbert space) and its effective condensate hydrodynamics (the GFT analogue of the Gross–Pitaevskii equation of real Bose condensates) can be recast in terms of cosmological dynamics of macroscopic observables like the volume of the universe, the density of emergent scalar matter, etc. While we are clearly just at the beginning in the development of this research direction, and in exploring the radical suggestions and potential new insights it provides, a number of important results have already been obtained: the emergent cosmological dynamics has the correct classical limit given by the Friedmann equation, but replaces the Big Bang singularity with a bouncing scenario, in which the universe passes from a previous contracting phase to the current expanding one [69]; the same dynamics has then been shown to have other interesting features, like the possibility of an accelerated inflation-like phase of expansion of purely quantum gravity origin (no need for additional degrees of freedom) [70], and it has recently been extended to a framework for studying cosmological perturbations [71], which appear to have a naturally close-to-scale invariant power spectrum and small amplitudes, as required to match cosmic microwave background (CMB) observations.

To conclude, the above examples show that we are at the beginning of a long road, shaped jointly by quantum gravity ideas, in turn coming from gravitational physics as well as high energy physics, and by concepts and methods from condensed matter theory and quantum many-body systems. Moreover, they provide support to the idea that the universe we live in, including its very foundations, i.e. its spatiotemporal properties, can itself be understood as a peculiar quantum many-body system. An admittedly radical, but certainly also fascinating hypothesis, which is only beginning to prove its fruitfulness.

References

1. S. Carlip, D.W. Chiou, W.T. Ni, R. Woodard, Int. J. Mod. Phys. D **24**(11), 1530028 (2015). https://doi.org/10.1142/S0218271815300281
2. C. Rovelli, in *Proceedings of the 9th Marcel Grossmann Meeting (MGIX MM): On Recent Developments, in Theoretical and Experimental General Relativity, Gravitation, and Relativistic Field Theories*, ed. by V.G. Gurzadyan, R.T. Jantzen, R. Ruffini (World Scientific, Singapore, 2000), pp. 742–768
3. D. Rickles, in *Integrating History and Philosophy of Science: Problems and Prospects (Boston Studies in the Philosophy and History of Science)*, ed. by S. Mauskopf, T. Schmaltz (Springer, New York, 2012), chap. 11, pp. 163–199. ISBN 9789400717442, 9789400717459. https://doi.org/10.1007/978-94-007-1745-9
4. D. Oriti (ed.), *Approaches to Quantum Gravity: Towards a New Understanding of Space, Time and Matter* (Cambridge University Press, Cambridge, 2009)
5. A. Ashtekar, J. Lewandowski, Class. Quantum Gravity **21**(15), R53 (2004). https://doi.org/10.1088/0264-9381/21/15/R01
6. C. Rovelli, *Quantum Gravity* (Cambridge University Press, Cambridge, 2004). ISBN 9780521837330
7. A. Ashtekar, J. Pullin, in *Loop Quantum Gravity: the First 30 Years* [72], chap. 1. ISBN9789813209923. https://doi.org/10.1142/10445
8. N. Bodendorfer (2016), arXiv:1607.05129 [gr-qc]
9. J. Ambjørn, A. Görlich, J. Jurkiewicz, R. Loll, Phys. Rep. **519**(4), 127 (2012). https://doi.org/10.1016/j.physrep.2012.03.007
10. A. Perez, Liv. Rev. Rel. **16**(1), 3 (2013). https://doi.org/10.12942/lrr-2013-3
11. D. Oriti, in *Foundations of Space and Time: Reflections on Quantum Gravity*, ed. by J. Murugan, A. Weltman, G.F.R. Ellis (Cambridge University Press, Cambridge, 2012), chap. 12. ISBN 9780521114400
12. T. Krajewski, PoS **QGQGS2011**, 005 (2011), *Proceedings of the 3rd Quantum Gravity and Quantum Geometry School (Zakopane, Poland, February 28–March 13, 2011)*, arXiv:1210.6257 [gr-qc]
13. A. Baratin, D. Oriti, J. Phys.: Conf. Ser. **360**(1), 012002 (2012). https://doi.org/10.1088/1742-6596/360/1/012002
14. D. Oriti, in *Ashtekar and Pullin* [72], chap. 5. ISBN9789813209923. https://doi.org/10.1142/10445
15. F. Dowker, Gen. Relat. Gravit. **45**(9), 1651 (2013). https://doi.org/10.1007/s10714-013-1569-y
16. F. Lizzi, in *Proceedings of the Workshop on Geometry, Topology, QFT and Cosmology, Paris, May 28–30,2008* (Observatoire de Paris, Paris, 2008), arXiv:0811.0268 [hep-th]
17. S. Majid, in *Oriti*, [4], chap. 24, p. 466. ISBN 9780521860451
18. M. Niedermaier, M. Reuter, Living Rev. Relativ. **9**(1), 5 (2006). https://doi.org/10.12942/lrr-2006-5
19. K.S. Stelle, in *Quantum Gravity and Quantum Cosmology*. Lecture Notes in Physics, vol. 863, ed. by G. Calcagni, L. Papantonopoulos, G. Siopsis, N. Tsamis (Springer, Berlin, 2013), pp. 3–30. ISBN 978-3-642-33036-0. https://doi.org/10.1007/978-3-642-33036-0_1
20. M. Blau, S. Theisen, Gen. Relativ. Gravit. **41**(4), 743 (2009). https://doi.org/10.1007/s10714-008-0752-z
21. S. Carlip, Int. J. Mod. Phys. D **23**(11), 1430023 (2014). https://doi.org/10.1142/S0218271814300237
22. S.D. Mathur, Class. Quantum Gravity **26**(22), 224001 (2009). https://doi.org/10.1088/0264-9381/26/22/224001
23. G. Chirco, S. Liberati, Phys. Rev. D **81**, 024016 (2010). https://doi.org/10.1103/PhysRevD.81.024016
24. T. Padmanabhan, Curr. Sci. **109**, 2236 (2015). https://doi.org/10.18520/v109/i12/2236-2242

25. L. Freidel, M. Geiller, J. Ziprick, Class. Quantum Gravity **30**(8), 085013 (2013). https://doi.org/10.1088/0264-9381/30/8/085013
26. D. Oriti, Class. Quantum Gravity **33**(8), 085005 (2016). https://doi.org/10.1088/0264-9381/33/8/085005
27. N. Seiberg, in *The Quantum Structure of Space and Time*, ed. by D. Gross, M. Henneaux, A. Sevrin (World Scientific, Singapore, 2007), pp. 163–213. ISBN 9789812569523. https://doi.org/10.1142/9789812706768_0005
28. O. Hohm, D. Lüst, B. Zwiebach, Fortschr. Phys. **61**(10), 926 (2013). https://doi.org/10.1002/prop.201300024
29. D. Oriti, Stud. Hist. Philos. Sci. B **46**(2), 186 (2014)
30. K. Crowther, Appearing out of nowhere: the emergence of spacetime in quantum gravity. PhD thesis, University of Sydney (2014), arXiv:1410.0345 [physics.hist-ph]
31. C. Barceló, S. Liberati, M. Visser, Living Rev. Relativ. **8**(1), 12 (2005). https://doi.org/10.12942/lrr-2005-12
32. S. Finazzi, S. Liberati, L. Sindoni, in *Proceedings of the 2nd Amazonian Symposium on Physics* (2012), arXiv:1204.3039 [gr-qc]
33. G. Volovik, Ann. Phys. **14**(1–3), 165 (2005). https://doi.org/10.1002/andp.200410123
34. B.L. Hu, Int. J. Theor. Phys. **44**(10), 1785 (2005). https://doi.org/10.1007/s10773-005-8895-0
35. D. Oriti, PoS **QG-Ph**, 030 (2007), *Proceedings of the Conference 'From Quantum to Emergent Gravity: Theory and Phenomenology' (Trieste, Italy, June 11–15 2007)*, arXiv:0710.3276 [gr-qc]
36. G.T. Horowitz, J. Polchinski, in *Oriti* [4], chap. 10, pp. 169–186. ISBN 9780521860451
37. J. de Boer, in *Theoretical Physics to Face the Challenge of LHC*. Lecture Notes of the Les Houches Summer School, vol. 97, ed. by L. Baulieu, K. Benakli, M.R. Douglas, B. Mansoulié, E. Rabinovici, L.F. Cugliandolo (Oxford University Press, Oxford, 2015), chap. 7. ISBN 9780198727965. https://doi.org/10.1093/acprof:oso/9780198727965.003.0007
38. J. Zaanen, Y. Liu, Y. Sun, K. Schalm, *Holographic Duality in Condensed Matter Physics* (Cambridge University Press, Cambridge, 2015)
39. N.H. March, G.G.N. Angilella, *Exactly Solvable Models in Many-body Theory* (World Scientific, Singapore, 2016)
40. A. Kitaev, A simple model of quantum holography (2005). Talks at KITP, April 7 and May 27, 2015
41. R. Gurau, J.P. Ryan, SIGMA **8**, 020 (2012), Contribution to the Special Issue on 'Loop Quantum Gravity and Cosmology', arXiv:1109.4812 [hep-th]. https://doi.org/10.3842/SIGMA.2012.020
42. R. Gurau, SIGMA **12**, 094 (2016), Contribution to the Special Issue on 'Tensor Models, Formalism and Applications', arXiv:1609.06439 [hep-th]. https://doi.org/10.3842/SIGMA.2016.094
43. R. Gurau, SIGMA **12**, 069 (2016), Contribution to the Special Issue on 'Tensor Models, Formalism and Applications', arXiv:1603.07278 [math-ph]. https://doi.org/10.3842/SIGMA.2016.069
44. V. Bonzom, L. Lionni, A. Tanasa, J. Math. Phys. **58**(5), 052301 (2017). https://doi.org/10.1063/1.4983562
45. M. Van Raamsdonk, in *New Frontiers in Fields and Strings. Proceedings of the 2015 Theoretical Advanced Study Institute in Elementary Particle Physics (TASI 2015, Boulder, Colorado, 1–26 June 2015)* (World Scientific, Singapore, 2016), chap. 5, pp. 297–351. arXiv:1609.00026 [hep-th]. https://doi.org/10.1142/9789813149441_0005
46. C. Cao, S.M. Carroll, S. Michalakis, Phys. Rev. D **95**, 024031 (2017). https://doi.org/10.1103/PhysRevD.95.024031
47. P. Hayden, S. Nezami, X.L. Qi, N. Thomas, M. Walter, Z. Yang, J. High Energy Phys. **2016**(11), 9 (2016). https://doi.org/10.1007/JHEP11(2016)009
48. M. Van Raamsdonk, Gen. Relativ. Gravit. **42**(10), 2323 (2010), Reprinted as Ref. [73]. https://doi.org/10.1007/s10714-010-1034-0

49. E.R. Livine, D.R. Terno, Reconstructing quantum geometry from quantum information: area renormalization, coarse-graining and entanglement on spin networks (2006), arXiv:gr-qc/0603008
50. P.A. Höhn, J. Phys.: Conf. Ser. **880**(1), 012014 (2017). https://doi.org/10.1088/1742-6596/880/1/012014
51. G. Chirco, F.M. Mele, D. Oriti, P. Vitale, Fisher metric, geometric entanglement and spin networks (2017), arXiv:1703.05231 [gr-qc]
52. E.R. Livine, Intertwiner entanglement on spin networks (2017), arXiv:1709.08511 [gr-qc]
53. V. Bonzom, F. Costantino, E.R. Livine, Commun. Math. Phys. **344**(2), 531 (2016). https://doi.org/10.1007/s00220-015-2567-6
54. B. Bahr, B. Dittrich, F. Hellmann, W. Kaminski, Phys. Rev. D **87**, 044048 (2013). https://doi.org/10.1103/PhysRevD.87.044048
55. B. Dittrich, F.C. Eckert, M. Martin-Benito, New J. Phys. **14**(3), 035008 (2012). https://doi.org/10.1088/1367-2630/14/3/035008
56. B. Dittrich, S. Mizera, S. Steinhaus, New J. Phys. **18**(5), 053009 (2016). https://doi.org/10.1088/1367-2630/18/5/053009
57. C. Delcamp, B. Dittrich, Towards a phase diagram for spin foams (2016), arXiv:1612.04506 [gr-qc]
58. G. Chirco, D. Oriti, M. Zhang, Group field theory and tensor networks: towards a ryu-takayanagi formula in full quantum gravity (2017), arXiv:1701.01383 [gr-qc]
59. M. Han, L.Y. Hung, Phys. Rev. D **95**, 024011 (2017). https://doi.org/10.1103/PhysRevD.95.024011
60. S. Carrozza, D. Oriti, V. Rivasseau, Commun. Math. Phys. **330**(2), 581 (2014). https://doi.org/10.1007/s00220-014-1928-x
61. D. Benedetti, J.B. Geloun, D. Oriti, J. High Energy Phys. **2015**(3), 84 (2015). https://doi.org/10.1007/JHEP03(2015)084
62. S. Carrozza, V. Lahoche, Class. Quantum Gravity **34**(11), 115004 (2017). https://doi.org/10.1088/1361-6382/aa6d90
63. S. Carrozza, SIGMA **12**, 070 (2016), Contribution to the special issue on 'Tensor Models, Formalism and Applications', arXiv:1603.01902 [gr-qc]. https://doi.org/10.3842/SIGMA.2016.070
64. S. Carrozza, V. Lahoche, D. Oriti, Phys. Rev. D **96**, 066007 (2017). https://doi.org/10.1103/PhysRevD.96.066007
65. S. Gielen, D. Oriti, L. Sindoni, Phys. Rev. Lett. **111**, 031301 (2013). https://doi.org/10.1103/PhysRevLett.111.031301
66. S. Gielen, D. Oriti, L. Sindoni, J. High Energy Phys. **2014**(6), 13 (2014). https://doi.org/10.1007/JHEP06(2014)013
67. S. Gielen, L. Sindoni, SIGMA **12**, 082 (2016), Contribution to the Special Issue on 'Tensor Models, Formalism and Applications', arXiv:1602.08104 [gr-qc]. https://doi.org/10.3842/SIGMA.2016.082
68. D. Oriti, Comptes Rendus Physique **18**(3), 235 (2017). https://doi.org/10.1016/j.crhy.2017.02.003
69. D. Oriti, L. Sindoni, E. Wilson-Ewing, Class. Quantum Gravity **33**(22), 224001 (2016). https://doi.org/10.1088/0264-9381/33/22/224001
70. M. de Cesare, A.G.A. Pithis, M. Sakellariadou, Phys. Rev. D **94**, 064051 (2016). https://doi.org/10.1103/PhysRevD.94.064051
71. S. Gielen, D. Oriti, Cosmological perturbations from full quantum gravity (2017), arXiv:1709.01095 [gr-qc]
72. A. Ashtekar, J. Pullin (eds.), *100 Years of General Relativity*, vol. 4 (World Scientific, Singapore, 2004). ISBN 9789813209923. https://doi.org/10.1142/10445
73. M. Van Raamsdonk, Int. J. Mod. Phys. D **19**(14), 2429 (2010), Reprint of Ref. [48]. https://doi.org/10.1142/S0218271810018529

Author Index

G. G. N. Angilella and C. Amovilli (eds.), *Many-body Approaches at Different Scales*, https://doi.org/10.1007/978-3-319-72374-7

Subject Index

G. G. N. Angilella and C. Amovilli (eds.), *Many-body Approaches at Different Scales*, https://doi.org/10.1007/978-3-319-72374-7

Printed in the United States
By Bookmasters